Ideal Magnetohydrodynamics

MODERN PERSPECTIVES IN ENERGY

Series Editors: Richard K. Lester and John Andelin
Founding Editor: David J. Rose†

BIOFUELS, AIR POLLUTION, AND HEALTH: A Global Review
Kirk R. Smith

BIOMASS ENERGIES: Resources, Links, Constraints
Vaclav Smil

CARBON-NITROGEN-SULFUR: Human Interference in Grand Biospheric Cycles
Vaclav Smil

ENERGY: The Conservation Revolution
John H. Gibbons and William U. Chandler

IDEAL MAGNETOHYDRODYNAMICS
Jeffrey P. Freidberg

LEARNING ABOUT ENERGY
David J. Rose

STRUCTURAL MATERIALS IN NUCLEAR POWER SYSTEMS
J. T. Adrian Roberts

Ideal Magnetohydrodynamics

Jeffrey P. Freidberg
Massachusetts Institute of Technology
Cambridge, Massachusetts

Plenum Press • New York and London

Library of Congress Cataloging in Publication Data

Freidberg, Jeffrey P.
Ideal magnetohydrodynamics.

(Modern perspectives in energy)
Includes bibliographies and index.
1. Magnetohydrodynamics. 2. Plasma (Ionized gases). 3. Fusion reactors. I. Title. II. Series.
QC718.5.M36F74 1987 538′.6 87-6999
ISBN 0-306-42512-2

A Division of Plenum Publishing Corporation
233 Spring Street, New York, N.Y. 10013

Printed in the United States of America

To Karen

Preface

Over the past two decades ideal magnetohydrodynamics (MHD) has developed into a relatively mature theory within the field of plasma physics. MHD represents the simplest self-consistent model describing the macroscopic equilibrium and stability properties of a plasma. Indeed, it is MHD considerations that provide the primary motivation for the unique and often ingenious magnetic geometries appearing on the fusion landscape.

Because of its importance, a two-semester sequence in MHD theory is offered at MIT. The first semester is devoted to ideal magnetohydrodynamics and is largely involved with the study of equilibrium and fast, often catastrophic, instabilities. The second semester treats slower, nonideal effects (such as resistivity, pressure anisotropy, and finite Larmor radius effects) and their influence on plasmas that are stable or at most, weakly unstable against the ideal modes.

This textbook is concerned with the first semester's material. Its goal is to provide an in-depth introduction to the subject of ideal magnetohydrodynamics. The style of presentation has a theoretical emphasis, indicative of the author's background. There is, nonetheless, considerable discussion of the application of theory to experiments and of the experiments themselves.

The intended audience is new graduate students, postdoctoral fellows, and scientists and engineers interested in plasma physics, particularly as related to magnetic fusion. The book has been written with the need to satisfy the pedagogical requirements of the student as a primary goal. As such, there are many descriptive explanations and virtually every equation is derived, either in detail, in outline form, or in the Appendices. Considerable effort has been devoted to making the derivations as simple as possible. While the subject matter of the book is almost entirely self-contained, an introductory plasma physics course is nonetheless recommended as a prerequisite.

One of the most enjoyable aspects of teaching is learning and introducing new material each time a course is taught. Since MHD has been offered for many years at MIT, there is, not surprisingly, a "surplus" of topics in the book. Instructors are thus free to emphasize basics versus applications, as well as to devote attention to specific configurations of choice. During one semester it is normally possible to cover approximately one half of the material in the book in reasonable detail.

Over the years most of the subject matter has been presented to students several times. However the topics of three-dimensional equilibrium and stability, appearing in Chapter 7 and part of Chapter 10, are rather complex and are usually covered sparingly in class. However, this material has been included for the sake of completeness and because of its importance to magnetic fusion.

Included in the book are the homework problems that have been developed during the history of the MHD course. Most of the problems are quite "substantial," requiring at least several hours to complete. They have been specifically formulated to provide both enlightenment about the underlying plasma physics and the development of mathematical skills.

Jeffrey P. Freidberg

Cambridge

Acknowledgments

The material for this book has evolved over many years of research and teaching. My colleagues in the fusion community, too numerous to mention, have contributed in a substantial way to virtually every topic covered in the book. For their collaboration, cooperation, and camaraderie, I acknowledge my deep appreciation.

I would like to express particular gratitude to John Wesson, Hans Goedbloed, and Gerhard Berge. Our collaborations, extending over many years of research, have proven invaluable as we attempted to educate ourselves in the subject of magnetohydrodynamics. I hope they have profited as much as I.

I am also indebted to Ian Hutchinson and my students Pekka Hakkarainen and Scott Haney for reading the completed manuscript and for the valuable suggestions that followed.

Preparing the manuscript has been a rather ambitious task. In this regard I would like to acknowledge Patricia Fina, Anna Kotsopoulos, and Vonya Perham for TEXing the manuscript and Cathy Lydon for drawing the figures. I also thank Albe Dawson for her help in proofreading.

A generous grant from the Bernard M. Gordon (1948) Engineering Curriculum Development Fund at MIT was provided for the text preparation and for this I am most grateful.

Finally, I would like to thank my wife Karen for her support and encouragement while I prepared the manuscript.

J.P.F

Contents

1

Introduction

1.1. IDEAL MHD AND MAGNETIC FUSION

The development of magnetic fusion as a commercial source of electricity requires the solutions to a number of challenging physics as well as technological problems. The physics problems are traditionally separated into three basic areas: equilibrium and stability, heating, and transport. The goal is the discovery of magnetic geometries that are capable of stably confining a sufficiently high density of plasma at a sufficiently high temperature for a sufficiently long time to produce net thermonuclear power. In its most elementary form the power balance in a magnetic fusion reactor is described in terms of the Lawson parameter $n\tau_E$ with n the plasma number density and τ_E the energy confinement time (Lawson, 1957). At reactor conditions $n\tau_E \gtrsim f(T)$, where typically $f(T) \approx 3 \times 10^{20}\,\mathrm{m}^{-3}\,\mathrm{s}$ at the optimized temperature $T \approx 15\,\mathrm{keV}$.

From a somewhat oversimplified view of the physics, the problem of maximizing $n\tau_E$ separates into two relatively independent parts. First, the maximum energy confinement time τ_E is determined by the microscopic behavior of the plasma: collisions and microinstabilities. This behavior ultimately leads to macroscopic energy transport which can be either classical or anomalous depending on the processes involved. Since these phenomena require a knowledge of individual particle motion on short length and time scales they are usually treated by kinetic models, but including only limited geometry because of the complexity of the physics. Such models are also used to determine the energy deposition and plasma temperature T associated with various methods of heating.

Second, the maximum density n is determined by macroscopic equilibrium and stability limits set by the magnetic geometry. Actually, at a given T, the critical figure of merit is $\beta \equiv 2\mu_0 nT/B^2$, the ratio of plasma energy to magnetic energy. Because of the relatively slow time and large length scales involved (compared to kinetic behavior), these phenomena are best described by fluid models.

Ideal magnetohydrodynamics (MHD) is the most basic single-fluid model for determining the macroscopic equilibrium and stability properties of a plasma. The model describes how magnetic, inertial, and pressure forces interact within an ideal perfectly conducting plasma in an arbitrary magnetic geometry. There is general consensus that any configuration meriting consideration as a fusion reactor must satisfy the equilibrium and stability limits set by ideal MHD. If not, catastrophic termination of the plasma on a very short time scale (compared to experimental times) is the usual consequence.

Thus, the role of ideal MHD in magnetic fusion is the discovery of magnetic geometries that possess attractive equilibrium and stability properties for fusion reactors.

It should be noted that while favorable ideal MHD behavior is a necessary condition in a reactor, it may not, even in the restrictive context of low-frequency macroscopic behavior, be sufficient. Nonideal effects such as electrical resistivity sometimes allow the development of slower, weaker instabilities, which nevertheless may still lead to enhanced transport or even, in certain circumstances, violent termination of the plasma. A complete evaluation of any given concept requires a knowledge of both the strong, ideal MHD behavior as well as the weaker, but still potentially dangerous nonideal effects. This textbook is concerned primarily with ideal MHD behavior.

Another point to note is that the separation between the two basic physics questions (i.e., ideal MHD $\rightarrow$ maximum n and kinetic theory $\rightarrow$ maximum τ_E) is only approximately correct and often there is significant overlap. For instance, ideal MHD often set limits on the largest current that can flow as well as on β. Clearly, a current limit will have a direct effect on the maximum possible ohmic heating. As another example, the pressure and current profiles in a plasma are determined by transport processes that are often anomalous. In certain magnetic configurations the maximum pressure and current limits are strong functions of the profiles, again indicating overlap in the physics. By studying only ideal MHD behavior such self-consistent interactions are not included.

With these considerations as background, it is the goal of this textbook to present an in-depth discussion of ideal MHD theory as applied to magnetic fusion with emphasis on developing an understanding of why certain geometries are more favorable than others. Since a main virtue of ideal MHD is its ability to distinguish different magnetic geometries, a number of specific examples are discussed. These are chosen from the currently most promising concepts in the international magnetic fusion program and include the tokamak, stellarator, reversed field pinch (RFP), and the Elmo bumpy torus (EBT). Each of these is a toroidal configuration. For reasons discussed later, ideal MHD provides a very poor description of the behavior of open-ended devices. Thus, there is no discussion of the mirror concept.

In presenting the material the basic approach adopted is one that relies heavily on analytic theory. Specifically, once the general properties of the ideal MHD model are established, each magnetic configuration is distinguished by a separate characteristic expansion of the model (where the expansion parameters are usually geometric or field ratios). Although such expansions are sometimes only marginally valid in actual experiments, they are extremely useful in delineating the basic physics issues of a given concept. Extensive numerical computations have been carried out for many ideal MHD configurations and these results are summarized in the context of the analytic theory.

1.2. UNITS

The basic units used throughout the textbook are MKS units. The one exception is the temperature. Whenever the Boltzmann constant K appears in the text, it is absorbed into the temperature. Thus, the symbol T refers to temperature measured in units of energy.

In the course of the text, a number of relations are derived in practical units defined as follows:

Number density	n	$10^{20}\ \mathrm{m}^{-3}$
Temperature	T	keV
Magnetic field	B	T (tesla)
Current	I	MA (megamperes)
Minor radius	a	m
Major radius	R	m

REFERENCES

Lawson, J. D. (1957). *Proc. Phys. Soc. London, Sec. B* **70**, 6.

2

The MHD Model

2.1. INTRODUCTION

The goal of Chapter 2 is to provide a physical understanding of the ideal MHD model. This includes (1) a basic description of the model, (2) a derivation starting from a more fundamental kinetic model, and, most important, (3) a discussion of its range of validity.

Whereas there is a straightforward and rigorous derivation of the model, there are, nonetheless, a number of surprises and subtleties involved. It is essential to understand these issues since ideal MHD, the simplest self-consistent model providing a tractable description of magnetic geometry, is so widely used in the design and interpretation of fusion experiments.

Questions arise for two reasons. First, one of the basic assumptions used in the derivation, i.e., that the plasma is collision dominated, is virtually never satisfied in plasmas of fusion interest. Even so, there is overwhelming empirical evidence that ideal MHD provides a very accurate description of most macroscopic plasma behavior. This apparent good fortune is not a coincidence but the consequence of some subtle physics; namely, those parts of the MHD model that are not valid are not directly involved in many if not most phenomena of interest. An attempt is made to clarify the issue by introducing a heuristic model known as collisionless MHD, which makes predictions almost identical to ideal MHD but is valid over a much wider parameter range. This model can be justified by rigorous calculation. However, substantial analysis is required, beyond the scope of the present textbook.

The second subtle MHD question concerns the following. Ideal MHD must be viewed as an asymptotic model in that specific length and time scales must be assumed for the derivation to be valid. Even when the collision-dominated assumption is not involved, there are many situations where MHD is used to describe phenomena on time scales far beyond its strict range of applicability. The issue is that the overall limits of validity arise from simultaneously satisfying every assumption required in the derivation. However, certain portions of the model are unaffected by certain assumptions, and consequently have a much wider range of validity. This important and useful information is discussed as the calculation progresses.

With these subtleties in mind attention is now focused on providing an in-depth description of the ideal MHD model.

2.2. DESCRIPTION OF THE IDEAL MHD MODEL

The ideal MHD model provides a single fluid description of long-wavelength, low-frequency, macroscopic plasma behavior. To put the model in perspective, it is perhaps useful to discuss first those plasma phenomena *not* described by ideal MHD.

Regarding physics in general, it has been pointed out that the three major discoveries of modern physics during the last century or so, namely,

1. Maxwell's equations with wave propagation.
2. Relativity.
3. Quantum mechanics.

are each eliminated in the derivation of MHD.

Within the narrower confines of plasma physics itself, there are a variety of processes important in fusion plasmas. Among them are

1. Radiation.
2. Rf heating.
3. Resonant particle effects.
4. Microinstabilities.
5. Classical transport.
6. Anomalous transport.
7. Resistive instabilities.
8. α-Particle behavior.

Similarly, none of these processes is adequately described by ideal MHD.

Although the apparent lack of physical content is humbling, the one crucial phenomenon simply but accurately described by the model is the effect of magnetic geometry on the macroscopic equilibrium and stability of fusion plasmas. Specifically, ideal MHD answers such basic questions as: How does a given magnetic geometry provide forces to hold a plasma in equilibrium? Why are certain magnetic geometries much more stable against macroscopic disturbances than others? Why do fusion configurations have such technologically undesirable shapes as a torus, a helix, or a baseball seam?

One should beware that despite the simplicity implied by the limited physical content, the ideal MHD model nevertheless is still too difficult to solve in most geometries of interest. This becomes evident by noting the many sophisticated expansions required throughout the text. Attempts to solve similar problems using more comprehensive kinetic models are virtually impossible, even numerically, in realistic two- and three-dimensional geometries.

With this perspective the ideal MHD model is given by

$$\frac{\partial \rho}{\partial t} + \nabla \cdot \rho \mathbf{v} = 0 \tag{2.1}$$

$$\rho \frac{d\mathbf{v}}{dt} = \mathbf{J} \times \mathbf{B} - \nabla p \tag{2.2}$$

$$\frac{d}{dt}\left(\frac{p}{\rho^{\gamma}}\right) = 0 \tag{2.3}$$

$$\mathbf{E} + \mathbf{v} \times \mathbf{B} = 0 \tag{2.4}$$

$$\nabla \times \mathbf{E} = -\frac{\partial \mathbf{B}}{\partial t} \tag{2.5}$$

$$\nabla \times \mathbf{B} = \mu_0 \mathbf{J} \tag{2.6}$$

$$\nabla \cdot \mathbf{B} = 0 \tag{2.7}$$

In these equations, the electromagnetic variables are the electric field $\mathbf{E}$, the magnetic field $\mathbf{B}$, and the current density $\mathbf{J}$. The fluid variables are the mass density ρ, the fluid velocity $\mathbf{v}$, and the pressure p. Also, $\gamma = 5/3$ is the ratio of specific heats and $d/dt = \partial/\partial t + \mathbf{v} \cdot \nabla$ is the convective derivative.

Equations (2.5)–(2.7) indicate that in ideal MHD the electromagnetic behavior is governed by the low-frequency, pre-Maxwell equations. Equations (2.1)–(2.3) govern the time evolution of mass, momentum, and energy, respectively. The mass equation implies that the total number of plasma particles is conserved; phenomena such as ionization recombination, or charge exchange do not play a role. The basic physics of the momentum equation corresponds to that of a fluid with three interacting forces: the pressure gradient force ∇p, the magnetic force $\mathbf{J} \times \mathbf{B}$, and the inertial force $\rho\, d\mathbf{v}/dt$. The energy equation expresses an adiabatic evolution characterized by a ratio of specific heats, $\gamma = 5/3$. The remaining relation, Eq. (2.4), is Ohm's law, which implies that in a reference frame moving with plasma the electric field is zero; that is, the plasma is a perfect conductor. It is the perfect conductivity assumption of Ohm's law that gives rise to the name "ideal" MHD.

As stated previously, the conditions for validity of the ideal MHD model imply that the phenomena of interest correspond to certain length and time scales. For macroscopic behavior the characteristic length scale is that of the overall plasma dimension. Denoting this dimension by a, then typically $a \sim 1$ m. The characteristic speed with which MHD phenomena occur is the thermal velocity of the plasma ions: $V_{T_i} = (2T_i/m_i)^{1/2}$ with T_i the ion temperature and m_i the ion mass. This gives rise to a characteristic time $\tau_M \equiv a/V_{T_i}$. For m_i equivalent to deuterium and $T_i = 2$ keV then $\tau_M \sim 2\ \mu$s. The MHD length and time scales can be compared with those of other basic plasma physics phenomena as shown in Tables 2.1 and 2.2. In computing these values it has been assumed that $a = 1$ m, $T_e = T_i = 2$ keV, $B = 5$ T, $n = 2 \times 10^{20}\ \mathrm{m}^{-3}$ (particle number density) and m_i equivalent to deuterium. Also the Coulomb logarithm has been set to $\ln \Lambda = 15$.

The richness of plasma physics is clearly evidenced by the large number and wide range of length and time scales. Among these, ideal MHD lies midway between a variety of high-frequency microscopic phenomena and the very low-frequency transport phenomena. This is the regime of macroscopic equilibrium and stability.

Observe that while typical MHD times (of the order of microseconds)

Table 2.1. Comparison of the Characteristic MHD Time with That of Other Basic Plasma Physics Phenomena. In Practical Formulas T is Expressed in keV and n in $10^{20}\ \mathrm{m}^{-3}$

Plasma physics time scales	Formulas	Numerical values (s)
Electron gyro period	$\tau_{ce} = 2\pi/\omega_{ce} = 2\pi m_e/eB_0$	7.1×10^{-12}
Electron plasma period	$\tau_{pe} = 2\pi/\omega_{pe} = 2\pi(m_e\epsilon_0/ne^2)^{1/2}$	7.9×10^{-12}
Ion plasma period	$\tau_{pi} = (m_i/m_e)^{1/2}\tau_{pe}$	4.8×10^{-10}
Ion gyro period	$\tau_{ci} = (m_i/m_e)\tau_{ce}$	2.6×10^{-8}
MHD time	$\tau_M = a/V_{T_i}$	2.3×10^{-6}
Electron–electron collision time	$\tau_{ee} = 7.4 \times 10^{-6} T_e^{3/2}/n$	1.0×10^{-5}
Ion–ion collision time	$\tau_{ii} = (2m_i/m_e)^{1/2}\tau_{ee}$	8.9×10^{-4}
Energy equilibration time	$\tau_{eq} = (m_i/2m_e)\tau_{ee}$	1.9×10^{-2}
Ignition time	$\tau_{ig} = 2.0/n$	1.0
Resistive diffusion time	$\tau_D = \mu_0 a^2/\eta = 51 a^2 T_e^{3/2}$	1.4×10^2

Table 2.2. Comparison of the Characteristic MHD Length with That of Other Basic Plasma Physics Phenomena

Plasma physics length scales	Formulas	Numerical values (m)
Electron gyro radius	$r_{Le} = (m_e T_e)^{1/2}/eB_0$	2.1×10^{-5}
Debye length	$\lambda_d = (\epsilon_0 T_e/e^2 n)^{1/2}$	2.4×10^{-5}
Electron skin depth	$\delta_e = c/\omega_{pe}$	3.8×10^{-4}
Ion gyro radius	$r_{Li} = (m_i/m_e)^{1/2} r_{Le}$	1.3×10^{-3}
Ion skin depth	$\delta_i = (m_i/m_e)^{1/2} \delta_e$	2.3×10^{-2}
MHD length	a	1.0
Ion–ion mfp	$\lambda_{ii} = V_{T_i} \tau_{ii}$	2.8×10^2
Electron–electron mfp	$\lambda_{ee} = \lambda_{ii}$	2.8×10^2

are much shorter than typical experimental times (of the order of seconds), this does not imply that MHD can be applied only during a small fraction of the time of interest. The widest use of MHD involves the repeated calculation of equilibrium and stability over many small time increments during the evolution of a discharge. For example, illustrated in Fig. 2.1 is a curve of the time evolution of the current in a tokamak. Also shown is a narrow time increment Δt during which MHD is valid. If at the time t_1 the external fields and the plasma current and pressure are given, then the MHD model can be used to calculate the equilibrium and stability of the system. This procedure is repeated many times as the pressure and current evolve on the transport time scale, an evolution whose physics is not contained within ideal MHD. There is an implication, however, that during each successive time increment Δt, the plasma is MHD stable. If not, the discharge would likely be destroyed within several MHD times because of the potency of the modes.

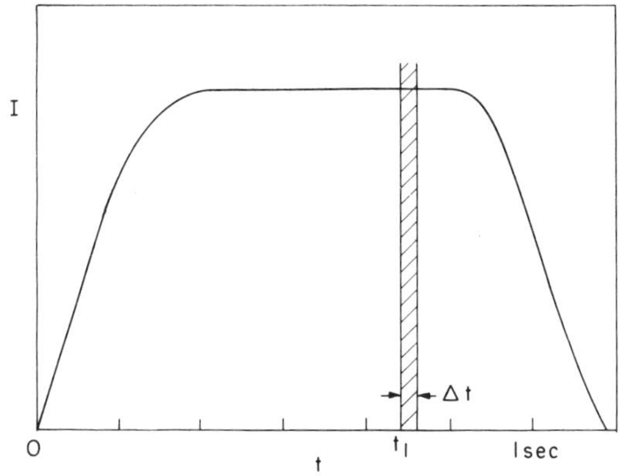

Figure 2.1. Schematic diagram of a 1-s current pulse for a typical tokamak. The increment $\Delta t \sim 50\ \mu s$ corresponds to an interval where ideal MHD is valid.

In summary, although many important plasma physics processes are neglected in the derivation of ideal MHD, the one critical phenomenon that remains is the self-consistent treatment of multidimensional magnetic geometry.

2.3. DERIVATION OF THE IDEAL MHD MODEL

2.3.1. Starting Equations

In order to appreciate more fully the physics content of ideal MHD as well as the subtleties involved, a derivation of the model is presented starting from basic principles. A number of such derivations exist in the literature and each, including the one presented here, follows the same general procedure: fluid moments are calculated from a general kinetic model, and then a number of assumptions is made to obtain closure of the system.

Many would agree that the most complete and rigorous derivation is described in the classic work by Braginskii (1965). The analysis presented here is noticeably simpler because the end goal is much narrower in scope; the derivation is focused solely on the ideal MHD model with little attention given to the general question of determining transport coefficients in different collisionality regimes. Even so, some of the results of Braginskii are quoted to demonstrate the smallness of certain terms neglected in the ideal MHD model.

The starting point for the present derivation is the full set of Maxwell's equations coupled with a kinetic model of the plasma, described by a Boltzmann equation for each species. [A discussion of the starting equations themselves can be found in Montgomery and Tidman (1964) and Krall and Trivelpiece (1973).] These equations are given by

$$\frac{\partial f_\alpha}{\partial t} + \mathbf{u} \cdot \nabla f_\alpha + \frac{q_\alpha}{m_\alpha}(\mathbf{E} + \mathbf{u} \times \mathbf{B}) \cdot \nabla_u f_\alpha = \left(\frac{\partial f_\alpha}{\partial t}\right)_c \tag{2.8}$$

$$\nabla \times \mathbf{E} = -\frac{\partial \mathbf{B}}{\partial t} \tag{2.9}$$

$$\nabla \cdot \mathbf{B} = 0 \tag{2.10}$$

$$\nabla \times \mathbf{B} = \mu_0 \mathbf{J} + \frac{1}{c^2}\frac{\partial \mathbf{E}}{\partial t} \tag{2.11}$$

$$\nabla \cdot \mathbf{E} = \frac{\sigma}{\epsilon_0} \tag{2.12}$$

with the current and charge densities defined as

$$\mathbf{J} \equiv \sum_{\alpha} q_{\alpha} \int \mathbf{u} f_{\alpha} \, d\mathbf{u} \tag{2.13}$$

$$\sigma \equiv \sum_{\alpha} q_{\alpha} \int f_{\alpha} \, d\mathbf{u} \tag{2.14}$$

Here, $f_\alpha(\mathbf{r}, \mathbf{u}, t)$ is the distribution function for each species α, and is in general a function of spatial coordinate, velocity, and time. It shall be assumed that the plasma is fully ionized and consists of two species, electrons and ions. Hence, $\alpha = e, i$.

In the Boltzmann description there are two types of forces that act on the particles. First there are the long-range Lorentz forces, $q_\alpha(\mathbf{E} + \mathbf{u} \times \mathbf{B})$ in which $\mathbf{E}$ and $\mathbf{B}$ are smoothly behaved fields calculated from the averaged current and charge densities as indicated in Eqs. (2.13) and (2.14). Second, the right-hand side of Eq. (2.8) represents the forces due to short-range interactions, or collisions. In the derivation presented here, the details of the collision operator are not important, only certain global conservation relations. For plasmas of fusion interest, the dominant collisions are elastic Coulomb collisions between both like and unlike particles. The conservation laws for elastic collisions can be summarized as follows. If the collision operator is defined in the usual way

$$\left(\frac{\partial f_\alpha}{\partial t}\right)_c = \sum_{\beta} C_{\alpha\beta} \tag{2.15}$$

where $C_{\alpha\beta}$ represents collisions of particles of species α with particles of species β, then

1. Conservation of particles between like and unlike particle collisions implies

$$\int C_{ee} \, d\mathbf{u} = \int C_{ii} \, d\mathbf{u} = \int C_{ei} \, d\mathbf{u} = \int C_{ie} \, d\mathbf{u} = 0 \tag{2.16}$$

2. Conservation of momentum and energy between like particle collisions implies

$$\int m_e \mathbf{u} C_{ee} \, d\mathbf{u} = \int m_i \mathbf{u} C_{ii} \, d\mathbf{u} = 0 \tag{2.17}$$

$$\int \frac{1}{2} m_e u^2 C_{ee} \, d\mathbf{u} = \int \frac{1}{2} m_i u^2 C_{ii} \, d\mathbf{u} = 0 \tag{2.18}$$

3. Conservation of total momentum and energy between unlike particle collisions implies

$$\int (m_e \mathbf{u} C_{ei} + m_i \mathbf{u} C_{ie})\, d\mathbf{u} = 0 \tag{2.19}$$

$$\int \frac{1}{2}(m_e u^2 C_{ei} + m_i u^2 C_{ie})\, d\mathbf{u} = 0 \tag{2.20}$$

[More information on collisions in a plasma can be obtained from Spitzer (1956), Braginskii (1965), Rose and Clark (1961), and Krall and Trivelpiece (1973).]

The full set of Boltzmann–Maxwell equations provides a very detailed and complete description of plasma behavior. At one end of the spectrum, it contains microscopic information about the orbits of individual charged particles on the very short cyclotron time scale and gyro radius length scale. At the other end, it accurately describes the macroscopic behavior of large plasma experiments including MHD equilibrium and stability as well as very slow transport phenomena. Not surprisingly, the complexity arising from this breadth of information makes it virtually impossible to solve, even numerically, the Boltzmann–Maxwell system of equations in any nontrivial geometry. This realization has led to the development of several simpler models with narrower physics content. Ideal MHD is such a model.

2.3.2. Two-Fluid Equations

The derivation of the ideal MHD equations begins by taking the appropriate moments of the Boltzmann equation. This calculation transforms a single equation for f_α in seven variables $(\mathbf{r}, \mathbf{u}, t)$ into an infinite set of fluid equations in four variables $(\mathbf{r}, t)$. The procedure is straightforward although admittedly somewhat tedious. Once the fluid equations are obtained, physical variables such as density, velocity, and pressure are defined. Then, depending upon the physical regime of interest, various approximations and expansions can be introduced, ultimately leading to a truncation of the infinite set of equations. The truncated system is by design focused on a specific regime of physics; as such, its total information content is much less than that of the original Boltzmann equation. The virtue of the fluid equations is that they are enormously simpler to solve.

To derive the ideal MHD model, the moments that are required correspond to mass, momentum, and energy; that is, starting with Eq.

(2.8) one evaluates

$$\int Q_i\left[\frac{df_\alpha}{dt} - \left(\frac{\partial f_\alpha}{\partial t}\right)_c\right] d\mathbf{u} = 0 \tag{2.21}$$

for $i = 1$–3 with $Q_i(\mathbf{u})$ given by

$$\begin{aligned} Q_1 &= 1 && \text{(mass)} \\ Q_2 &= m_\alpha \mathbf{u} && \text{(momentum)} \\ Q_3 &= m_\alpha u^2/2 && \text{(energy)} \end{aligned} \tag{2.22}$$

After a straightforward calculation the fluid equations for each species can be written as

$$\begin{gathered} \frac{\partial n_\alpha}{\partial t} + \nabla \cdot n_\alpha \mathbf{v}_\alpha = 0 \\ \frac{\partial}{\partial t}(m_\alpha n_\alpha \mathbf{v}_\alpha) + \nabla \cdot (m_\alpha n_\alpha \langle \mathbf{uu} \rangle) - q_\alpha n_\alpha (\mathbf{E} + \mathbf{v}_\alpha \times \mathbf{B}) = \int m_\alpha \mathbf{u} C_{\alpha\beta}\, d\mathbf{u} \\ \frac{\partial}{\partial t}\left(\frac{1}{2} m_\alpha n_\alpha \langle u^2 \rangle\right) + \nabla \cdot \left(\frac{1}{2} m_\alpha n_\alpha \langle u^2 \mathbf{u} \rangle\right) - q_\alpha n_\alpha \mathbf{v}_\alpha \cdot \mathbf{E} = \int \frac{1}{2} m_\alpha u^2 C_{\alpha\beta}\, d\mathbf{u} \end{gathered} \tag{2.23}$$

Here, n_α and $\mathbf{v}_\alpha$ are the macroscopic number density and fluid velocity defined by

$$n_\alpha(\mathbf{r}, t) \equiv \int f_\alpha\, d\mathbf{u} \tag{2.24}$$

$$\mathbf{v}_\alpha(\mathbf{r}, t) \equiv \frac{1}{n_\alpha}\int \mathbf{u} f_\alpha\, d\mathbf{u} \tag{2.25}$$

and the only nonzero contribution to the collision terms is from unlike particle interactions, $\alpha \neq \beta$. The quantities $\langle \mathbf{uu} \rangle$, $\langle u^2 \rangle$, and $\langle u^2 \mathbf{u} \rangle$ are higher moments of the distribution function defined by the general relation

$$\langle Q \rangle \equiv \frac{1}{n_\alpha}\int Q f_\alpha\, d\mathbf{u} \tag{2.26}$$

The next step in the derivation is to introduce a new velocity variable $\mathbf{w} = \mathbf{u} - \mathbf{v}_\alpha(\mathbf{r}, t)$ representing the random thermal motion of the particles, and defined so that $\langle \mathbf{w} \rangle = 0$. By introducing this variable into Eq. (2.23) one can write the fluid equations in terms of more physical macroscopic quantities. In particular, the quantities of interest are the scalar pressure,

$$p_\alpha = \tfrac{1}{3} n_\alpha m_\alpha \langle w^2 \rangle \tag{2.27}$$

the total pressure tensor,

$$\mathsf{P}_\alpha \equiv n_\alpha m_\alpha \langle \mathbf{ww} \rangle \tag{2.28}$$

the anisotropic part of the pressure tensor,

$$\boldsymbol{\Pi}_\alpha \equiv \mathsf{P}_\alpha - p_\alpha \mathsf{I} \tag{2.29}$$

the temperature,

$$T_\alpha \equiv p_\alpha / n_\alpha \tag{2.30}$$

the heat flux due to random motion,

$$\mathbf{h}_\alpha \equiv \tfrac{1}{2} n_\alpha m_\alpha \langle w^2 \mathbf{w} \rangle \tag{2.31}$$

the mean momentum transfer between unlike particles due to the friction of collisions,

$$\mathbf{R}_\alpha \equiv \int m_\alpha \mathbf{w} C_{\alpha\beta} \, d\mathbf{w} \tag{2.32}$$

and the heat generated due to collisions between unlike particles,

$$Q_\alpha \equiv \int \frac{1}{2} m_\alpha w_\alpha^2 C_{\alpha\beta} \, d\mathbf{w} \tag{2.33}$$

Substituting these definitions into Eq. (2.23) leads to the following form of the moment equations:

$$\left(\frac{dn_\alpha}{dt} \right)_\alpha + n_\alpha \nabla \cdot \mathbf{v}_\alpha = 0 \tag{2.34}$$

$$m_\alpha n_\alpha \left(\frac{d\mathbf{v}_\alpha}{dt} \right)_\alpha - q_\alpha n_\alpha (\mathbf{E} + \mathbf{v}_\alpha \times \mathbf{B}) + \nabla \cdot \mathsf{P}_\alpha = \mathbf{R}_\alpha \tag{2.35}$$

$$n_\alpha \left[\frac{d}{dt} \left(\frac{m_\alpha v_\alpha^2}{2} + \frac{3}{2} T_\alpha \right) \right]_\alpha + \nabla \cdot (\mathbf{v}_\alpha \cdot \mathsf{P} + \mathbf{h}_\alpha) - q_\alpha n_\alpha \mathbf{v}_\alpha \cdot \mathbf{E} = Q_\alpha + \mathbf{v}_\alpha \cdot \mathbf{R}_\alpha \tag{2.36}$$

where

$$\left(\frac{d}{dt} \right)_\alpha \equiv \frac{\partial}{\partial t} + \mathbf{v}_\alpha \cdot \nabla \tag{2.37}$$

is the convective derivative for the species α.

A further reduction can be made leading to a simpler form of the energy equation [Eq. (2.36)]. This form is obtained by evaluating the dot product of Eq. (2.35) with $\mathbf{v}_\alpha$ and subtracting the result from Eq. (2.36). Upon replacing the energy equation with the simplified form one obtains the final set of two-fluid equations given by

$$\begin{gathered}
\left(\frac{dn_\alpha}{dt}\right)_\alpha + n_\alpha \nabla \cdot \mathbf{v}_\alpha = 0 \\
n_\alpha m_\alpha \left(\frac{d\mathbf{v}_\alpha}{dt}\right)_\alpha - q_\alpha n_\alpha (\mathbf{E} + \mathbf{v}_\alpha \times \mathbf{B}) + \nabla \cdot \mathsf{P}_\alpha = \mathbf{R}_\alpha \\
\frac{3}{2} n_\alpha \left(\frac{dT_\alpha}{dt}\right)_\alpha + \mathsf{P}_\alpha : \nabla \mathbf{v}_\alpha + \nabla \cdot \mathbf{h}_\alpha = Q_\alpha \\
\nabla \times \mathbf{E} = -\frac{\partial \mathbf{B}}{\partial t} \\
\nabla \times \mathbf{B} = \mu_0 e (n_i \mathbf{v}_i - n_e \mathbf{v}_e) + \frac{1}{c^2} \frac{\partial \mathbf{E}}{\partial t} \\
\nabla \cdot \mathbf{E} = \frac{e}{\epsilon_0} (n_i - n_e) \\
\nabla \cdot \mathbf{B} = 0
\end{gathered} \tag{2.38}$$

where it has been assumed that the ions are singly charged so that $q_i = -q_e = e$.

This set of equations is exact if not very useful since there is as yet no prescription for closing the system: there are more unknowns than equations. The particular prescription that leads to the ideal MHD equations consists of the following steps. First, certain asymptotic orderings are introduced which eliminate the very high-frequency, short-wavelength information in the model. These orderings are well satisfied for the phenomena of interest, i.e., those involving the macroscopic behavior of fusion plasmas. Next, the equations are rewritten as a set of single-fluid equations by the introduction of appropriate single-fluid variables. In the final and crucial step the plasma is assumed to be collision dominated. This enables the higher-order moments to be approximated from standard transport theory. It is then possible to determine a set of validity conditions such that the remaining equations are those of ideal MHD.

A remarkable feature of the derivation is that while the model is derived on the assumption of a collision-dominated plasma, the final ideal MHD equations do not depend in any explicit way on the details of the collisions. The main purpose of the collisional transport theory is to provide the appropriate scaling of the transport coefficients which are then systematically ordered out of the model by virtue of the length and time scales associated with macroscopic MHD behavior.

2.3.3. Low-Frequency, Long-Wavelength, Asymptotic Expansions

There are two important approximations that can be made, leading to substantial simplifications in the two-fluid equations. These approximations serve to eliminate the high-frequency, short-wavelength information from the model. Each approximation has the form of an asymptotic expansion, eliminating a leading-order time derivative and thereby altering the basic mathematical structure of the model.

The first approximation represents the transformation of the full Maxwell's equations to the low-frequency Maxwell's equations. This limit is formally accomplished by letting $\epsilon_0 \to 0$. As a result, the displacement current $\epsilon_0\, \partial \mathbf{E}/\partial t$ and the net charge $\epsilon_0 \nabla \cdot \mathbf{E}$ can both be neglected. The resulting set of equations is easily shown to be Gallilean invariant.

The neglect of the displacement current requires that the electromagnetic waves of interest have phase velocities much slower than the speed of light, $\omega/k \ll c$ and that the characteristic thermal velocities be nonrelativistic, V_{T_e}, $V_{T_i} \ll c$, $V_{T_\alpha} \equiv (2T_\alpha/m_\alpha)^{1/2}$. The neglect of the net charge restricts attention to plasma behavior whose characteristic frequency is much less than the electron plasma frequency $\omega \ll \omega_{pe}$, $\omega_{pe} = (n_0 e^2/m_e \epsilon_0)^{1/2}$ and whose characteristic length is much longer than the Debye length, $a \gg \lambda_d$, $\lambda_d = V_{T_e}/\omega_{pe}$. An examination of Tables 2.1 and 2.2 shows that these assumptions are satisfied by a large margin when considering the macroscopic behavior of fusion plasmas.

The neglect of $\epsilon_0 \nabla \cdot \mathbf{E}$ implies that

$$n_i = n_e \equiv n \tag{2.39}$$

This relationship is called the quasineutral approximation. It should be emphasized that Eq. (2.39) does not imply that $\mathbf{E} = 0$ or $\nabla \cdot \mathbf{E} = 0$, only that $\epsilon_0 \nabla \cdot \mathbf{E}/en \ll 1$. For any low-frequency macroscopic charge separation that tends to develop, the electrons have more than an adequate time to respond, thus creating an electric field which maintains the plasma in local quasineutrality.

As an example, consider an electrostatic problem with $\mathbf{E} = -\nabla\phi$ and $n_e = n_e(\phi, \mathbf{r})$ and $n_i = n_i(\phi, \mathbf{r})$. Equating n_e to n_i and inverting the relationship then gives $\phi = \phi(\mathbf{r})$. Quasineutrality implies that the electric field calculated from $\mathbf{E} = -\nabla\phi$ satisfies $\epsilon_0 \nabla \cdot \mathbf{E}/en_e \ll 1$.

The second asymptotic assumption neglects electron inertia in the electron momentum equation and is accomplished formally by letting $m_e \to 0$. This implies that on time scales of MHD interest the electrons have an infinitely fast response time because of their small mass. Specifically, time scales long compared to those of the electron plasma frequency, ω_{pe}, and the electron cyclotron frequency, $\omega_{ce} = eB/m_e$, are required. Similarly, the length scales must be long compared to the Debye length, λ_d, and the electron gyro radius, $r_{Le} = V_{T_e}/\omega_{ce}$. As before, an examination of Tables 2.1 and 2.2 shows that these conditions are easily satisfied for macroscopic phenomena in fusion plasmas.

One somewhat subtle point should be made regarding the neglect of electron inertia. There is a class of low-frequency, long-wavelength plasma modes whose length and time scales are not too different from those of MHD. These modes, called drift waves, are strongly affected by electron dynamics along the magnetic field lines. Thus, even with their small mass, the electrons do not respond "instantaneously" since they have to move a long distance to transverse one wavelength. By neglecting electron inertia, the physics of the drift wave is no longer accurately treated. One should keep in mind that a complete study of low-frequency plasma stability may require several separate treatments of different phenomena.

As a general comment, note that ideal MHD invariably treats motion parallel to the field lines incorrectly. However, for most problems involving MHD equilibrium and stability these motions are found to be unimportant. Parallel motion does play an important rôle in more sophisticated models, such as those describing drift waves, but the corresponding instabilities when they occur are much weaker than MHD modes.

2.3.4. The Single-Fluid Equations

Using the asymptotic approximations just described, one can derive a set of single-fluid equations by introducing appropriate fluid variables. To begin, it is customary to introduce a mass density rather than a number density. Since $m_e \to 0$ and $n_i = n_e \equiv n$, the mass density is defined as

$$\rho = m_i n \tag{2.40}$$

Likewise, the momentum of the fluid is carried by the ions, so that the appropriate definition of the fluid velocity is given by

$$\mathbf{v} = \mathbf{v}_i \tag{2.41}$$

The current density is proportional to the difference in flow velocity between electrons and ions

$$\mathbf{J} = en(\mathbf{v}_i - \mathbf{v}_e) \tag{2.42}$$

or

$$\mathbf{v}_e = \mathbf{v} - \mathbf{J}/en \tag{2.43}$$

The final definitions required are for the total pressure and temperature

$$\begin{aligned} p &= nT = p_e + p_i \\ T &= T_e + T_i \end{aligned} \tag{2.44}$$

Equations (2.40)–(2.44) relate the single-fluid variables ρ, $\mathbf{v}$, $\mathbf{J}$, p, T to the two-fluid variables n, $\mathbf{v}_i$, $\mathbf{v}_e$, p_i, p_e, T_e, T_i.

Two points require discussion before proceeding with the single-fluid equations. First, care must be exercised to conserve the total information content of the starting equations. For instance, several of the single-fluid equations are obtained by combining various equations of the two-fluid model. If these manipulations involve two equations from the starting model, then two equivalent equations must appear in the final model. This is not as trivial as it might seem, as evidenced by the second point of discussion: namely, that if one counts fluid variables then the two-fluid model has 11 unknowns while the single-fluid model has nine unknowns. A careful accounting of the information content in the two-fluid equations is shown to explain and eliminate this imbalance.

Consider now the derivation of the single-fluid model. The first equation is obtained from the mass conservation equations. Multiplying the ion mass equation by m_i one finds

$$\frac{\partial \rho}{\partial t} + \nabla \cdot \rho \mathbf{v} = 0 \tag{2.45}$$

which is identical to Eq. (2.1), the ideal MHD mass conservation relation. The other information contained in these equations is obtained by multiplying the electron and ion equations by e and then subtracting. The result is

$$\nabla \cdot \mathbf{J} = 0 \tag{2.46}$$

Equation (2.46) is consistent with charge conservation in the low-frequency form of Maxwell's equations.

The next set of single-fluid equations follows from the momentum equations. The electron and ion equations are first added together. Making use of the fact that $\mathbf{R}_e = -\mathbf{R}_i$ leads to the relationship

$$\rho \frac{d\mathbf{v}}{dt} - \mathbf{J} \times \mathbf{B} + \nabla p = -\nabla \cdot (\mathbf{\Pi}_i + \mathbf{\Pi}_e) \tag{2.47}$$

Here, $d/dt = \partial/\partial t + \mathbf{v} \cdot \nabla$ now represents the convective derivative moving with the (ion) fluid. The left-hand side of Eq. (2.47) corresponds to the ideal MHD momentum equation. The condition under which the right-hand side is negligible determines one range of validity for the MHD equations and will be discussed shortly. Note that there is no electric force, $\sigma\mathbf{E}$, acting on the fluid since $\sigma = 0$ as a result of the quasineutral approximation.

The second piece of information contained in the two-fluid momentum equations is obtained by simply rewriting the electron equation in terms of the single-fluid variables. This leads to the following relation:

$$\mathbf{E} + \mathbf{v} \times \mathbf{B} = \frac{1}{en}(\mathbf{J} \times \mathbf{B} - \nabla p_e - \nabla \cdot \mathbf{\Pi}_e + \mathbf{R}_e) \tag{2.48}$$

The left-hand side of Eq. (2.48) corresponds to Ohm's law. The conditions under which the right-hand side is negligible determine another range of validity to be discussed shortly.

The third set of single-fluid equations follows from the two-fluid energy equations. After some straightforward algebra one obtains

$$\frac{d}{dt}\left(\frac{p_i}{\rho^\gamma}\right) = \frac{2}{3\rho^\gamma}(Q_i - \nabla \cdot \mathbf{h}_i - \mathbf{\Pi}_i : \nabla \mathbf{v}) \tag{2.49}$$

$$\frac{d}{dt}\left(\frac{p_e}{\rho^\gamma}\right) = \frac{2}{3\rho^\gamma}\left[Q_e - \nabla \cdot \boldsymbol{h}_e - \mathbf{\Pi}_e : \nabla\left(\mathbf{v} - \frac{\mathbf{J}}{en}\right)\right] + \frac{1}{en}\mathbf{J} \cdot \nabla\left(\frac{p_e}{\rho^\gamma}\right) \tag{2.50}$$

where $\gamma = 5/3$. The last term in Eq. (2.50) results from the relation $(d/dt)_e = (d/dt) - (\mathbf{J}/en) \cdot \nabla$. The left-hand sides of Eqs. (2.49) and (2.50) are closely related to the ideal MHD equation of state. There are a number of terms on the right-hand side, which in order to be neglected define additional conditions on the region of validity. Furthermore, since MHD is a single-fluid model, still other assumptions must be made to couple the individual electron and ion temperatures and pressures into single-fluid variables.

The final equations of the single-fluid model are the remaining low-frequency Maxwell equations given by

$$\begin{aligned} \nabla \times \mathbf{E} &= -\frac{\partial \mathbf{B}}{\partial t} \\ \nabla \times \mathbf{B} &= \mu_0 \mathbf{J} \\ \nabla \cdot \mathbf{B} &= 0 \end{aligned} \tag{2.51}$$

As stated previously, Eq. (2.46) is consistent with the low-frequency form of Ampere's law.

The single-fluid model is described by Eqs. (2.45)–(2.51). No assumptions other than the asymptotic approximations have been made at this point. Although the left-hand side of the fluid equations is identical to the MHD model, the full set of equations is still incomplete because of the presence of the as yet undefined higher moments and the unresolved coupling of the electron and ion energies.

2.3.5. The Ideal MHD Limit

This section contains a description of the critical assumptions required for the single-fluid model to reduce to ideal MHD; in particular, quantitative conditions are derived which determine when the right-hand sides of Eqs. (2.47)–(2.50) are negligible.

The basic requirement for the validity of ideal MHD is that both the electrons and ions be collision dominated. This is the usual requirement for a fluid model to be useful. If there are sufficient collisions, a given particle remains reasonably close to its neighboring particles during the time scales of interest. In this case the division of the plasma into small identifiable fluid elements provides a good description of the physics.

The question as to whether a given model describes collision dominated or collision-free behavior rests largely in the evolution of the pressure tensor; that is, if one considers the exact mass and momentum equations it is only the pressure tensor that remains as an undefined higher moment. A comparison of Eqs. (2.47) and (2.2) indicates that in ideal MHD, $\mathbf{\Pi}$ must be negligible, implying that the full pressure tensor P reduces to a simple scalar isotropic pressure p. One intuitively expects an isotropic pressure to arise in systems where many collisions take place on a time scale short compared to those of interest. The collisions rapidly randomize the distribution function giving rise to an isotropic pressure. This is the basic assumption of the scalar pressure ideal MHD model. In collisionless models closure of the pressure tensor is often more complicated, requiring the solution of reduced kinetic equations. A remarkable feature of MHD is that for most problems involving equilibrium and stability the specific model describing the evolution of P plays only a minor role in the physics. It is this fact that is ultimately responsible for the unexpectedly reliable predictions of ideal MHD well outside its regime of validity.

The collision-dominated assumption enters into the single-fluid equations as follows. In this limit the distribution function for both electrons and ions is nearly locally Maxwellian in form. As a consequence, one can refer to well-established theories, such as given by Braginskii (1965), in order to obtain expressions for the higher-order fluid moments in terms of appropriate transport coefficients. By defining the characteristic length and time scales of ideal MHD it is then possible to compare the MHD terms with the transport terms. This then determines the conditions under which the right-hand sides of the single-fluid equations are negligible and the model reduces to ideal MHD.

A convenient place to begin the analysis is with the specification of the characteristic MHD length and time scales. Since the main goal of ideal MHD is the investigation of macroscopic phenomena, the length scales of interest correspond to the macroscopic dimensions of the plasma denoted by a. The typical time scale of MHD interest corresponds to a/V_{T_i}, $V_{T_i} = (2T_i/m_i)^{1/2}$, the ion thermal transit time across a macroscopic plasma dimension. This time scale is characteristic of many MHD plasma instabilities and represents the fastest time scale in which macroscopic plasma motion can occur. It should be noted that certain MHD waves and other phenomena can have time scales somewhat faster or slower than a/V_{T_i}. For the present purposes, however, it is not crucial to make these distinctions. In determining the scaling relations it is helpful to introduce the characteristic MHD frequency ω and wave number k as follows:

$$\omega \sim \frac{\partial}{\partial t} \sim \frac{V_{T_i}}{a}$$

$$k \sim \nabla \sim \frac{1}{a} \tag{2.52}$$

and, similarly, the resulting velocity

$$\frac{\omega}{k} \sim v \sim V_{T_i} \tag{2.53}$$

The next step in the analysis is to consider the conditions for the collision-dominated transport theory to be valid. There are two such conditions. The first requires that during the MHD time scale of interest, each species has sufficient collisions to make the distribution function nearly Maxwellian. For the ions the dominant collision mechanism is due to ion–ion interactions, characterized by a collision time τ_{ii}. The electrons become Maxwellian by colliding with either other electrons or ions. Since $\tau_{ee} \sim \tau_{ei}$ it is not important to make this distinction. Hence, the condition that each species be collision dominated is given by

$$\text{Ions } \omega\tau_{ii} \sim V_{T_i}\tau_{ii}/a \ll 1 \tag{2.54a}$$

$$\text{Electrons } \omega\tau_{ee} \sim (m_e/m_i)^{1/2} V_{T_i}\tau_{ii} \ll 1 \tag{2.54b}$$

Here, use has been made of the fact that $\tau_{ee} \sim (m_e/m_i)^{1/2}\tau_{ii}$ when $T_e \sim T_i$. As might be expected, this condition is more restrictive for ions than electrons.

The second condition for the collision-dominated theory to be valid requires that the macroscopic scale length be much longer than the mean free path. Noting that the mean free path for each species is given by $\lambda_\alpha \sim V_{T\alpha}\tau_{\alpha\alpha}$, the condition $\lambda_\alpha \ll a$ reduces to Eq. (2.54a) for both electrons and ions; that is, for the ions both the time scale and length scale requirements yield the same condition for high collisionality. For the electrons, the length scale requirement is more restrictive and yields the same condition as for the ions. Thus, the conditions for a collision-dominated plasma can be summarized as follows:

$$V_{T_i}\tau_{ii}/a \sim V_{T_e}\tau_{ee}/a \ll 1 \tag{2.55}$$

The collision-dominated assumption is now used to estimate the higher-order moments, which enter as transport terms, in the single-fluid equations. (Readers unfamiliar with the evaluation of transport coefficients should attempt Problem 2.2, which demonstrates a simple procedure for deriving such coefficients.)

The analysis begins with Eq. (2.47), the momentum equation. The matrix elements for $\mathbf{\Pi}_i$ and $\mathbf{\Pi}_e$ are a rather complicated series of terms. The leading-order effect, however, is viscosity. Moreover, the ion viscosity coefficient is larger than that of the electrons by a factor $(m_i/m_e)^{1/2}$. Using the results of Braginskii (1965) it follows that the largest elements of the $\mathbf{\Pi}_i$ tensor have the form (in rectangular coordinates)

$$\Pi_{jj} \sim \mu(2\nabla_{\|} \cdot \mathbf{v}_{\|} - \tfrac{2}{3}\nabla \cdot \mathbf{v}) \sim \mu V_{T_i}/a \tag{2.56}$$

where the viscosity coefficient μ is given by

$$\mu \sim nT_i\tau_{ii} \tag{2.57}$$

If the right-hand side of Eq. (2.47) is now compared with the ∇p term one finds

$$|\nabla \cdot \mathbf{\Pi}_i/\nabla p| \sim V_{T_i}\tau_{ii}/a \ll 1 \tag{2.58}$$

Therefore, if the collision-dominated assumption is satisfied, the viscosity is negligible and the momentum equation reduces to that of ideal MHD.

The next equation to consider is the Ohm's law, given by Eq. (2.48). From the argument just given it is clear that the $\mathbf{\Pi}_e$ term can be neglected compared to the ∇p_e term. From the momentum equation it also follows that the $\mathbf{J} \times \mathbf{B}$ and ∇p_e terms are comparable.The $\mathbf{J} \times \mathbf{B}$ term represents the Hall effect while the ∇p_e term represents the effect of the electron diamagnetic drift. Comparing either of these terms with the $\mathbf{v} \times \mathbf{B}$ term yields

$$\frac{|\nabla p_e/en|}{|\mathbf{v} \times \mathbf{B}|} \sim \frac{r_{Li}}{a} \tag{2.59}$$

where $r_{Li} = V_{T_i}/\omega_{ci}$ is the ion gyro radius. Thus, if it is assumed that

$$r_{Li}/a \ll 1 \tag{2.60a}$$

the $\mathbf{J} \times \mathbf{B}$ and ∇p_e terms can be neglected in the Ohm's laws. Note that

Eq. (2.60a) implies that

$$\omega/\omega_{ci} \sim r_{Li}/a \ll 1 \tag{2.60b}$$

MHD frequencies are much slower than the ion gyro frequency.

The remaining term on the right-hand side of Eq. (2.48), $\mathbf{R}_e/en$, represents the momentum transfer due to the friction of collision between electrons and ions. The dominant contribution of $\mathbf{R}_e$ is electrical resistivity. Using the results from transport theory, $\mathbf{R}_e$ can be expressed as

$$\frac{1}{en}\mathbf{R}_e \sim \eta\mathbf{J} \tag{2.61}$$

where the electrical resistivity η is given by

$$\eta \sim \frac{m_e}{ne^2\tau_{ei}} \tag{2.62}$$

Substituting the scaling relation (from the momentum equation) $|\mathbf{J}| \sim |\nabla p|/|\mathbf{B}|$ leads to the following requirement for the $\eta\mathbf{J}$ term to be negligible compared to the $\mathbf{v} \times \mathbf{B}$ term:

$$\frac{|\eta\mathbf{J}|}{|\mathbf{v} \times \mathbf{B}|} \sim \frac{(m_e/m_i)^{1/2}}{\omega\tau_{ii}}\left(\frac{r_{Li}}{a}\right)^2 \ll 1 \tag{2.63}$$

Equation (2.63) implies that, while τ_{ii} must be sufficiently large for the theory to be valid, there should not be too many collisions or else the plasma will be dominated by resistive diffusion. Alternatively, one can view Eq. (2.63) as a requirement to make the macroscopic dimension a large enough so that the resistive diffusion time is long compared to the characteristic MHD time. Thus, in order for the ideal MHD Ohm's law to apply, the small gyro radius and small resistivity conditions must both be satisfied.

The last equations to consider are the energy relations given by Eqs. (2.49) and (2.50). With the assumptions already made, most of the terms on the right-hand sides are negligible. In particular,

$$\begin{aligned}
\frac{\mathbf{\Pi}_e : \nabla(\mathbf{J}/en)}{\partial p_e/\partial t} &\sim \left(\frac{m_e}{m_i}\right)^{1/2}\left(\frac{r_{Li}}{a}\right)\left(\frac{V_{T_i}\tau_{ii}}{a}\right) \ll 1 \\
\frac{(\mathbf{J}\cdot\nabla p_e)/en}{\partial p_e/\partial t} &\sim \frac{r_{Li}}{a} \ll 1 \\
\frac{\mathbf{\Pi}_e : \nabla\mathbf{v}}{\partial p_e/\partial t} &\sim \left(\frac{m_e}{m_i}\right)^{1/2}\left(\frac{V_{T_i}\tau_{ii}}{a}\right) \ll 1 \\
\frac{\mathbf{\Pi}_i : \nabla\mathbf{v}}{\partial p_i/\partial t} &\sim \left(\frac{V_{T_i}\tau_{ii}}{a}\right) \ll 1
\end{aligned} \tag{2.64}$$

The remaining terms contain the heat flux $\mathbf{h}$ and the collisional heating Q. The largest contribution to $\mathbf{h}$ is due to thermal conductivity. There are separate conductivity coefficients along and perpendicular to the field and for electrons and ions. By far, the largest coefficients are those along the field so that $\mathbf{h}_\alpha \approx -\kappa_{\|\alpha}\nabla_\| T_\alpha$. The main contributions to the collisional heating Q are joule heating and electron and ion energy equilibration. If the condition to neglect resistive diffusion, Eq. (2.63), is satisfied, then joule heating is also negligible. Therefore, what remains of the energy equations is as follows:

$$\begin{aligned}\frac{d}{dt}\left(\frac{p_i}{\rho^\gamma}\right) &= \frac{2}{3\rho^\gamma}\left[\nabla_\| \cdot (\kappa_{\|i}\nabla T_i) + \frac{n(T_e - T_i)}{\tau_{\text{eq}}}\right] \\ \frac{d}{dt}\left(\frac{p_e}{\rho^\gamma}\right) &= \frac{2}{3\rho^\gamma}\left[\nabla_\| \cdot (\kappa_{\|e}\nabla T_e) - \frac{n(T_e - T_i)}{\tau_{\text{eq}}}\right]\end{aligned} \tag{2.65}$$

where τ_{eq} is the energy equilibration time. As they now stand, these equations describe the separate time evolution of both the electrons and ions. In order to obtain the MHD energy equation a further assumption is required which couples the electron and ion energies together. This assumption corresponds to the condition that the energy equilibration time be short compared to the characteristic time of interest, so that $T_e \approx T_i$. This can be expressed as $\omega\tau_{\text{eq}} \ll 1$, or, using the relation $\tau_{\text{eq}} \sim (m_i/m_e)^{1/2}\tau_{ii}$,

$$\left(\frac{m_i}{m_e}\right)^{1/2}\frac{V_{T_i}\tau_{ii}}{a} \ll 1 \tag{2.66}$$

Because the energy equilibration time is long compared to the momentum exchange time, Eq. (2.66) is even more restrictive in terms of collisionality than Eq. (2.55), the basic condition for the collision-dominated expansion to be valid.

If Eq. (2.66) is satisfied, then

$$\begin{aligned}T_i &\approx T_e = T/2 \\ p_i &\approx p_e = p/2\end{aligned} \tag{2.67}$$

Note that the strong collisionality assumption resolves the problem of the unequal number of fluid variables in the two-fluid and ideal MHD models.

Both the electron and ion energy equations yield the identical piece of information—namely, that $T_i \approx T_e$, in the limit of small τ_{eq}. The second piece of independent information follows from annihilating the $T_i \approx T_e$ redundancy by adding the equations and setting $T_e = T_i = T/2$.

The result is

$$\frac{d}{dt}\left(\frac{p}{\rho^{\gamma}}\right) = \frac{1}{3\rho^{\gamma}}\nabla_{\parallel}\cdot[(\kappa_{\parallel e} + \kappa_{\parallel i})\nabla_{\parallel}T] \tag{2.68}$$

The ideal MHD equation of state follows when the right-hand side of Eq. (2.68) can be neglected. Transport theory shows that the parallel electron thermal conductivity, $\kappa_{\parallel e} \sim nT_e\tau_{ee}/m_e$, is larger by $(m_i/m_e)^{1/2}$ than that of the ions. Consequently, the right-hand side is negligible when

$$\frac{\nabla_{\parallel}\cdot(\kappa_{\parallel e}\nabla_{\parallel}T)}{\partial p/\partial t} \sim \left(\frac{m_i}{m_e}\right)^{1/2}\frac{V_{T_i}\tau_{ii}}{a} \ll 1 \tag{2.69}$$

which is identical to Eq. (2.66).

This completes the derivation of the ideal MHD equations. In summary, the validity of ideal MHD imposes several conditions on the plasma: (1) high collisionality, (2) characteristic dimensions much larger than an ion gyro radius, and (3) sufficient size that resistive diffusion is negligible, despite the high collisionality.

2.4. REGION OF VALIDITY

2.4.1. Overall Criteria

With the somewhat tedious analysis completed, the next important issue is to focus on the most restrictive assumptions required for the derivation and to discuss their regions of validity. Furthermore, there is additional useful information to be extracted by examining the conditions for validity one equation at a time. This provides insight into which specific phenomena are not accurately described by ideal MHD; perhaps more important, it also indicates those phenomena that will be reliably treated even when the overall validity conditions are violated.

These goals are accomplished in several steps. The starting point is the introduction of dimensionless variables

$$\begin{aligned} y &= \frac{r_{Li}}{a} \\ x &= \left(\frac{m_i}{m_e}\right)^{1/2}\frac{V_{T_i}\tau_{ii}}{a} \end{aligned} \tag{2.70}$$

As just discussed, there are three independent conditions which must be satisfied for ideal MHD to be valid. They are

$$\begin{aligned}&(1)\ \text{High collisionality } x \ll 1\\&(2)\ \text{Small gyro radius } y \ll 1\\&(3)\ \text{Small resistivity } y^2/x \ll 1\end{aligned}\tag{2.71}$$

These conditions are illustrated in Fig. 2.2. The shaded region on each curve indicates that the given condition has been satisfied. In the region labeled "ideal MHD" all three conditions are simultaneously satisfied. Although a significant number of approximations has been made in the derivation, there is a substantial region of parameter space where all the assumptions are satisfied and the ideal MHD equations are valid.

The next question to be asked is whether plasmas of fusion interest lie in the region of MHD validity. This can be answered by transforming the (x, y) diagram into an (n, T) diagram and observing whether values of n and T of fusion interest lie in the region of validity.

The first step is to define the parameter range of fusion interest. Past experiments and extrapolations to future fusion reactors indicate that the densities and temperatures of fusion plasmas lie in the range

$$\begin{aligned}10^{18}\,\text{m}^{-3} < n < 10^{22}\,\text{m}^{-3}\\0.5\,\text{keV} < T < 50\,\text{keV}\end{aligned}\tag{2.72}$$

These conditions describe a rectangle in the (n, T) diagram.

The next step is to rewrite the conditions in Eq. (2.71) in terms of n and T. Since the B field explicitly appears, a prescription is needed to specify how B varies with n and T. A reasonable choice is to assume

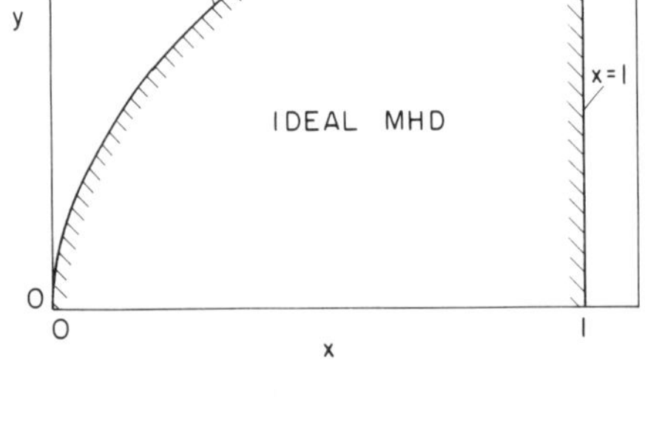

Figure 2.2. Region of validity of the ideal MHD model in terms of the normalized variables $y = r_{Li}/a$ vs. $x = (m_i/m_e)^{1/2}(V_{T_i}\tau_{ii}/a)$.

$\beta = 2\mu_0 nT/B^2$ is held fixed. The parameter β measures the ratio of plasma energy to magnetic energy. Its value is important in fusion reactor designs and is often limited by MHD instabilities. Consequently, treating β and the scale length a as parameters leads to the following expressions for the three MHD criteria of validity:

$$\begin{aligned} &(1)\ \text{High collisionality } x = 3.0 \times 10^3 (T^2/an) \ll 1 \\ &(2)\ \text{Small gyro radius } y = 2.3 \times 10^{-2} (\beta/na^2)^{1/2} \ll 1 \\ &(3)\ \text{Small resistivity } y^2/x = 1.8 \times 10^{-7} \beta/aT^2 \ll 1 \end{aligned} \tag{2.73}$$

In these expressions the units are a (m), T(keV), n ($10^{20}\,\mathrm{m}^{-3}$). Also, the Coulomb logarithm appearing in the expression for τ_{ii} has been set to 15, and the ion mass corresponds to deuterium.

Equation (2.73) is illustrated in Fig. 2.3 for the case $a = 1$ m and $\beta = 0.05$. Observe that the conditions of small gyro radius and small resistivity are well satisfied for plasmas of fusion interest. Note, however, that the high collisionality assumption is never satisfied. *Thus, the region in which ideal MHD is valid completely excludes plasmas of fusion interest*! This disconcerting conclusion is, however, inconsistent with the overwhelming empirical evidence demonstrating that ideal MHD provides a very accurate description of most macroscopic plasma behavior.

The question may now be asked whether or not this is coincidence or, alternatively, the result of some subtle and perhaps unexpected physics? The resolution follows from an examination of the validity

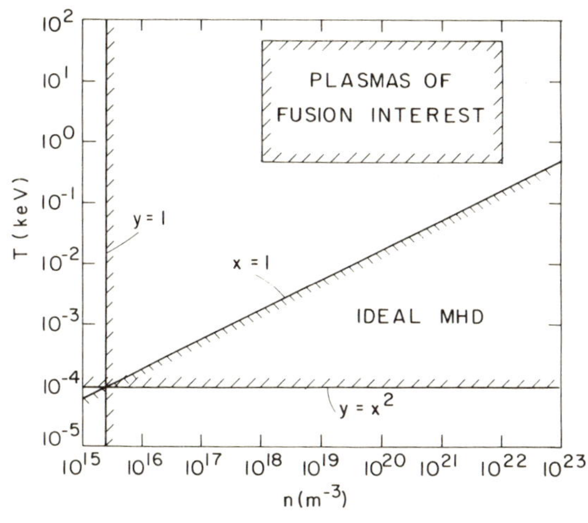

Figure 2.3. Region of validity of the ideal MHD model in (n, T) parameter space for fixed $\beta = 0.05$ and $a = 1$ m.

conditions, one equation at a time. The results show that there are indeed subtle physics issues at play which, in addition to providing an explanation, suggest the introduction of a modified MHD model, collisionless MHD.

2.4.2. Conservation of Mass

To begin, consider the conservation of mass relation, Eq. (2.1). Since none of the MHD validity conditions are required to obtain this equation, its range of applicability extends far beyond that of ideal MHD. It is valid for collisionless or collision-dominated systems regardless of the size of the ion gyro radius or the influence of plasma resistivity. Modifications are required when phenomena such as recombination, ionization, charge exchange, or source terms due to gas puffing or pellet injection are present.

2.4.3. Momentum Equation

The validity of the momentum equation is also much wider than the ideal MHD conditions would imply, although the reasons involve considerably more subtle physics. Recall that the collision-dominated assumption is required to neglect $\mathbf{\Pi}$. In a collisionless plasma the magnetic field in a certain sense plays the role of collisions for the perpendicular motion; that is, perpendicular to the field, particles are confined to the vicinity of a given field line executing nearly (two-dimensional) isotropic motion if their gyro radius is much smaller than the characteristic plasma dimension. In fact, a calculation of $\mathbf{\Pi}_{\perp i}$ in the collisionless regime [see, for instance, Bowers and Haines (1971)] shows that $\mathbf{\Pi}_{\perp i}/p_i \sim r_{Li}/a \ll 1$. Therefore, the perpendicular motion is fluid-like, implying that the perpendicular components of the momentum equation provide an excellent description of plasma behavior in either the collision-dominated or collision-free regimes.

The general situation parallel to the field is much more complicated and it is here that ideal MHD treats the physics very inaccurately. In a collisionless plasma there is no reason *a priori* to assume that any simple relationship exists between the perpendicular and parallel pressures, $p_\perp$ and $p_\parallel$; perpendicular to the field, particles execute well-confined gyro orbits, while along the field they execute free-streaming kinetic motion.

The practical situation parallel to the field simplifies considerably for the following subtle reason. Consider the application of the MHD model over the narrow increment Δt as discussed in connection with Fig. 2.1. By assumption, the plasma has evolved through a series of macroscopically stable states over the long period of time $0 < t < t_1$. During this period there is more than adequate time for collisions to isotropize the pressure. Consequently, at $t = t_1$, it is a good approximation to set $p_\perp = p_\parallel$ as *initial conditions*. During the MHD increment Δt, the two pressures will in general evolve quite differently because of the different physics involved perpendicular and parallel to the field. However, it is a property of most MHD instabilities that the plasma pressure neither compresses nor expands as the modes grow; instead, the pressure is convected with the fluid motion. Compression and expansion correspond to very stable motions and would suppress most instabilities if excited to any appreciable level. This crucial fact of plasma behavior, demonstrated explicitly in Chapter 8, implies that neither $p_\perp$ nor $p_\parallel$ changes significantly from their initial values during the increment Δt.

Under the situation just described, it is accurate to use the ideal MHD momentum equation in both the collision-dominated and collisionless regimes. In the collisionless regime that part of the physics treated inaccurately by ideal MHD does not matter: the nature of MHD instabilities is such that motions where the inaccuracies would be important are not excited.

2.4.4. Energy Equation

The ideal MHD energy equation [Eq. (2.3)] is never valid for plasmas of fusion interest. Like the parallel motion in the momentum equation, it too, for similar reasons, has little effect on the applicability of the model.

Even assuming that evolution over the long period $0 < t < t_1$ leads to temperature equilibration, one still requires negligible parallel thermal conductivity for the adiabatic energy relation to be valid. Substituting numerical values indicates that it would be far more accurate to treat the parallel thermal conductivity as infinite rather than zero. Therefore, a more appropriate equation of state follows from setting $\nabla_\parallel \cdot \kappa_\parallel \nabla_\parallel T = 0$, yielding

$$\mathbf{B} \cdot \nabla T = 0 \tag{2.74}$$

An alternate interpretation of Eq. (2.74) in the collisionless regime is that highly mobile particles are free to stream along magnetic field lines, thereby rapidly phase mixing and equilibrating any parallel gradients that tend to arise.

The situation is as follows. The ideal MHD equation of state in general allows compressions and expansions as well as convection. However, because of the nature of MHD instabilities, the plasma behaves incompressibly. Replacing the ideal equation of state, Eq. (2.3), with the more appropriate collisionless relation, Eq. (2.74), further reinforces the incompressible nature of MHD instabilities in that parallel temperature gradients are not permitted.

The equivalence of the two equations of state can be shown analytically if one starts with the assumption that the motions of interest for MHD instabilities are incompressible. By definition, incompressibility implies

$$\frac{d\rho}{dt} = 0 \tag{2.75}$$

indicating that the plasma mass is convected with the fluid. From the conservation of mass relation it then follows that incompressibility is equivalent to

$$\nabla \cdot \mathbf{v} = 0 \tag{2.76}$$

The adiabatic equation of state, before incompressibility is assumed, can be written as

$$\frac{dp}{dt} + \gamma p \nabla \cdot \mathbf{v} = 0 \tag{2.77}$$

Invoking incompressibility gives

$$\frac{dp}{dt} = 0 \tag{2.78}$$

implying that the energy as well as the mass is convected with the fluid.

The same result follows from the collisionless equation of state. Taking the total time derivative of Eq. (2.74) and using Faraday's law yields

$$\frac{d}{dt}\left[\mathbf{B} \cdot \nabla\left(\frac{p}{\rho}\right)\right] = \mathbf{B} \cdot \nabla \frac{d}{dt}\left(\frac{p}{\rho}\right) = 0 \tag{2.79}$$

Invoking incompressibility and assuming $\mathbf{B} \cdot \nabla \neq 0$ again leads to

$$\frac{dp}{dt} = 0 \tag{2.80}$$

The collision-dominated and collisionless equations of state are identical when the plasma motions of interest are incompressible.

A corollary of this analysis is the following. Any calculation using the ideal MHD model in which the results depend explicitly upon the ratio of specific heats γ is unreliable. Equation (2.77) implies that such results must involve compressibility. Therefore, they make explicit use of the collision-dominated assumption, which is not satisfied in fusion plasmas.

2.4.5. Ohm's Law

The final equation to consider is Ohm's law [Eq. (2.4)]. Here, the main terms neglected (i.e., Hall effect, electron diamagnetic drift, and electrical resistivity) do not depend upon the collision-dominated assumption. Thus, Ohm's law is valid in the collisionless regime for phenomena on the MHD time scale. Even so, there are two subtleties that should be noted.

First, for systems with substantial equilibrium flows, $v \sim V_{T_i}$, Ohm's law implies $E \sim V_{T_i}B$. The corrections to $\mathbf{E}$ from the Hall effect and the electron diamagnetic drift are smaller by r_{Li}/a and hence negligible. However, for systems in static equilibrium (i.e., $\mathbf{v} = 0$), Ohm's law predicts $\mathbf{E} = 0$. While the corrections to $\mathbf{E}$ are "small," of order r_{Li}/a, they must now be compared to zero; specifically, the corrections are given by

$$\mathbf{E} \approx \frac{m_i}{e\rho}(\mathbf{J} \times \mathbf{B} - \nabla p_e) \tag{2.81}$$

which for static equilibria with $T_e = T_i$ reduces to

$$\mathbf{E} \approx \frac{m_i}{2e}\nabla\left(\frac{\gamma}{\gamma - 1}\frac{p}{\rho}\right) \tag{2.82}$$

Consistent with the assumption of static equilibrium, the electric field is electrostatic: $\mathbf{E} = -\nabla\phi$. Since the electric field in the low-frequency form of Maxwell's equations appears only in Faraday's law, the error associated with not including the contribution to $\mathbf{E}$ given by Eq. (2.82) has no effect on the evaluation of the remainder of the fluid and field variables. This is not an issue for the perturbed fields since they vary on the MHD time scale. The perturbed electric field has a large inductive component which dominates the Hall effect and electron diamagnetic drift corrections.

The second subtlety involves the resistivity. Although its magnitude is very small it nonetheless plays an important role. This follows because the ideal MHD Ohm's law implies $E_\parallel = 0$ for any situation. With resistivity included, $E_\parallel = \eta J_\parallel$, which, while small, again must be compared to zero. The critical feature is that unlike the Hall and diamagnetic effects discussed above, the resistivity contribution does not arise from an electrostatic potential. As such, it has the important property that, while small, it allows a new class of motions to occur that are entirely excluded from ideal MHD: the tearing and reconnecting of magnetic field lines. These phenomena occur on a hybrid time scale related to the geometric mean of the short ideal MHD and long resistive diffusion time scales. Since the hybrid scale is much longer than the characteristic MHD time, resistivity has little effect on ideal MHD behavior. However, the hybrid time scale sets a well-defined upper limit to the incremental width of time Δt over which ideal MHD is valid. Typical numerical values for the hybrid scale are on the order of 1 ms.

2.4.6. Summary of Validity Conditions

The ideal MHD model follows from a straightforward, self-consistent closure of the two-fluid moment equations. The conditions for its validity are (1) high collisionality, (2) small ion gyro radius, and (3) small resistivity. Conditions (2) and (3) are well satisfied in fusion plasmas. In contrast, condition (1) is never satisfied.

A more subtle examination of ideal MHD, one equation at a time, shows that the model is still reliable even though the collision-dominated assumption is not satisfied. Crucial to this argument is the claim that for problems involving MHD equilibrium and stability, the plasma motions of interest are incompressible. Specifically, the conservation of mass and the perpendicular momentum equation are shown to be valid in the collisionless limit. However, both the parallel momentum equation and the energy equation are incorrect in this regime. Nevertheless, for incompressible motions neither of these equations plays an important role.

An analysis of Ohm's law demonstrates that ideal MHD accurately calculates the inductive part of the electric field, but often incorrectly predicts the electrostatic part. Even so, for the low-frequency Maxwell's equations, this has no effect on the evaluation of the remaining fluid and field variables. Finally, resistivity, while small, plays an important role in that it allows new motions to occur that are prohibited in ideal MHD. These phenomena occur on the slower MHD-diffusive hybrid time scale and thus set a practical upper limit on the maximum time period Δt over which ideal MHD can be considered reliable.

2.5. COLLISIONLESS MHD

The intuition just developed suggests the introduction of a modified MHD model known as collisionless MHD. This model has the property of making predictions that are nearly identical to those of ideal MHD in almost all cases of interest, but which assumes collisionless rather than collision-dominated physics; the regime of validity for collisionless MHD encompasses plasmas of fusion interest. In those special situations where the predictions do not coincide, neither model is in general reliable.

It is worth pointing out that other models also exist attempting to improve the reliability of ideal MHD. One such model is the Guiding Center Fluid model (sometimes known as the Double Adiabatic theory) in which the pressure is allowed to be anisotropic and there are separate equations of state for $p_\perp$ and $p_\|$. Its validity, like that of ideal MHD, should be questioned in the collisionless regime where behavior is not fluidlike along the field lines.

A second model is the Guiding Center Plasma model (sometimes known as zeroth-order Drift Kinetic theory). Here, the perpendicular motion is fluidlike, while a one-dimensional (in $u_\|$) kinetic equation governs the behavior parallel to the field. This model also allows pressure anisotropy. It is in general far more difficult to solve than ideal MHD because of the complexities associated with the parallel kinetic motion.

The collisionless MHD model actually represents a special simplifying limit of the Guiding Center Plasma theory. However, to demonstrate this requires a substantial amount of analysis, beyond the scope of the present textbook. Instead, a simpler heuristic derivation is presented based on single-particle guiding center orbit theory. The derivation of the mass conservation and momentum equations is, in point of fact, rigorous. The heuristic part of the analysis involves the closure of the system. Here, the critical assumption invoked is that the motions of interest are incompressible. This gives rise to an equation state where the plasma pressure is neither compressed nor expanded, but is simply convected with the fluid motion.

The calculation proceeds as follows. As discussed previously, the conservation of mass equation is exact requiring no assumptions about collisionality or gyro radius size. Hence, the ideal MHD mass equation also applies in the collisionless MHD model:

$$\frac{\partial \rho}{\partial t} + \nabla \cdot \rho \mathbf{v} = 0 \tag{2.83}$$

The major part of the derivation involves the momentum equation. The basic idea is to calculate the self-consistent perpendicular plasma current induced by the single-particle guiding center drifts in the low-frequency ($\omega \ll \omega_{ci}$), small gyro radius ($r_{Li} \ll a$), collisionless limit. The cross product of this current with **B** then gives the collisionless MHD momentum equation.

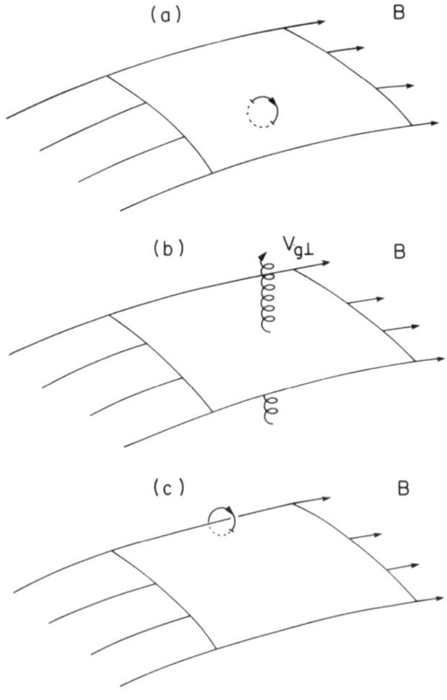

Figure 2.4. Mechanism of current generation due to single particle guiding center motion: (a) pure gyro motion: no current; (b) guiding center drifts: $\mathbf{J}_{g\perp}$; (c) magnetization current: $\mathbf{J}_M$.

The perpendicular currents that flow can be understood by examining Fig. 2.4. To zeroth order in r_{Li}/a each particle executes a small circular gyro orbit about a given field line (Fig. 2.4a). Since each orbit enters and leaves the surface one time, no current is generated by this motion.

There are two first-order (in r_{Li}/a) contributions to the perpendicular current. The first is due to the single-particle guiding center drifts: the $\mathbf{E} \times \mathbf{B}$ drift, the ∇B drift, the curvature drift, and the polarization drift. These drifts are perpendicular to the magnetic field and thereby produce a perpendicular current (Fig. 2.4b). The second contribution is due to the magnetization current. This is the current created by those gyro orbits that just intersect the surface once because of their proximity to the edge (Fig. 2.4c).

Turning to specifics, one can calculate the current due to the guiding center drifts by noting that the velocity of a particle gyrating about a field line can be expressed as $\mathbf{u} = \mathbf{u}_\perp + u_\parallel \mathbf{b} + \mathbf{V}_{g\perp}$. Here $\mathbf{u}_\perp$ is the fast gyration velocity about the field line, $u_\parallel$ is the velocity along the field line, $\mathbf{b}$ is a unit vector parallel to $\mathbf{B}$, and $\mathbf{V}_{g\perp}$ is the slower single-particle drift velocity of the particle's guiding center. [For those not completely familiar with single-particle motion in electric and magnetic fields, an excellent discussion has been given by Chen (1984).] In the low-frequency small gyro radius limit, $\mathbf{V}_{g\perp}$ is given by

$$\mathbf{V}_{g\perp} = \mathbf{V}_E + \mathbf{V}_{\nabla B} + \mathbf{V}_\kappa + \mathbf{V}_p \tag{2.84}$$

where $\mathbf{V}_E$ is the $\mathbf{E} \times \mathbf{B}$ velocity

$$\mathbf{V}_E = \frac{\mathbf{E} \times \mathbf{B}}{B^2} \tag{2.85}$$

$\mathbf{V}_{\nabla B}$ is the ∇B drift

$$\mathbf{V}_{\nabla B} = \frac{u_\perp^2}{2\omega_c} \frac{\mathbf{B} \times \nabla B}{B^2} \tag{2.86}$$

$\mathbf{V}_p$ is the polarization drift

$$\mathbf{V}_p = \frac{\mathbf{b}}{\omega_c} \times \frac{d}{dt}\left(\frac{\mathbf{E} \times \mathbf{B}}{B^2}\right) \tag{2.87}$$

and $\mathbf{V}_\kappa$ is the curvature drift

$$\mathbf{V}_\kappa = -\frac{u_\parallel^2}{\omega_c} \frac{\boldsymbol{\kappa} \times \mathbf{B}}{B} \tag{2.88}$$

with $\boldsymbol{\kappa} = \mathbf{b} \cdot \nabla \mathbf{b} = \mathbf{R}_c/R_c^2$ the curvature vector of the magnetic field line (i.e., $\mathbf{R}_c$ is the radius of curvature vector). These expressions are valid for both electrons and ions when one uses the appropriate sign and magnitude of q and m appearing in $\omega_c = qB/m$. Also, the fields are to be evaluated at the guiding center of the particle.

Combining these results leads to the following expressions for the ion and electron guiding center drifts:

$$\mathbf{V}_{g\perp i} = \frac{\mathbf{E} \times \mathbf{B}}{B^2} + \frac{1}{\omega_{ci}} \mathbf{b} \times \left[\frac{u_\perp^2}{2} \frac{\nabla B}{B} + u_\parallel^2 \boldsymbol{\kappa} + \frac{d}{dt}\left(\frac{\mathbf{E} \times \mathbf{B}}{B^2}\right)\right] \tag{2.89}$$

$$\mathbf{V}_{g\perp e} = \frac{\mathbf{E} \times \mathbf{B}}{B^2} + \frac{1}{\omega_{ce}} \mathbf{b} \times \left[\frac{u_\perp^2}{2} \frac{\nabla B}{B} + u_\parallel^2 \boldsymbol{\kappa} + \frac{d}{dt}\left(\frac{\mathbf{E} \times \mathbf{B}}{B^2}\right)\right] \tag{2.90}$$

In the MHD ordering the terms in the bracket are smaller by r_{Li}/a than the $\mathbf{E} \times \mathbf{B}$ velocity. Consequently,

$$\mathbf{V}_{g\perp e} \approx \mathbf{V}_{g\perp i} = \frac{\mathbf{E} \times \mathbf{B}}{B^2} \tag{2.91}$$

Since the dominant drift for each ion is identical (i.e., independent of u), this motion corresponds to the perpendicular component of the macroscopic fluid velocity; that is,

$$\mathbf{v}_\perp = \frac{1}{n} \int \mathbf{V}_{g\perp i} f_i(\mathbf{u})\, d\mathbf{u} = \frac{\mathbf{E} \times \mathbf{B}}{B^2} \tag{2.92}$$

Equation (2.92) can be rewritten as

$$\mathbf{E}_{\perp} + \mathbf{v} \times \mathbf{B} = 0 \tag{2.93}$$

which coincides with the perpendicular component of the ideal MHD Ohm's law.

The parallel electric field is assumed to satisfy

$$E_{\parallel} = 0 \tag{2.94}$$

For incompressible MHD motions in a collisionless plasma, no density or temperature gradients are allowed to develop parallel to the magnetic field. Hence, no parallel electric field is induced.

Although the $\mathbf{E} \times \mathbf{B}$ drifts are large, they are identical for electrons and ions and consequently produce no net current in a quasineutral plasma ($n_e = n_i = n$). The guiding center current, which is first order in r_{Li}/a, is proportional to the difference in drift velocities and is given by

$$\mathbf{J}_{g\perp} = e\int [\mathbf{V}_{g\perp i}f_i(\mathbf{u}) - \mathrm{V}_{g\perp e}f_e(\mathbf{u})]\, d\mathbf{u} \tag{2.95}$$

Carrying out the velocity integration yields

$$\mathbf{J}_{g\perp} = \frac{1}{B}\mathbf{b} \times \left(p_{\perp}\frac{\nabla B}{B} + p_{\parallel}\boldsymbol{\kappa} + \rho\frac{d\mathbf{v}_{\perp}}{dt}\right) \tag{2.96}$$

where

$$\begin{aligned} p_{\perp} &= \sum_{\alpha}\int \frac{m_{\alpha}u_{\perp}^2}{2} f_{\alpha}(\mathbf{u})\, d\mathbf{u} \\ p_{\parallel} &= \sum_{\alpha}\int m_{\alpha}u_{\parallel}^2 f_{\alpha}(\mathbf{u})\, d\mathbf{u} \\ \rho &= m_i\int f_i(\mathbf{u})\, d\mathbf{u} \end{aligned} \tag{2.97}$$

Note that the contribution of the electron polarization drift is small by m_e/m_i and is thus neglected.

The total perpendicular current is calculated by summing the guiding center current due to the particle drifts with the magnetization current due to the particles intersecting the edge of the surface. The magnetization current is given by

$$\mathbf{J}_M = \nabla \times \mathbf{M} \tag{2.98}$$

with

$$\mathbf{M} = -\frac{p_{\perp}}{B}\mathbf{b} \tag{2.99}$$

Note that the magnetization current arises even if there is no drift of the guiding centers. It occurs when there is a gradient in the density of magnetic moments. [For those unfamiliar with the magnetization current an excellent discussion is given in Longmire (1963). Also, see Problem 2.7.]

The total perpendicular current is then given by

$$\mathbf{J}_\perp = \mathbf{J}_M + \mathbf{J}_{g\perp} \tag{2.100}$$

After forming the cross product of Eq. (2.100) with **B** and carrying out some simple vector manipulations, one obtains

$$\rho\left(\mathbf{b} \times \frac{d\mathbf{v}_\perp}{dt}\right) \times \mathbf{b} = \mathbf{J} \times \mathbf{B} - \nabla_\perp p_\perp - (p_\parallel - p_\perp)\boldsymbol{\kappa} \tag{2.101}$$

Equation (2.101) is the perpendicular momentum equation for the general anisotropic guiding center model.

As stated previously the mass and momentum equations [Eqs. (2.83) and (2.101)] are exact in the small gyro radius collisionless limit. The final step in the derivation is to close the system, and it is here that the heuristic assumption is introduced. Specifically, it is assumed that the motions of interest are incompressible,

$$\nabla \cdot \mathbf{v} = 0 \tag{2.102}$$

implying that the density and pressures are convected with the fluid motion; no compressions or expansions are allowed. Under this assumption, the mass conservation relation reduces to

$$\frac{d\rho}{dt} = 0 \tag{2.103}$$

Applying Eq. (2.102) to the perpendicular and parallel pressures leads to closure of the system; the incompressible equations of state are given by

$$\frac{dp_\perp}{dt} = 0 \tag{2.104a}$$

$$\frac{dp_\parallel}{dt} = 0 \tag{2.104b}$$

Based on previous discussions it is a good approximation to assume $p_\perp = p_\parallel$ as initial conditions for an MHD time interval Δt. Under this assumption, Eq. (2.104) implies that $p_\perp = p_\parallel \equiv p$ for all times.

The derivation of the collisionless MHD model is now complete. The equations describing the model are summarized and compared with ideal

MHD:

Collisionless MHD	Ideal MHD
$(d\rho/dt) = 0$	$d\rho/dt = -\rho\nabla\cdot\mathbf{v}$
$\rho(d\mathbf{v}_\perp/dt)_\perp = \mathbf{J}\times\mathbf{B} - \nabla_\perp p$	$\rho(d\mathbf{v}/dt)_\perp = \mathbf{J}\times\mathbf{B} - \nabla_\perp p$
$\mathbf{B}\cdot\nabla(v_\parallel/B) = -\nabla\cdot\mathbf{v}_\perp$	$\rho(d\mathbf{v}/dt)_\parallel = -\mathbf{b}\cdot\nabla p$
$dp/dt = 0$	$dp/dt = -\gamma p\nabla\cdot\mathbf{v}$
$\mathbf{E}+\mathbf{v}\times\mathbf{B} = 0$	$\mathbf{E}+\mathbf{v}\times\mathbf{B} = 0$
$\nabla\times\mathbf{E} = -\partial\mathbf{B}/\partial t$	$\nabla\times\mathbf{E} = -\partial\mathbf{B}/\partial t$
$\nabla\times\mathbf{B} = \mu_0\mathbf{J}$	$\nabla\times\mathbf{B} = \mu_0\mathbf{J}$
$\nabla\cdot\mathbf{B} = 0$	$\nabla\cdot\mathbf{B} = 0$

Note that the condition $\nabla\cdot\mathbf{v} = 0$ plays the role of the parallel momentum equation in collisionless MHD in that it describes the relationship determining $v_\parallel$.

Finally, a comparison of the two models indicates that if one replaces the parallel momentum equation in ideal MHD with the incompressibility condition $\nabla\cdot\mathbf{v} = 0$, then both models becomes essentially identical. Consequently, to the extent that the motions of interest are indeed incompressible, both models make very similar predictions, even though ideal MHD is assumed collision dominated while collisionless MHD is assumed collision free.

2.6. SUMMARY

The ideal MHD model is a single-fluid model that describes the effects of magnetic geometry on the macroscopic equilibrium and stability properties of fusion plasmas. The length and time scales of interest are a, the macroscopic plasma dimension, and a/V_{T_i}, the ion thermal transit time across the plasma. The model is derived in a straightforward manner by forming the mass, momentum, and energy moments of the Boltzmann equation. The moment equations reduce to ideal MHD with the introduction of three critical assumptions: high collisionality, small ion gyro radius, and small resistivity.

An analysis of the validity conditions shows that the collision-dominated assumption is never satisfied in plasmas of fusion interest. The remaining two conditions are satisfied by a wide margin. A careful examination of the collision-dominated assumption shows that those particular parts of ideal MHD treated inaccurately (i.e., the parallel momentum and energy equations) play little if any practical role in MHD equilibrium and stability; that is, these equations primarily describe the compression and expansion of a plasma whereas most MHD instabilities involve incompressible motions. The model is incorrect only where it does not matter.

This realization leads to the introduction of a modified MHD model known as collisionless MHD which makes predictions nearly identical to ideal MHD; however, as its name implies, it does not depend on the collision-dominated assumption. It is thus valid for plasmas of fusion interest. The derivation follows from an analysis of single-particle guiding center motion in a collisionless plasma and the subsequent closure of the system by the heuristic assumption that the motions of interest are incompressible.

A comparison of the two models shows that for incompressible motions the equations describing each system nearly coincide. The implication is that for such motions both models make similar predictions even though one depends on the collision-dominated assumption while the other is assumed collision free.

The remainder of the text deals primarily with ideal MHD, mainly for historical reasons, but at certain critical points detailed comparisons are made to show the relationship between the two models.

REFERENCES

Bowers, E. C., and M. G. Haines (1971). *Phys. Fluids* **14,** 165.

Braginskii, S. I. (1965). In *Reviews of Plasma Physics,* edited by M. A. Leontovich, Consultants Bureau, New York, Vol. I.

Chen, F. F. (1974). *Introduction to Plasma Physics,* Plenum, New York.

Krall, N. A., and A. W. Trivelpiece (1973). *Principles of Plasma Physics,* McGraw-Hill, New York.

Longmire, C. L. (1963). *Elementary Plasma Physics,* Wiley-Interscience, New York.

Montgomery, D. C., and D. A. Tidman (1964). *Plasma Kinetic Theory,* McGraw-Hill, New York.

Rose, D. J., and M. Clark, (1961). *Plasmas and Controlled Fusion,* MIT, Cambridge, Massachusetts.

Spitzer, L. (1956). *Physics of Fully Ionized Gases,* Interscience, New York.

PROBLEMS

2.1. Starting from the Boltzmann–Maxwell equations, and following the steps outlined in the text, derive the general set of two-fluid equations given by Eq. (2.38).

2.2. This problem demonstrates a simple procedure for estimating the transport coefficients in a collision-dominated plasma. Keep in mind that the problem actually represents a form of dimensional analysis since the simple collision operator used below conserves neither mass, momentum, nor energy. Consider the one-dimensional kinetic model

$$u_x \frac{\partial f}{\partial x} = -\nu(f - f_0)$$

where f is the electron distribution function, $f_0 = N(m/2\pi T)^{3/2} \exp(-mu^2/2T)$ is a background Maxwellian, $u^2 = u_x^2 + u_y^2 + u_z^2$ and $T = T(x)$. For simplicity assume ν and N are constants.

a. Assume large collisionality (i.e., large ν) and solve for $f = f_0 + f_1 + f_2 + \cdots$ up to and including second order.
b. Calculate $\mathbf{J}_e$.
c. Calculate $\mathbf{R}_e$. Define $\mathbf{R}_e = eN\eta \mathbf{J}_e$ and compute the resistivity η.
d. Calculate the heat flux $\mathbf{h}_e$ and show that it is of the form $\mathbf{h}_e = -\kappa(\partial T/\partial x)\mathbf{e}_x$. Compute the thermal conductivity κ.
e. Calculate the anisotropic element of the pressure tensor Π_{xx} and show that it can be written as $\Pi_{xx} = -\mu(\partial v_x/\partial x)$ where v_x is the macroscopic fluid velocity. Compute the viscosity μ.
f. Note that the viscosity and thermal conductivity approach infinity as the collision frequency approaches zero. Is this a contradiction? Explain.

2.3. Consider a system in which each species of particles is acted upon by an additional force arising from an external potential; that is, Newton's law for each particle is given by

$$m_\alpha \frac{d\mathbf{u}_\alpha}{dt} = q_\alpha(\mathbf{E} + \mathbf{u}_\alpha \times \mathbf{B}) - \nabla\phi_\alpha$$

where $\phi_\alpha = \phi_\alpha(\mathbf{r})$ is a fixed potential function.

a. Using the same assumptions made in the derivation of ideal MHD, derive an equivalent set of closed, single-fluid equations which include the effects of ϕ_e and ϕ_i.
b. In order to complete step (a) some assumptions must be made regarding the size of ϕ_e and ϕ_i. Order ϕ_e and ϕ_i so that their contribution to the momentum equation is comparable to the $\mathbf{J} \times \mathbf{B}$ force. What ordering gives this result?
c. Assume the external potential represents a fixed gravitational force in the $\mathbf{e}_x$ direction of magnitude g. What are the corresponding functional dependences of ϕ_e and ϕ_i? For $m_e \ll m_i$ write down the modified MHD equations including gravity.

2.4. Does the local Maxwellian distribution function

$$f = n\left(\frac{m}{2\pi T}\right)^{3/2} \exp(-mu^2/2T)$$

$$n = n(\mathbf{r}, t), \qquad T = T(\mathbf{r}, t)$$

satisfy the Vlasov equation (Boltzmann equation without the collision term)?

2.5. Does a system in steady state $\partial/\partial t = 0$, having y symmetry $\partial/\partial y = 0$, with distribution function

$$f = n_0\left(\frac{m}{2\pi T_0}\right)^{3/2} \exp[-(\epsilon + \alpha p^2/m)/T_0]$$

$$n_0,\ T_0,\ \alpha \text{ constants}$$

$$\epsilon = mu^2/2 + q\phi(x, z)$$

$$p = mu_y + qA_y(x, z)$$

satisfy the Vlasov equation? Here, ϕ and A_y are the scalar and vector potential, respectively. For this distribution function calculate n, $\mathbf{v}$, p, $\mathbf{\Pi}$ and $\mathbf{h}$ as functions of ϕ and A_y.

2.6. In Chapter 2 it was shown that the ideal and collisionless MHD models are quite similar in many respects despite the opposing limits on collisionality. To further illustrate this point derive the following exact result. Consider a plasma described by a hybrid model. The ions satisfy the Vlasov equation. The electrons satisfy the massless fluid equation $0 = -e(\mathbf{E} + \mathbf{v}_e \times \mathbf{B}) - \nabla p_e/n$. Assume the plasma is quasineutral: $n_i = n_e \equiv n$.

a. For static equilibria $(\partial/\partial t = 0)$ show that any function $f_i = f_i(\epsilon)$, $\epsilon = m_i u^2/2 + e\phi(\mathbf{r})$ satisfies the Vlasov equation for arbitrary three-dimensional geometry.

b. Calculate $\mathbf{J} = e\int \mathbf{u} f_i\, d\mathbf{u} - en\mathbf{v}_e$ and show that

$$\mathbf{J} \times \mathbf{B} = \nabla p$$

where $p = p_e + (m_i/3)\int u^2 f_i\, d\mathbf{v}$. Note that this equation is identical to the ideal MHD static equilibrium relation.

c. Compare the equilibrium electric field in ideal MHD and the hybrid model.

2.7. The magnetization current used in the derivation of collisionless MHD can be derived as follows. Consider the surface S illustrated below. The magnetization current arises from those guiding center orbits that intersect the edge of the surface just once.

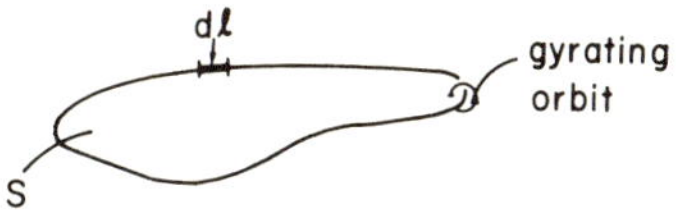

a. Show that the current arising from a single such particle is given by

$$\delta I = -\frac{e\omega_c}{2\pi}$$

Explain the minus sign.

b. Show that the total number of particles intersecting the edge one time can be written as

$$N = \int \pi r_L^2 f \, d\mathbf{u} \, dl$$

where $dl = \mathbf{b} \cdot d\mathbf{l}$. Appropriate subscripts should be substituted for electrons and ions.

c. Using Stokes' theorem show that the magnetization current can be expressed as

$$\mathbf{J}_M = \nabla \times \mathbf{M}$$

where

$$\mathbf{M} = -(p_\perp / B)\mathbf{b}$$

3

General Properties of Ideal MHD

3.1. INTRODUCTION

This chapter presents a discussion of the basic properties of the ideal MHD model. These properties include the conservation laws satisfied by the model as well as the boundary conditions that must be applied. The discussion demonstrates the physical foundations of ideal MHD while providing insight into its reliability in predicting experimental behavior.

The material is organized as follows: First, a short description is given of the three most common classes of boundary conditions that couple the plasma behavior to the externally applied magnetic fields: (1) plasma surrounded by a perfectly conducting wall, (2) plasma isolated from a perfectly conducting wall by an insulating vacuum region, and (3) plasma surrounded by external coils. The most complex of these provides a quite accurate description of realistic experimental conditions.

Second, it is shown that despite the significant number of approximations made, the ideal MHD model still conserves mass, momentum, and energy, both locally and globally. This is one basic reason for the reliability of the model.

Finally, a short calculation shows that as a consequence of the perfect conductivity assumption, the plasma and magnetic field lines are constrained to move together; that is, the field lines are "frozen" into the plasma. This leads to important topological constraints on the model.

3.2. BOUNDARY CONDITIONS

In order to formulate properly an MHD problem a set of appropriate boundary conditions must be specified to couple the plasma behavior to the externally applied magnetic fields. For problems involving MHD equilibrium and stability, there are three classes of boundary conditions, of varying complexity, which are often used. A discussion of each now follows.

3.2.1. Perfectly Conducting Wall

The first and simplest boundary condition assumes that the plasma extends out to a stationary, perfectly conducting wall, $r = R_W(\theta, z)$, as shown in Fig. 3.1. In this case the electromagnetic boundary conditions require that the tangential electric field and normal magnetic field vanish on the conducting wall:

$$\mathbf{n} \times \mathbf{E}|_{R_W} = 0 \tag{3.1}$$

$$\mathbf{n} \cdot \mathbf{B}|_{R_W} = 0 \tag{3.2}$$

Here, $\mathbf{n}$ is the outward-pointing normal vector. It then follows from the ideal Ohm's law that $\mathbf{n} \times \mathbf{E} + (\mathbf{n} \cdot \mathbf{B})\mathbf{v} - (\mathbf{n} \cdot \mathbf{v})\mathbf{B} = 0$; that is, the normal component of velocity also automatically vanishes on the wall:

$$\mathbf{n} \cdot \mathbf{v}|_{R_W} = 0 \tag{3.3}$$

Thus, once appropriate initial data and the shape of the wall are specified, any two of the conditions given above completely specify the problem. Note that there are no constraints imposed on the tangential velocity at the wall.

3.2.2. Insulating Vacuum Region

A somewhat more realistic set of boundary conditions assumes that the plasma is isolated from the conducting wall by a vacuum region as

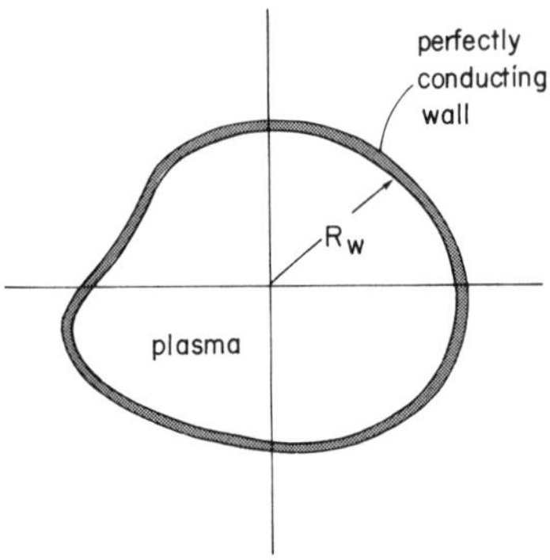

Figure 3.1. Boundary condition corresponding to a plasma surrounded by a perfectly conducting wall located at $r = R_w(\theta, z)$.

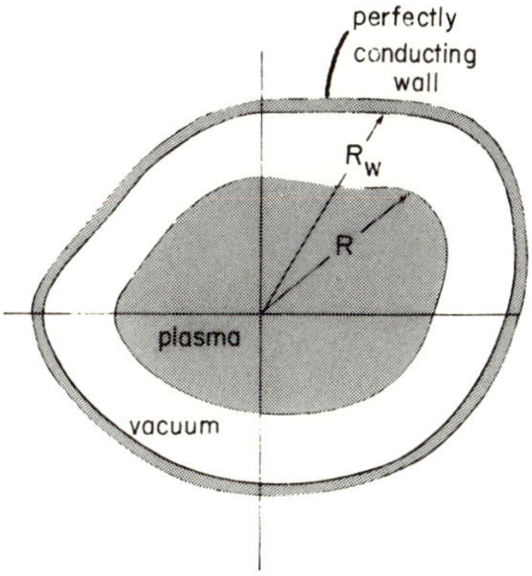

Figure 3.2. Boundary condition corresponding to a plasma whose surface is at $r = R(\theta, z)$ isolated from a perfectly conducting wall at $r = R_w(\theta, z)$ by a vacuum region.

shown in Fig. 3.2. In most cases this model is more appropriate than the previous one for describing "confined" plasmas.

In principle one solves the combined plasma–vacuum system as follows. In the plasma region the ideal MHD equations apply, while in the vacuum region, where the fluid variables are not defined, the relevant equations are those determining the magnetic field:

$$\nabla \times \hat{\mathbf{B}} = 0 \tag{3.4}$$

$$\nabla \cdot \hat{\mathbf{B}} = 0 \tag{3.5}$$

Here, quantities with a ^ denote vacuum variables.

Assume that the equations can be solved in each region and consider the boundary conditions. On the perfectly conducting wall, $r = R_W(\theta, z)$,

$$\hat{\mathbf{n}} \cdot \hat{\mathbf{B}}|_{R_W} = 0 \tag{3.6}$$

must be satisfied.

Unlike the previous case, the plasma surface, $r = R(\theta, z, t)$, is free to move since the plasma is surrounded by vacuum. Hence, $\mathbf{n} \cdot \mathbf{v}|_R$ is arbitrary. There are, however, three jump conditions that must be satisfied to connect the magnetic fields across the surface. These arise from the divergence **B** equation, Ampere's law, and the momentum equation. The other jump conditions that can be derived involve either the electric field or the fluid velocity and are not explicitly required to solve MHD problems.

Consider the magnetic jump conditions. In general, it is possible to sustain a jump in the pressure and tangential magnetic field if surface currents are allowed to flow. Upon integrating the relevant equations across the surface in the usual manner one obtains

$$[\![\mathbf{n}\cdot\mathbf{B}]\!]_R = 0 \tag{3.7}$$

$$[\![\mathbf{n}\times\mathbf{B}]\!]_R = \mu_0\mathbf{K} \tag{3.8}$$

$$[\![p + B^2/2\mu_0]\!]_R = 0 \tag{3.9}$$

where $[\![Q]\!]$ denotes $\hat{Q} - Q$ and $\mathbf{K}$ is the surface current density, assumed known. Also, in the integration of the momentum equation it has been assumed that the plasma acceleration is finite.

In the interesting situation where there are no surface currents and the pressure falls smoothly to zero at the plasma edge, the jump conditions reduce to

$$\mathbf{n}\cdot\mathbf{B}|_R = \mathbf{n}\cdot\hat{\mathbf{B}}|_R \tag{3.10}$$

$$\mathbf{n}\times\mathbf{B}|_R = \mathbf{n}\times\hat{\mathbf{B}}|_R \tag{3.11}$$

$$B^2|_R = \hat{B}^2|_R \tag{3.12}$$

The normal and tangential components of $\mathbf{B}$ and the magnetic pressure are continuous.

Although Eqs. (3.7)–(3.9) completely specify the boundary conditions, the plasma–vacuum problem is in practice difficult to solve. The reason is that a straightforward counting of boundary conditions indicates that there appears to be one too many: four conditions on three components of $\mathbf{B}$. This is not actually the case. If one wants to specify the surface currents and the shape of the outer conductor independently, it is necessary to self-consistently determine the correct shape of the plasma surface $R(\theta, z, t)$, and therein lies the extra degree of freedom. This is known as the "free boundary" problem. Often it is far simpler, but somewhat less relevant, to specify the shape of the plasma surface and then determine the appropriate shape of the outer perfect conductor. This is known as the "fixed boundary" problem.

There are two further points worth noting. First, since the ideal Ohm's law implies that the plasma is a perfect conductor, then the quantity $\mathbf{n}\cdot\mathbf{B}|_R$ is automatically zero. Therefore, in practice Eq. (3.11) reduces to

$$\mathbf{n}\cdot\hat{\mathbf{B}}|_R = 0 \tag{3.13}$$

Second, in many cases of interest, the metallic chamber surrounding the plasma may be a very good but not perfect conductor. In these situations, a slow time-varying magnetic field $\mathbf{B}_a$ may be applied and allowed to penetrate the chamber. Once established, the applied field $\mathbf{B}_a$ remains constant on the fast MHD time scale: $\mathbf{B}_a \approx \mathbf{B}_a(\mathbf{r})$. For problems of this type, the vacuum field can be written as $\hat{\mathbf{B}}(\mathbf{r}, t) = \mathbf{B}_a(\mathbf{r}) + \tilde{\mathbf{B}}(\mathbf{r}, t)$, where

$\nabla \times \mathbf{B}_a = \nabla \times \tilde{\mathbf{B}} = 0$ and $\nabla \cdot \mathbf{B}_a = \nabla \cdot \tilde{\mathbf{B}} = 0$ in the vacuum. The only difference that occurs is in the boundary condition at the wall, Eq. (3.6), which now becomes

$$\mathbf{n} \cdot \tilde{\mathbf{B}}|_{R_w} = 0 \tag{3.14}$$

The quantity $\mathbf{B}_a|_{R_w}$ is assumed given.

3.2.3. Plasma Surrounded by External Coils

The most difficult but realistic set of boundary conditions corresponds to the situation where the plasma is confined by the magnetic fields created by a fixed set of external current-carrying conductors as shown in Fig. 3.3. This problem is more difficult than the previous one because the current-carrying conductors must necessarily have gaps between them. Consequently, the problem must be solved in an infinite rather than finite domain.

In terms of the boundary conditions the requirements on the plasma–vacuum interface remain unchanged [i.e., Eqs. (3.7)–(3.9) still apply]. However, with external coils the wall condition, Eq. (3.6), is no longer required. Instead, the vacuum field is written as $\hat{\mathbf{B}} = \mathbf{B}_a + \tilde{\mathbf{B}}$, where $\mathbf{B}_a$ is the applied field due to the external coils and $\tilde{\mathbf{B}}$ is the induced field due to the plasma. The applied field is obtained by writing $\mathbf{B}_a = \nabla \times \mathbf{A}$ and then calculating $\mathbf{A}$ from the Biot–Savart law

$$\mathbf{A} = \frac{\mu_0}{4\pi} \sum_i \int \frac{\mathbf{J}_i}{|\mathbf{r} - \mathbf{r}'|} d\mathbf{r}' \tag{3.15}$$

where the sum is over the external conductors. The induced field satisfies

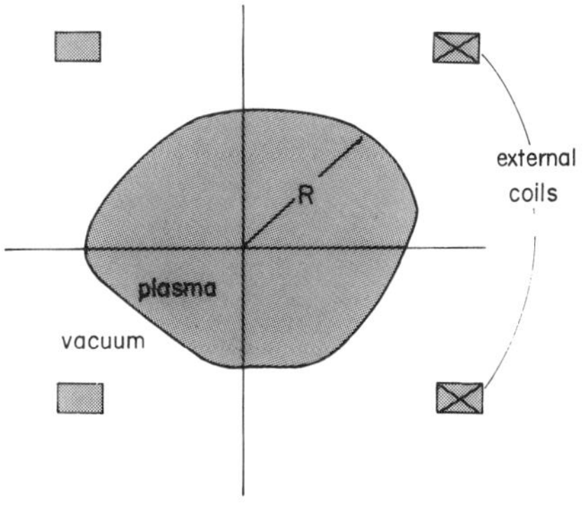

Figure 3.3. Boundary condition corresponding to a plasma whose surface is at $r = R(\theta, z)$ surrounded by a vacuum region with external current carrying coils.

$\nabla \times \tilde{\mathbf{B}} = \nabla \cdot \tilde{\mathbf{B}} = 0$ and must be regular throughout the entire vacuum region, including $r \to \infty$.

Solutions to the MHD equations satisfying the external coil boundary conditions provide an accurate description of plasma behavior in realistic experimental situations. Because of the complexity involved it is perhaps not surprising that most such applications require substantial numerical computations.

The MHD equations and any set of the boundary conditions just discussed form a complete basis to investigate the macroscopic equilibrium and stability of fusion plasmas.

3.3. LOCAL CONSERVATION RELATIONS

The original Boltzmann–Maxwell equations from which the MHD model was derived conserve mass, momentum, and energy, not only macroscopically, but microscopically as well. Since a considerable number of assumptions were made in the derivation of the MHD equations it is important to investigate whether the resulting model still satisfies the basic conservation laws. In this section the question of local conservation is treated by showing that both the ideal and collisionless MHD models can be written in conservation form. The implications regarding global conservation are discussed in the next section.

In order to show local conservation it is necessary to write the mass, momentum, and energy equations in the form

$$\frac{\partial}{\partial t}(\ ) + \nabla \cdot (\ ) = 0 \tag{3.16}$$

Once this is accomplished, it is straightforward to derive the global conservation relations.

After some standard manipulations the ideal MHD equations can be expressed in conservation form as follows:

$$\begin{aligned} \frac{\partial \rho}{\partial t} + \nabla \cdot \rho \mathbf{v} &= 0 \\ \frac{\partial \rho \mathbf{v}}{\partial t} + \nabla \cdot \mathsf{T} &= 0 \\ \frac{\partial w}{\partial t} + \nabla \cdot \mathbf{s} &= 0 \end{aligned} \tag{3.17}$$

where the stress tensor T, energy w, and energy flux $\mathbf{s}$ are given by

$$\begin{aligned}
\mathsf{T} &= \rho\mathbf{vv} + \left(p + \frac{B^2}{2\mu_0}\right)\mathsf{I} - \frac{\mathbf{BB}}{\mu_0} \\
w &= \frac{1}{2}\rho v^2 + \frac{B^2}{2\mu_0} + \frac{p}{\gamma - 1} \\
\mathbf{s} &= \left(\frac{1}{2}\rho v^2 + \frac{\gamma}{\gamma - 1}p\right)\mathbf{v} + \frac{1}{\mu_0}\mathbf{E} \times \mathbf{B}
\end{aligned} \tag{3.18}$$

Note that the stress tensor is composed of three terms. The $\rho\mathbf{vv}$ term is the Reynolds stress. In a local coordinate system in which $\mathbf{v}$ is in the z direction this part of the tensor has the form

$$\mathsf{T}_R = \begin{pmatrix} 0 & 0 & 0 \\ 0 & 0 & 0 \\ 0 & 0 & \rho v^2 \end{pmatrix} \tag{3.19}$$

Thus, the Reynolds stress gives rise to a pressure along $\mathbf{v}$. This term represents an important contribution to systems with large fluid flow but is usually not important in studies of plasma stability where the flows are either small or zero.

The remaining part of the stress tensor includes the effect of the pressure and the magnetic field. In a local coordinate system in which $\mathbf{B}$ lies along the z direction this part of the tensor can be expressed as follows:

$$\mathsf{T}_B = \begin{pmatrix} p_\perp & 0 & 0 \\ 0 & p_\perp & 0 \\ 0 & 0 & p_\parallel \end{pmatrix} \tag{3.20}$$

where

$$\begin{aligned}
p_\perp &= p + \frac{B^2}{2\mu_0} \\
p_\parallel &= p - \frac{B^2}{2\mu_0}
\end{aligned} \tag{3.21}$$

The quantities $p_\perp$ and $p_\parallel$ represent the total pressures perpendicular and parallel to the magnetic field, respectively. Equation (3.21) implies, as expected, that the plasma pressure p acts isotropically perpendicular and parallel to the field. In contrast, the magnetic pressure, $B^2/2\mu_0$, adds to the total pressure perpendicular to the field, while subtracting when parallel to the field. The "negative parallel magnetic pressure" in fact corresponds to a tension along the magnetic field lines. This anisotropic behavior of the magnetic field (i.e., pressure perpendicular to **B**, tension parallel to **B**) is fundamental to the understanding of the equilibrium and stability properties of magnetic geometries.

Consider now the energy equation. The quantity w represents the sum of the plasma kinetic energy w_K and the potential energy w_P:

$$
\begin{aligned}
w &= w_K + w_P \\
w_K &= \tfrac{1}{2}\rho v^2 \\
w_P &= \frac{B^2}{2\mu_0} + \frac{p}{\gamma - 1}
\end{aligned}
\tag{3.22}
$$

Note that the potential energy is made up of two contributions, the magnetic energy and the internal energy of the plasma, $[(3/2)nT$ for $\gamma = 5/3]$.

The total energy flux **s** can be expressed as

$$
\mathbf{s} = \left(\frac{1}{2}\rho v^2 + \frac{p}{\gamma - 1}\right)\mathbf{v} + p\mathbf{v} + \frac{1}{\mu_0}\mathbf{E} \times \mathbf{B} \tag{3.23}
$$

Here, the first term represents the flow of plasma kinetic plus internal energy. The second term corresponds to the mechanical work done on or by the plasma as it moves, and the last term, the Poynting flux, represents the flow of electromagnetic energy.

Summarizing, it has just been demonstrated that the ideal MHD equations can be written in local conservation form in which each of the terms has a simple physical interpretation.

A similar situation prevails for the collisionless MHD model. Here, an important feature is that the energy equation divides into two separate parts because of the decoupling of the parallel motion. The individual contributions are given by

$$
\begin{aligned}
&\frac{\partial w}{\partial t} + \nabla \cdot \mathbf{s} = 0 \\
&\frac{\partial p}{\partial t} + \nabla \cdot p\mathbf{v} = 0
\end{aligned}
\tag{3.24}
$$

where

$$
\begin{aligned}
w &= \frac{1}{2}\rho v_\perp^2 + \frac{B^2}{2\mu_0} \\
\mathbf{s} &= \frac{1}{2}\rho v_\perp^2 \mathbf{v} + p\mathbf{v} + \frac{1}{\mu_0}\mathbf{E} \times \mathbf{B}
\end{aligned}
\tag{3.25}
$$

The decoupling of the plasma internal energy from the kinetic and magnetic energies is a consequence of the incompressibility assumption; this assumption suppresses those parallel motions that would give rise to pressure gradients along the field. Thus, thermal sound waves cannot be excited and the internal energy of the plasma must remain constant.

3.4. GLOBAL CONSERVATION LAWS

By integrating the local conservation laws over the volumes appropriate to each of the three sets of boundary conditions previously discussed, it is possible to obtain a set of global conservation laws for ideal MHD. These laws are exact, and valid for general, nonlinear, multidimensional, time-dependent situations. Since they depend on the particular boundary conditions applied, each set of conditions must be treated separately.

3.4.1. Perfectly Conducting Wall

For this case it is appropriate to integrate the local conservation laws out to the conducting wall. Making use of the boundary conditions given by Eqs. (3.1)–(3.3) then leads to the following global conservation relations:

$$\frac{dM}{dt} = 0 \tag{3.26}$$

$$\frac{d\mathbf{P}}{dt} = -\int\left(p + \frac{B^2}{2\mu_0}\right)\mathbf{n}\,dS \tag{3.27}$$

$$\frac{dW}{dt} = 0 \tag{3.28}$$

where M is the total mass of plasma

$$M = \int \rho\, d\mathbf{r} \tag{3.29}$$

$\mathbf{P}$ is the mechanical momentum of the plasma

$$\mathbf{P} = \int \rho\mathbf{v}\, d\mathbf{r} \tag{3.30}$$

and W is the total kinetic and potential energy of the plasma and magnetic field

$$W = \int \left(\frac{1}{2}\rho v^2 + \frac{p}{\gamma - 1} + \frac{B^2}{2\mu_0}\right) d\mathbf{r} \tag{3.31}$$

Equation (3.26) demonstrates that the total mass of plasma is conserved. Equation (3.27) describes the conservation of momentum. The boundary term represents the total force exerted by the walls on the plasma. Assuming the system remains in place, this force must vanish identically and $d\mathbf{P}/dt = 0$; that is, mechanical momentum is conserved. The final equation, Eq. (3.28), shows that the total energy of the system is conserved. From the definition of W, Eq. (3.31), one observes that during a given plasma motion (an instability, for instance) energy can in general be transferred between the magnetic field, the plasma internal energy, and the plasma kinetic energy. This should be compared with the energy relations for collisionless MHD which have the form

$$\begin{aligned} \frac{dW}{dt} &= 0 \\ \frac{dU}{dt} &= 0 \end{aligned} \tag{3.32}$$

where

$$\begin{aligned} W &= \int \left(\frac{1}{2}\rho v_\perp^2 + \frac{B^2}{2\mu_0}\right) d\mathbf{r} \\ U &= \int p\, d\mathbf{r} \end{aligned} \tag{3.33}$$

In this model the suppression of parallel motion implies that the total internal energy of the plasma remains constant during any plasma motion. If there is to be a transfer of energy it can only occur between the magnetic field and the perpendicular kinetic energy.

3.4.2. Insulating Vacuum Region

When the plasma is surrounded by a vacuum region it is of particular interest to focus attention on the conservation of energy. This relation is slightly more complicated than for the case of the perfectly conducting wall since the plasma boundary is now allowed to move. The result is that it is not the individual, but the combined plasma–vacuum energy that is conserved.

To show this, first note that for a global quantity Z defined by

$$Z(t) = \int z(\mathbf{r}, t)\, d\mathbf{r} \tag{3.34}$$

the total time derivative in a volume whose boundary is moving with velocity $\mathbf{u}$ is given by (see Problem 3.1)

$$\frac{dZ(t)}{dt} = \int_V \frac{\partial z}{\partial t}\, d\mathbf{r} + \int_S z\mathbf{n} \cdot \mathbf{u}\, dS \tag{3.35}$$

Equation (3.35) is applied to the plasma energy by setting $Z(t) = W(t)$ [Eq. (3.31)] and $z(\mathbf{r}, t) = w(\mathbf{r}, t)$ [Eq. (3.22)]. The volume is chosen as that of the plasma, and the velocity $\mathbf{u} = \mathbf{v}$ corresponds to the motion of the plasma surface. Making use of the ideal Ohm's law, and the boundary condition $\mathbf{n} \cdot \mathbf{B}|_S = 0$, it follows that

$$\frac{dW}{dt} = -\int \left(p + \frac{B^2}{2\mu_0}\right)\mathbf{n} \cdot \mathbf{v}\, dS \tag{3.36}$$

Note that if the plasma surface is moving, the boundary term is in general nonzero.

Consider now the vacuum region. Here the total energy is given by

$$\hat{W} = \int \frac{\hat{B}^2}{2\mu_0}\, d\mathbf{r} \tag{3.37}$$

From Eq. (3.35) it follows that

$$\frac{d\hat{W}}{dt} = \int \frac{1}{\mu_0}\left(\hat{\mathbf{B}} \cdot \frac{\partial \hat{\mathbf{B}}}{\partial t}\right) d\mathbf{r} - \int \frac{\hat{B}^2}{2\mu_0}\mathbf{n} \cdot \mathbf{v}\, dS \tag{3.38}$$

The minus sign appears because $\mathbf{n}$ is an inward normal vector to the vacuum region. The first term can be simplified by substituting Faraday's law and converting the volume integral to a surface integral by the divergence theorem. Setting $\hat{\mathbf{E}} = -\mathbf{v} \times \hat{\mathbf{B}}$ on the surface and using the boundary condition $\mathbf{n} \cdot \hat{\mathbf{B}}|_S = 0$ one obtains

$$\frac{d\hat{W}}{dt} = \int \frac{\hat{B}^2}{2\mu_0}\mathbf{n} \cdot \mathbf{v}\, dS \tag{3.39}$$

The final energy relation is obtained by adding Eqs. (3.36) and (3.39)

$$\frac{d}{dt}(W + \hat{W}) = 0 \tag{3.40}$$

Here, the boundary terms cancel by virtue of Eq. (3.9), the pressure balance jump conditions.

Equation (3.40) implies that when an ideal MHD plasma is isolated from a conducting wall by a vacuum region, the combined energy of the plasma–vacuum system is conserved. The fact that only the total is conserved indicates that, in general, energy will flow from one region to the other as the plasma moves.

3.4.3. Plasma Surrounded by External Coils

If the conducting wall in the vacuum is replaced by a series of current-carrying conductors, the energy of the system is no longer conserved. The reason is that with external sources present, energy can be supplied to or extracted from the system. Nevertheless, it is still possible to derive a relatively simple energy relation for the region interior to the conductors. The procedure is almost identical to that given for the insulating vacuum region boundary condition. The main difference is that in the vacuum region the integration is carried out to any convenient surface S_W interior to the conductors. The boundary term no longer vanishes on this surface since it is not a perfect conductor. Instead one obtains

$$\frac{d}{dt}(W + \hat{W}) = -\int \frac{1}{\mu_0}(\mathbf{E} \times \mathbf{B}) \cdot \mathbf{n}_W \, dS_W \tag{3.41}$$

Equation (3.41) can be simply interpreted as showing that the rate of increase of the total energy in the combined plasma–vacuum system is equal to the electromagnetic power flowing into the region.

3.5. CONSERVATION OF FLUX: "FROZEN" FIELD LINE PICTURE

The final conservation law to be considered concerns the magnetic flux. The basic result, a consequence of the perfect conductivity Ohm's law, is that the magnetic flux passing through any arbitrary open surface area moving with the plasma is constant; flux is conserved locally as well as globally.

To show this, note that the time rate of change of the flux passing through any moving surface, S, is given by (see Problem 3.2)

$$\frac{d\psi}{dt} = \int \frac{\partial \mathbf{B}}{\partial t} \cdot \mathbf{n} \, dS - \oint \mathbf{u} \times \mathbf{b} \cdot d\mathbf{l} \tag{3.42}$$

Here, **u** is the velocity of the surface and

$$\psi = \int \mathbf{B} \cdot \mathbf{n} \, dS \tag{3.43}$$

is the magnetic flux.

Substituting Faraday's law for $\partial \mathbf{B}/\partial t$ and converting the surface integral into a line integral by applying Stokes' theorem yields

$$\frac{d\psi}{dt} = -\oint (\mathbf{E} + \mathbf{u} \times \mathbf{B}) \cdot d\mathbf{l} \tag{3.44}$$

If the velocity **u** is chosen to coincide with the plasma velocity **v**, and the ideal Ohm's law is assumed valid, then

$$\frac{d\psi}{dt} = 0 \tag{3.45}$$

Since the derivation of Eq. (3.45) applies to any arbitrary surface area, it follows immediately, by considering the entire plasma cross section, that the total flux in an ideal MDH plasma is conserved. A more interesting case is to allow the surface area to coincide with the cross section of a long thin flux tube (see Fig. 3.4). Application of Eq. (3.45) then leads to the well-known intuitive picture that in the ideal MHD model, magnetic lines move with the plasma; they are "frozen" into the fluid.

The conservation of flux relation has very important implications about the structure of the magnetic field. This follows because any allowable physical motion of the plasma requires that neighboring fluid elements remain adjacent to one another; fluid elements are not allowed to tear or break into separate pieces. Since the magnetic lines move with the plasma, the field line topology must thus be preserved during any physically allowable MHD motion. This is a very strict requirement on the structure of the magnetic fields.

There are many configurations in plasma physics in which it is intuitively clear that it is energetically favorable for field lines to break

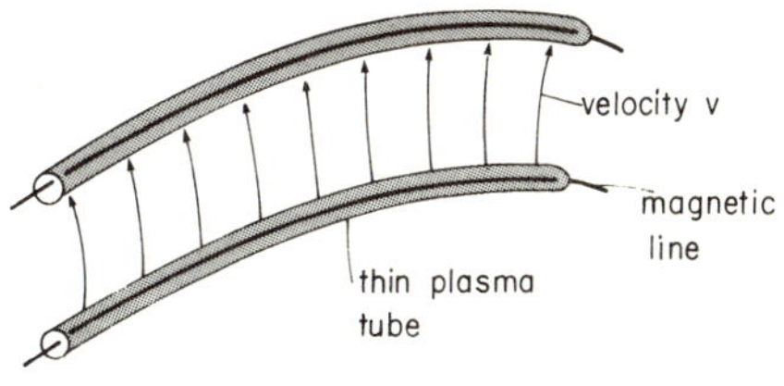

Figure 3.4. Thin tube of plasma containing a magnetic field line. As the plasma moves with velocity **v**, the field line is "frozen" to the plasma motion.

and reconnect into new configurations with lower potential energy. Such transitions are not allowed in ideal MHD because of the constraint on the topology. It is for this reason that the introduction of even a small resistivity can have a dramatic effect on plasma stability, much larger than indicated by simple dimensional arguments. A small dissipation removes the topological constraint and allows a much wider class of motions to take place, although admittedly on a much slower time scale. These new motions allow the plasma to access the lower energy states which are prohibited in the ideal model. As stated earlier, the effects of resistivity lie beyond the scope of this textbook.

3.6. SUMMARY

The results of this chapter show that the ideal MHD model conserves mass, momentum, energy, and magnetic flux. These conservation laws apply to general nonlinear, time-dependent, multidimensional systems. The existence of such laws is a nontrivial consequence in view of the many assumptions made in the derivation of the model. However, having shown the existence of the conservation laws, one can proceed with some confidence to the problems of equilibrium and stability of magnetic fusion configurations knowing that the model should prove reliable and provide valuable insight because of its inherently sound foundation.

REFERENCES

Boundary Conditions and Conservation Laws

Bernstein, I. B., E. A. Frieman, M. D. Kruskal, and R. M. Kulsrud (1958). *Proc. R. Soc. London, Ser. A* **244,** 17.

Goedbloed, J. P. (1979). *Lecture Notes on Ideal Magnetohydrodynamics,* Fom-Instituut voor Plasmafysica, Nieuwegein, Netherlands.

Structure of Magnetic Fields

Bateman, G. (1978). *MHD Instabilities,* MIT, Cambridge, Massachusetts.

Grad, H., and H. Rubin (1958). In *Proceedings of the Second United Nations International Conference on the Peaceful Uses of Atomic Energy,* United Nations, Geneva, Vol. 31, p. 190.

Morozov, A. I., and L. S. Solov'ev (1966). In *Reviews of Plasma Physics,* edited by M. A. Leontovich, Consultants Bureau, New York, Vol. II.

Newcomb, W. A. (1958). *Ann. Phys.* (NY) **3,** 347.

PROBLEMS

3.1. Consider a quantity $Z(t)$ defined as

$$Z(t) = \int_V z(\mathbf{r}, t)\, d\mathbf{r}$$

where V is a closed volume. If the surface S surrounding the volume is moving with a velocity $\mathbf{v}$ show that

$$\frac{dZ}{dt} = \int_V \frac{\partial z}{\partial t} d\mathbf{r} + \int_S z\mathbf{v}\cdot\mathbf{n}\, dS$$

3.2. Consider a quantity $C(t)$ defined by

$$C(t) = \int_S \mathbf{c}\cdot\mathbf{n}\, dS$$

where S is an open surface and $\mathbf{n}$ is the surface normal. If the boundary of the surface is moving with a velocity $\mathbf{v}$ show that

$$\frac{dC}{dt} = \int \frac{\partial \mathbf{c}}{\partial t}\cdot \boldsymbol{n}\, dS + \oint \mathbf{c}\times\mathbf{v}\cdot d\mathbf{l}$$

3.3. Derive the exact nonlinear local energy conservation laws for the ideal and collisionless MHD models. Calculate the global conservation laws assuming the presence of a vacuum region between the plasma and the perfectly conducting wall.

3.4. This problem provides an alternate demonstration that in an ideal MHD plasma the field lines are frozen into the plasma. It can be easily shown that any divergence-free magnetic field can be written as

$$\mathbf{B} = \nabla\alpha \times \nabla\beta$$

where $\alpha = \alpha(\mathbf{r}, t)$, $\beta = \beta(\mathbf{r}, t)$ represent field line coordinates; that is, the intersection of the surfaces $\alpha = \alpha_0$ and $\beta = \beta_0$ defines a line that is everywhere tangent to $\mathbf{B}$, thus defining the field line. Compute $\partial\mathbf{B}/\partial t$ and substitute the result into Faraday's law using the ideal MHD Ohm's law $\mathbf{E} + \mathbf{v}\times\mathbf{B} = 0$. Show that the resulting relation is satisfied when the field lines move with the plasma:

$$\frac{d\alpha}{dt} = \frac{d\beta}{dt} = 0$$

with $d/dt = \partial/\partial t + \mathbf{v}\cdot\nabla$.

3.5. The requirement $d\alpha/dt = d\beta/dt = 0$ in Problem 3.4 is not unique, as it is possible to add various homogeneous solutions to α and β. To illustrate this point consider a magnetic field $\mathbf{B} = B_\theta(r)\mathbf{e}_\theta + B_z(r)\mathbf{e}_z$. Write $\mathbf{B} = \nabla\alpha\times\nabla\beta$ with $\alpha = r^2/2$ and $\beta = \theta f_1(r) + (z/r)f_2(r)$.

a. Express $f_1(r)$ and $f_2(r)$ in terms of $B_\theta(r)$ and $B_z(r)$.
b. On any surface $\alpha = \alpha_0$ show that replacing $\beta(r, \theta, z)$ by $\beta(r, \theta, z) + g(r, t)$ leaves the magnetic field unchanged but corresponds to either a rotation in θ or a translation in z of the field line. Note that such rotations and translations have no physical significance because of the cylindrical symmetry.
c. Assume the plasma moves with a velocity $\mathbf{v} = v_\theta(r)\mathbf{e}_\theta + v_z(r)\mathbf{e}_z$ and $v_r = 0$. Is this velocity field consistent with Faraday's law? Could the given magnetic field be consistent with any $v_r \neq 0$? Explain.

The conclusion is that the concept of "frozen in" field lines, while correct, is not unique because of the ambiguity in identifying a field line from one instant of time to the next. The "frozen in" concept represents an extremely useful, although not unique, interpretation of the motion of plasma and field lines.

4

Equilibrium General Considerations

4.1. INTRODUCTION

The goal of MHD equilibrium theory is the discovery of magnetic geometries that (1) confine and isolate hot plasmas from material walls and (2) provide macroscopic stability at sufficiently high values of β ($\beta \equiv$ plasma energy/magnetic energy) to be attractive as potential fusion reactors.

Since most of the remainder of the textbook focuses on the problem of discovering geometries with high equilibrium and stability β limits, it is worthwhile demonstrating the crucial role that β plays in fusion ignition experiments and reactors. In this connection consider the following simple calculation of the plasma parameters required to achieve ignition.

At ignition the plasma is heated by fusion-produced α particles whose power density is given by

$$p_\alpha = \tfrac{1}{4} Q_\alpha n^2 \overline{\sigma v}(T_i) \text{ watts/m}^3 \tag{4.1}$$

where $Q_\alpha = 3.5\,\text{MeV}$ is the α-particle energy and $\overline{\sigma v}$ is the velocity-averaged fusion cross section. This power must exceed the power density loss rate (primarily due to thermal conduction), which in the zero-dimensional approximation is given by

$$p_L = \frac{3}{2}\frac{n(T_e + T_i)}{\tau_E} = \frac{3nT_i}{\tau_E} \text{ watts/m}^3 \tag{4.2}$$

where τ_E is the energy confinement time and $T_e \approx T_i$. Ignition requires $p_\alpha > p_L$, leading to the well-known condition

$$n\tau_E > \frac{12}{Q_\alpha}\left(\frac{T_i}{\overline{\sigma v}}\right) \text{ m}^{-3}\,\text{s} \tag{4.3}$$

From the point of view of plasma physics the quantity τ_E is limited by transport, which may be classical, neoclassical, or, as is often the case, anomalous. This is the domain of kinetic theory. While MHD equilibrium considerations play an important role in defining the magnetic geometry, MHD β limits are not usually critical in the determination of τ_E as long as they are not exceeded.

This is not the case for n; here, the maximum allowable density is directly related to MHD equilibrium and instability limits. More precisely, it is not n, but $\beta \equiv 2\mu_0 n(T_e + T_i)/B^2$ that is limited by MHD. Thus, plasma physics considerations suggest that a more appropriate form of Eq. (4.3) can be written as

$$\beta\tau_E > \frac{48\mu_0}{Q_\alpha}\left(\frac{T_i^2}{\overline{\sigma v}}\right)\frac{1}{B^2}\,\mathrm{s} \tag{4.4}$$

The quantity $T_i^2/\overline{\sigma v}$ has a minimum at $T_i \approx 15$ keV. Substituting numerical values, one obtains

$$\beta\tau_E > \frac{2.3}{B^2}\,\mathrm{s} \tag{4.5}$$

Since B has a maximum allowable value set by technological considerations, Eq. (4.5) illustrates the importance of achieving high β in fusion ignition experiments and reactors. It is here that MHD equilibrium and stability limits play a crucial role.

Fusion research since the 1950s has led to the discovery of a number of concepts possessing attractive MHD properties. Chapter 4 focuses on the common features of these configurations. Included are (1) a description of the basic equilibrium equations, (2) the need for toroidicity, (3) the concept of magnetic flux surfaces, (4) the definition of the basic plasma parameters and figures of merit describing an MHD equilibrium, and (5) the fundamental conflict between the requirements for equilibrium and stability in toroidal geometry.

Thus, a framework and perspective are provided from which to view the problem of obtaining MHD equilibria. This background, once established, is the basis for the discussions of the specific applications presented in the chapters that follow.

4.2. BASIC EQUATIONS

To begin, consider the MHD equilibrium equations given by

$$\mathbf{J} \times \mathbf{B} = \nabla p \tag{4.6}$$

$$\nabla \times \mathbf{B} = \mu_0 \mathbf{J} \tag{4.7}$$

$$\nabla \cdot \mathbf{B} = 0 \tag{4.8}$$

These are just the time-independent form of the full MHD equations with $\mathbf{v} = 0$; the equilibria of interest are static.

Equilibrium is achieved by balancing the magnetic force $\mathbf{J} \times \mathbf{B}$ with the pressure gradient force ∇p. It is interesting to note that while the full MHD equations including dynamics are valid for times approaching the hybrid MHD-transport time scale, the equilibrium equations themselves are actually valid over the much longer transport time scale. This follows because for static or very slowly moving systems, the issues of $E_{\|}$, collisions, and viscosity are either negligible or exactly cancel when adding the individual electron and ion momentum equations to obtain the single-fluid MHD momentum equation.

The question of static ($\mathbf{v} = 0$) vs. stationary ($\mathbf{v} \neq 0$) equilibria is also worth discussing. In general, stationary equilibria with nonzero flows are possible, but within the context of ideal MHD such flows should satisfy $\mathbf{v} \sim V_{T_i}$. This is much larger than is normally observed experimentally except in certain high-energy neutral-beam-injected tokamaks. The equilibrium flows that are observed are usually quite small and are associated with either the radial transport velocity or the azimuthal drifts arising from small ambipolar electric fields. A self-consistent treatment of these flows requires a more sophisticated model than ideal MHD. A second point is that plasmas with flows are often more unstable; for example, plasma rotation gives rise to centrifugal forces which represent an additional source of free energy to drive instabilities. Thus, from the confinement point of view, systems with flows are less attractive. For these reasons, attention is focused on static ideal MHD equilibria.

Finally, note that for static equilibria the density $\rho(\mathbf{r})$ is arbitrary.

4.3. THE VIRIAL THEOREM

The ideal MHD equilibrium equations satisfy a particular integral relation known as the virial theorem (Shafranov, 1966). A consequence of this theorem is the basic requirement that any MHD equilibrium must be supported by externally supplied currents; it is not possible to create a configuration confined solely by the currents flowing within the plasma itself. If this were not the case, one could design very attractive fusion reactors requiring only minimal "start-up"coils which could be shut off once the plasma currents were established. See Fig. 4.1.

To demonstrate the virial theorem, recall that the equilibrium equations can be written in conservation form,

$$\nabla \cdot \mathsf{T} = 0 \tag{4.9}$$

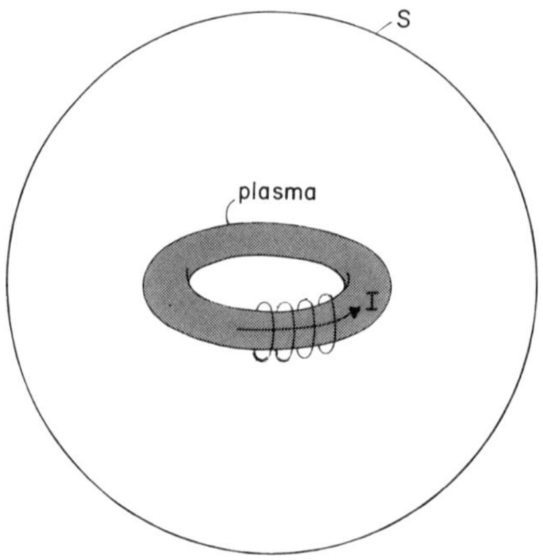

Figure 4.1. Geometry illustrating a plasma attempting to confine itself by its own currents. The surface S is the one referred to in the virial theorem.

where the stress tensor T is given by

$$\mathsf{T} = p_{\perp}(\mathsf{I} - \mathbf{bb}) + p_{\parallel}\mathbf{bb} \tag{4.10}$$

$$p_{\perp} = p + B^2/2\mu_0 \tag{4.11}$$

$$p_{\parallel} = p - B^2/2\mu_0 \tag{4.12}$$

$$\mathbf{b} = \mathbf{B}/B \tag{4.13}$$

Note that in a local rectangular coordinate system, with $\mathbf{B}$ defined in the z direction, T has the form

$$\mathsf{T} = \begin{pmatrix} p_{\perp} & 0 & 0 \\ 0 & p_{\perp} & 0 \\ 0 & 0 & p_{\parallel} \end{pmatrix} \tag{4.14}$$

indicating that $p_{\perp}$ and $p_{\parallel}$ represent the pressures perpendicular and parallel to $\mathbf{B}$, respectively.

The next step in the derivation is to integrate the identity

$$\nabla \cdot (\mathbf{r} \cdot \mathsf{T}) = \mathbf{r} \cdot (\nabla \cdot \mathsf{T}) + \mathrm{Trace}(\mathsf{T}) \tag{4.15}$$

over an arbitrary volume V, bounded by the surface S. Setting $\nabla \cdot \mathsf{T} = 0$ for equilibrium, one obtains

$$\int \left(3p + \frac{B^2}{2\mu_0}\right) dV = \int \left[\left(p + \frac{B^2}{2\mu_0}\right)(\mathbf{n} \cdot \mathbf{r}) - \frac{B^2}{2\mu_0}(\mathbf{r} \cdot \mathbf{b})(\mathbf{n} \cdot \mathbf{b})\right] dS \tag{4.16}$$

where $d\mathbf{S} = \mathbf{n}\, dS$ and $\mathbf{n}$ is the outward-pointing normal vector to S.

Assume now that the virial theorem is false; confined equilibria do exist without external currents. Let S lie outside the confined plasma so that $p(S) = 0$ (see Fig. 4.1). If no external currents are present, S can extend to infinity. Furthermore, if the equilibrium currents are indeed confined to the plasma, then **B** must decrease with radius at least as fast as a dipole field for large r: $B(S) \leq K/r^3$ as $r \to \infty$. Consequently, the integrand on the right-hand side of Eq. (4.16) scales as follows for large r:

$$B^2 r\, dS \propto \frac{1}{r^3} \tag{4.17}$$

Under these circumstances, as S extends to infinity, the right-hand side of Eq. (4.16) vanishes while the left-hand side approaches a nonzero positive constant. The resulting contradiction thus proves the virial theorem. When confined equilibria are surrounded by external conductors, Eq. (4.16) is not violated, since the right-hand side must now be evaluated over the surface of the conductors.

4.4. TOROIDICITY

The most obvious common feature in current magnetic fusion concepts is that, with one important exception, each is constructed in the shape of a torus. The intent is to create configurations in which magnetic field lines remain contained within the toroidal volume; lines should not intersect either the external current-carrying conductors or the vacuum chamber, whichever is closer to the plasma.

The reason for building relatively complicated toroidal systems rather than simpler, linear, open-ended systems is associated with the enormous difference in the energy loss rates perpendicular and parallel to a magnetic field. In an open-ended device, magnetic field lines leave the system and ultimately make contact with material walls (see Fig. 4.2a). Both density and energy can be lost very quickly because charged

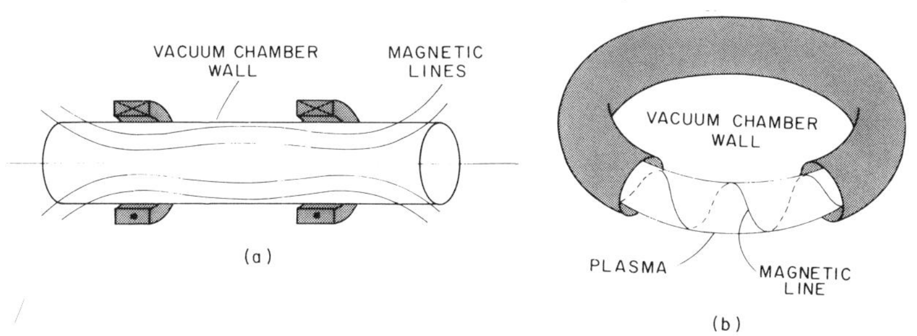

Figure 4.2. Magnetic field line trajectories in (a) open-ended systems and (b) toroidal systems.

particles move freely along magnetic field lines. Assuming that the potentially faster particle loss is eliminated by magnetic trapping, the dominant remaining energy loss mechanism is parallel thermal conduction by electrons. In a toroidal device with contained lines, the dominant energy loss mechanism is cross-field ion thermal conduction (see Fig. 4.2b). The ratio of the classical thermal conductivities is given by [Braginskii (1965)]

$$\frac{\kappa_{\|e}}{\kappa_{\perp i}} = 1.12\left(\frac{m_i}{m_e}\right)^{1/2}(\omega_{ci}\tau_{ii})^2 \approx 6.2 \times 10^{10}\left(\frac{B^2T_i^3}{n^2}\right) \tag{4.18}$$

where ω_{ci} is the ion cyclotron frequency and τ_{ii} is the ion–ion collision time for momentum exchange. It is also assumed that (1) the ions are singly charged deuterons, ($Z = 1$, $m_i = 2m_{proton}$), (2) the electrons and ions are at the same temperature ($T_e = T_i$), and (3) the Coulomb logarithm appearing in τ_{ii} has been set to 15 ($\ln \Lambda = 15$). For plasma parameters corresponding either to current experimental operation or to reactor conditions this ratio is enormous. As an example, consider a tokamak with $T_i = 2\,\text{keV}$, $B = 5\,\text{T}$, and $n = 2 \times 10^{20}\,\text{m}^{-3}$. In this case $\kappa_{\|e}/\kappa_{\perp i} = 3.1 \times 10^{12}$.

In practice the actual values of the perpendicular thermal conductivities in toroidal devices can be substantially larger than their classical values. The reasons for this are twofold. First, the orbits of a special class of particles are strongly modified by toroidal effects, leading to a disproportionately large contribution to $\kappa_{\perp i}$ (neoclassical transport). Second, anomalous effects due to plasma microinstabilities increase the effective collision frequency leading to greatly enhanced values of $\kappa_{\perp e}$. On the other hand, experimental measurements of $\kappa_{\|e}$ indicate that its value is approximately classical. Even with these larger values of $\kappa_\perp$, the ratio of $\kappa_\|/\kappa_\perp$ remains so large that there exists a common consensus that to avoid such losses, the magnetic configuration must be toroidal.

The one major exception to this consensus is the open-ended mirror confinement concept. Here, by a combination of innovative ideas and clever plasma operation, large gains have been made in reducing endloss. However, even with the improvements in confinement, mirrors, as of this writing, have yet to achieve the comparable performance of tokamaks.

4.5. MAGNETIC FLUX SURFACES

In fusion devices of interest the magnetic field lines lie in general on a set of closed nested toroidal surfaces. This follows from the equilibrium

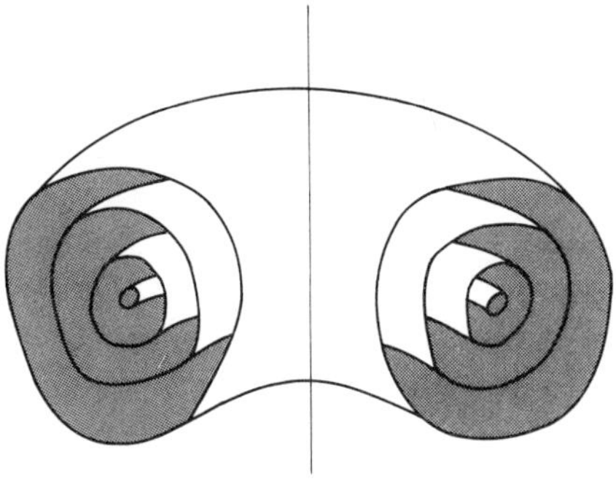

Figure 4.3. Contours of constant pressure in a well-confined toroidal equilibrium.

relation

$$\mathbf{B} \cdot \nabla p = 0 \tag{4.19}$$

That is, for well-confined equilibria the pressure is maximum near the center of the poloidal cross section and is approximately constant as a function of toroidal angle. For such profiles the contours of constant pressure are nested toroidal surfaces. See Fig. 4.3. From Eq. (4.19) it follows that the magnetic lines lie on the p = const contours, and, consequently, these contours are usually referred to as either magnetic flux surfaces or simply flux surfaces. The limiting flux surface, which approaches a single magnetic line where the pressure is maximum, is called the magnetic axis. See Fig. 4.4.

Similarly, the equilibrium relation

$$\mathbf{J} \cdot \nabla p = 0 \tag{4.20}$$

implies that the current lines also lie on the surfaces of constant pressure; the current flows between and not across flux surfaces. Note that while

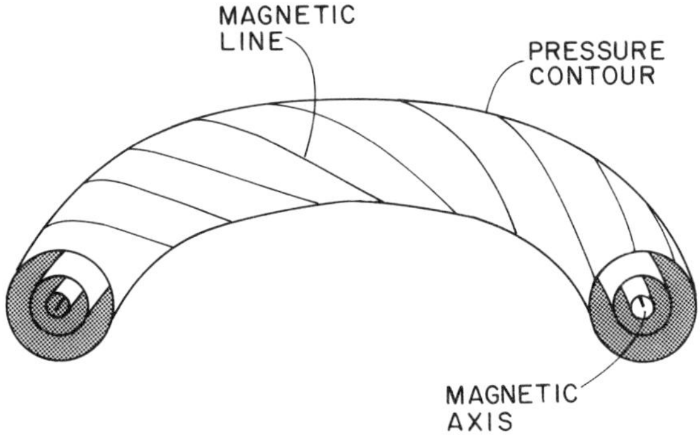

Figure 4.4. Toroidal flux surfaces showing the magnetic axis and the magnetic lines lying on a surface.

both **J** and **B** lie on constant p contours, this does not imply that their directions are parallel. In general, the angle between **J** and **B** is arbitrary, and the limit of being either purely parallel or perpendicular indeed corresponds to a special case. The general situation is illustrated in Fig. 4.5.

There are three important classes of magnetic field line trajectories that must be distinguished: rational, ergodic, and stochastic. A constant p contour in which all the lines exactly close on themselves after a finite number of toroidal circuits is known as a rational surface. If instead the field lines do not close, but cover the entire constant p surface, this corresponds to an ergodic surface. In those situations where the field line actually fills a volume, this is known as a region of stochasticity.

In general, toroidal magnetic geometries with ergodic or rational surfaces are the ones of interest and importance to MHD. Most configurations with desirable MHD confinement properties have both ergodic and rational surfaces. Usually the ratio of rational to ergodic surfaces is of measure zero. The tokamak, stellarator, reversed field pinch, and spheromak are configurations of this type. Counterexamples that have only closed lines are the Elmo bumpy torus and the toroidal multipole.

Systems with regions of stochasticity are undesirable for MHD equilibrium and stability. Over such regions the equilibrium relation $\mathbf{B} \cdot \nabla p = 0$ implies that $\nabla p = 0$; this is equivalent to a region with infinite transport. Stochastic regions can occur only in systems without geometric symmetry, for instance, in the vicinity of a separatrix in a three-dimensional stellarator or, more importantly, over substantial regions of plasma where there exists multidimensional turbulence involving a large number of different field harmonics. Because of their unattractive confinement properties, systems with regions of stochasticity are not treated here.

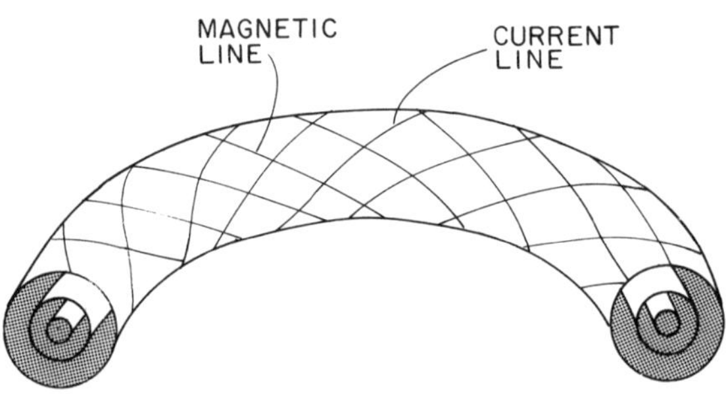

Figure 4.5. Toroidal flux surface showing magnetic lines and current lines lying on the surface.

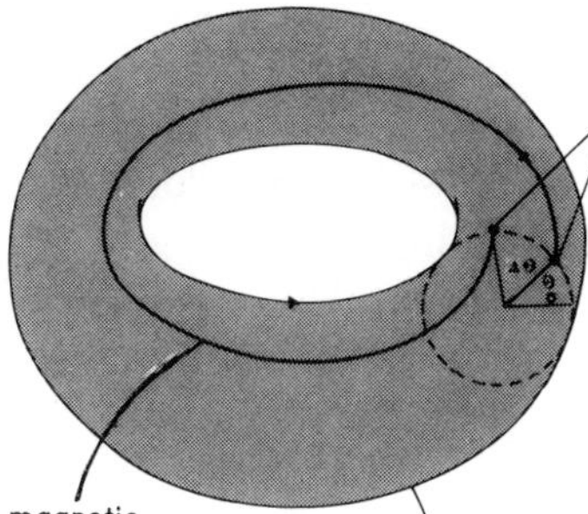

Figure 4.6. Magnetic field line projections used in the definition of rotational transform.

The usual way to characterize the two important classes of surfaces is in terms of the rotational transform, defined as follows. Imagine the projection of a magnetic line on a given poloidal cross section as shown in Fig. 4.6. After one transit around the torus the magnetic line returns to a slightly different angle, $\theta_0 + \Delta\theta$, on the flux surface. In general, $\Delta\theta$ depends upon the poloidal angle, θ_0, where the line started. The rotational transform, ι, is the average value of the angle $\Delta\theta$ after an infinite number of transits:

$$\iota \equiv \lim_{N\to\infty} \frac{1}{N} \sum_{1}^{N} \Delta\theta_n \tag{4.21}$$

If ι is a rational fraction of 2π the line is closed. If it is not, the line is ergodic. The rotational transform plays an important role in both equilibrium and stability, and practical procedures for its evaluation are discussed in later sections. Note that ergodic surfaces can be mapped out either by plotting p = const contours or by tracing magnetic field-line trajectories many circuits of the torus. For closed-line systems, however, the flux surfaces are defined as p = const surfaces since closed lines do not trace out complete surfaces.

4.6. SURFACE QUANTITIES: BASIC PLASMA PARAMETERS AND FIGURES OF MERIT

By carrying out appropriate integrals over the flux surfaces one can define a number of quantities that are of importance to the application of MHD theory. These quantities, known as "surface quantities," describe the global properties of the equilibria and as such are essential to distinguish different configurations and to interpret experimental data.

Several quantities also serve as important figures of merit measuring MHD stability. Furthermore, as discussed in the next section, a precise number of surface quantities must be specified *a priori* in order to formulate a well-posed equilibrium problem.

Surface quantities are, in general, one-dimensional functions depending only upon the flux surface label. For example, each contour of the set of nested toroidal surfaces illustrated in Fig. 4.3 is labeled by a different value of the pressure p. If one now calculates the poloidal flux ψ_p contained within any given pressure contour (see Fig. 4.7)

$$\psi_p \equiv \int \mathbf{B} \cdot d\mathbf{S}_p \qquad \text{poloidal flux} \tag{4.22}$$

then

$$\psi_p = \psi_p(p) \tag{4.23}$$

The amount of poloidal flux contained within any given pressure contour depends only upon the value of p labeling that contour.

Note that it makes physical sense for ψ_p to be a surface quantity since the magnetic lines lie in the constant p contours. Were the magnetic lines to cross the pressure surfaces the definition of "contained flux" would become ambiguous. In fact, since $\mathbf{J}$ and $\mathbf{B}$ both lie on constant p

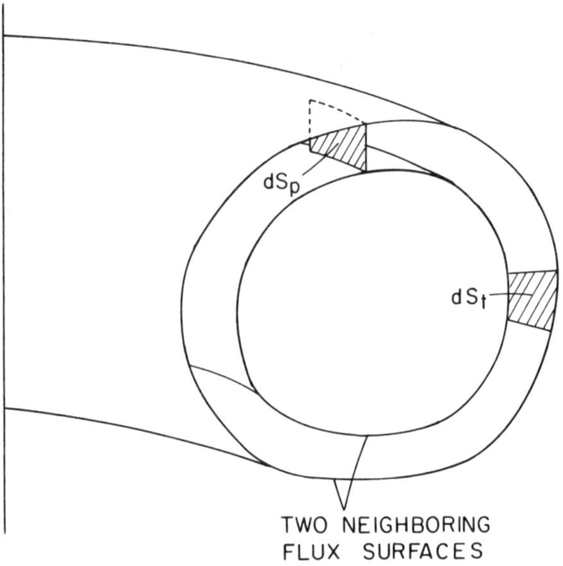

Figure 4.7. Illustration of the poloidal and toroidal surface elements $d\mathbf{S}_p$ and $d\mathbf{S}_t$.

contours, it follows that each of the quantities (see Fig. 4.7)

$$\begin{aligned} \psi_t &= \int \mathbf{B} \cdot d\mathbf{S}_t \qquad \text{toroidal flux} \\ I_t &= \int \mathbf{J} \cdot d\mathbf{S}_t \qquad \text{toroidal current} \\ I_p &= \int \mathbf{J} \cdot d\mathbf{S}_p \qquad \text{poloidal current} \end{aligned} \tag{4.24}$$

is also a surface quantity. In contrast, the components of magnetic field and current density vary with poloidal and toroidal angle, θ and ϕ, around a flux surface and thus do not qualify as surface quantities, i.e., B_p is a function $B_p(p, \theta, \phi)$ and not $B_p(p)$.

A further point to discuss concerns the labeling of the flux surfaces. Up until now it has been assumed that each flux surface has been labeled by a value of p. Clearly this is not necessary. One could just as well label the surfaces with the values of ψ_p, or with any other surface quantity for that matter. In many MHD calculations appearing in the literature, either ψ_p, ψ_t, or a combination thereof is used as the label. The key point is that a properly defined surface quantity remains a surface quantity regardless of how the flux surfaces are labeled.

With this as background, one can now define several basic plasma parameters and figures of merit which measure the quality of an MHD equilibrium with respect to confinement efficiency and stability. Each of the quantities introduced is discussed in detail in later sections of the textbook. For present purposes of convenience, they are simply introduced without proof and with little discussion. Also, keep in mind that the figures of merit related to stability are neither necessary nor sufficient; instead, in combination they serve as approximate guidelines for achieving overall favorable stability properties.

In order to define useful figures of merit, several requirements must be met: (1) the quantities must be relatively easy to evaluate but still accurately reflect the physics issue under consideration; (2) the quantities must be expressible in terms of simple physical parameters, easily related to experiment; and (3) the quantities must be defined in a sufficiently general manner so as to apply to all configurations of interest. This last requirement is greatly facilitated by using the volume V contained within a flux surface as the flux surface label. Thus, surface quantities such as the pressure, poloidal flux, etc., can be written as $p = p(V)$, $\psi_p = \psi_p(V)$, etc.

The figures of merit are now given as follows.

Plasma β. The plasma β is a parameter that measures the efficiency of plasma confinement by the magnetic field. It is defined as the ratio of the averaged plasma energy to the averaged magnetic energy. Specifically,

$$\bar{\beta} \equiv \frac{2\mu_0 \langle p \rangle}{\langle B_t^2 + B_p^2 \rangle} \tag{4.25}$$

where

$$\langle p \rangle = \frac{1}{V_0} \int_0^{V_0} p(V)\, dV \tag{4.26}$$

V_0 is the total plasma volume and B_t and B_p are the toroidal and poloidal magnetic fields, respectively. Similar averages can be written for $\langle B_t^2 \rangle$ and $\langle B_p^2 \rangle$. However, it is convenient to replace these averages with simpler but similar expressions more easily related to experiments. Thus, the average toroidal magnetic energy is replaced by

$$\langle B_t^2 \rangle \to B_0^2 \tag{4.27}$$

Here, B_0 is the applied toroidal magnetic field at the edge of the plasma, evaluated at the midplane $R = R_0$. See Fig. 4.8 for the geometry.

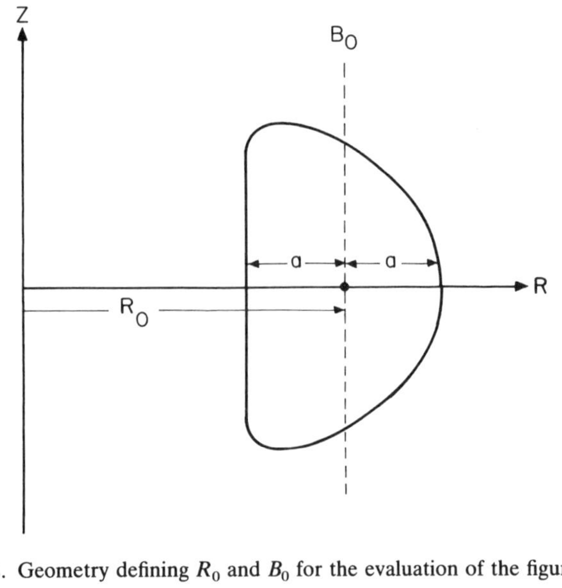

Figure 4.8. Geometry defining R_0 and B_0 for the evaluation of the figures of merit.

Similarly, the poloidal magnetic field is replaced by

$$\langle B_p^2 \rangle \to \bar{B}_p^2 \tag{4.28}$$

where

$$\bar{B}_p = \frac{\mu_0 I_0}{2\pi a \kappa} \tag{4.29}$$

I_0 is the total toroidal plasma current, A is the area of the cross section, a is the minor radius of the plasma, and $\kappa = A/\pi a^2$ is the plasma "elongation."

With these replacements, the definition of $\bar{\beta}$, used throughout the textbook, becomes

$$\bar{\beta} \equiv \frac{2\mu_0 \langle p \rangle}{B_0^2 + \bar{B}_p^2} \tag{4.30}$$

It is often useful to define separate toroidal and poloidal β's measuring plasma confinement efficiency with respect to each component of the magnetic field. These definitions are given by

$$\beta_t \equiv \frac{2\mu_0 \langle p \rangle}{B_0^2} \tag{4.31}$$

and

$$\beta_p = \frac{2\mu_0 \langle p \rangle}{\bar{B}_p^2} = \frac{8\pi^2 a^2 \kappa^2 \langle p \rangle}{\mu_0 I_0^2} \tag{4.32}$$

Note that

$$\frac{1}{\bar{\beta}} = \frac{1}{\beta_t} + \frac{1}{\beta_p} \tag{4.33}$$

indicating that the smaller of the two quantities β_t or β_p dominates the overall magnetic confinement efficiency.

In general high values of $\bar{\beta}$ are desirable for fusion reactor economics and technology. However, there is a maximum allowable value of $\bar{\beta}$ set by MHD equilibrium requirements and by instabilities driven by the pressure gradient.

Kink Safety Factor. The kink safety factor is a parameter that measures stability against long-wavelength modes driven by the toroidal plasma current. It is given by

$$q_* \equiv \frac{aB_0}{R_0 \bar{B}_p} = \frac{2\pi a^2 \kappa B_0}{\mu_0 R_0 I_0} \tag{4.34}$$

In many fusion applications it is desirable to operate at high current, i.e., low q_*, for improved confinement and ohmic heating. However, kink instabilities prevent this type of operation and set a lower limit on the minimum achievable q_*.

Rotational Transform and MHD Safety Factor. The rotational transform and MHD safety factor are two simply related surface quantities that play an important role in both MHD equilibrium and stability. To evaluate these quantities, assume the configurations of interest consist of a set of nested toroidal ergodic flux surfaces, labeled by the volume V enclosed within each surface. In principle, one can then calculate the toroidal and poloidal fluxes contained within any given surface:

$$\begin{aligned}\psi_t &= \psi_t(V)\\ \psi_p &= \psi_p(V)\end{aligned} \tag{4.35}$$

It is shown in Section 6.7 that the rotational transform defined in Eq. (4.21) can be written as

$$\iota(V) \equiv 2\pi\left(\frac{d\psi_p/dV}{d\psi_t/dV}\right) \tag{4.36}$$

The MHD safety factor q is related to ι by

$$q(V) \equiv \frac{2\pi}{\iota} = \frac{d\psi_t/dV}{d\psi_p/dV} \tag{4.37}$$

Equilibria can be designed with small, moderate, or large ι. Large ι usually implies a robust toroidal equilibrium, but can also indirectly give rise to MHD instabilities. Note that while Eqs. (4.35)–(4.37) are general, there are in most cases of practical interest more efficient ways of calculating $q(V)$. Also, in certain special limits one can show that $q(V_0) = q_*$, a result that sometimes leads to confusion as to the role each plays. Finally, the relations presented here can be extended to include rational surfaces and purely closed-line configurations if the flux surfaces labeled by V are recognized to coincide with the contours of constant pressure.

Shear. The shear is a quantity that measures the change in pitch angle of a magnetic field line from one flux surface to the next. It plays an important role in stabilizing MHD instabilities, particularly those driven by the pressure gradient. The shear is a surface quantity defined on each flux surface as follows:

$$s(V) \equiv 2\frac{V}{q}\frac{dq}{dV} \tag{4.38}$$

As a general rule, large shear is favorable for stability.

Magnetic Well. The magnetic well is a quantity that measures plasma stability against short perpendicular wavelength modes driven by the plasma pressure gradient. In Section 10.3 this quantity is shown to be closely related to the average curvature of a magnetic field line. In the limit $\mathbf{J} \to 0$ the vacuum magnetic well is defined as

$$\hat{W} = 2\frac{V}{\langle B^2 \rangle}\frac{d}{dV}\left\langle \frac{B^2}{2} \right\rangle \tag{4.39}$$

where

$$\langle Q \rangle \equiv \int_0^L \frac{Q\,dl}{B} \bigg/ \int_0^L \frac{dl}{B} \tag{4.40}$$

and dl is the incremental arc length along the field line. On ergodic surfaces $\langle B^2 \rangle$ is evaluated in the limit $L \to \infty$. For closed magnetic lines L corresponds to one complete circuit. Also, in Eq. (4.40) Q is to be evaluated on the field line.

At finite $\mathbf{J}$ the stability condition from which $\hat{W}$ is derived becomes considerably more complicated. In this regime there is a simple modification to $\hat{W}$ that accounts for one important finite $\mathbf{J}$ effect, namely, plasma diamagnetism. The modified form which serves as a useful figure of merit is given by

$$W = 2\frac{V}{\langle B^2 \rangle}\frac{d}{dV}\langle \mu_0 p + B^2/2 \rangle \tag{4.41}$$

On ergodic surfaces W is a surface quantity: $W = W(V)$. On rational surfaces the situation is more involved. In many configurations, particularly those with some degree of symmetry, there is a smooth transition from ergodic to rational surfaces with W remaining a surface quantity. However, in sufficiently general geometry (e.g., toroidal helix) W can be a function of both V and θ_0 on rational surfaces. In these situations $W(V, \theta_0)$ is no longer a surface quantity. (This is in contrast to the rotational transform, which is the same for all magnetic lines on a rational surface since magnetic lines are not allowed to cross.) Even so, the quantity $W(V, \theta_0)$ still plays an important role in plasma stability.

In general, a configuration has favorable stability properties if $W(V)$ is large and positive. Such systems tend to confine plasma in regions of low B; thus, instabilities driven by the pressure gradient are suppressed since the plasma has difficulty expanding into a high-B region. In practice, there are important topological constraints inherent in any toroidal magnetic geometry which limit the magnitude and even the sign of $W(V)$.

Summary. A number of basic plasma parameters and figures of merit have been introduced whose purpose is to provide intuition as to the quality and desirability of a given MHD equilibrium for fusion applications. Of particular interest to the equilibrium confinement properties is the value of $\bar{\beta}$ that can be supported, as well as the value of ι or, equivalently, q. The guidelines for stability test whether q_* is large enough and whether or not there is adequate shear s and magnetic well W.

4.7. EQUILIBRIUM DEGREES OF FREEDOM

In addition to their importance in describing the global properties of fusion plasmas, the surface quantities play a crucial role in the proper formulation of MHD equilibrium problems. The reason is that while many different surface quantities can be defined, only two are independent. One can see this intuitively by recognizing that there are two independent controls that can be exercised through the external circuits producing any given equilibrium: one corresponding to the toroidal field circuit (e.g., regulate the toroidal flux evolution) and the other to the poloidal field circuit (e.g., regulate the toroidal current evolution). The existence of two arbitrary-free surface functions has been demonstrated mathematically for one-, two-, and general three-dimensional configurations.

The implication is that in order to formulate a well-posed equilibrium problem it must be both possible and necessary to specify *a priori* two and only two independent surface quantities.

The existence of two free functions is on the one hand a great advantage of MHD theory. This freedom allows the theory to describe the equilibria of a wide range of configurations. On the other hand, one should appreciate that there is no physics contained within the ideal MHD model that predicts the specific functional dependence of the two free functions. Hence, for a given configuration the accuracy of the results depends upon the accuracy with which the free functions can be specified. For example, if $p(V)$ and $q(V)$ are chosen as the free functions, their functional dependence must be specified at the outset of the problem, based on physical intuition, experimental data, or the results of a transport code. Only after these functions are specified is the MHD equilibrium problem properly formulated, permitting a solution for the fields, flux surfaces, etc.

4.8. THE BASIC PROBLEM OF TOROIDAL EQUILIBRIUM

To a very good approximation, the problem of achieving toroidal equilibrium separates into two parts. First, the magnetic configuration must provide radial confinement (i.e., radial pressure balance) in the poloidal plane so that the pressure contours form closed nested surfaces. See Fig. 4.9. Although the manners in which they do so are quite different, either toroidal or poloidal fields can readily accomplish this. Second, the configuration must compensate the radially outward expansion force (i.e., toroidal force balance) inherent in all toroidal geometries without sacrificing stability. See Fig. 4.10. Although the forces associated with toroidal force balance are usually smaller than those corresponding to radial pressure balance, they are nevertheless more difficult to compensate. With this in mind it is helpful to examine two opposing

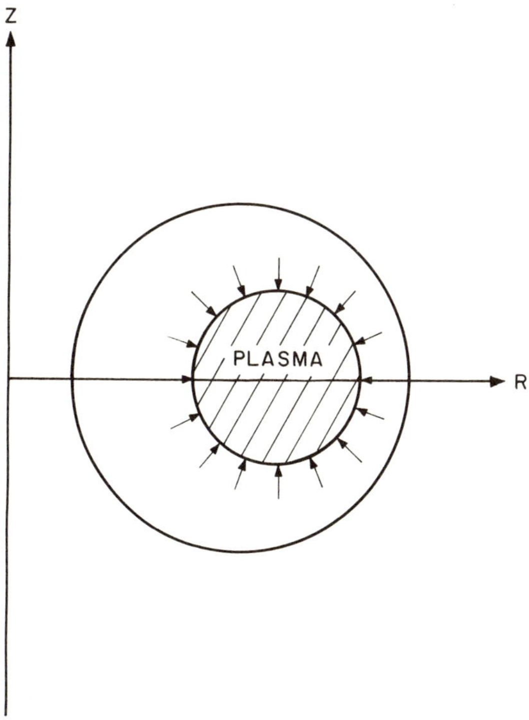

Figure 4.9. Schematic diagram of the forces required for radial pressure balance.

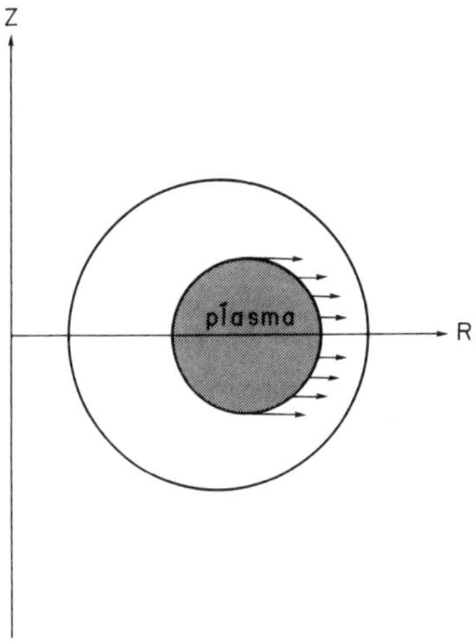

Figure 4.10. Schematic diagram of the outward toroidal expansion force inherent in all toroidal configurations.

limits that serve to illustrate the basic nature of the toroidal force-balance problem.

First, consider a configuration with a purely poloidal magnetic field as shown in Fig. 4.11. There are two forces that cause the plasma to expand radially outward in the $+R$ direction: (1) the hoop force and (2) the tire tube force.

The hoop force arises from the toroidal current and is similar to the force produced by a current-carrying loop of wire. One can understand this force intuitively by examining the field pattern illustrated in Fig. 4.12a. Conservation of flux implies that $\psi_1 = \psi_2$. Since ψ_1 passes through a much smaller area than ψ_2, one expects the average magnetic

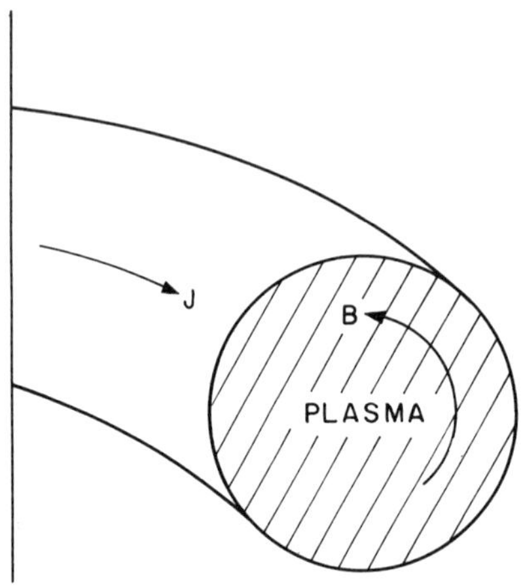

Figure 4.11. Toroidal configuration with a purely poloidal magnetic field.

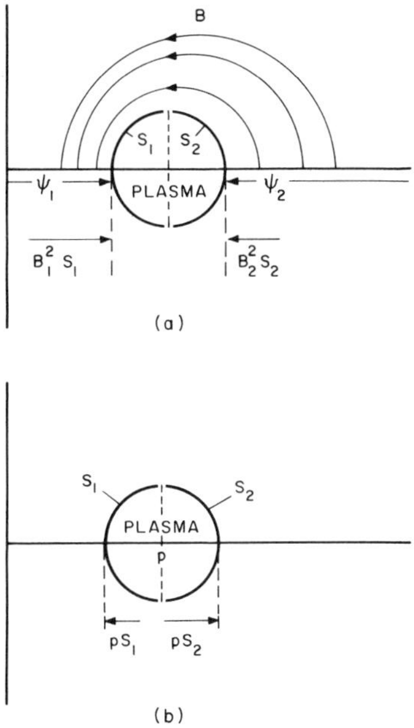

Figure 4.12. Diagram showing the toroidal expansion forces in a system with purely poloidal fields (a) hoop force and (b) tire tube force.

fields on the plasma surfaces S_1 and S_2 to satisfy $B_1 > B_2$. The net outward force is proportional to $\mathbf{F} \sim \mathbf{e}_R(B_1^2 S_1 - B_2^2 S_2)/2\mu_0$. Even though S_1 is smaller than S_2, the quadratic dependence on B dominates, indicating that $F_R > 0$: F_R corresponds to an outward hoop force.

The tire tube forces arises because the area S_1 on the inside of a constant pressure contour is smaller than the outer area S_2. See Fig. 4.12b. If the average pressure of the plasma is p there is a net outward force $\mathbf{F} \sim -\mathbf{e}_R(pS_1 - pS_2)$. Since $S_2 > S_1$, then $F_R > 0$: F_R corresponds to an outward force very similar to that found in a rubber tire tube.

The combined outward force can be compensated in two different ways. If a perfectly conducting shell surrounds the plasma, then as the plasma expands along R in response to the outward force, the poloidal magnetic flux outside the plasma is compressed, thereby increasing the magnetic pressure (Fig. 4.13a). Equilibrium is achieved when the plasma shifts sufficiently far so that the outward force balances the inward force due to the increased magnetic pressure.

A second way to balance the outward force is to replace the conducting wall with a set of fixed current-carrying vertical field coils (Fig. 4.13b). This is an important practical option. In actual experiments with conducting walls the material used is usually copper, which has a high but not infinite conductivity. Hence, flux can only remain compressed for

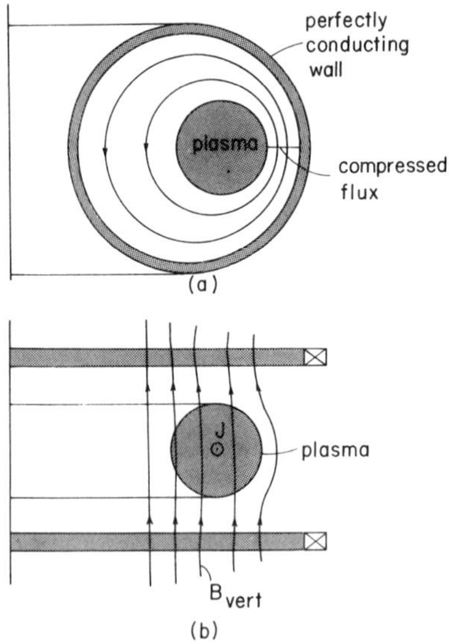

Figure 4.13. Toroidal equilibrium in a configuration with purely poloidal fields by (a) a perfectly conducting wall and (b) externally applied vertical field.

about a skin time, which is often shorter than the experimental times of interest. By a proper choice of magnitude and sign, the vertical field generated by the external coils produces an inward compensating $\mathbf{J} \times \mathbf{B}_{\text{vert}}$ force for equilibrium; $F_R = 2\pi R_0 I_0 B_{\text{vert}}$. Thus, configurations with purely poloidal magnetic fields can easily be designed with good toroidal equilibrium properties.

This favorable conclusion is negated by the fact that such configurations develop disastrous MHD instabilities leading to the destruction of the plasma. (The stability of configurations with purely poloidal fields is discussed in detail in Chapter 9.)

As a consequence, one is motivated to examine the opposite limit, a configuration with a purely toroidal field as shown in Fig. 4.14. In Chapter 9, it is demonstrated that this configuration has inherently better stability properties than the purely poloidal case. Nevertheless, from the simple calculation outlined below, it follows that a purely toroidal configuration cannot be held in magnetohydrodynamic equilibrium.

To show this, assume that all the current flows solely in a thin layer at the plasma surface and that the pressure p is a constant p_0 in the plasma and zero outside. (Note that $\mathbf{J} \times \mathbf{B} = \nabla p$ is trivially satisfied in the plasma.) Assume the plasma is completely diamagnetic so that the magnetic field in the plasma is zero. Outside, from Maxwell's equations, it follows that the toroidal field decreases inversely with R; that is, $\mathbf{B} = B_\phi \mathbf{e}_\phi$, where $B_\phi = B_0(R_0/R)$, $B_0 = \mu_0 I_c/2\pi R_0$, and I_c is the current

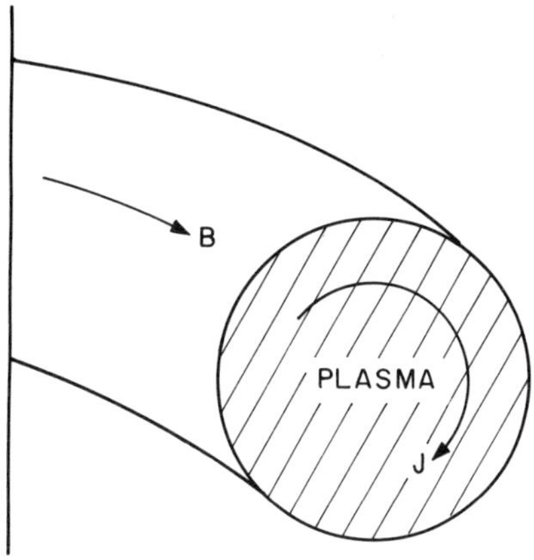

Figure 4.14. Toroidal configuration with a purely toroidal magnetic field.

flowing in the toroidal field coils. See Fig. 4.15a. Because of the $1/R$ dependence it is clear that the magnetic pressure on the inside of the torus ($\pi/2 < \theta < 3\pi/2$) is always greater than that on the outside: $B_1^2 > B_2^2$. This effect is partially compensated by the slightly smaller area on the inside $S_1 > S_2$, but the quadratic dependence on B dominates and there remains a net outward force. There is an additional outward force arising from the tire tube effect, identical to that in the purely poloidal case. The combined outward force F_R (in the $+R$ direction) can easily be calculated by integrating the $\mathbf{e}_R$ component of the total pressure jump over the plasma surface

$$F_R = -\int [\![p + B^2/2\mu_0]\!]\mathbf{e}_r \cdot \mathbf{e}_R \, dS \tag{4.42}$$

where

$$\begin{aligned} [\![p + B^2/2\mu_0]\!] &= \frac{B_0^2}{2\mu_0}\left(\frac{R_0}{R}\right)^2 - p \\ \mathbf{e}_r \cdot \mathbf{e}_R &= \cos\theta \\ dS &= 2\pi a R \, d\theta \end{aligned} \tag{4.43}$$

and a is the plasma radius. Writing $R = R_0 + a\cos\theta$ and taking the large aspect ratio limit, $R_0/a \gg 1$, one obtains the following expression for the outward expansion force:

$$F_R = 2\pi^2 a^2\left(p + \frac{B_0^2}{2\mu_0}\right) \tag{4.44}$$

The first and second terms in Eq. (4.44) represent the tire tube force and the $1/R$ force, respectively.

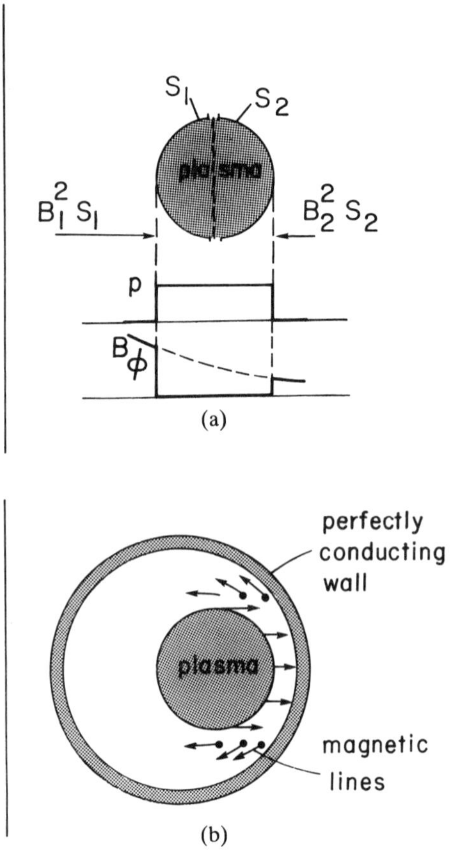

Figure 4.15. Toroidal force balance in a system with a purely toroidal field: (a) sharp boundary model showing the $1/R$ outward force and (b) lack of toroidal equilibrium.

Since the magnetic field is in the toroidal direction, a conducting wall is not able to compensate this force; that is, as the plasma moves outward, the magnetic lines simply slip around and let the plasma drift through (Fig. 4.15b); because of the topology, toroidal flux does not get trapped between the plasma and the wall. Likewise, vertical fields do not help because by symmetry they cannot produce any net force.

The conclusion from this calculation is that a configuration with a purely toroidal magnetic field cannot be held in equilibrium because of the $1/R$ dependence of the magnetic field resulting from the toroidal geometry. One question to be asked is whether or not the outward drift can be negligible on the time scale of interest if the aspect ratio is made sufficiently large. The answer is negative. If a vacuum chamber were located at a minor radius b, then the time required for the plasma to make contact is given by

$$\tau = \left(\frac{2b}{F_R/M}\right)^{1/2} = \left(\frac{2bR_0\rho_0}{p + B_0^2/2\mu_0}\right)^{1/2} \tag{4.45}$$

where F_R/M is the plasma acceleration, $M = \rho_0 V_0$ is the total mass, and ρ_0 is the mass density. Even in an extreme case, $b = 1\,\text{m}$, $R_0 = 1\,\text{km}$, $B_0 = 5\,\text{T}$, $n_0 = 10^{21}\,\text{m}^{-3}$, $p = B_0^2/2\mu_0$, and deuterium m_i, one finds $\tau = 18\,\mu\text{s}$, which is much too short to be of practical interest.

The basic problem of finding satisfactory magnetic geometries for confining fusion plasmas can thus be summarized as follows. On the one hand, toroidal systems with a predominantly poloidal magnetic field have good toroidal equilibrium properties but poor stability properties. On the other hand, when the toroidal field is dominant, stability is inherently much better, but serious equilibrium problems exist. Attempts to resolve this dilemma have led to the discovery of a number of different configurations, often managing to combine the desirable features of both toroidal and poloidal systems while to a reasonable extent suppressing the undesirable ones. Chapters 5–7 discuss the equilibrium properties of the currently most promising magnetic configurations.

Before proceeding, however, it is worthwhile to develop some intuition into the origin of these configurations by examining the toroidal equilibrium problem from the single-particle point of view.

Consider first the equilibrium problem in the purely toroidal field case. In a low β situation where the plasma diamagnetism is negligible the dominant drifts experienced by the particles are due to ∇B and curvature. These drifts arise because of the $1/R$ dependence of B and the centrifugal force, both toroidal effects. The drift velocity is in the Z direction, as illustrated in Fig. 4.16a, and is given by

$$\mathbf{V}_{g\perp} = \frac{1}{\omega_{c0}R_0}\left(\frac{u_\perp^2}{2} + u_\parallel^2\right)\mathbf{e}_Z \tag{4.46}$$

where $\omega_{c0} = qB_0/m$. Note that the drift is in opposite directions for electrons and ions. As a result, positive and negative charges collect at the top and bottom of the plasma, respectively, as shown in Fig. 4.16b. This charge sets up a Z-directed electric field, which then propels the entire plasma outward along R with the $\mathbf{E} \times \mathbf{B}/B^2$ drift velocity; hence, there is no toroidal equilibrium for the configuration with purely toroidal field.

Consider the situation now where a small poloidal field is superimposed on the toroidal field. See Fig. 4.17a. The magnetic field lines wrap

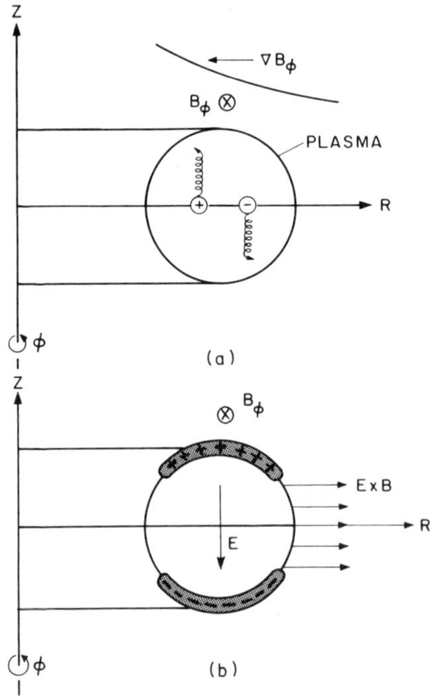

Figure 4.16. Single-particle picture of the lack of toroidal equilibrium in a system with purely toroidal field (a) guiding center drifts and (b) induced outward $\mathbf{E} \times \mathbf{B}$ drift.

slowly around the plasma on approximately helical trajectories. If $B_\theta \ll B_\phi$, the particles still experience the same up–down drift given by Eq. (4.46). One is then faced with the apparently paradoxical question of how a small poloidal field can lead to single-particle confinement for particles that have a constant vertical drift. The resolution is illustrated in Fig. 4.17b. The dashed lines represent flux surfaces. To zeroth order, a particle's parallel motion simply follows a field line; thus, the projection of the particle's orbit on a poloidal cross section would coincide with the flux surface if there was no up–down drift. If the drift is now included, as the particle spirals from point 1 to point 2, it actually arrives at point 2′ because of the upward drift. From point 2′ to 3′ the drift off the surface continues to increase. However, from point 3′ to 4′ and then back to 1, the upward drift causes the particle to drift back toward the surface, eventually returning to its starting position. The point is that a constant upward drift corresponds to motion away from the surface half the time and toward the surface the other half of the time; on average the net motion due to the drift cancels and the particles remain confined.

This simple picture demonstrates how a poloidal field, or rotational transform in general leads to single-particle confinement and macroscopic

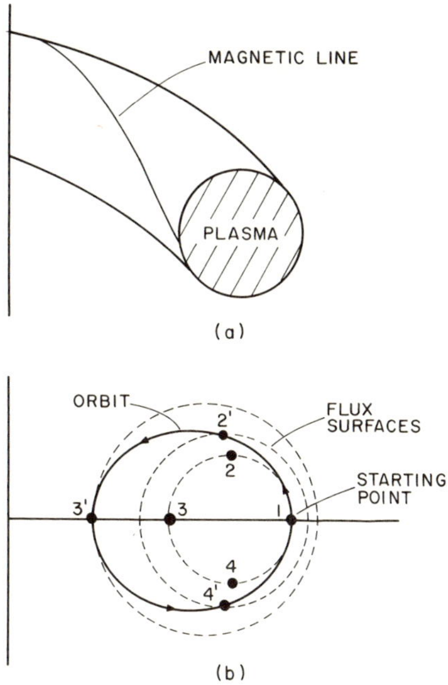

Figure 4.17. Single-particle picture of toroidal confinement in the presence of a poloidal magnetic field: (a) field line wrapping around the torus. To leading order the particle's parallel motion follows this line; (b) particle orbit including a uniform upward drift.

toroidal force balance. While rotational transform is effective in this role it is not a necessary requirement. Closed-line equilibria with zero transform are possible, but modulations in the field are required to produce guiding center drifts in the poloidal direction so that a particle samples the entire flux surface during its trajectory.

4.9. SUMMARY

Ideal MHD equilibrium theory has the goal of discovering magnetic geometries that stably confine hot plasmas at sufficiently high values of β to be interesting as fusion reactors. Over years of research many different concepts have been developed to achieve this goal. These different concepts, nevertheless, share a number of common features, which are the subject of Chapter 4. A summary is given as follows.

Basic Equations. The equilibrium equations of interest correspond to the time-independent $\partial/\partial t = 0$, static $\mathbf{v} = 0$, limit of the MHD model, describing a balance between the pressure gradient force ∇p and the magnetic force $\mathbf{J} \times \mathbf{B}$. The equilibrium model by itself is actually valid from the MHD time scale up until the very long transport time scale.

Virial Theorem. The virial theorem demonstrates that, as enticing as it may seem, it is not possible to construct a magnetic geometry in which the plasma is confined solely by its own currents; an external coil system must be supplied.

Toroidicity. Because the thermal conduction loss rate parallel to the magnetic field is enormous compared to that perpendicular to the field, there is near consensus in the fusion community that magnetic geometries of fusion interest should be toroidal. The one important exception to this philosophy is the mirror concept.

Magnetic Flux Surfaces. In a well-confined toroidal MHD equilibrium, the contours of constant pressure form a set of nested tori. Both the magnetic and currrent lines lie on these contours, which are known as flux surfaces.

Surface Quantities: Figures of Merit. Given an MHD equilibrium, it is possible to define a number of global quantities depending only upon the flux surface label. These surface quantities represent important physical parameters which can be related to experiment and be used as figures of merit for measuring stability. Included are average beta $\bar{\beta}$, toroidal beta β_t, poloidal beta β_p, kink safety factor q_*, rotational transform ι, *MHD* safety factor q, shear s, and magnetic well W.

Equilibrium Degrees of Freedom. For a well-posed MHD equilibrium problem it must be both possible and necessary to specify *a priori* two and only two independent surface quantities.

The Basic Problem of Toroidal Equilibrium. The problem of MHD equilibrium separates into two parts: radial pressure balance and toroidal force balance. The latter is the more difficult problem and herein lies the basic dilemma of magnetic geometry. Systems with predominantly poloidal magnetic fields are easy to maintain in toroidal force balance but have poor MHD stability properties. In contrast, systems with predominantly toroidal fields have poor toroidal force balance but much more favorable stability properties (in equivalent straight systems). The challenge then is to discover configurations that combine the favorable and minimize the unfavorable features of these two limiting configurations.

REFERENCES

Braginskii, S. I. (1965). In *Reviews of Plasma Physics,* edited by M. A. Leontovich, Consultants Bureau, New York, Vol. I.
Shafranov, V. D. (1966). In *Reviews of Plasma Physics,* edited by M. A. Leontovich, Consultants Bureau, New York, Vol. II.

GENERAL REFERENCES

Basic Properties of MHD Equilibria

Morozov, A. I., and L. S. Solov'ev (1966). In *Reviews of Plasma Physics,* edited by M. A. Leontovich, Consultants Bureau, New York, Vol. II.
Solov'ev, L. S., and V. D. Shafranov (1970). In *Review of Plasma Physics,* edited by M. A. Leontovich, Consultants Bureau, New York, Vol. V.
Shafranov, V. D. (1966). In *Reviews of Plasma Physics,* edited by M. A. Leontovich, Consultants Bureau, New York, Vol. II.

Effects of Plasma Rotation and Finite Larmor Radius Effects on MHD Stability

Bowers, E. C., and M. G. Haines (1971). *Phys. Fluids* **14,** 165.
Freidberg, J. P., and L.D. Pearlstein (1978). *Phys. Fluids* **21,** 1207.
Freidberg, J. P., and J. A. Wesson (1970). *Phys. Fluids* **13,** 1117.
Rosenbluth, M. N., N. A. Krall, and N. Rostoker (1962). *Nucl. Fusion Suppl., Part I,* 143.
Taylor, J. B. (1962). *J. Nucl. Energy, Part* **C 4,** 406.

PROBLEMS

4.1. For each of the magnetic configurations given below, determine the magnetic field line trajectories and the shape of the flux surfaces.

a. $B_r = 0$
$B_\theta = B_{\theta 0}[ar/(r^2 + a^2)]$
$B_z = B_{z0}$
with $B_{\theta 0}$, B_{z0}, and a constants.

b. $B_\theta = 0$
$B_z = B_{z0}(1 + \delta \cos kz)$
with B_{z0} and δ constants. (You will need to calculate B_r.)

c. $B_\theta = B_{\theta 0}(r + \Delta \cos \theta)/a$
$B_z = B_{z0}$
with B_{z0}, $B_{\theta 0}$, a, and Δ constants. (You will need to calculate B_r.)

4.2. Following the discussion associated with Fig. 4.17, sketch the orbit of a particle starting off at point 1 with a parallel velocity in the opposite direction of that illustrated.

4.3. Show that the MHD equilibrium equations for an arbitrary three-dimensional geometry can be written as

$$\nabla_\perp\left(p + \frac{B^2}{2\mu_0}\right) - \frac{B^2}{\mu_0}\boldsymbol{\kappa} = 0$$

where $\boldsymbol{\kappa} = \mathbf{b}\cdot\nabla\mathbf{b}$ is the field line curvature and $\mathbf{b} = \mathbf{B}/B$.

4.4. For a prescribed magnetic field $\mathbf{B}$ and corresponding unit vector $\mathbf{b} = \mathbf{B}/B$, prove that the field line curvature is given by

$$\boldsymbol{\kappa} = \mathbf{b}\cdot\nabla\mathbf{b}$$

where $\boldsymbol{\kappa} = \mathbf{R}_c/R_c^2$ and $\mathbf{R}_c$ is the radius of curvature. Show that $\boldsymbol{\kappa}$ is perpendicular to $\mathbf{b}$: $\boldsymbol{\kappa}\cdot\mathbf{b} = 0$.

4.5. The purpose of this surprisingly lengthy problem is to demonstrate a procedure for calculating the outward toroidal hoop force inherent in all toroidal Z pinches, tokamaks, and reversed field pinches. Consider the geometry illustrated below:

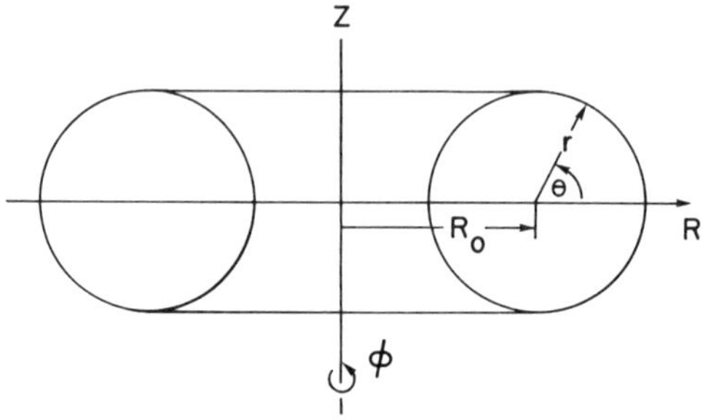

Assume the plasma is axisymmetric (i.e., $\partial/\partial\phi = 0$) and carries a uniform current density, J_0 in the cross section.

a. Using the Biot–Savart law, show that the vector potential, for arbitrary aspect ratio, can be written as

$$A_\phi(R, Z) = \frac{\mu_0 J_0}{4\pi}\int R'\,dR'\,dZ'\,G(R, Z; R', Z') \qquad (1)$$

where

$$G = \frac{2A}{RR'}[(1 - k^2/2)K - E]$$

$$A^2 = (R + R')^2 + (Z - Z')^2$$

$$k^2 = 4RR'/A^2$$

and K and E are the complete elliptic integrals defined by

$$E(k) = \int_0^{\pi/2} (1 - k^2 \sin^2 y)^{1/2}\, dy \tag{2}$$

$$K(k) = \int_0^{\pi/2} \frac{dy}{(1 - k^2 \sin^2 y)^{1/2}} \tag{3}$$

(Hint: be careful using the Biot–Savart law in nonrectangular geometries.)

b. Consider now the fields in the vicinity of the plasma (i.e., $r \sim a$) in the large aspect ratio limit $a/R_0 \ll 1$. Convert from $(R, Z; R', Z')$ coordinates to $(r, \theta; r', \theta')$ coordinates. Show that the leading-order contribution to $k'^2 \equiv 1 - k^2$ is of order (a^2/R_0^2).

c. Show that as $k'^2 \to 0$, $E(k) \to 1$.

d. Show that as $k'^2 \to 0$, $K(k) \to \ln(4/k')$. [Hint: In Eq. (3) make the substitution $\cos y = (k'/k)\sinh(x/2)$. Show that in the transformed integral $\sinh(x/2)$ can be accurately approximated by $(1/2)\exp(x/2)$ when $k'^2 \ll 1$. With this simplification, the resulting integral can be evaluated by standard techniques.]

e. Using the large aspect ratio limits of E and K, evaluate the leading-order contribution to A_ϕ [from Eq. (1)] on the plasma surface. Relate $A_\phi(a, \theta)$ to the flux ψ enclosed within the plasma ring. Define the external self-inductance L as $L \equiv \psi/I$ and show that

$$L = \mu_0 R_0[\ln(8R_0/a) - 2] \tag{4}$$

[Hint: To complete this step you will need to show that

$$I(\alpha) = \int_0^{\pi} \ln(1 + \alpha^2 - 2\alpha \cos z)\, dz = 0 \qquad \text{for} \quad \alpha^2 < 1$$

This can be demonstrated by noting that $I(0) = 0$ and then showing that $dI/d\alpha = 0$ for any $\alpha^2 < 1$.]

f. Show that the magnetic energy stored in the volume surrounding the plasma is given by

$$W = \tfrac{1}{2} L I^2$$

[Hint: This can easily be shown from general considerations by converting the volume integral of magnetic energy to a surface integral using Gauss' theorem and the fact that the field in this region is a vacuum field.]

g. Assume now that the outward hoop force F_H causes a virtual displacement of the plasma from R_0 to $R_0 + dR_0$. Equate the work done during this displacement to the decrease in potential (i.e., magnetic) energy. Note that since the plasma is a perfect conductor, the displacement must be made subject to the constraint that the flux enclosed by the plasma remains constant. Calculate the outward hoop force F_H.

5

Equilibrium
One-Dimensional Configurations

5.1. INTRODUCTION

Although the magnetic configurations of fusion interest are toroidal, one can begin to develop physical intuition by first investigating their one-dimensional cylindrically symmetric analogs: the θ pinch, the Z pinch, and the general screw pinch. This serves to separate the two basic problems of MHD equilibrium, radial pressure balance and toroidal force balance, so that each can be studied individually.

The one-dimensional model focuses entirely on radial pressure balance. The question of toroidal force balance does not enter since by definition the geometry is linear. For most configurations, once radial pressure balance is established, toroidicity can be introduced by means of an aspect ratio expansion, from which one can then investigate toroidal force balance.

Chapter 5 provides a description of the basic one-dimensional configurations and how they provide radial pressure balance in a plasma. In particular, it is shown that both toroidal and poloidal fields as well as combinations thereof can easily satisfy this objective.

5.2. THE θ PINCH

The θ pinch represents the one-dimensional analog of the toroidal configuration with purely toroidal field. The "equivalent" torus consists of a section $2\pi R_0$ in length of the infinitely long cylindrically symmetric

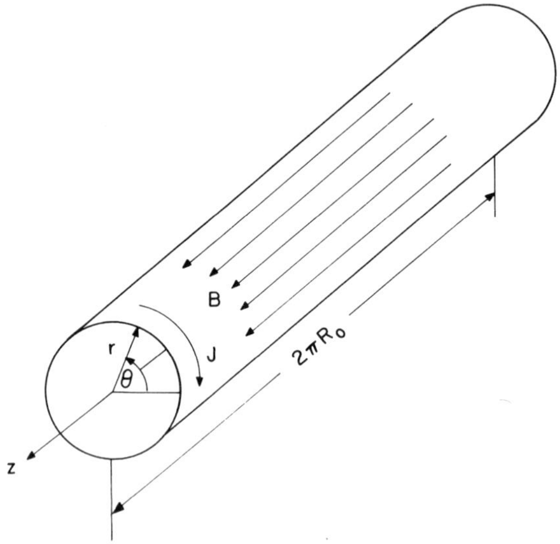

Figure 5.1. Linear θ-pinch geometry.

linear plasma column. In a θ pinch the only nonzero component of $\mathbf{B}$ is in the z direction. It is applied externally and induces a large diamagnetic θ current J_θ in the plasma as shown in Fig. 5.1 (hence the name θ pinch).

The basic equilibrium relation for the θ pinch is easily obtainable from the radial component of the momentum equation. However, a slightly more formal derivation is presented which demonstrates the general procedure used throughout the textbook to solve MHD equilibrium problems. Specifically, there is a natural sequence in which the MHD equations can be solved that provides a simple and direct path to the final equations of interest. This sequence is given by (1) the $\nabla \cdot \mathbf{B}$ equation, (2) Ampere's law, and (3) the momentum equation.

$\nabla \cdot \mathbf{B}$ Equation. The equation $\nabla \cdot \mathbf{B} = 0$ is trivially satisfied for the θ pinch because of symmetry; all quantities are only a function of the radius r:

$$\frac{\partial B_z}{\partial z} = 0 \tag{5.1}$$

Ampere's Law. Ampere's law shows that only $J_\theta(r)$ is nonzero and is given by

$$J_\theta = -\frac{1}{\mu_0}\frac{dB_z}{dr} \tag{5.2}$$

Momentum Equation. The only nontrivial component of the momentum equation is the radial component which reduces to

$$J_\theta B_z = \frac{dp}{dr} \tag{5.3}$$

where the pressure $p = p(r)$. Substituting Eq. (5.2) yields

$$\frac{d}{dr}\left(p + \frac{B_z^2}{2\mu_0}\right) = 0 \tag{5.4}$$

or, upon integrating,

$$p + \frac{B_z^2}{2\mu_0} = \frac{B_0^2}{2\mu_0} \tag{5.5}$$

where B_0 is the externally applied magnetic field.

Equation (5.4) is the basic radial pressure balance relation for a θ pinch. Its integrated form, Eq. (5.5), indicates that at any local value of r the sum of the particle pressure and the magnetic pressure is a constant, equal to the externally applied magnetic pressure; the plasma is confined radially by the pressure of the externally applied magnetic field.

Illustrated in Fig. 5.2 is a set of typical profiles given, for example, by

$$\begin{aligned} p &= p_0 \exp(-r^2/a^2) \\ B_z &= B_0[1 - \beta_0 \exp(-r^2/a^2)]^{1/2} \end{aligned} \tag{5.6}$$

where $\beta_0 \equiv 2\mu_0 p_0/B_0^2$ is the β on axis. The pressure is peaked at $r = 0$ and decreases rapidly for large r, thus isolating itself from the containing wall. This indicates good radial confinement and closed-nested pressure contours.

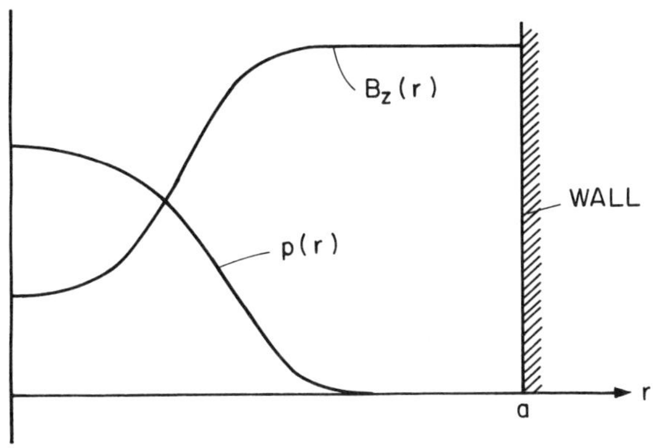

Figure 5.2. Equilibrium profiles for a θ pinch.

Note that Eq. (5.4) implies that the θ pinch has one free surface function, e.g., $B_z(r)$, from which it is then possible to calculate $p(r)$. The second free function, e.g., $B_\theta(r)$, has been set to zero because of the special symmetry associated with the θ pinch.

The basic plasma parameters and figures of merit can now be easily calculated. Since $B_\theta = 0$, $\bar{\beta} = \beta_t$ with the toroidal β given by

$$\beta_t = \frac{4\mu_0}{a^2 B_0^2} \int_0^a pr\,dr = \frac{2}{a^2} \int_0^a \left(1 - \frac{B_z^2}{B_0^2}\right) r\,dr \tag{5.7}$$

where a is the radius of the containing vessel. Equation (5.7) implies that

$$0 < \beta_t < 1 \tag{5.8}$$

This wide range of variation including access to high values of β_t indicates that the θ pinch is an excellent radial container of plasma.

Since there is no Z current in a θ pinch ($I_0 = 0$) the quantities β_p, q_*, ι, and s approach limiting values. The poloidal β approaches infinity $\beta_p = \infty$, and as such is not a relevant quantity. Similarly, $q_* = \infty$, indicating favorable stability against current-driven kinks, a trivial result since there is no longitudinal current. In a θ pinch, $B_\theta = 0$ everywhere, implying that the magnetic field lines are straight (Fig. 5.1). Consequently, $\iota = 0$. The shear is not well defined when $\iota = 0$; however, since the magnetic lines point in the same direction on each surface (i.e., the z direction), the natural limit has $s = 0$. Shear is not available to stabilize MHD instabilities.

The magnetic well is also easily evaluated. Identifying the volume $V = 2\pi^2 R_0 r^2$, one obtains the following expression for the vacuum magnetic well:

$$\hat{W} = \frac{2V}{\langle B^2 \rangle} \frac{d}{dV} \left\langle \frac{B^2}{2} \right\rangle = \frac{r}{B_z} \frac{dB_z}{dr} \tag{5.9}$$

In a θ pinch the vacuum field corresponds to $B_z = \text{const}$, implying that $\hat{W} = 0$: the configuration is neutral with respect to magnetic well stabilization.

At finite β, the pressure modified form of W, given by Eq. (4.41), must be used. One finds

$$W = \frac{2V}{\langle B^2 \rangle} \frac{d}{dV} \left\langle \mu_0 p + \frac{B^2}{2} \right\rangle = \frac{\mu_0 r}{B_z^2} \frac{d}{dr} \left(p + \frac{B_z^2}{2\mu_0} \right) = 0 \tag{5.10}$$

Thus, the pressure contribution to W exactly cancels $\hat{W}$, again implying a neutral result with respect to magnetic well stabilization.

This cancellation is quite general and is responsible for the often-

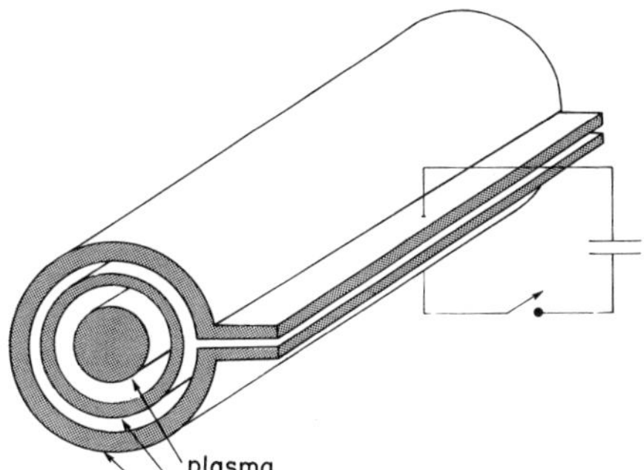

Figure 5.3. Schematic diagram of a θ-pinch experiment.

heard statement that "a plasma cannot dig its own magnetic well by means of its diamagnetic confinement currents."

The θ pinch is more than a theoretical concept and in the early history of the fusion program a number of such devices were built. The typical experimental situation is illustrated in Fig. 5.3. The plasma fills an insulating discharge tube (usually quartz or Pyrex), which is surrounded by a single turn coil connected to a large capacitor bank. The gas is initially preionized and the switch on the capacitor bank closed. As the capacitor discharges, a current flows in the coil, producing a B_z field in the discharge chamber. The parameters of the high-voltage capacitor bank are chosen so that the rise time of the field is quite short, on the order of $2\,\mu$s, while the overall pulse length is typically 10–$50\,\mu$s. The rapidly rising magnetic field acts like a piston, imparting a large impulse of momentum and energy to the particles as they are reflected. This energy is ultimately converted to heat after repeated reflections off the converging piston.

The performance of θ pinches represents one of the early successes of the fusion program. Using the implosion heating method, ion temperatures of 1–4 keV are routinely obtained at very high density, $n \sim 1$–$2 \times 10^{22}\,\mathrm{m}^{-3}$. The peak value of β on axis is typically 0.7–0.9, quite a high value, and corresponds to volume-averaged $\beta_t \sim 0.05$. As further evidence of good performance, radial profile measurements made by holographic interferometry demonstrate excellent radial confinement, nested circular flux surfaces, and no indications of macroscopic instability. Because of the high temperature, the θ pinch is the first laboratory device to have successfully produced a substantial number of thermonuclear neutrons.

As might be expected, the biggest problem with the θ pinch is end loss. Typically, little if anything is done to provide axial confinement of particles or energy; the plasma simply flows out of the device along field lines in a characteristic time $\tau \sim L/V_{T_i}$, where L is the length of the magnet. For a 5-m device, $\tau \sim 10\,\mu$s, a very short time indeed. Because of the end-loss problem the θ pinch has been replaced by more advanced toroidal concepts capable of much longer confinement times.

Before closing the discussion of θ pinches, several comments should be made regarding toroidal equilibrium. The θ pinch represents a highly degenerate configuration. A complete and detailed stability analysis indicates that the straight θ pinch is neutrally stable, and as has been previously shown, cannot be bent into a toroidal equilibrium. Therefore, additional fields must be added to provide equilibrium. Since the basic configuration is neutrally stable, the overall stability will depend sensitively on the additional fields. Thus, the important problems have yet to be faced. Nevertheless, the θ pinch represents the radial pressure balance properties of a number of interesting fusion concepts: in particular, high and low β stellarators, high β tokamaks, the central cell of the tandem mirror, and the Elmo bumpy torus.

5.3. THE Z PINCH

The Z pinch is a one-dimensional model of the toroidal configuration with purely poloidal field and is in many ways orthogonal to the θ pinch. In a Z pinch, only B_θ is nonzero and consists entirely of the self-field induced by the longitudinal current J_z flowing in the plasma (hence the name Z pinch). See Fig. 5.4. Unlike the θ pinch, when the plasma current vanishes there is no background B field remaining. The basic equation describing radial pressure balance in a Z pinch is determined as follows.

$\nabla \cdot \mathbf{B}$ Equation. As for the θ pinch the $\nabla \cdot \mathbf{B} = 0$ equation is trivially satisfied because of symmetry; all quantities are only functions of r:

$$\frac{1}{r}\frac{\partial B_\theta}{\partial \theta} = 0 \tag{5.11}$$

Ampere's Law. Application of Ampere's law shows that only $J_z(r)$ is nonzero.

$$J_z = \frac{1}{\mu_0 r}\frac{d}{dr}(rB_\theta) \tag{5.12}$$

Momentum Equation. Only the radial component of the momentum equation is nontrivial:

$$J_z B_\theta = -\frac{dp}{dr} \tag{5.13}$$

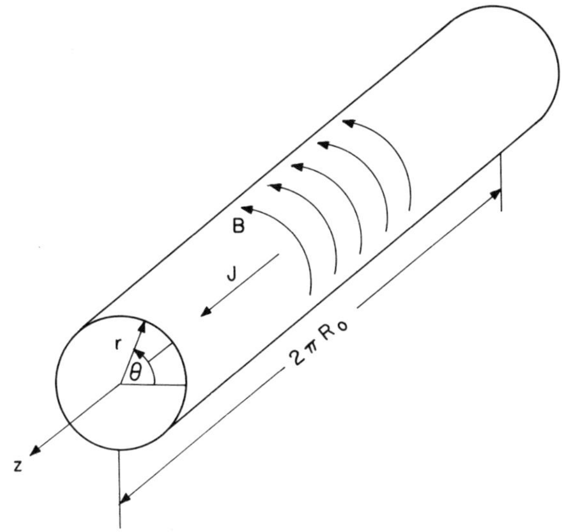

Figure 5.4. Linear Z-pinch geometry.

Substituting Eq. (5.12) yields

$$\frac{dp}{dr} + \frac{B_\theta}{\mu_0 r}\frac{d}{dr}(rB_\theta) = 0 \tag{5.14}$$

which can be rewritten as

$$\frac{d}{dr}\left(p + \frac{B_\theta^2}{2\mu_0}\right) + \frac{B_\theta^2}{\mu_0 r} = 0 \tag{5.15}$$

Equation (5.15) is the basic radial pressure-balance relation for a Z pinch. The two terms in the derivative represent the particle pressure and the magnetic pressure. The last term represents the tension force generated by the curvature of the magnetic field lines. The connection between the tension force and the curvature is a general result obtained by substituting **J** from Ampere's law directly into the momentum equation for arbitrary magnetic configurations. Noting that $(\nabla \times \mathbf{B}) \times \mathbf{B} = -\nabla(B^2/2) + \mathbf{B}\cdot\nabla\mathbf{B}$ and $\boldsymbol{\kappa} = \mathbf{b}\cdot\nabla\mathbf{b}$ one finds

$$\nabla_\perp\left(p + \frac{B^2}{2\mu_0}\right) - \frac{B^2}{\mu_0}\boldsymbol{\kappa} = 0 \tag{5.16}$$

Since $\boldsymbol{\kappa} = -\mathbf{e}_r/r$ in a Z pinch the connection is now apparent.

The tension force is very important. In contrast to the θ pinch, for a Z pinch it is the tension force and not the magnetic pressure gradient that provides radial confinement of the plasma. This can be seen explicitly by examining a specific example. It can be easily verified that the well-known Bennett profiles (Bennett, 1934) given below satisfy the Z-pinch pressure-balance relation:

$$\begin{aligned} B_\theta &= \frac{\mu_0 I_0}{2\pi} \frac{r}{r^2 + r_0^2} \\ J_z &= \frac{I_0}{\pi} \frac{r_0^2}{(r^2 + r_0^2)^2} \\ p &= \frac{\mu_0 I_0^2}{8\pi^2} \frac{r_0^2}{(r^2 + r_0^2)^2} \end{aligned} \tag{5.17}$$

Here, I_0 is the total plasma current and r_0 is the plasma scale length. These profiles are illustrated in Fig. 5.5. Also illustrated are curves of p', $(B_\theta^2/2\mu_0)'$ and $B_\theta^2/\mu_0 r$. Note that in the outer region of the plasma, $r > r_0$, the plasma and magnetic pressure gradients are both pointing outward; magnetic pressure is not confining the plasma. Instead it is the tension force that acts inwards, thereby providing radial pressure balance. If one thinks of the magnetic field lines as rubber bands surrounding a column of plasma, the tension force is then obvious. In analogy with the θ pinch, the Z-pinch equilibrium has one free function,

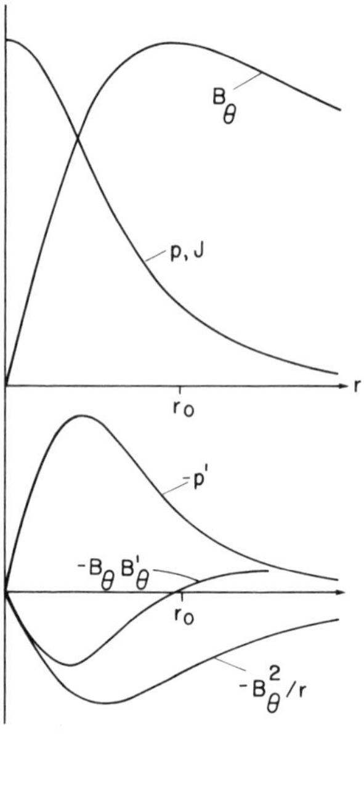

Figure 5.5. Equilibrium profiles for a Z pinch.

e.g., B_θ, which then determines p. The other free function, e.g., B_z, has been set to zero because of the special Z-pinch symmetry.

Using the basic Z-pinch relationships, Eqs. (5.12)–(5.15), it is straightforward to calculate the basic plasma parameters and figures of merit. The confinement efficiency is measured by $\bar{\beta} = \beta_p$ [Eq. (4.32)] and is obtained by multiplying Eq. (5.14) by r^2 and then integrating $0 < r < a$ with a the plasma edge. If one identifies $V = 2\pi^2 R_0 r^2$ and assumes that $B_\theta \to \mu_0 I_0/2\pi r$ for large r, then a simple calculation yields

$$\beta_p = \frac{16\pi^2}{\mu_0 I_0^2} \int_0^a pr\,dr = 1 \tag{5.18}$$

This result is known as the Bennett Pinch Relation and is valid for any pure Z-pinch profile. While high β_p is desirable for confinement efficiency, the lack of flexibility in achieving small to moderate β_p is a disadvantage; that is, some classes of potentially dangerous MHD modes can be stabilized if β_p is sufficiently low. The fact that this cannot occur in a pure Z pinch is one reason why its stability properties are so poor.

Since $B_z = 0$ for a Z pinch the quantities β_t, q_*, and q approach limiting values. First, $\beta_t = \infty$ and as such is not a relevant quantity. Second, $q_* = 0$, indicating poor stability against current-driven kinks. Strictly speaking, q_* refers to modes driven by currents flowing parallel to **B**. Since **J** and **B** are perpendicular, the value of q_* should not be relevant. Still, $q_* = 0$ might be considered ominous. Next, since the magnetic line trajectories never progress along z, it follows that $q = 0$. As in the θ pinch, the lines from one surface to the next are parallel, so the natural limit for the shear in a Z pinch is $s = 0$.

The final quantity of interest is the magnetic well. A simple calculation yields the following expressions for the vacuum magnetic well $\hat{W}$ and the pressure-corrected magnetic well W:

$$\begin{aligned} \hat{W}(r) &= \frac{r}{B_\theta}\frac{dB_\theta}{dr} \\ W(r) &= -1 \end{aligned} \tag{5.19}$$

Since $W(r) < 0$, this implies unfavorable stability properties with respect to pressure-driven MHD instabilities.

A number of linear Z-pinch experiments were constructed during the early years of the fusion program. A typical device is illustrated schematically in Fig. 5.6. Here, a capacitor bank is discharged across two electrodes located at each end of a discharge tube. The high voltage ionizes the gas and produces a Z current flowing along the discharge. These early experiments exhibited disastrous instabilities, often leading

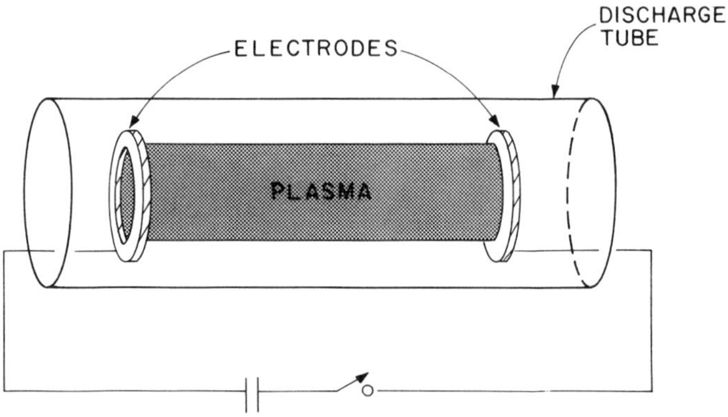

Figure 5.6. Schematic diagram of a linear Z-pinch experiment.

to a complete quenching of the plasma after $\sim$1–2 μs. This behavior, discussed in detail in Section 9.3, is consistent with the unfavorable figures of merit derived above.

Finally, with regard to toroidal equilibrium, recall that the Z pinch can easily be bent into a torus, although its stability properties remain poor. Nevertheless, the Z pinch represents the radial pressure balance properties of several interesting fusion concepts including the ohmically heated tokamak and the reversed field pinch.

5.4. THE GENERAL SCREW PINCH

It should come as no great surprise that in the search for satisfactory toroidal equilibria, one approach of interest is to consider configurations with both toroidal and poloidal fields. The hope, of course, is to combine the favorable and suppress the unfavorable features of each field. The one-dimensional analogs of such systems have both $B_\theta(r)$ and $B_z(r)$ nonzero. See Fig. 5.7. The basic pressure balance relation for a general screw pinch follows in a straightforward manner.

$\nabla \cdot \mathbf{B}$ *Equation.* Even with two field components present, cylindrical symmetry guarantees that $\nabla \cdot \mathbf{B} = 0$ is trivially satisfied:

$$\frac{1}{r}\frac{\partial B_\theta}{\partial \theta} + \frac{\partial B_z}{\partial z} = 0 \tag{5.20}$$

Ampere's Law. Application of Ampere's law shows that two

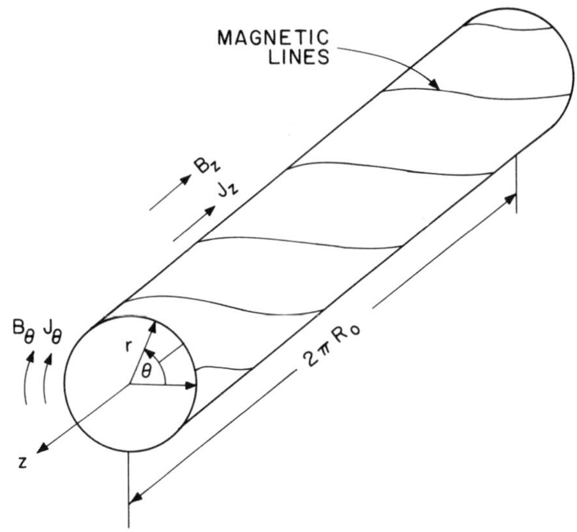

Figure 5.7. General screw-pinch geometry.

components of current, J_θ and J_z, are nonzero:

$$\begin{aligned} J_\theta &= -\frac{1}{\mu_0}\frac{dB_z}{dr} \\ J_z &= \frac{1}{\mu_0 r}\frac{d}{dr}(rB_\theta) \end{aligned} \tag{5.21}$$

Momentum Equation. As before, only the radial component of the momentum equation is nontrivial:

$$J_\theta B_z - J_z B_\theta = \frac{dp}{dr} \tag{5.22}$$

Substituting Eq. (5.21) yields

$$\frac{d}{dr}\left(p + \frac{B_\theta^2 + B_z^2}{2\mu_0}\right) + \frac{B_\theta^2}{\mu_0 r} = 0 \tag{5.23}$$

Equation (5.23) is the basic radial pressure-balance relation for a general screw pinch. Note that even though the momentum equation is nonlinear, the θ-pinch and Z-pinch forces add as a linear superposition, a consequence of the high degree of symmetry. Although Eq. (5.23) is a relatively simple relation, it nevertheless exhibits many of the features and flexibility expected in more realistic, multidimensional, toroidal calculations. This is demonstrated by examining Eq. (5.23) and evaluating the corresponding plasma parameters and figures of merit.

Free Functions. There are two free functions available to specify the equilibrium; for example, $B_\theta(r)$ and $B_z(r)$. The θ pinch and Z pinch are special choices where either B_θ or B_z is set to zero, respectively. In the general case, once both profiles are specified, the pressure is calculated from Eq. (5.23).

Flux Surfaces. The contours of constant pressure are given by $r =$ const. Thus, the magnetic flux surfaces consist of closed concentric circles.

β_t and β_p. Both β_t and β_p can be varied over a wide range if B_z is not equal to zero. This can be seen from the general one-dimensional macroscopic energy balance relation obtained by multiplying Eq. (5.23) by r^2 and integrating $0 < r < a$ with a the plasma edge:

$$2\pi \int_0^a pr\,dr = \frac{\mu_0 I_0^2}{8\pi} + 2\pi \int_0^a \frac{B_0^2 - B_z^2}{2\mu_0} r\,dr \tag{5.24}$$

The first term represents the plasma energy line density, the second term is a measure of the energy in the poloidal magnetic field, and the last term represents the magnetic energy associated with the diamagnetic part of the toroidal field.

Introducing the definitions

$$\begin{aligned} \beta_t &= \frac{4\mu_0}{a^2 B_0^2} \int_0^a pr\,dr \\ \beta_p &= \frac{16\pi^2}{\mu_0 I_0^2} \int_0^a pr\,dr \end{aligned} \tag{5.25}$$

Eq. (5.24) can be rewritten as

$$\frac{1}{\beta_p} + \frac{\alpha_t}{\beta_t} = 1 \tag{5.26}$$

where

$$\alpha_t = \frac{2}{a^2} \int_0^a \left(1 - \frac{B_z^2}{B_0^2}\right) r\,dr \tag{5.27}$$

represents the plasma diamagnetism. α_t lies in the range $-\infty < \alpha_t < 1$, implying a wide range of variation for β_p and β_t.

Kink Safety Factor. The kink safety factor is readily evaluated from Eq. (4.34) and is given by

$$q_* = \frac{2\pi a^2 B_0}{\mu_0 R_0 I_0} \tag{5.28}$$

q_* can vary over a wide range, depending on the ratio of current to toroidal field and the geometry a^2/R_0.

Rotational Transform. The magnetic lines wrap around the column along helical paths. See Fig. 5.7. This gives rise to a nonzero rotational transform which can be calculated as follows. The average value of the angle $\Delta\theta$ [see Eq. (4.21)] is independent of θ_0, the starting angle, because of the circular symmetry. Thus, to calculate ι it is only necessary to integrate the field line trajectory a distance $2\pi R_0$ in z, corresponding to one transit around the "equivalent" torus.

$$\iota \equiv \Delta\theta = \int_0^{\Delta\theta} d\theta = \int_0^{2\pi R_0} \frac{d\theta}{dz} dz \tag{5.29}$$

Using the equations for the field line trajectories

$$\begin{aligned} \frac{dr}{dz} &= \frac{B_r(r)}{B_z(r)} = 0 \\ \frac{d\theta}{dz} &= \frac{B_\theta(r)}{rB_z(r)} \end{aligned} \tag{5.30}$$

then gives $r = \text{const}$, indicating circular flux surfaces and

$$\iota(r) = \frac{2\pi R_0 B_\theta(r)}{rB_z(r)} \tag{5.31}$$

The MHD safety factor $q(r) = 2\pi/\iota(r)$ has the form

$$q(r) = \frac{rB_z}{R_0 B_\theta} \tag{5.32}$$

Again, with two free functions available there is a wide range of flexibility in the $q(r)$ profile. Also, note that $q(a) = q_*$.

Shear. The shear is evaluated from Eq. (4.38) after making the identification $V = 2\pi^2 R_0 r^2$. One finds

$$s(r) = \frac{r}{q}\frac{dq}{dr} \tag{5.33}$$

For a screw pinch, the shear is in general not zero, thereby providing a measure of safety against a variety of MHD instabilities.

Magnetic Well. The magnetic well is easily evaluated in a manner analogous to that of the rotational transform. Substituting in Eq. (4.41) yields

$$W(r) = \frac{\mu_0 r}{B^2}\frac{d}{dr}\left(p + \frac{B^2}{2\mu_0}\right) = -\frac{B_\theta^2}{B^2} \tag{5.34}$$

Clearly $W(r) < 0$ corresponding to instability. However, the screw pinch has the flexibility of making the magnitude of $W(r)$ small by requiring $B_\theta^2 \ll B_z^2$. Equally important, the destabilizing influence of $W(r) < 0$ can be compensated for by shear.

Because of its flexibility the general screw-pinch relation describes a wide variety of configurations. Several toroidal configurations in particular are reasonably well described (at least in leading order) for stability as well as equilibrium by the linear model because they are characterized by circular cross sections. These include circular low-β tokamaks and the reversed field pinch.

Finally, note that toroidal equilibrium is in general not difficult to achieve in a screw pinch since $B_\theta \neq 0$.

5.5. SUMMARY

An examination of MHD equilibria in one-dimensional cylindrically symmetric configurations focuses attention on the problem of radial pressure balance. Both toroidal and poloidal fields can provide radial plasma confinement. This has been demonstrated for the limiting cases of the θ pinch and Z pinch, as well as the general screw pinch, which contains a combination of B_θ and B_z. A summary of these configurations is given as follows.

The θ Pinch. The θ pinch is the one-dimensional analog of the torus with purely toroidal fields: $B_z(r)$, $J_\theta(r)$, $p(r)$. In this configuration, radial confinement is provided by the externally applied magnetic pressure. The θ pinch is capable of a wide range of toroidal β, $0 < \beta_t < 1$. It has zero rotational transform, shear and magnetic well: $\iota(r) = s(r) = W(r) = 0$, thus indicating neutral stability.

The Z Pinch. The Z pinch is the one-dimensional analog of the torus with purely poloidal fields: $B_\theta(r)$, $J_z(r)$, $p(r)$. Here, radial pressure balance is provided by the tension in the magnetic field lines. In a Z pinch the poloidal β is always unity: $\beta_p = 1$. The configuration has zero MHD safety factor and shear: $q(r) = s(r) = 0$. However, the magnetic well $W(r) < 0$, indicating instability with no other stabilizing factor to compensate.

The General Screw Pinch. The screw pinch is the one-dimensional analog of toroidal configurations with both toroidal and poloidal magnetic fields; for example, the tokamak and reversed field pinch. Screw-pinch geometries are capable of a wide range of variation in both β_t and β_p. In general, the kink safety factor, the MHD safety factor, the shear, and the magnetic well are nonzero. The magnetic well $W(r)$ is always negative, indicating instability. However, this can be compensated for by the shear and by making the magnitude of $W(r)$ small.

REFERENCES

Bennett, W. H. (1934). *Phys. Rev.* **45**, 980.

PROBLEMS

5.1. Consider a θ pinch with a magnetic field $\mathbf{B} = B(r)\mathbf{e}_z$ and pressure $p = p(r)$. Assume $p(r) = p_0 \exp(-r^2/a^2)$, $B(\infty) = B_0$, $n(0) = 3 \times 10^{22}\,\mathrm{m}^{-3}$, $T_i = 2\,\mathrm{keV}$, $T_e = 0.8\,\mathrm{keV}$ and $B_0 = 8\,\mathrm{T}$, and $a = 0.01\,\mathrm{m}$. The temperatures are uniform in space.

a. Calculate $\beta_0 \equiv \beta(0)$.

b. Draw a diagram indicating the direction of $\mathbf{B}$ and the plasma current density $\mathbf{J}$.

c. Calculate the magnitude and sign of the curvature drift of a thermal ion and a thermal electron at $r = a$.

d. Calculate the magnitude and sign of the ∇B drift of a thermal ion and a thermal electron at $r = a$.

e. Aren't the signs in step (d) inconsistent with step (b)? Explain.

5.2. Consider a static ideal MHD Z-pinch equilibrium with

$$J = c_1 \frac{r^2/a^2}{(1 + r^2/a^2)^3}$$

where c_1 is a constant.

a. Calculate $B_\theta(r)$ and $p(r)$. Express your answers in terms of I, the total current. Sketch the fields and the currents.

b. Since $J(r)$ vanishes for large r the Z pinch is apparently confined by its own current. Doesn't this violate the virial theorem? Explain.

5.3. Often times in MHD equilibrium theory one must specify $p(\psi)$ rather than $p(r)$. This problem demonstrates how the specification of $p = p(\psi)$ leads to a determination of the equilibrium fields.

a. Consider a one-dimensional θ pinch. Show that the flux contained within a given radius r is equal to $2\pi r A_\theta$, where A_θ is the vector potential.

b. Define $\psi = rA_\theta$ and assume p is specified as follows: $p(\psi) = p_0 \exp(-\psi/\psi_0)$ with p_0 and ψ_0 constants. Using the one-dimensional radial pressure-balance relation for a θ pinch, derive a differential equation for $\psi(r)$. This equation should be a first-order, nonlinear, ordinary differential equation. For convenience, write the equation in terms of the normalized variables and parameters:

$$H = \psi/\psi_0, \qquad \lambda = a^2 B_0/\psi_0$$
$$x = r^2/2a^2, \qquad \beta = 2\mu_0 p_0/B_0^2$$

Here B_0 is the magnetic field far from the plasma, and a is a scale factor representing the plasma radius.

c. Solve this equation for H, subject to the condition $\psi(0) = 0$. Calculate and sketch $p(r)$ and $B_z(r)$.

d. Calculate $\lambda = \lambda(\beta)$ by defining the scale length a such that

$$\pi p_0 a^2 \equiv \int p(r) r \, dr \, d\theta$$

5.4. In tokamaks and reversed field pinches the safety factor $q(r) \equiv rB_z/R_0B_\theta$ is an important quantity of physical interest. Here, $R_0 = \text{const}$ represents the major radius of an equivalent torus.

a. Assuming a force-free configuration ($\nabla p = 0$) derive a differential equation for $B_z(r)$ assuming $q(r)$ is a known specified function.

b. Calculate the fields for the case $q = \text{const}$ assuming $B_z(0) = B_0$. Evaluate q if the total Z current flowing in the plasma is I.

5.5. A rigid rotor θ pinch is characterized by a current density $\mathbf{J} = J_\theta \mathbf{e}_\theta$, where $J_\theta = -e\Omega r n(r)$ and the angular rotation velocity $\Omega = \text{const}$. Assume for simplicity that the temperature $T(r) = T_0 n(r)/n_0$.

a. Derive the differential equation determining $n(r)$.

b. Determine $n(r)$ assuming $n(0) = n_0$ and $n(r_0) = 0$, where the edge of the plasma r_0 is a specified quantity.

5.6. Prove that any force-free screw-pinch configuration is paramagnetic; that is,

$$\int_0^a (B_0^2 - B_z^2) r \, dr < 0$$

where B_0 is the applied B_z field at the edge of the plasma.

5.7. This problem investigates the use of "gravity" in a slab geometry to model plasma rotation and magnetic field line curvature.

a. Derive the one-dimensional radial pressure-balance relation for a general screw pinch including the effects of a stationary angular velocity $\Omega(r)\mathbf{e}_\theta$.

b. Derive the one-dimensional pressure-balance relation for a slab model of a plasma including a gravity $g(x)\mathbf{e}_x$. Assume $\rho(x)$, $B_y(x)$, $B_z(x)$, and $p(x)$ are nonzero. Also assume there is no equilibrium flow: $\mathbf{v} = 0$.

c. By comparing the results in (a) and (b) determine a correspondence between B_θ, Ω, and g so that gravity models the effects of rotation and poloidal field line tension.

5.8. This problem treats the adiabatic compression of a θ pinch. Consider the simple surface current model of an infinitely long, straight θ pinch as shown. Here n_i, p_i, B_i, a_i, and B_0 represent the initial, known state of the plasma. At $t = 0$, the applied field at $r = b$ is slowly (i.e., adiabatically) increased until B_0 reaches the value $B_0' \equiv \lambda B_0$. The quantity $\lambda > 1$ is the multiplication factor of the applied field and is assumed known. Also, the plasma obeys an adiabatic equation of state.

a. Derive, but do not solve, an algebraic equation which relates the final β of the plasma (β_f) to the initial β of the plasma (β_i). Express this equation as $F(\beta_f, \beta_i, \lambda, \gamma,) = 0$ where γ is the ratio of specific heats.

b. Solve this equation for β_f for the special case $\gamma = 1$, $\beta_i = 0.5$, and $\lambda = 2$.

c. If you did the problem correctly you should find that $\beta_f < \beta_i$. Explain why the final β decreases in spite of the fact that the plasma has been compressed.

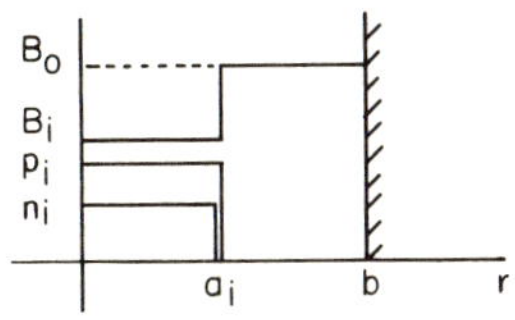

5.9. Temperature and magnetic profile measurements on a straight screw pinch indicate that the B_θ profile is given by

$$B_\theta(r) = \frac{\mu_0 I}{2\pi} \frac{r}{r^2 + a^2}$$

where I is the plasma current and a is the average radius of the plasma. For simplicity assume that the pressure and B_z field are related by

$$B_z^2(r) = B_0^2 - 2\mu_0 \lambda p(r).$$

Here B_0 is the externally applied field and λ is a constant to be determined. Diamagnetic loop measurements indicate that the "excluded flux" is given by

$$\Delta\psi = 0.02(\pi B_0 a^2).$$

The "excluded flux" represents the difference between the toroidal flux contained within the conducting wall (of radius $b \gg a$) for the case of no plasma present and the case where plasma is present, for the same value of

B_0:

$$\Delta\psi = \psi_{\text{tor}}\ (\text{no plasma present}) - \psi_{\text{tor}}\ (\text{plasma present}).$$

a. Calculate λ. To simplify the analysis assume $\mu_0 I/2\pi a B_0 = 0.2 \ll 1$.
b. Calculate the volume averaged toroidal β if $b/a = 3$.

6

Equilibrium

Two-Dimensional Configurations

6.1. INTRODUCTION

Chapter 5 has shown that a one-dimensional, cylindrically symmetric magnetic geometry accurately models the problem of radial pressure balance in many configurations. In Chapter 6 the problem of toroidal force balance is addressed in the simplest nontrivial two-dimensional geometry, that of an axisymmetric torus.

A derivation is presented of the Grad–Shafranov equation, the basic equation describing axisymmetric toroidal equilibrium. The solutions to this equation provide a complete description of ideal MHD equilibria: radial pressure balance, toroidal force balance, equilibrium β limits, rotational transform, shear, magnetic well, etc. A wide number of configurations are accurately modeled by the Grad–Shafranov equation. Among them are all types of tokamaks, the spheromak, the reversed field pinch, and toroidal multipoles.

An important aspect of the analysis is the use of asymptotic expansions, with the inverse aspect ratio $\epsilon = a/R_0$ serving as the expansion parameter. For many configurations the asymptotic analysis mathematically separates the problems of radial pressure balance and toroidal force balance. It thus provides a method for obtaining analytic results as well as developing physical intuition. In a certain sense, each different asymptotic expansion serves to mathematically identify a specific MHD concept.

In addition, an equation similar to the Grad–Shafranov equation, but for helically symmetric equilibria, is presented. This equation, which is at least as difficult to solve, represents the leading-order description of low-β and high-β stellarators, heliacs, and the Elmo bumpy torus. The solutions all correspond to infinitely long straight helices. Bending such a configuration into a torus requires a full three-dimensional calculation and is discussed in Chapter 7.

6.2. THE GRAD–SHAFRANOV EQUATION

6.2.1. Derivation

The Grad–Shafranov equation is a two-dimensional, nonlinear, elliptic partial differential equation obtained from the reduction of the ideal MHD equations [Eqs. (4.6)–(4.8)] for the case of toroidal axisymmetry (Grad and Rubin, 1958; Shafranov, 1960). The geometry of interest is illustrated in Fig. 6.1.

Here, R, ϕ, Z describe the usual right-handed cylindrical coordinate system and the assumption of toroidal axisymmetry implies that $\partial S/\partial\phi = 0$, where S is any scalar. The derivation proceeds as follows.

$\nabla \cdot \mathbf{B}$ *Equation.* The axisymmetric assumption implies that the $\nabla \cdot \mathbf{B} = 0$ equation can be written as

$$\frac{1}{R}\frac{\partial(RB_R)}{\partial R} + \frac{\partial B_Z}{\partial Z} = 0 \tag{6.1}$$

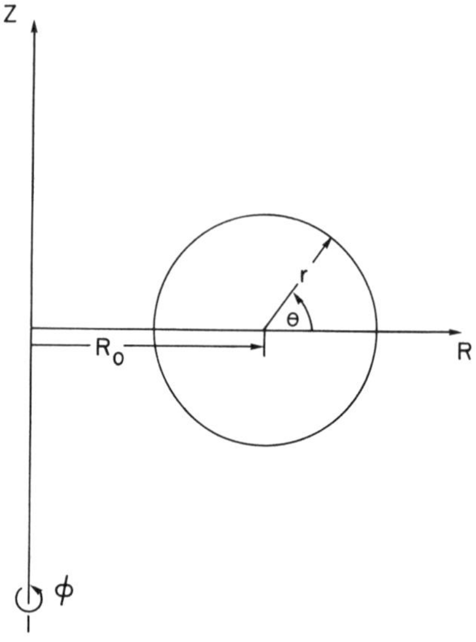

Figure 6.1. Geometry for axisymmetric toroidal equilibrium.

It is useful to introduce a stream function ψ for the poloidal magnetic field:

$$B_R = -\frac{1}{R}\frac{\partial \psi}{\partial Z}, \qquad B_Z = \frac{1}{R}\frac{\partial \psi}{\partial R} \tag{6.2}$$

where $\psi = RA_\phi$ and A_ϕ is the toroidal component of vector potential. Note that B_ϕ is unaffected by Eq. (6.1). In more compact notation

$$\begin{aligned} \mathbf{B} &= B_\phi \mathbf{e}_\phi + \mathbf{B}_p \\ \mathbf{B}_p &= \frac{1}{R}\nabla\psi \times \mathbf{e}_\phi \end{aligned} \tag{6.3}$$

The stream function ψ is closely related to the poloidal flux in the plasma. ψ_p is given by

$$\psi_p = \int \mathbf{B}_p \cdot d\mathbf{A} \tag{6.4}$$

If one chooses for the area a washer-shaped surface (see Fig. 6.2) lying in the $Z = 0$ plane, extending from the magnetic axis $R = R_a$ to an arbitrary ψ contour defined by $\psi = \psi(R_b, 0)$ then

$$\psi_p = \int_0^{2\pi} d\phi \int_{R_a}^{R_b} dR\, RB_Z(R, Z = 0) \tag{6.5}$$

or

$$\psi_p = 2\pi\psi \tag{6.6}$$

Here, the arbitrary integration constant associated with ψ has been chosen so that $\psi(R_a, 0) = 0$; the poloidal flux on axis is zero.

In axisymmetric tori it is convenient to label the flux surfaces with ψ rather than p or V.

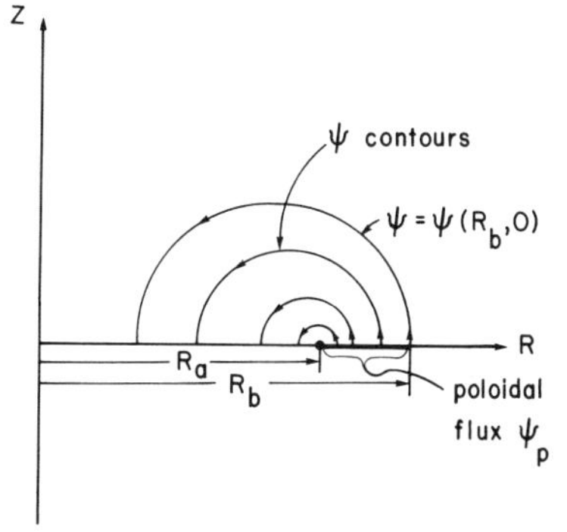

Figure 6.2. Washer-shaped surface through which the poloidal flux ψ_p passes.

Ampere's Law. Substituting Eq.(6.3) into Ampere's law leads to the following expression for the current density:

$$\mu_0\mathbf{J} = \mu_0 J_\phi \mathbf{e}_\phi + \frac{1}{R}\nabla(RB_\phi) \times \mathbf{e}_\phi$$
$$\mu_0 J_\phi = -\frac{1}{R}\Delta^*\psi \tag{6.7}$$

where the elliptic operator Δ^* is given by

$$\Delta^*\psi \equiv R^2\nabla\cdot\left(\frac{\nabla\psi}{R^2}\right) = R\frac{\partial}{\partial R}\left(\frac{1}{R}\frac{\partial\psi}{\partial R}\right) + \frac{\partial^2\psi}{\partial Z^2} \tag{6.8}$$

Momentum Equation. The final step in the derivation is to substitute the expressions for **B** and **J** [Eqs. (6.3) and (6.7)] into the momentum equation [Eq. (4.6)]. A convenient way to do this is to decompose Eq. (4.6) into three components, along **B**, **J**, and $\nabla\psi$ (normal to the flux surface). The **B** component gives $\mathbf{B}\cdot\nabla p = 0$ or

$$\mathbf{e}_\phi \cdot \nabla\psi \times \nabla p = 0 \tag{6.9}$$

As expected, this implies that p is a surface quantity

$$p = p(\psi) \tag{6.10}$$

Similarly, the **J** component gives $\mathbf{J}\cdot\nabla p = 0$ or

$$\mathbf{e}_\phi \cdot \nabla\psi \times \nabla(RB_\phi) = 0 \tag{6.11}$$

Hence, RB_ϕ is also a surface quantity

$$RB_\phi = F(\psi) \tag{6.12}$$

The quantity $F(\psi)$ is related to the net poloidal current flowing in the plasma and the toroidal field coils; specifically, the current flowing through a disk-shaped surface lying in the $Z = 0$ plane extending out to an arbitrary ψ contour defined by $\psi \equiv \psi(R_b, 0)$. See Fig. 6.3. One finds

$$I_p = \int \mathbf{J}_p \cdot d\mathbf{A} = -\int_0^{2\pi} d\phi \int_0^{R_b} dR\, RJ_Z(R, Z = 0) \tag{6.13}$$

or

$$I_p(\psi) = -2\pi F(\psi) \tag{6.14}$$

The Grad–Shafranov equation is obtained by substituting Eqs. (6.10) and (6.12) into the $\nabla\psi$ component of the momentum equation.

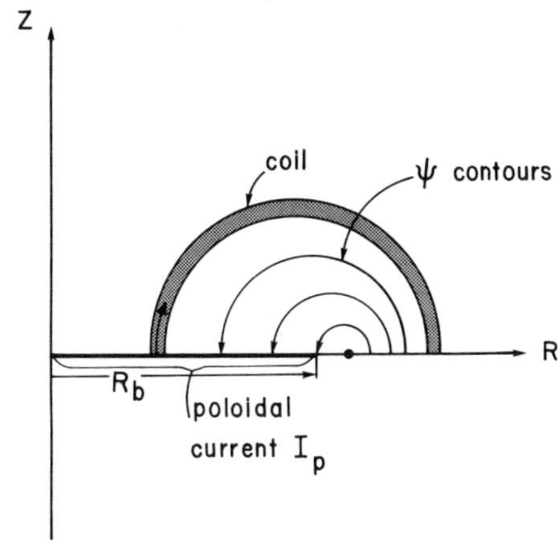

Figure 6.3. Disk-shaped surface through which the total (plasma plus coil) poloidal current I_p flows.

The result, along with a summary of the field relations, is given by

$$\Delta^* \psi = -\mu_0 R^2 \frac{dp}{d\psi} - F\frac{dF}{d\psi} \tag{6.15}$$

where

$$\mathbf{B} = \frac{1}{R}\nabla\psi \times \mathbf{e}_\phi + \frac{F}{R}\mathbf{e}_\phi \tag{6.16}$$

$$\mu_0 \mathbf{J} = \frac{1}{R}\frac{dF}{d\psi}\nabla\psi \times \mathbf{e}_\phi - \frac{1}{R}\Delta^*\psi\mathbf{e}_\phi \tag{6.17}$$

and $p(\psi)$ and $F(\psi)$ are two free functions.

Equation (6.15) is the Grad–Shafranov equation, a second-order nonlinear partial differential equation describing axisymmetric toroidal equilibria. The nature of the equilibria (i.e., tokamak, reversed field pinch, etc.) is to a large extent determined by the choices of the two free functions $p(\psi)$ and $F(\psi)$ and, of course, the boundary conditions. These issues are discussed in detail on an individual basis in the sections that follow.

6.2.2. Plasma Parameters and Figures of Merit

Consider now the evaluation of the basic plasma parameters and the figures of merit. By exploiting the axisymmetry it is possible to obtain general expressions for these quantities in terms of the flux function ψ. To begin, assume $p(\psi)$ and $F(\psi)$ have been specified and a solution has been obtained for ψ and the resulting magnetic fields. For purposes of intuition assume the calculation has been carried out in the toroidal coordinates r, θ, ϕ, shown in Fig. 6.1, and defined by

$$\begin{aligned} R &= R_0 + r\cos\theta \\ Z &= r\sin\theta \end{aligned} \tag{6.18}$$

Furthermore, assume that the flux function $\psi = \psi(r, \theta)$ can be inverted to give

$$r = \hat{r}(\theta, \psi) \tag{6.19}$$

This is the equation for the flux surfaces, $\psi = \text{const.}$

The Volume V. The volume is evaluated as follows:

$$V(\psi) = \int_0^{2\pi} \int_0^{2\pi} \int_0^{\hat{r}} Rr\,dr\,d\theta\,d\phi \tag{6.20}$$

which reduces to

$$V(\psi) = \pi R_0 \int_0^{2\pi} d\theta\,\hat{r}^2\left[1 + \frac{2}{3}\left(\frac{\hat{r}}{R_0}\right)\cos\theta\right] \tag{6.21}$$

The Safety Factor q. The safety factor is slightly more complicated. The equations describing the field line trajectories are given by

$$\frac{R\,d\phi}{B_\phi} = \frac{dr}{B_r} = \frac{r\,d\theta}{B_\theta} = \frac{dl_p}{B_p} \tag{6.22}$$

where $dl_p = [(dr)^2 + (r\,d\theta)^2]^{1/2}$ is the poloidal arc length and $B_p = (B_r^2 + B_\theta^2)^{1/2} = |\nabla\psi|/R$ is the poloidal magnetic field. As the magnetic line wraps exactly once around the poloidal cross section, i.e., $\Delta\theta = 2\pi$, then the toroidal angle $\Delta\phi$ along which it travels is given by

$$\Delta\phi = \int_0^{\Delta\phi} d\phi = \int_0^{2\pi}\left(\frac{rB_\phi}{RB_\theta}\right)_S d\theta \tag{6.23}$$

Here, the integrand I is to be evaluated on the flux surface on which the magnetic line lies: $I(r, \theta)|_S \equiv I[\hat{r}(\theta, \psi), \theta)]$. The average fraction of $\Delta\theta$ traversed during a single toroidal transit is by definition the rotational transform and can be found by taking simple proportions:

$$\frac{\iota}{2\pi} = \frac{\Delta\theta}{\Delta\phi} = \frac{2\pi}{\Delta\phi} \tag{6.24}$$

The safety factor $q(\psi) = 2\pi/\iota$ then has the form

$$q(\psi) = \frac{1}{2\pi}\int_0^{2\pi}\left(\frac{rB_\phi}{RB_\theta}\right)_S d\theta = \frac{F(\psi)}{2\pi}\oint \frac{dl_p}{R^2 B_p} \tag{6.25}$$

where the line integral in the last expression corresponds to one poloidal circuit.

The Magnetic Well and Shear. To calculate the shear and magnetic well note that Eq. (6.21) implies that

$$\frac{dV}{d\psi} = 2\pi \int_0^{2\pi} d\theta \, R\hat{r} \frac{\partial \hat{r}}{\partial \psi} \tag{6.26}$$

At fixed θ, $\partial\hat{r}/\partial\psi = (\partial\psi/\partial\hat{r})^{-1} = (1/RB_\theta)_S$. Thus,

$$\frac{dV}{d\psi} = 2\pi \int_0^{2\pi} \left(\frac{r}{B_\theta}\right)_S d\theta = 2\pi \oint \frac{dl_p}{B_p} \tag{6.27}$$

Using this result, the shear and magnetic well can be expressed in terms of ψ derivatives as follows:

$$s(\psi) = 2\left(\frac{V}{V'}\right)\left(\frac{q'}{q}\right) \tag{6.28}$$

$$W(\psi) = 2\left(\frac{V}{V'}\right)\left(\frac{\langle B^2/2 + \mu_0 p\rangle'}{\langle B^2\rangle}\right) \tag{6.29}$$

where the prime denotes differentiation with respect to ψ, $\langle Q\rangle$ is defined as

$$\langle Q\rangle = \frac{\int (rQ/B_\theta)_S \, d\theta}{\int (r/B_\theta)_S \, d\theta} = \frac{\oint Q \, dl_p/B_p}{\oint dl_p/B_p} \tag{6.30}$$

and

$$\langle B^2\rangle = \langle [F^2 + (\nabla\psi)^2]/R^2\rangle \tag{6.31}$$

There exists a simple correspondence between the magnetic well W and the field line curvature $\boldsymbol{\kappa}$ which is useful in developing intuition about the stability of different magnetic geometries. Recall that $W > 0$ is favorable for stability. In low β, Eq. (6.29) implies that the sign of W is proportional to $\langle B^2\rangle'/\langle B^2\rangle$; that is, $\langle B^2\rangle$ increasing with radius corresponds to stability. This behavior is easily correlated to field line curvature as shown in Fig. 6.4. The regions labeled F and U represent favorable and unfavorable curvature, respectively. As one moves radially outward in a region labeled U, the field tends to becomes smaller because of the large radius of curvature: $\langle B^2\rangle$ decreases indicating instability. Conversely, for regions labeled F, $\langle B^2\rangle$ increases with increasing radius because of the smaller radius of curvature, implying stability. This simple intuition regarding field line curvature is a useful guideline for testing stability.

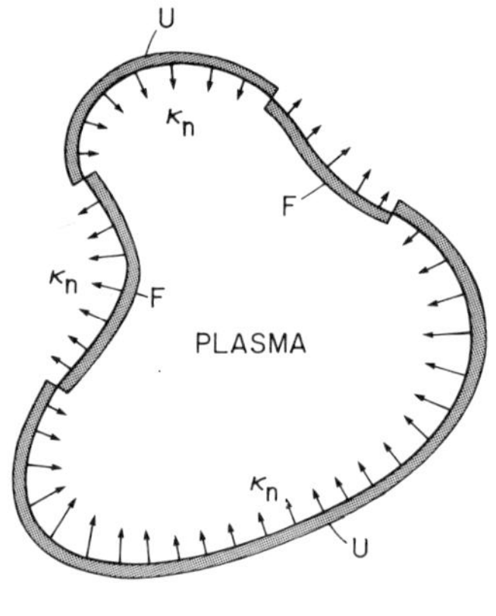

Figure 6.4. Schematic diagram of outward-pointing favorable curvature regions F and inward-pointing unfavorable curvature regions U.

β_t, β_p and q_*. Finally, the quantities β_t, β_p, and q_* are obtained by substituting into Eqs. (4.31), (4.32), and (4.34). One finds

$$\beta_t = \frac{2\mu_0\langle p\rangle}{B_0^2} \tag{6.32}$$

$$\beta_p = \frac{2\mu_0\langle p\rangle}{\bar{B}_p^2} \tag{6.33}$$

$$q_* = \frac{aB_0}{R_0\bar{B}_p} \tag{6.34}$$

where

$$\langle p\rangle = \frac{1}{V_0}\int_0^{V_0} p(V)\,dV = \frac{1}{V_0}\int_0^{\psi_b} pV'\,d\psi \tag{6.35}$$

$$\bar{B}_p = \frac{\mu_0 I_0}{2\pi a\kappa} = \frac{1}{2\pi a\kappa}\oint B_p\,dl_p \tag{6.36}$$

$$\kappa = \frac{1}{2\pi}\int_0^{2\pi} d\theta\left[\frac{\hat{r}(\psi_b\theta)}{a}\right]^2 \tag{6.37}$$

$V_0 = V(\psi_b)$ and ψ_b is the value of ψ at the plasma boundary.

The quantities β_t, β_p, q_*, $q(\psi)$, and $W(\psi)$ represent the basic plasma parameters and figures of merit for axisymmetric tori as described by the Grad–Shafranov equation. In many of the specific applications that follow, these expressions simplify considerably.

Because of the nonlinearity, the Grad–Shafranov equation must in general be solved numerically. Most of the major fusion laboratories around the world have developed such codes. They can be used for essentially arbitrary $p(\psi)$ and $F(\psi)$ as well as for rather sophisticated boundary conditions. Examples of these numerical equilibria are presented in the sections that follow.

However, from the point of view of obtaining physical understanding, it is perhaps more instructive to perform asymptotic analyses of the Grad–Shafranov equation. The simplifications that arise often permit analytic solutions that are of great aid in developing intuition. The basic expansion parameter is the inverse aspect ratio, r/R. When $r/R \ll 1$ the toroidal effects are weak, allowing many configurations to be viewed as perturbations about a straight system. Extensive use of the inverse aspect ratio expansion is made in the remainder of Chapter 6 as well as Chapter 7.

Before proceeding with the applications it is useful to note that a "Grad-Shafranov" equation also exists for systems with helical symmetry. In such systems all quantities are functions of the two independent variables $(r,\ \alpha \equiv l\theta + hz)$ where $2\pi/h$ is the helical period and l describes the order of the poloidal periodicity. The derivation of the equation is similar to that of the axisymmetric torus, and the results are as follows (Solov'ev, 1967):

$$
\begin{aligned}
B_r &= -\frac{1}{r}\frac{\partial\psi}{\partial\alpha} \\
lB_\theta + hrB_z &= \frac{\partial\psi}{\partial r} \\
lB_z - hrB_\theta &= F(\psi) \\
\Delta^*\psi &= -\mu_0\frac{dp}{d\psi} - \frac{F}{l_0^2}\frac{dF}{d\psi} + \frac{2hlF}{l_0^4} \\
\Delta^*\psi &\equiv \frac{1}{r}\frac{\partial}{\partial r}\left(\frac{r}{l_0^2}\frac{\partial\psi}{\partial r}\right) + \frac{1}{r^2}\frac{\partial^2\psi}{\partial\alpha^2} \\
l_0^2 &\equiv l^2 + h^2r^2
\end{aligned}
\tag{6.38}
$$

The equation for ψ is of comparable difficulty to solve as the axisymmetric Grad–Shafranov equation. It provides a good zeroth-order description of stellarators, heliacs, and the Elmo bumpy torus but does not include the effects of toroidal force balance. This requires a three-dimensional calculation and is discussed in Chapter 7.

6.3. THE REVERSED FIELD PINCH (RFP)

It is convenient to begin discussions of individual concepts with the reversed field pinch (RFP) because it, of all others, is the one most accurately described as a perturbation about a linear system. Application of the appropriate aspect ratio expansion leads to a relatively simple analytic solution to the Grad–Shafranov equation.

RFP Expansion. The asymptotic expansion is defined in terms of the inverse aspect ratio

$$\epsilon \equiv a/R_0 \tag{6.39}$$

where a is the characteristic scale length of the minor plasma radius (i.e., $r \sim a$) and it is assumed that $\epsilon \ll 1$. The reversed field pinch is distinguished from other configurations by means of a specific ordering of the fields and pressure. This ordering, designated as the RFP expansion, is given by

$$\begin{aligned} \frac{B_\phi}{B_p} &\sim 1 \\ \beta_p \sim \frac{2\mu_0 p}{B_p^2} &\sim 1 \\ \beta_t \sim \frac{2\mu_0 p}{B_\phi^2} &\sim 1 \\ q \sim \frac{rB_\phi}{RB_p} &\sim \epsilon \end{aligned} \tag{6.40}$$

Thus, an RFP is a high-β device, operating with a small safety factor, in which both poloidal and toroidal fields play an important role in radial pressure balance. Specific profiles are discussed shortly.

The boundary conditions are also quite simple. An accurate approximation to RFP experiments is to assume the plasma is surrounded by a perfectly conducting, circular wall of minor radius b. On this surface, $\mathbf{n} \cdot \mathbf{B} = 0$, implying that the wall is a flux surface. Hence, the boundary conditions on $\psi(r, \theta)$ require

$$\begin{aligned} &(1)\ \psi \text{ regular for } r \leq b \\ &(2)\ \psi(b, \theta) = \text{const} \end{aligned} \tag{6.41}$$

The solution to the Grad–Shafranov equation proceeds as follows. First, the flux function is expanded in ϵ:

$$\begin{aligned} \psi(r, \theta) &= \psi_0(r) + \psi_1(r, \theta) \ldots \\ \psi_0/rRB_p &\sim 1 \\ \psi_1/\psi_0 &\sim \epsilon \end{aligned} \tag{6.42}$$

Here, ψ_0 represents the leading order cylindrically symmetric component of ψ, and ψ_1 represents a small correction arising from toroidal effects.

The next step is to substitute Eq. (6.42) into the Grad–Shafranov equation. This step is accomplished by transforming the coordinates from (R, ϕ', Z) to (r, θ, ϕ). The transformation is given by

$$\begin{aligned} R &= R_0 + r\cos\theta \\ Z &= r\sin\theta \\ \phi' &= -\phi \end{aligned} \tag{6.43}$$

In the (r, θ, ϕ) system the basic vector operations can be written as

$$\begin{aligned} \nabla\psi &= \nabla_c\psi \\ \nabla\cdot\mathbf{B} &= \nabla_c\cdot\mathbf{B} + \frac{\mathbf{e}_R\cdot\mathbf{B}}{R} \\ \nabla\times\mathbf{B} &= \nabla_c\times\mathbf{B} - \frac{B_\phi}{R}\mathbf{e}_Z \\ \mathbf{e}_R &= \cos\theta\,\mathbf{e}_r - \sin\theta\,\mathbf{e}_\theta \\ \mathbf{e}_Z &= \sin\theta\,\mathbf{e}_r + \cos\theta\,\mathbf{e}_\theta \\ \mathbf{e}_{\phi'} &= -\mathbf{e}_\phi \end{aligned} \tag{6.44}$$

Here, ∇_c is the gradient operator in a linear (r, θ, z) right-handed cylindrical coordinate system with $\partial/\partial z$ replaced by $(1/R)\partial/\partial\phi$. These relations imply that the Δ^* operator appearing in the Grad–Shafranov equation can be expressed as

$$\Delta^*\psi = \frac{1}{r}\frac{\partial}{\partial r}\left(r\frac{\partial\psi}{\partial r}\right) + \frac{1}{r^2}\frac{\partial^2\psi}{\partial\theta^2} - \frac{1}{R}\left(\cos\theta\frac{\partial\psi}{\partial r} - \frac{\sin\theta}{r}\frac{\partial\psi}{\partial\theta}\right) \tag{6.45}$$

Zeroth-Order Solution (Radial Pressure Balance). Substituting the RFP expansion into Eq. (6.45) leads to the following zeroth-order contribution:

$$\frac{1}{r}\frac{d}{dr}\left(r\frac{d\psi_0}{dr}\right) = -\mu_0 R_0^2\frac{dp}{d\psi_0} - F\frac{dF}{d\psi_0} \tag{6.46}$$

If one notes that the zeroth-order fields are given by

$$\begin{aligned} B_{r0}(r) &= 0 \\ B_{\theta 0}(r) &= \frac{1}{R_0}\frac{d\psi_0}{dr} \\ B_{\phi 0}(r) &= \frac{F(\psi_0)}{R_0} \\ p_0(r) &= p(\psi_0) \end{aligned} \tag{6.47}$$

then Eq. (6.46) reduces to

$$\frac{d}{dr}\left(p_0 + \frac{B_{\phi 0}^2}{2\mu_0}\right) + \frac{B_{\theta 0}}{\mu_0 r}\frac{d}{dr}(rB_{\theta 0}) \tag{6.48}$$

As expected, this is just the radial pressure balance relation for the straight general screw pinch [Eq. (5.23)]. In place of $p(\psi)$, $F(\psi)$ two of the four quantities $p_0(r)$, $B_{\phi 0}(r)$, $B_{\theta 0}(r)$, $\psi_0(r)$ can be specified as arbitrary free functions. For convenience, assume that $B_{\theta 0}$ and $B_{\phi 0}$ are free. Then p_0' is determined from Eq. (6.48) and ψ_0 from $\psi_0 = R_0 \int B_{\theta 0}\, dr$.

The typical radial profiles for an RFP are illustrated in Fig. 6.5. These somewhat unusual profiles were first observed in the Zeta experiment (Butt *et al.*, 1958; Bodin and Newton, 1980) and later shown theoretically (Robinson, 1971) to have favorable ideal MHD stability properties. RFP stability is discussed in Chapter 9. As far as equilibrium is concerned, observe that the profiles have comparable $B_{\theta 0}$ and $B_{\phi 0}$ and that the pressure is either hollow or very flat rather than peaked. A crucial feature is the reversal of $B_{\phi 0}$ toward the outside of the plasma; hence, the name reversed field pinch. In an approximate sense radial pressure balance can be described as follows: the plasma is compressed between the $B_{\theta 0}$ tension on the outside and the $B_{\phi 0}$ pressure on the inside.

First-Order Solution (Toroidal Effects). Consider now the first-order toroidal corrections to the Grad–Shafranov equation. Substitution

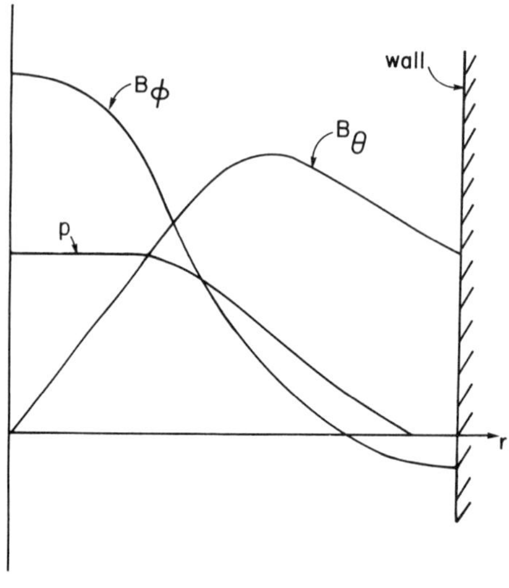

Figure 6.5. Schematic diagram of the B_ϕ, B_θ, and p profiles providing radial pressure balance in an RFP.

of the RFP expansion yields

$$\nabla^2\psi_1 = -\psi_1 \frac{d^2}{d\psi_0^2}\left(\mu_0 R_0^2 p + \frac{F^2}{2}\right) - \left(2\mu_0 R_0 r \frac{dp}{d\psi_0} - \frac{1}{R_0}\frac{d\psi_0}{dr}\right)\cos\theta \tag{6.49}$$

where ∇^2 is the cylindrical Laplacian (the subscript c on ∇_c has been dropped for convenience) and p and F are functions of ψ_0. Equation (6.49) can be solved by observing that the inhomogeneous terms all have a θ dependence proportional to $\cos\theta$. This, coupled with the assumption that the boundary is circular, implies that $\psi_1(r, \theta)$ can be written as

$$\psi_1(r, \theta) = \psi_1(r)\cos\theta \tag{6.50}$$

After some algebra, where use is made of the zeroth-order radial pressure balance relation, one obtains the following simplified equation for $\psi_1(r)$

$$\left[rB_\theta^2\left(\frac{\psi_1}{B_\theta}\right)'\right]' = rB_\theta^2 - 2\mu_0 r^2 p' \tag{6.51}$$

where a prime denotes differentiation with respect to r and the subscript 0 has been dropped from the zeroth-order quantities. The solution to Eq. (6.51) satisfying the boundary conditions is given by

$$\psi_1(r) = B_\theta(r)\int_r^b \frac{dx}{xB_\theta^2(x)}\int_0^x \left[2\mu_0 y^2 \frac{dp(y)}{dy} - yB_\theta^2(y)\right] dy \tag{6.52}$$

The contribution of ψ_1 to the total flux represents the toroidal correction to reversed field pinch equilibria.

Plasma Parameters and Figures of Merit. Using the zeroth- and first-order solutions for ψ one can investigate the basic plasma parameters and figures of merit for the RFP configuration. In fact, since B_θ, B_ϕ, and p are each of zeroth order, the values of β_t, β_p, q_*, q, s, and W are, to within corrections of order ϵ, identical to those in a straight system.

The quantities β_t and β_p are obtained from Eq. (5.25). The results are

$$\begin{aligned} \beta_t &= \frac{4\mu_0}{b^2 B_\phi^2(b)}\int_0^b pr\,dr \sim 1 \\ \beta_p &= \frac{16\pi^2}{\mu_0 I_0^2}\int_0^b pr\,dr \sim 1 \end{aligned} \tag{6.53}$$

where $\mu_0 I_0 = 2\pi b B_\theta(b)$.

Although both β_t and β_p are formally of order unity, the typical RFP profile (see Fig. 6.5) has $|B_\phi(b)| \ll B_\theta(b)$; that is, $\beta_t \gg \beta_p$, indicating that the toroidal field does not provide much external magnetic pressure for radial pressure balance. The plasma is almost entirely confined by the tension in the poloidal magnetic field. In this sense, RFP radial pressure balance is similar to that in a Z pinch.

The kink safety factor, obtained from Eq. (5.28), is given by

$$q_* = \frac{2\pi b^2 B_\phi(b)}{\mu_0 R_0 I_0} \sim \epsilon \tag{6.54}$$

This low value of q_* is an indication that RFPs are potentially very unstable to current-driven kink instabilities. In Chapter 9 it is shown that it is only the presence of the perfectly conducting wall that suppresses these modes.

The MHD safety factor, shear, and magnetic well are determined from Eqs. (5.32), (5.33), and (5.34):

$$\begin{aligned} q(r) &= \frac{rB_\phi}{R_0 B_\theta} \sim \epsilon \\ s(r) &= \frac{r}{q}\frac{dq}{dr} \sim 1 \\ W(r) &= -\frac{B_\theta^2}{B^2} \sim 1 \end{aligned} \tag{6.55}$$

and are illustrated in Fig. 6.6 for typical RFP profiles. Note that $W(r)$ is large and negative, indicating strong unfavorable curvature. This, however, is counteracted by high shear associated with the B_ϕ reversal. The balance between shear and unfavorable curvature is discussed in Chapter 9.

The Shafranov Shift. Consider next the effects of toroidicity. Since $q_* \sim 1/I_0$ is small, the toroidal current is large. This has the consequence of requiring only a small outward shift of the plasma to compress sufficient flux for toroidal equilibrium, even with $\beta_p \sim 1$. To demonstrate the point, one assumes a realistic plasma model in which all the plasma currents vanish outside a particular flux surface. This is the plasma boundary. From the expansion for the flux function, it follows that the boundary is a slightly shifted circle, denoted by

$$\begin{aligned} r_a &= a + \Delta\cos\theta \\ \Delta/a &\sim \epsilon \end{aligned} \tag{6.56}$$

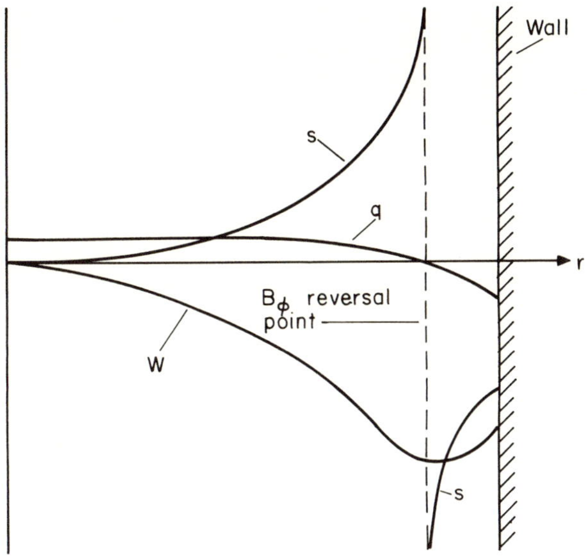

Figure 6.6. Typical profiles of $q(r)$, $s(r)$, and $W(r)$ for an RFP.

See Fig. 6.7. The shift Δ is determined from the condition $\psi[r_a(\theta), \theta] =$ const, or

$$\Delta = -\psi_1(a)/\psi_0'(a) \tag{6.57}$$

The quantity $\psi_1(a)$ is obtained from Eq. (6.52):

$$\psi_1(a) = B_\theta(a)\int_a^b \frac{dx}{xB_\theta^2(x)} K(x) \tag{6.58}$$

$$K(x) = \int_0^x \left[2\mu_0 y^2 \frac{dp}{dy} - yB_\theta^2(y)\right] dy \tag{6.59}$$

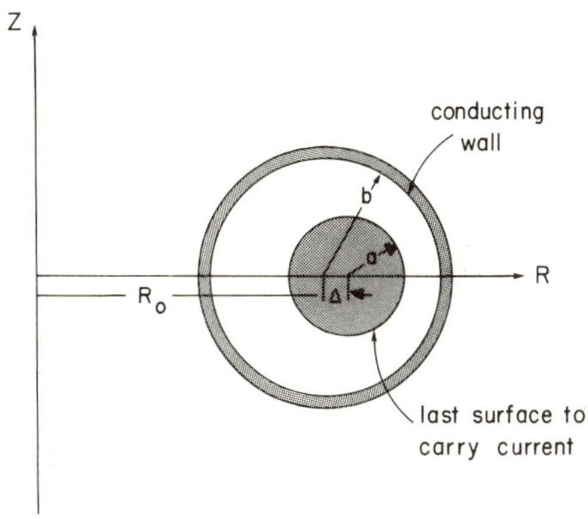

Figure 6.7. Geometry illustrating the shift Δ of the last flux surface to carry current, measured with respect to the perfectly conducting wall.

This expression is simplified as follows. First, the limit of integration x in Eq. (6.59) always lies in the range $a \le x \le b$. Since $p = 0$ for $x > a$, the pressure term in Eq.(6.59) can be integrated by parts, yielding

$$2\mu_0 \int_0^a y^2 \frac{dp}{dy}\, dy = -a^2 B_\theta^2(a)\beta_p \tag{6.60}$$

The second term in $K(x)$ is evaluated by breaking the interval $0 < y < x$ into two subintervals, $0 < y < a$ and $a < y < x$. Since $J_\phi = 0$ for $r > a$, then $B_\theta(r) = B_\theta(a)(a/r)$ in this region. Thus, the second term is given by

$$\int_0^x y B_\theta^2(y)\, dy = a^2 B_\theta^2(a)\left[\frac{l_i}{2} + \ln\left(\frac{x}{a}\right)\right] \tag{6.61}$$

where l_i is the normalized internal inductance per unit length associated with the toroidal current flowing in the plasma:

$$l_i \equiv \left(\frac{L_i}{2\pi R_0}\right)\Big/\left(\frac{\mu_0}{4\pi}\right) = \frac{2}{a^2 B_\theta^2(a)} \int_0^a B_\theta^2 r\, dr \tag{6.62}$$

Here L_i is the internal plasma inductance defined by $(1/2)L_i I_0^2 = \int (B_\theta^2/2\mu_0)\, d\mathbf{r}$. Equations (6.60) and (6.61) are substituted in Eq.(6.58) and the remaining integral evaluated. This ultimately leads to the following expression for Δ, first obtained by Shafranov (1966):

$$\frac{\Delta}{b} = \frac{b}{2R_0}\left[\left(\beta_p + \frac{l_i - 1}{2}\right)\left(1 - \frac{a^2}{b^2}\right) + \ln\frac{b}{a}\right] \tag{6.63}$$

Note that Δ is positive, indicating an outward shift. Since $b/R_0 \sim \epsilon \ll 1$ in the asymptotic expansion, Eq. (6.63) verifies that the equilibrium shift is indeed small.

The response to the outward shift can be seen explicitly by calculating the poloidal magnetic field on the plasma surface:

$$\begin{aligned} B_p[r_a(\theta), \theta] &\approx \frac{\mu_0 I_0}{2\pi a}\left[1 + \left(\frac{\psi_1'(a)}{R_0 B_a} - \frac{\Delta}{a} - \frac{a}{R_0}\right)\cos\theta\right] \\ &\approx \frac{\mu_0 I_0}{2\pi a}\left[1 + \frac{a}{R_0}\left(\beta_p + \frac{l_i}{2} - 1\right)\cos\theta\right] \end{aligned} \tag{6.64}$$

where $B_a = \mu_0 I_0/2\pi a$. The first term, independent of θ, represents the straight cylindrical contribution due to the toroidal current. Of the $\cos\theta$ terms, the (-1) contribution is a result of the $1/R$ dependence of the toroidal geometry which causes B_p to decrease slightly on the outside $(\theta = 0)$ and increase on the inside $(\theta = \pi)$. The $l_i/2$ term represents the compression of the outer poloidal flux in response to the outward hoop force. Hence, it causes B_p to increase slightly on the outside and decrease on the inside. The β_p term contains the combined response of the "tire tube" force and the $1/R$ toroidal field force. If the toroidal field were zero, then $\beta_p = 1$ [Eq. (5.26)]. This term then represents a compression of the outer poloidal flux caused by the outward force of the plasma pressure. If the toroidal field has a typical RFP profile, then β_p is somewhat reduced, thereby decreasing the compression of the flux. Stated another way, the B_ϕ field in a reversed field pinch is paramagnetic (i.e., large when the pressure is high and small when the pressure is low). Thus, in contrast to the θ pinch, which is diamagnetic, the $1/R$ force due to the toroidal field is actually inward. It partially compensates the pressure force, resulting in a reduced compression of the outer poloidal flux.

Experimental Discussion/Numerical Solutions. "Exact" numerical solutions of the Grad–Shafranov equation indicate that the asymptotic expansion described above represents an excellent approximation of reversed field pinch equilibria. A typical example of a numerical equilibrium is shown in Fig. 6.8. Here the functions $p(\psi)$ and $F(\psi)$ were chosen to simulate closely experimental profiles for the Los Alamos ZT-40 RFP. Note the small outward shift of the plasma pressure and the compression of the poloidal flux on the outside of the torus.

In recent years the experimental performance of RFP devices has improved dramatically. Higher densities $n \gtrsim 10^{20}\,\mathrm{m}^{-3}$, temperatures $T \gtrsim 0.5\,\mathrm{keV}$, and confinement times $\tau_E \gtrsim 0.1$ ms have been achieved, although performance is not yet at the same level of similar-sized tokamaks.

The typical operation of an RFP experiment is illustrated in Fig. 6.9. A relatively low toroidal bias field, $B_\phi \sim 0.5\,\mathrm{T}$, is established initially throughout the discharge tube. Gas is then injected into the chamber. The experiment starts when the current is switched on in the primary circuit. The plasma represents the secondary of the transformer. As the primary current rises, the secondary voltage ionizes the gas and a secondary plasma current begins to flow. The plasma current continues to increase until it reaches a level of about 500 kA.

During the rise of current, the increasing poloidal field compresses the plasma. Since the plasma is a good conductor, the toroidal flux is trapped and compressed with the plasma. In this manner $B_\phi(0)$ increases and a typical $B_\phi(r)$ RFP profile is created. Strictly speaking, if the plasma behaved in a classical manner, then $B_\phi(b)$ could never become negative, no matter how strong the compression. In fact, experimental observations indicate that in certain parameter ranges the B_ϕ field spontaneously

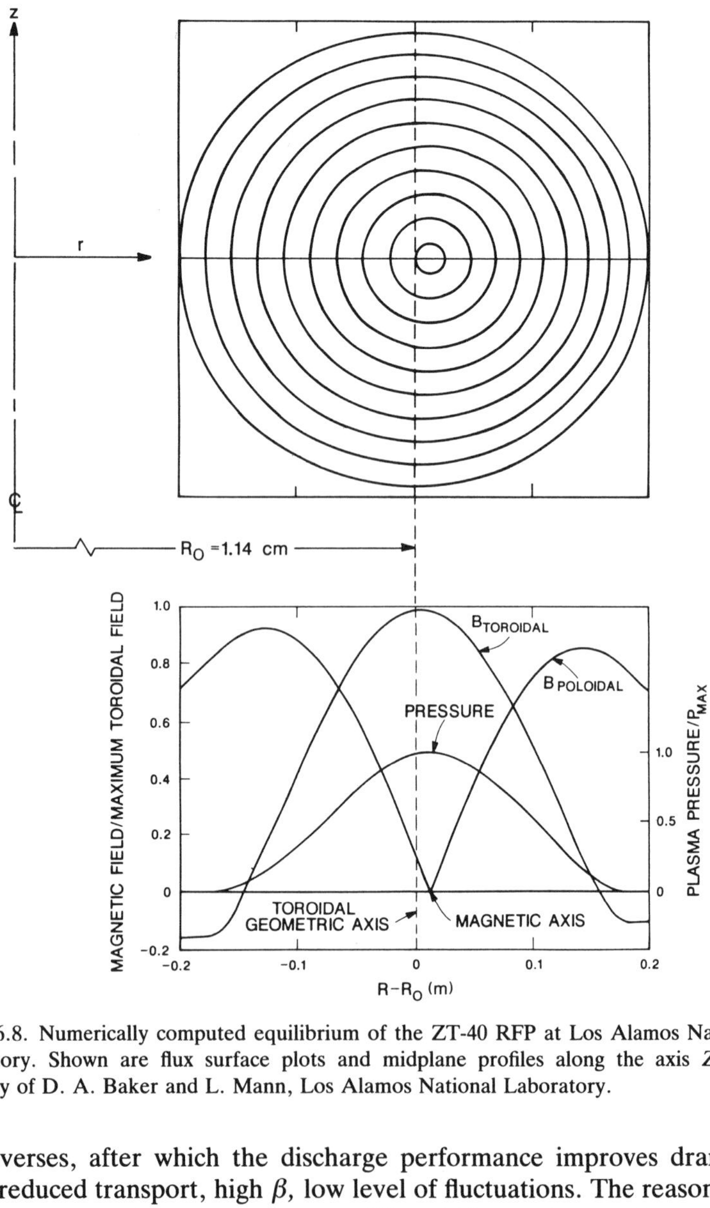

Figure 6.8. Numerically computed equilibrium of the ZT-40 RFP at Los Alamos National Laboratory. Shown are flux surface plots and midplane profiles along the axis $Z = 0$. Courtesy of D. A. Baker and L. Mann, Los Alamos National Laboratory.

self-reverses, after which the discharge performance improves dramatically; reduced transport, high β, low level of fluctuations. The reasons for this unexpected behavior involve RFP stability and are discussed in Chapter 9.

Summary. In summary, the reversed field pinch is a high-β, low-q

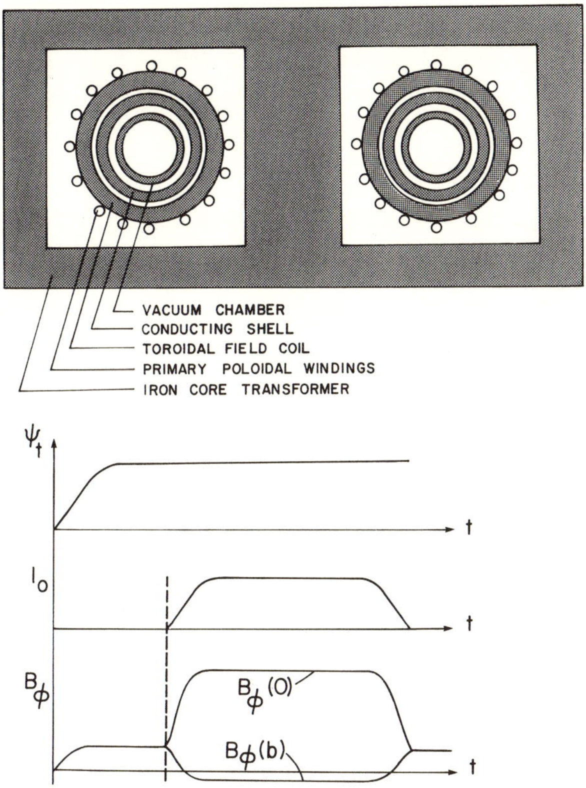

Figure 6.9. Schematic diagram of an RFP experiment. The curves illustrate a typical time history of toroidal flux ψ_t, toroidal current I_0, and toroidal field B_ϕ.

configuration in which the plasma is contained radially between the toroidal magnetic pressure on the inside and the poloidal magnetic pressure on the outside. A large toroidal current provides toroidal force balance by compressing the poloidal flux between the conducting wall and the slightly outward-shifted plasma.

6.4. THE OHMICALLY HEATED TOKAMAK

For about a decade starting in the early 1970s, the ohmically heated tokamak has been the leader of the international magnetic fusion program in terms of overall performance. Temperatures on the order of several keV are routinely achieved, as well as long confinement times at reasonably high densities. In fact, a major milestone in the magnetic fusion program was accomplished in an ohmically heated tokamak. Specifically, the first experiment to exceed the requirements for driven breakeven (i.e., thermonuclear neutron power $\geq$ applied input power) $n\tau_E \geq 6 \times 10^{19}\,\mathrm{m}^{-3}\,\mathrm{s}$, is Alcator C, a high-field, ohmically heated tokamak at MIT.

Although the basic field configuration is quite different from the reversed field pinch, the analysis is nevertheless quite similar. The underlying reason is that the pressure is sufficiently small so that only a small outward shift is required for equilibrium, thus implying that the inverse aspect ratio expansion should be very accurate.

In applying the expansion, it is initially assumed for simplicity that the plasma is surrounded by a perfectly conducting shell. This assumption is then relaxed and replaced by the condition that toroidal force balance is achieved by an external vertical field. Since the typical experimental time scales of tokamak operation are much greater than the skin penetration time of the surrounding walls, the vertical field boundary condition is a much more realistic one.

Ohmically Heated Tokamak Expansion. The asymptotic ordering in ϵ that defines the ohmically heated tokamak is given by

$$\begin{aligned}
\frac{B_p}{B_\phi} &\sim \epsilon \\
\beta_t &\sim \frac{2\mu_0 p}{B_\phi^2} \sim \epsilon^2 \\
\beta_p &\sim \frac{2\mu_0 p}{B_p^2} \sim 1 \\
q &\sim \frac{rB_\phi}{RB_p} \sim 1
\end{aligned} \tag{6.65}$$

The configuration is dominated by a large vacuum toroidal field. As its name implies, the plasma is heated and supported in radial pressure balance by the ohmic current, a confinement situation analogous to that in a pure Z pinch. In practice, one wants the ohmic current as large as possible for ease of toroidal equilibrium, for maximizing β, and for maximizing the ohmic heating. However, the current is limited by the stability constraint that the safety factor be of order unity. The combination of Z-pinch radial pressure balance and finite safety factor gives rise to a low plasma pressure, $\beta_t \sim \epsilon^2$. While this is not a desirable scaling for fusion applications, it is very convenient for theoretical analysis in that the implied toroidal shift is small, thus validating the use of the asymptotic expansion. In actual experimental situations the toroidal field can also enter the radial pressure balance relation either diamagnetically or paramagnetically. However, in an ohmically heated tokamak the deviation from the vacuum field is very small, of order ϵ^2, leaving the Z pinch nature of radial pressure balance unchanged. From the point of view of MHD theory, the technologically expensive large toroidal field plays little or no role in either radial pressure balance or toroidal force balance. Its primary function is to provide stability at high ohmic currents by generating safety factors on the order of unity.

Under the ohmically heated tokamak expansion the Grad–Shafranov equation is solved by formally expanding as follows:

$$\begin{aligned} \psi(r, \theta) &= \psi_0(r) + \psi_1(r)\cos\theta \ldots \\ p(\psi) &= p_2(\psi_0) + \frac{dp_2(\psi_0)}{d\psi_0}\psi_1 \cos\theta \cdots \\ F(\psi) = RB_\phi &= R_0[B_0 + B_{\phi 2}(\psi_0) \ldots] \end{aligned} \tag{6.66}$$

where $B_0 = \text{const}$ is the vacuum toroidal field at $R = R_0$, $B_{\phi 2}(\psi)$ is a new free function replacing $F(\psi)$, and

$$\begin{aligned} \psi_0/rRB_0 &\sim \epsilon \\ \psi_1/\psi_0 &\sim \epsilon \\ 2\mu_0 p_2/B_0^2 &\sim \epsilon^2 \\ B_{\phi 2}/B_0 &\sim \epsilon^2 \end{aligned} \tag{6.67}$$

Zeroth-Order Solution (Radial Pressure Balance). In analogy with the RFP, the leading-order contribution to the Grad–Shafranov equation gives

$$\frac{d}{dr}\left(p_2 + \frac{B_0 B_{\phi 2}}{\mu_0}\right) + \frac{B_{\theta 1}}{\mu_0 r}\frac{d}{dr}(rB_{\theta 1}) \tag{6.68}$$

where

$$\begin{aligned} B_{\theta 1}(r) &= \frac{1}{R_0}\frac{d\psi_0}{dr} \\ B_{\phi 2}(r) &= B_{\phi 2}[\psi_0(r)] \end{aligned} \tag{6.69}$$

As expected, Eq. (6.68) represents the expanded form of the one-dimensional radial pressure balance relation for a general screw-pinch. For convenience one can think of $B_{\theta 1}$ and $B_{\phi 2}$ as the two free functions.

A typical set of radial profiles for an ohmically heated tokamak is

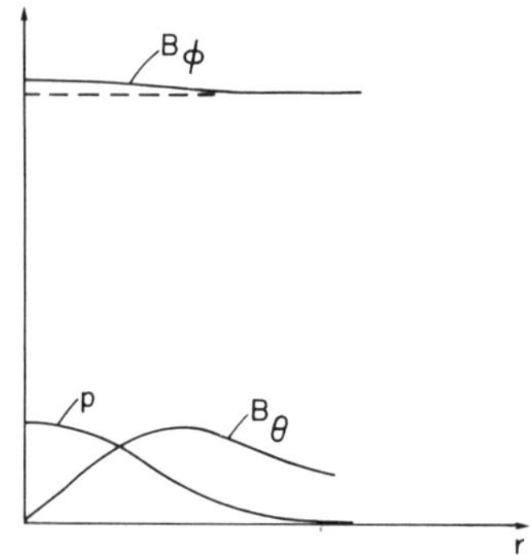

Figure 6.10. Typical radial profiles for an ohmically heated tokamak.

illustrated in Fig. 6.10. Observe the large, slightly paramagnetic toroidal field. Both the toroidal current density and plasma pressure are peaked on axis and decrease to zero at the edge of the plasma. Since B_ϕ is slightly paramagnetic, the entire plasma pressure is held in radial pressure balance by the poloidal magnetic field.

First-Order Solution (Toroidal Force Balance). Consider now the toroidal corrections to $\psi(r, \theta)$. The equation that determines $\psi_1(r)$ is obtained from the first-order contribution to the Grad–Shafranov equation. Not surprisingly, this equation is identical to Eq. (6.51). If the plasma were surrounded by a perfectly conducting shell located at $r = b$, then the solution would be identical to Eq. (6.52), repeated here for convenience:

$$\psi_1(r) = B_{\theta 1}(r)\int_r^b \frac{dx}{xB_{\theta 1}^2(x)}\int_0^x \left[2\mu_0 y^2 \frac{dp_2(y)}{dy} - yB_{\theta 1}^2(y)\right] dy \quad (6.70)$$

To simulate more realistic tokamak boundary conditions, assume that a set of vertical field coils has been placed outside the conducting shell. In vacuum these coils produce an additional magnetic field, $B_v \mathbf{e}_Z$, with $B_v / B_0 \sim \epsilon^2$.

In actual tokamak experiments, it is the interaction of this field with the toroidal current that produces the compensating force required for toroidal force balance; that is, the typical penetration time of a copper shell or vacuum chamber is sufficiently short compared to experimental times that it does not behave like a perfect conductor. One must thus use a vertical field for toroidal equilibrium.

Assume that the vertical field coils have been energized over a long period of time so that their fields have diffused through the conducting boundary. Keep in mind that once this has occurred, the wall still acts as a perfect conductor on the faster MHD time scale. In this situation, the boundary condition on the flux function is modified so that $\psi(b, \theta) = \text{const} + \psi_v(b, \theta)$ where $\psi_v(r, \theta) = R_0 B_v r \cos\theta$ is the flux function due to the vertical field. [See Eq. (3.14) for a discussion of the boundary conditions.]

The full toroidal correction to ψ is obtained by adding an appropriate homogeneous solution to ψ_1 [Eq. (6.70)]. Writing $\psi_1(\text{total}) = \psi_{1T}(r)\cos\theta$ one finds

$$\psi_{1T}(r) = \psi_1(r) + \left[\frac{bR_0B_v}{B_{\theta 1}(b)}\right]B_{\theta 1}(r) \tag{6.71}$$

The main effect of adding the vertical field is to change the toroidal shift of the plasma, which is now given by

$$\begin{aligned}\Delta &= -\psi_{1T}(a)/\psi_0'(a) \\ &= -\psi_1(a)/\psi_0'(a) - \Delta_v\end{aligned} \tag{6.72}$$

where

$$\Delta_v = bB_v/B_{\theta 1}(b) \tag{6.73}$$

Δ can be written as

$$\frac{\Delta}{b} = \frac{b}{2R_0}\left[\left(\beta_p + \frac{l_i - 1}{2}\right)\left(1 - \frac{a^2}{b^2}\right) + \ln\frac{b}{a}\right] - \frac{B_v}{B_{\theta 1}(b)} \tag{6.74}$$

where $B_{\theta 1}(b) = \mu_0 I_0/2\pi b$, l_i is given by Eq. (6.62), and typically $l_i \gtrsim 1/2$.

Basic Plasma Parameters and Figures of Merit. This completes the formal analysis of the ohmically heated tokamak. Using the relations just derived one can investigate the basic plasma parameters and figures of merit. As with the RFP, many of the quantities of interest are accurately given by the leading-order fields alone, corresponding to the straight system. This is a consequence of the small toroidal shift. Included among these quantities are β_t, β_p, q_*, q, and s but, as will be shown, not W.

The quantities β_t and β_p can be written as

$$\beta_t = \frac{4\mu_0}{a^2B_0^2}\int_0^a pr\,dr \sim \epsilon^2 \tag{6.75}$$

$$\beta_p = \frac{16\pi^2}{\mu_0 I_0^2}\int_0^a pr\,dr \sim 1 \tag{6.76}$$

Since the plasma pressure is supported by the poloidal field, which is small compared to the toroidal field for purposes of stability, the value of β_t is of order ϵ^2. Whatever other virtues the ohmically heated tokamak may have, it is not very efficient in its confinement of plasmas.

Examination of the general, one-dimensional, equilibrium integral relation [Eq. (5.24)] indicates that when $B_{\phi 2} = 0$ (i.e., pure vacuum toroidal field) then $\beta_p = 1$. If $B_{\phi 2} < 0$, the plasma is diamagnetic and $\beta_p > 1$, since part of the pressure is now confined by the toroidal field. Similarly, for a paramagnetic configuration, $B_{\phi 2} > 0$ and $\beta_p < 1$.

The kink safety factor is given by

$$q_* = \frac{2\pi a^2 B_0}{\mu_0 I_0 R_0} \sim 1 \tag{6.77}$$

Circular tokamaks normally operate with sufficiently high values of $q_* \sim 3$ that current-driven kink modes are suppressed without the need of a conducting wall as required by the RFP.

Further useful equilibrium information can be obtained from the general relationship between β_t, β_p, and q_*, implied by Eqs. (4.31)–(4.34):

$$\beta_t = \frac{\epsilon^2 \beta_p}{q_*^2} \tag{6.78}$$

for the ohmically heated tokamak. Here, $\epsilon \equiv a/R_0$. Since $\beta_p \approx 1$, Eq. (6.78) implies that the total pressure that can be contained is proportional to I_0^2.

The MHD safety factor and shear are given by their straight screw-pinch values

$$\begin{aligned} q(r) &= \frac{rB_0}{R_0 B_{\theta 1}} \sim 1 \\ s(r) &= \frac{r}{q}\frac{dq}{dr} \sim 1 \end{aligned} \tag{6.79}$$

Note that the safety factor is of order unity and is an increasing function of radius, a property common to all tokamaks. Also, $q(a) = q_*$. The shear is zero near the origin and increases monotonically with the radius. At larger radii, where the fields approach vacuum fields, then $s(r) \to 2$.

The Magnetic Well. The magnetic well in a tokamak is different from the straight screw pinch value. This is a consequence of the small poloidal field and the curvature of the toroidal field. Recall that the normal curvature $\kappa_n \equiv \mathbf{n} \cdot \boldsymbol{\kappa}$ with $\boldsymbol{\kappa} = \mathbf{b} \cdot \nabla \mathbf{b}$. Upon substituting the tokamak expansion, one obtains, after a straightforward calculation, an expression for κ_n correct to order ϵ^2

$$\kappa_n(r, \theta) = -\frac{B_{\theta 1}^2}{rB_0^2} - \left(1 - \frac{r\cos\theta}{R_0}\right)\frac{\cos\theta}{R_0} + \frac{\Delta \sin^2\theta}{rR_0} \tag{6.80}$$

where $\Delta(r) = -\psi_1(r)/\psi_0'(r)$ is the shift of the flux surfaces. The first term represents the unfavorable curvature associated with the poloidal field. The second term represents the curvature of the toroidal field, which is favorable on the inside ($\theta = \pi$) and unfavorable on the outside ($\theta = 0$). The toroidal curvature averages to zero in leading order but gives rise to a small net stabilizing contribution in first order. This follows because $|\kappa_n|_{\text{tor}} \propto 1/R$; the magnitude of κ_n is slightly larger for small R where the curvature is favorable than for large R where it is unfavorable. The last term represents the modification of the curvature due to the shift in the flux surfaces. Since the surfaces are shifted outward, $\psi_1 < 0$, implying that this term is also stabilizing.

Note that each of the terms in Eq. (6.80) is present for the RFP as well as the ohmic tokamak. However, in an RFP $B_\theta \sim B_\phi$, implying that the toroidal corrections are negligible. In a tokamak $B_\theta \sim \epsilon B_\phi$ and all the terms are competitive.

The magnetic well is evaluated by substituting the toroidal tokamak equilibrium into Eq. (6.29). After a slightly tedious calculation (see Problem 6.9) one finds

$$W(r) = -\frac{B_{\theta 1}^2}{B_0^2} - \frac{r}{2R_0}\left[\frac{(r^2\Delta)'}{r}\right]' + \frac{r^2}{2R_0^2} \tag{6.81}$$

The first term is the straight cylindrical screw pinch contribution which is destabilizing. The second and third terms represent the effect of toroidicity and are always stabilizing. A sketch of $W(r)$ is illustrated in Fig. 6.11. For sufficiently large q, $W(r)$ can be made positive (i.e., stabilizing) over the entire plasma. This result is also suggested by examining $W(r)$ in the vicinity of $r = 0$ and $r = a$:

$$\begin{aligned} W(r) &\approx -\frac{r^2}{R_0^2 q_0^2}\left[1 - q_0^2\left(\frac{5}{4} - \frac{3}{2}\frac{p_2''}{B_{\theta 1}'^2}\right)\right]_{r=0}, & r \ll a \\ W(r) &\approx -\frac{a^2}{R_0^2 q_a^2}\left[1 - q_a^2\left(1 + \beta_p + \frac{l_i}{2}\right)\right], & r = a \end{aligned} \tag{6.82}$$

Thus, in an ohmically heated tokamak toroidicity plays an important role in stabilizing MHD modes driven by the pressure gradient.

The Shafranov Shift. The final topic of interest concerns the Shafranov shift, given by Eq.(6.74). This equation expresses the relationship between the shift, the vertical field, and the plasma parameters. It is of particular interest to determine the vertical field required to make the plasma shift zero. This is easily accomplished by setting $\Delta = 0$ in Eq.

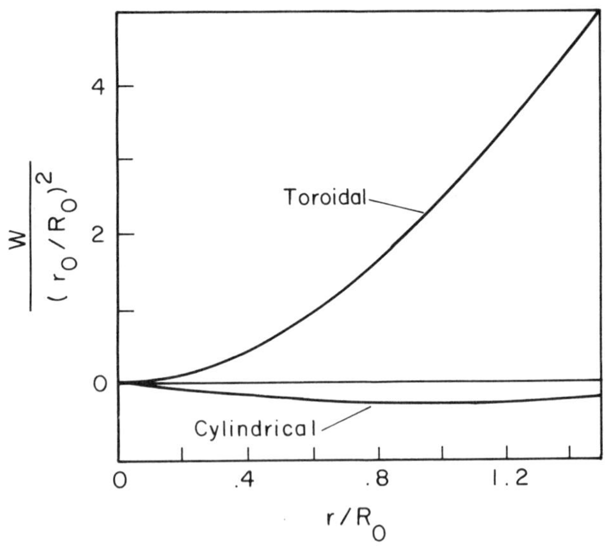

Figure 6.11. Plot of the magnetic well versus radius for an ohmically heated tokamak. The profiles correspond to $B_\phi = B_0 =$ const and B_θ and p satisfying radial pressure balance with amplitudes adjusted so that $q_0 = 1$. Also shown is $W/(r_0/R_0)^2 = -(r/qR_0)^2$, the magnetic well for the straight cylinder.

(6.74) and solving for B_v:

$$B_v = \frac{\mu_0 I_0}{4\pi R_0}\left[\left(\beta_p + \frac{l_i - 1}{2}\right)\left(1 - \frac{a^2}{b^2}\right) - \ln\frac{b}{a}\right] \tag{6.83}$$

Even more important is the question of determining the vertical field required to hold the plasma in equilibrium with its center at $R = R_0$ when no conducting shell is present. To obtain this result it is necessary to set $\Delta = 0$ and take the limit $b/a \gg 1$. This limit is smooth for the first group of terms in Eq. (6.83). The logarithmic term is more complicated since as b gets larger it eventually reaches its geometric limit $b = R_0$. In fact, a plausible estimate of the required vertical field is obtained by setting $\Delta = 0$ and $b = R_0$ in the logarithmic term [i.e., $\ln(b/a) \to \ln(R_0/a)$]. Note, however, that setting $b = R_0$ violates the large aspect ratio expansion, which requires $b \ll R_0$.

The exact value of B_v requires a surprisingly lengthy calculation (Shafranov, 1966). However, a more accurate value than the heuristic one just given is obtained by observing that the $\ln(b/a)$ term represents the force due to the change in magnetic energy contained between the plasma and wall as the plasma is shifted outward. It is the analog of the l_i

term, except that it is applied to the external flux. This force can be written as

$$F_R = -\mathbf{e}_R \cdot \nabla U \tag{6.84}$$

where the potential energy U is given by

$$U = \int \frac{B_p^2}{2\mu_0}\, d\mathbf{r} \tag{6.85}$$

and the integration is over the volume outside the plasma. Equation (6.84) can be rewritten as follows:

$$F_R = -\frac{1}{2}\frac{\partial}{\partial R_0}(L_e I_0^2) \tag{6.86}$$

where L_e is the external inductance associated with the poloidal flux ψ_e between the plasma and the wall. Since $\psi_e = L_e I_0$ remains constant as the plasma shifts outward,

$$F_R = \frac{I_0^2}{2}\frac{\partial L_e}{\partial R_0} \tag{6.87}$$

If the wall is reasonably near the plasma, then

$$L_e = \mu_0 R_0 \ln\frac{b}{a} \tag{6.88}$$

If the wall is removed from the plasma, then the value of L_e to be used should correspond to that of a circular current-carrying loop of wire with major radius R_0 and minor radius a. This value of L_e can be found in standard electromagnetic textbooks (Stratton, 1941). See Problem 4.5. For $a/R_0 \ll 1$

$$L_e = \mu_0 R_0\left(\ln\frac{8R_0}{a} - 2\right) \tag{6.89}$$

After carrying out the differentiation with respect to R_0 and equating the forces, it follows that the appropriate limit for $\ln(b/a)$ as $b \to R_0$ is given by

$$\ln\frac{b}{a} \to \ln\frac{8R_0}{a} - 1 = \ln\frac{R_0}{a} + 1.079 \tag{6.90}$$

Note that this differs from the heuristic assumption only by the factor of 1.079. Substituting Eq. (6.90) into Eq. (6.83) yields the value of B_v required for toroidal equilibrium with no conducting shell:

$$B_v = \frac{\mu_0 I_0}{4\pi R_0}\left(\beta_p + \frac{l_i - 3}{2} + \ln\frac{8R_0}{a}\right) \tag{6.91}$$

While Eq. (6.91) is simple in appearance, it is somewhat unintuitive in that it is not easily recognizable in terms of the forces holding the plasma in toroidal equilibrium. A more intuitive form is obtained by multiplying by $2\pi R_0 I_0$ and then performing some standard manipulations. See Problem 6.10. The result is

$$2\pi R_0 I_0 B_v = \frac{I_0^2}{2} \frac{\partial}{\partial R_0}(L_e + L_i) + 4\pi^2 \int_0^a \left(p - \frac{B_0 B_{\phi 2}}{\mu_0}\right) r\, dr \qquad (6.92)$$

The left-hand side represents the vertical field force. The first term on the right-hand side represents the hoop force and includes the total, internal plus external, poloidal flux. The pressure term represents the tire tube force, and the last term, the $1/R$ force due to the toroidal field.

Experimental Discussion/Numerical Results. The results described above have proven to be a very reliable guide for the operation of tokamak experiments. They have also been shown to be quite accurate when compared to exact numerical solutions of the Grad–Shafranov equation. In fact, because of the importance of tokamaks to the international fusion program, a number of sophisticated numerical codes have been developed to solve the Grad–Shafranov equation. One important feature incorporated in these codes is the ability to solve the more difficult but realistic problem in which the plasma is surrounded by a set of external coils rather than a perfect conductor. An example of such a numerical calculation is illustrated in Fig. 6.12 for the Alcator C experiment at MIT. Note the small shift of the flux surfaces and the

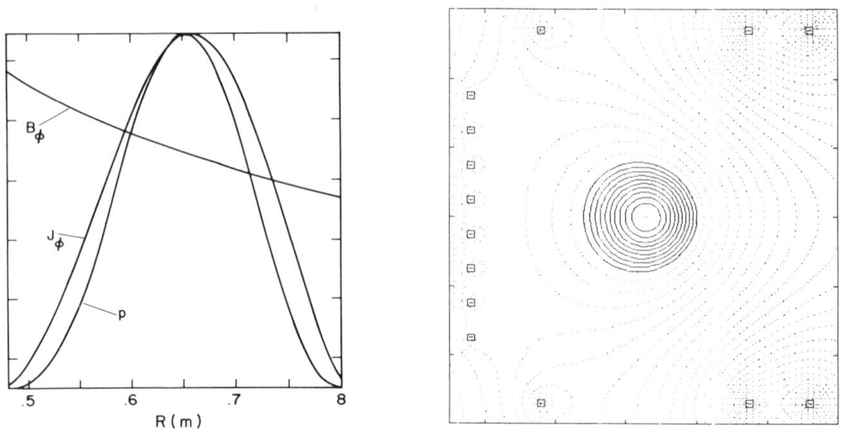

Figure 6.12. Numerically computed equilibrium of the ohmically heated Alcator C tokamak at MIT. Shown are flux surface plots and midplane profiles along the axis $Z = 0$. Courtesy J. Ramos, MIT.

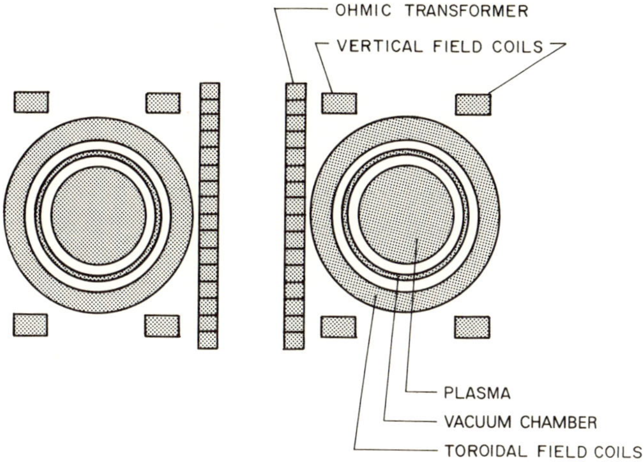

Figure 6.13. Schematic diagram of an ohmically heated tokamak.

compression of the poloidal flux on the outside of the torus, in agreement with the asymptotic theory.

There are many ohmically heated tokamak experiments currently in existence in the international magnetic fusion program. A schematic diagram of such an experiment is illustrated in Fig. 6.13. Tokamak operation proceeds as follows. A large toroidal field, typically $B_\phi \sim 3$–$10\,\mathrm{T}$, is applied and allowed to penetrate through the vacuum chamber. Neutral gas is then injected, followed by initiation of the current in the primary of the air core ohmic transformer. The plasma, which acts as the secondary, becomes ionized and begins to carry the induced transformer current. The current rises to its maximum value, after which it is clamped. Most of the important experimental measurements take place during the flattop portion of the current pulse. Because of the large toroidal field there is only a weak compression of the plasma (compared to an RFP) as the current rises. The final state thus exhibits a slightly paramagnetic B_ϕ, corresponding to $\beta_p \sim 0.2$–0.5.

The peak toroidal current in an ohmically heated tokamak varies from several hundred kA to several MA, depending on machine size. Densities usually lie in the range of $n \sim 2 \times 10^{19}$–$5 \times 10^{20}\,\mathrm{m}^{-3}$, with $n \sim 10^{20}\,\mathrm{m}^{-3}$ a typical operating value. For ohmic operation the electron and ion temperatures reach values of $T_i \approx T_e \sim 1$–$2\,\mathrm{keV}$. Most impressive are the energy confinement times, which as of 1986 range from tens of milliseconds in small devices to nearly a second in the largest machines. Tokamak experiments are continually resupplied with energy through ohmic heating and with particles through gas puffing at the plasma edge or pellet injection, thus allowing the evolution to a quasisteady state. In this way, the overall experimental pulse length can greatly exceed the energy confinement time. Finally, note that the typical value of β_t in an ohmic tokamak is very low: $\beta_t \lesssim 1\%$.

Summary. An ohmically heated tokamak is a device in which both radial pressure balance and toroidal force balance are achieved almost entirely by the poloidal field as in a Z pinch. The ratio of toroidal current to toroidal field is limited in amplitude by the requirement that the safety factor q be of order unity for stability. The sole purpose (from the ideal MHD point of view) of the large, technically and economically expensive, toroidal field is stability. Overall, stability is provided by a combination of high q_*, shear, and a favorable magnetic well. Since all the heating and confinement are provided by the relatively small poloidal field, the resulting pressure is small ($\beta_t \sim \epsilon^2$), requiring only a small shift or small vertical field for toroidal force balance.

6.5. THE HIGH-β TOKAMAK

The progress in the performance of ohmically heated tokamaks has been impressive in recent years, particularly in the achievement of large values of the $n\tau_E$ parameter. In fact, there are a number of features that have earned the ohmically heated tokamak its high ranking in the magnetic fusion program: good equilibrium, good stability, good confinement, and good heating. Even so, the low value of β_t remains as an important disadvantage from the point of view of fusion reactor requirements. It is thus worthwhile to investigate variations of the basic tokamak configuration which attempt to correct this problem. That is the goal of the high-β tokamak.

From a general point of view there are two modifications to the ohmically heated tokamak which serve to raise β_t. The first modification requires the application of external heating sources such as neutral beams or rf power. Even putting the question of β_t aside, such external heating will almost certainly be required since ohmic heating alone, while good, probably cannot raise the temperature in a tokamak sufficiently high to cause ignition. With external heating the plasma β_t can be raised substantially.

The second modification assumes the cross section of the plasma becomes noncircular. Optimizing over this new degree of freedom leads to a higher value of β_t for the same value of q_*. The circular high-β

tokamak is described in this section, followed in Section 6.6 by a discussion of the effects of noncircularity.

The basic idea of the high-β tokamak is as follows. Application of an independent source of external heating at fixed toroidal current and toroidal field causes an increase in the plasma pressure. The higher β_t that results is confined by newly generated poloidal diamagnetic currents induced in the plasma. When sufficient heating is applied, radial pressure balance is provided almost entirely by the toroidal field. This is the regime of the high-β tokamak. The plasma pressure is decoupled from the ohmic heating current (in contrast to the ohmically heated tokamak) and confinement is more closely related to that in a θ pinch rather than a Z pinch. The key feature of the high-β tokamak is that β_t is externally raised to the maximum value that can be supported in toroidal force balance by the toroidal current while satisfying the stability condition, $q_* \gtrsim 1$.

High-β Tokamak Expansion. Taking these requirements into account leads to the following inverse aspect ratio expansion for the high-β tokamak:

$$\begin{aligned} \frac{B_p}{B_\phi} &\sim \epsilon \\ \beta_t &\sim \frac{2\mu_0 p}{B_\phi^2} \sim \epsilon \\ \beta_p &\sim \frac{2\mu_0 p}{B_p^2} \sim \frac{1}{\epsilon} \\ q &\sim \frac{rB_\phi}{RB_p} \sim 1 \end{aligned} \tag{6.93}$$

Note that β_t is one order larger than in a conventional tokamak. Also, $\beta_p \sim 1/\epsilon$ indicates that the poloidal field has only a minor role in radial pressure balance. The confinement is provided by a small diamagnetic depression in the toroidal field. Even though the depression is small, of order ϵ, it still confines $1/\epsilon$ more plasma than the poloidal field. Thus, in a high-β tokamak the toroidal field provides both radial pressure balance and stability, while the poloidal field is required primarily for toroidal force balance, with a small contribution to the heating.

Because of the higher β's, the toroidal shift of the flux surfaces is of order unity and the flux surfaces themselves are no longer small deviations from circles. Consequently, although the inverse aspect ratio expansion does somewhat simplify the Grad–Shafranov equation, the leading-order equilibrium remains inherently two-dimensional; that is, radial pressure balance and toroidal force balance enter simultaneously.

As a result of this feature only the leading-order contribution to the Grad–Shafranov equation is required. The appropriate formal expansion for the high-β tokamak is given by

$$\begin{aligned} \psi(r, \theta) &= \psi_0(r, \theta) \ldots \\ p(\psi) &= p_1(\psi_0) \ldots \end{aligned} \tag{6.94}$$

where

$$\begin{aligned} \psi_0/rRB_\phi &\sim \epsilon \\ 2\mu_0 p_1/B_\phi^2 &\sim \epsilon \end{aligned} \tag{6.95}$$

The expansion for $F(\psi)$ is slightly subtle because the Grad–Shafranov equation basically determines the poloidal field, whereas the primary radial pressure balance is dominated by the toroidal field. The expansion for $F(\psi)$ must automatically take this into account. The procedure is to introduce a new function $B_{\phi2}(\psi)$ in place of $F(\psi)$ as follows:

$$F^2(\psi) = R_0^2 B_0^2\left[1 - \frac{2\mu_0 p_1(\psi)}{B_0^2} + \frac{2B_{\phi2}(\psi)}{B_0}\right] \tag{6.96}$$

where, as before, B_0 is the vacuum toroidal field at $R = R_0$. Equation (6.96) implies that the deviation from a θ-pinch pressure-balance relation is of order $B_{\phi2}/B_0 \sim \epsilon^2$, which is one order smaller than the pressure itself. Substituting into the Grad–Shafranov equation leads to a two-dimensional partial differential equation for high-β tokamak equilibria. This equation is given by

$$\nabla^2\psi_0 = -R_0^2 B_0^2\left[\frac{B'_{\phi2}}{B_0} + \frac{2\mu_0 p'_1}{B_0^2}\left(\frac{r}{R_0}\right)\cos\theta\right] \tag{6.97}$$

where prime denotes differentiation with respect to ψ_0.

Zeroth-Order Solution (Radial Pressure Balance and Toroidal Force Balance). Since Eq. (6.97) is a nonlinear two-dimensional partial differential equation, it must in general be solved numerically. However, for special choices of $B_{\phi2}(\psi_0)$ and $p_1(\psi_0)$, Eq. (6.97) can be solved analytically, and it is such solutions that provide a great deal of insight into the nature of high-β tokamak equilibria. To implement this idea a simple model is discussed below, first investigated by Haas (1972).

Consider the case where $B_{\phi2}(\psi_0)$ and $p_1(\psi_0)$ are linear in ψ_0 so that $R_0^2 B_0 B'_{\phi2}(\psi_0) = -A$, $2\mu_0 R_0 p'_1(\psi_0) = -C$. Assume the plasma surface is circular with radius a, and is surrounded by vacuum. The linear dependence of $B_{\phi2}$ implies that the toroidal current density is nearly uniform in the plasma (at least for low β) and abruptly jumps to zero across the plasma surface. The function $\psi_0(r, \theta)$ must be regular for $r \le a$ and constant on the surface $r = a$. The constant is normalized so $\psi_0(a, \theta) = 0$. Equation (6.97) reduces to

$$\nabla^2\psi_0 = A + Cr\cos\theta \tag{6.98}$$

The solutions for ψ_0, B_θ, and p are easily found and are given by

$$\begin{aligned}
\psi_0(r, \theta) &= \frac{A}{4}(r^2 - a^2) + \frac{C}{8}(r^3 - a^2 r)\cos\theta \\
B_\theta(r, \theta) &= \frac{1}{2R_0}\left[Ar + \frac{C}{4}(3r^2 - a^2)\cos\theta\right] \\
p(r, \theta) &= \frac{C}{8\mu_0 R_0}\left[A(a^2 - r^2) + \frac{C}{2}(a^2 r - r^3)\cos\theta\right]
\end{aligned} \tag{6.99}$$

At this point it is useful to express the constants A and C in terms of the more physical parameters, q_* and β_t. The results are

$$\begin{aligned}
\frac{1}{q_*} &= \frac{\mu_0 R_0 I_0}{2\pi a^2 B_0} = \frac{R_0}{2\pi a^2 B_0}\int_0^{2\pi} B_\theta(a, \theta) a\, d\theta = \frac{A}{2B_0} \\
\beta_t &= \frac{2\mu_0\langle p\rangle}{B_0^2} = \frac{2\mu_0}{\pi a^2 B_0^2}\int_0^{2\pi}\int_0^a pr\, dr\, d\theta = \frac{a^2 AC}{8R_0 B_0^2}
\end{aligned} \tag{6.100}$$

Substituting these expressions leads to the following equilibrium solution for the high-β tokamak:

$$\begin{aligned}
\frac{\psi_0}{a^2 B_0} &= \frac{1}{2q_*}[\rho^2 - 1 + \nu(\rho^3 - \rho)\cos\theta] \\
\frac{B_\theta}{\epsilon B_0} &= \frac{1}{q_*}\left[\rho + \frac{\nu}{2}(3\rho^2 - 1)\cos\theta\right] \\
\frac{B_r}{\epsilon B_0} &= -\frac{\nu}{2q_*}(\rho^2 - 1)\sin\theta \\
\frac{2\mu_0 p}{B_0^2} &= 2\beta_t(1 - \rho^2)(1 + \nu\rho\cos\theta) \\
\frac{\mu_0 R_0 J_\phi}{B_0} &= -\frac{2}{q_*}(1 + 2\nu\rho\cos\theta) \\
\frac{B_\phi}{B_0} &= 1 - \epsilon\rho\cos\theta - \beta_t(1 - \rho^2)(1 + \nu\rho\cos\theta)
\end{aligned} \tag{6.101}$$

where $\rho = r/a$, $\nu = \beta_t q_*^2/\epsilon$ and $\epsilon = a/R_0$.

Basic Plasma Parameters and Figures of Merit. Consider now the equilibrium properties of the high-β tokamak. The situation is more complicated than for the RFP and ohmically heated tokamak since radial pressure balance and toroidal force balance occur in the same order. Intuition into the nature of the radial pressure balance can be developed by plotting the curves of B_ϕ, p, and B_θ along the midplane; specifically, one defines a rectangular coordinate system centered about R_0, $x = \rho \cos \theta$, $y = \rho \sin \theta$, and illustrates B_ϕ, p, and B_θ vs. x for $y = 0$. This is shown in Fig. 6.14. Note that there is a small diamagnetic depression of order ϵ superimposed on the $1/R$ decay of the toroidal field. It is in this depression that the plasma pressure $p \sim \epsilon B_\phi^2/\mu_0$ is held in radial pressure balance. Since $B_\theta/B_\phi \sim \epsilon$ for $q \sim 1$, the poloidal field can at most confine a pressure of order $B_\theta^2 \sim \epsilon^2 B_\phi^2$. This is one order smaller in ϵ than the plasma pressure, indicating that the poloidal field has only a very small effect on radial pressure balance.

The general equilibrium relation between β_t, β_p, and q_* is the same as for the ohmically heated tokamak

$$\beta_t = \epsilon^2 \beta_p / q_*^2 \tag{6.102}$$

except that β_t and β_p are each one order larger in ϵ: $\beta_t \sim \epsilon$, $\beta_p \sim 1/\epsilon$, and $q_* \sim 1$. Furthermore, from the definition of ν given above it follows that

$$\nu = \epsilon \beta_p \tag{6.103}$$

An examination of Eq. (6.101) indicates that it is useful to view ν as the tokamak expansion parameter with $\nu \ll 1$ corresponding to the ohmically heated system and $\nu \sim 1$, the high-β system.

The toroidal high-β effects enter the evaluation of the remaining plasma parameters and the toroidal force balance relation in a number of interesting ways. To begin, observe that the flux surfaces $\psi_0 = \text{const}$,

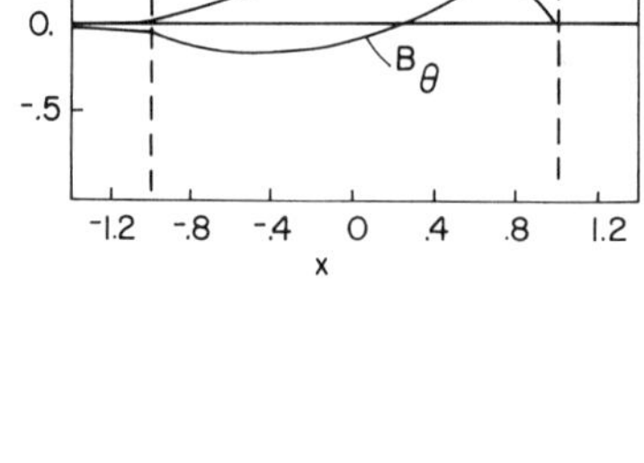

Figure 6.14. B_ϕ/B_0, B_θ/B_0, and $2\mu_0 p/B_0^2$ vs. $x = (R - R_0)/a$ along the midplane $y = Z/a = 0$. The plasma parameters are $\epsilon = 0.35$, $\nu = 0.8$, $q_* = 1.2$, and $\beta_t = 0.2$.

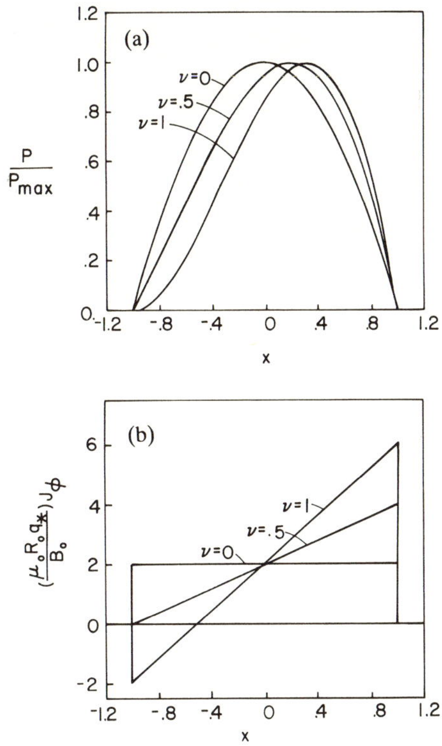

Figure 6.15. Midplane profiles for the high-β tokamak for different values of ν: (a) plasma pressure and (b) toroidal current density.

while round, are no longer pure circles, except for $r = a$. The pressure profile is essentially parabolic for small ν but becomes quite distorted, shifting to the outside for $\nu \sim 1$. This is illustrated in Fig. 6.15. Also shown is the J_ϕ profile, which for small ν is constant across the plasma jumping to zero at the boundary. For $\nu \sim 1$, the current accumulates on the outside in order to provide a stronger toroidal restoring force to balance the increased tire tube and $1/R$ expansion forces. In fact, for $\nu > 1/2$ the current reverses on the cylindrical surface

$$R = R_0(1 - \epsilon/2\nu) \tag{6.104}$$

A current reversal may be difficult to realize in systems with finite conductivity, but this question lies outside the scope of ideal MHD.

The next quantity of interest is the shift of the magnetic axis Δ_0 with respect to the plasma surface. This quantity, defined as the radius where $\partial\psi(r = \Delta_0, \theta = 0)/\partial r = 0$ is easily calculated and is given by

$$\frac{\Delta_0}{a} = \frac{\nu}{1 + (1 + 3\nu^2)^{1/2}} \tag{6.105}$$

Since $\nu \sim 1$ in the high-β ordering, the shift of the magnetic axis is finite and not small as in the ohmically heated tokamak.

The evaluation of the MHD safety factor and shear is quite complicated, even in this simple model. The reason is that the flux surfaces are not simple circles. For nonzero ν the q profile is an increasing function of radius. In this regard it is sufficient to calculate q on the magnetic axis and q at the edge of the plasma.

It can be shown, by expanding the flux function about the magnetic axis and substituting into Eq. (6.25), that

$$q_0 = \Delta_0 B_\phi \left(\frac{\partial^2 \psi}{\partial r^2} \frac{\partial^2 \psi}{\partial \theta^2} \right)^{-1/2} \tag{6.106}$$

where q_0 is the MHD safety factor on axis and B_ϕ, ψ_{rr}, and $\psi_{\theta\theta}$ are to be evaluated at $r = \Delta_0$, $\theta = 0$. (See Problem 6.5.) Equation (6.106) is exact; no expansion in ϵ or β is required. Substitution of the equilibrium solution into Eq. (6.106) yields

$$\begin{aligned} q_0 &= q_* \left[\frac{3}{\eta(2 + \eta)} \right]^{1/2} \\ \eta &= (1 + 3\nu^2)^{1/2} \end{aligned} \tag{6.107}$$

Since η is always greater than unity then $q_0 < q_*$. In the low-β limit, $\nu \to 0$ and $q_0 \to q_*$. For higher β the ratio of q_0/q_* becomes progressively smaller.

The next point to note is that in the high-β ordering the variation of B_θ around the cross section is of order unity in order to support the higher pressure in toroidal force balance. This variation by definition averages to zero when calculating $1/q_*$ so that $q_* \sim 1/I$. However, when calculating q_a as defined by Eq. (6.25) the integrand is more strongly weighted on the inside of the torus where $B_\theta(a, \theta)$ is small. This leads to the result that $q_a > q_*$. Specifically, on the flux surface $r = a$

$$q_a = \frac{aB_0}{2\pi R_0} \int_0^{2\pi} \frac{d\theta}{B_\theta(a, \theta)} = \frac{q_*}{(1 - \nu^2)^{1/2}} \tag{6.108}$$

In the limit of low β, $\nu \to 0$ and $q_a \to q_*$. Nevertheless, in the high-β regime it is important to distinguish between q_* and q_a, particularly since $1/q_a$ is not simply proportional to I. The apparent breakdown of the solution when $\nu \geq 1$, as evidenced by Eq. (6.108), is an important aspect of high-β tokamak behavior and will be discussed shortly, after evaluation of the remaining plasma parameters.

The Shear. Having calculated q_0 and q_a, one can now focus attention on the shear. Since q_0 is finite, it follows that the shear on axis is zero: $s_0 = 0$. The shear just inside the plasma boundary can be evaluated by expanding the solution for ψ_0 about $r = a$ and assuming that ψ_0 is a small quantity. Writing $\rho = 1 - \rho_1$ with $\rho_1 \ll 1$, one finds that the equation for the flux surface $\rho_1(\theta)$, corresponding to the surface $\psi_0 = $ const, is given by

$$\rho_1(\theta) = -\frac{q_* \psi_0}{a^2 B_0} \frac{1}{1 + \nu \cos\theta} \tag{6.109}$$

Substituting into Eqs. (6.21), (6.25), and (6.28) yields the following expression for s_a, the shear at the edge of the plasma:

$$s_a = \frac{3\nu^2}{(1 - \nu^2)^{3/2}} \tag{6.110}$$

At low β, $\nu \rightarrow 0$ and $s_a \rightarrow 0$. As β is increased, the edge shear increases.

The trends q_0/q_* decreasing, q_a/q_* increasing, and s_a increasing as $\nu = \epsilon\beta_p$ increases are typical of many high-β tokamak profiles. The vanishing of the shear as $\nu \rightarrow 0$ is a peculiarity of the simple model treated here. In this limit the current profile becomes flat. For more general profiles, the shear at the edge is finite, at any β including $\beta \rightarrow 0$.

The Magnetic Well. The magnetic well is the next quantity of interest. On axis the magnetic well is zero. In the high-β limit, the well is dominated by the toroidal curvature and is one order larger in ϵ than for the ohmically heated tokamak. Recall that the magnetic well is proportional to $\langle B^2 \rangle'$. The averaging process involves a weighted integral with $1/B_p$ the weight factor. The largest contribution occurs on the inside of the torus where B_p is smallest. This is the region of favorable curvature where B has its largest gradient. After a straightforward but slightly tedious calculation one can show that the magnetic well on the plasma surface is given by

$$W_a = \frac{\epsilon\nu}{1 - \nu^2}[3 - (1 - \nu^2)^{1/2}] \tag{6.111}$$

W_a becomes increasingly positive with β and approaches zero as $\nu \rightarrow 0$. Actually, when ν decreases the ohmically heated contributions to $W_a \sim \epsilon^2$ become important. The sign of W_a corresponds to stability.

The High-β Tokamak Equilibrium Limit. Perhaps the most interesting feature of high-β tokamak equilibria is the appearance of an upper limit on the value of β that can be contained for a given current. This can be seen in Eq. (6.108), which requires $\nu < 1$ for existence of the solutions. At the limiting value $\nu = 1$, the plasma parameters are given by

$$\begin{aligned} \beta_p &= 1/\epsilon \\ \beta_t &= \epsilon/q_*^2 \\ \Delta_0/a &= 1/3 \\ q_0 &= (3/8)^{1/2} q_* \\ q_a &\to \infty \end{aligned} \tag{6.112}$$

For a given inverse aspect ratio ϵ, the actual numerical value of β_t is obtained from the minimum allowable q_* as determined from stability considerations.

The origin of the equilibrium limit can be understood by examining the fields in the vacuum region. The solution that matches onto Eq. (6.101) and satisfies the vacuum equation $\nabla^2 \hat{\psi}_0 = 0$ is given by

$$\frac{\hat{\psi}_0}{a^2 B_0} = \frac{1}{q_*} \left[\ln \rho + \frac{\nu}{2} \left(\rho - \frac{1}{\rho} \right) \cos \theta \right] \tag{6.113}$$

The first term represents the equivalent "straight" contribution of the toroidal current. The last two terms represent the vertical field and the corresponding diamagnetic response, respectively. Note that the vacuum field has a separatrix at the point defined by $B_\theta(r_s, \theta_s) = B_r(r_s, \theta) = 0$, where $\theta_s = \pi$ and

$$\frac{r_s}{a} = \rho_s = \frac{1 + (1 - \nu^2)^{1/2}}{\nu} \tag{6.114}$$

For low β, $\nu \ll 1$ and the separatrix $\rho_s \approx 2/\nu$ is far from the plasma surface. As β (i.e., ν) increases, the vertical field increases and the separatrix moves in toward the plasma. When $\nu = 1$, the separatrix moves onto the plasma surface. Since it can move no further, this corresponds to the equilibrium limit. The existence of the separatrix is shown graphically in Fig. 6.16, by summing the contributions of the vertical field and the $1/r$ field of the toroidal current.

From the results presented above, one might conjecture that all high-β tokamaks have an equilibrium β limit with similar scaling in ϵ. This is indeed true, but only under the condition that increasing values of β are held in toroidal force balance by *increasing the vertical field, keeping the toroidal current fixed.* Such a procedure is convenient for numerical calculations but does not in general represent the physical evolution of a given experimental discharge. This question is addressed in

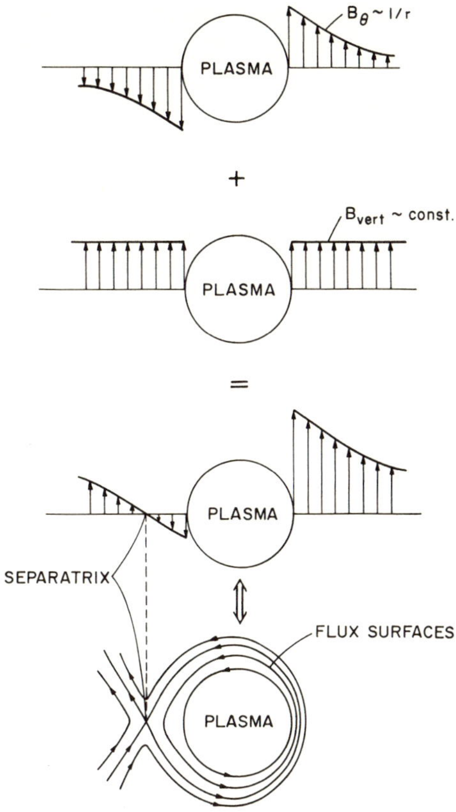

Figure 6.16. The appearance of a separatrix in a high-β tokamak.

the section on "Flux-Conserving Tokamaks," where it is shown that an equilibrium limit of the type discussed can be avoided for interesting regimes of tokamak operation although the nature of such equilibria changes qualitatively when $\beta q_*^2/\epsilon \gtrsim 1$.

The Equilibrium Vertical Field. The final question of interest concerns the vertical field required for toroidal force balance. This quantity is easily obtainable by examining Eq. (6.113). Since the $\rho \cos\theta$ term explicitly represents the vertical field source, it follows that

$$B_v = \frac{\mu_0 I_0}{4\pi R_0}\beta_p \tag{6.115}$$

This result coincides with the corresponding result for the conventional tokamak given by Eq. (6.91) if one keeps in mind that the β_p term is $1/\epsilon$ times larger than other terms in the high-β ordering.

Experimental Discussion/Numerical Results. Many numerical studies have been carried out that investigate the equilibrium properties of high-β tokamaks. A particular example is illustrated in Fig. 6.17 corresponding to the circular cross section TFTR experiment at Princeton Plasma Physics Laboratory. Note the large shift of the peak plasma pressure as well as the buildup of the toroidal current on the outside of the torus. The B_ϕ profile exhibits a clear diamagnetic depression which provides the radial pressure balance in the plasma.

Experimentally, high-β tokamak operation is achieved by the addition of auxiliary heating power supplies to an ohmically heated tokamak. This power usually corresponds to one of two types. First, high-energy neutral beams are injected into the plasma imparting their energy through collisions. Second, high-power rf heating sources are applied in which the frequency is carefully chosen to resonate with one of

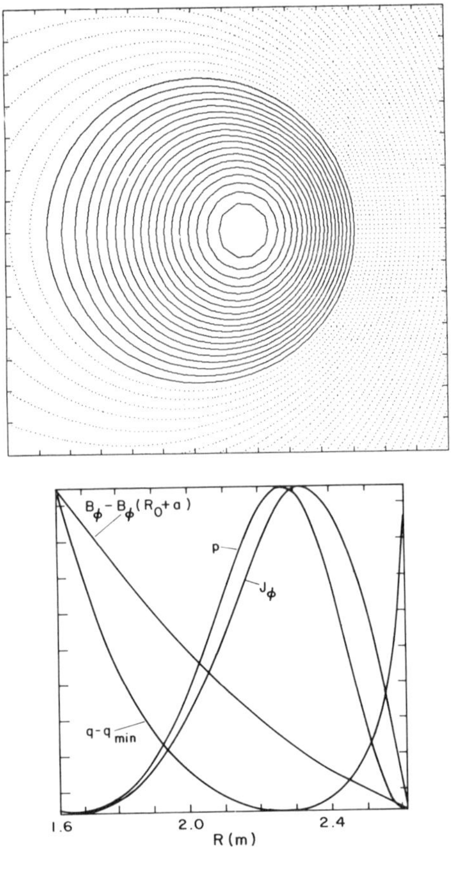

Figure 6.17. Numerically computed equilibrium of the circular cross section TFTR at Princeton Plasma Physics Laboratory. The flux surface plot and midplane profiles correspond to auxiliary heated, high-β tokamak operation. Courtesy J. Manickam, Princeton Plasma Physics Laboratory.

the basic characteristic frequencies of the plasma: the ion gyro frequency $\omega = \omega_{ci}$, the lower hybrid frequency $\omega \approx \omega_{pi}$, and the electron gyro frequency $\omega = \omega_{ce}$.

Auxiliary heating power supplies are in general quite expensive and complicated technologically. Even so, they have been successful in providing power to raise the β in tokamaks. Auxilliary power levels from factors of 2–10 larger than the ohmic power have been supplied leading to β values of about $\beta_t \sim 4\%–5\%$. The DIII experiment at GA Technologies is the first large tokamak to have demonstrated such high values of β.

One undesirable feature observed in many high-β experiments is a decrease in the energy confinement time as the auxiliary power is increased. At the present time there is no clear explanation of this phenomenon, and how important it will be to future tokamaks remains to be seen.

Summary. In summary, a high-β tokamak is a configuration in which radial pressure balance is provided almost entirely by the toroidal field as in a θ pinch. The toroidal current provides toroidal force balance but is limited in value by the stability condition $q \gtrsim 1$. The large, nearly vacuum, toroidal field, is responsible for stability as in the ohmic tokamak. In the regime of the high-β tokamak, β_t is raised to the largest value that can be held in toroidal force balance by a current satisfying $q \gtrsim 1$. This leads to a scaling $\beta_t \sim \epsilon$, $\beta_p \sim 1/\epsilon$ which is one order larger in ϵ than in an ohmic tokamak. There is in general an upper limit on the achievable β_t's if, as β_t is raised, toroidal force balance is maintained by increasing the applied vertical field but keeping the toroidal current and toroidal field fixed. In this case a separatrix forms and ultimately moves on to the plasma surface.

A more realistic description of the plasma behavior as β_t increases is given in Section 6.7 on "the flux-conserving tokamak." This discussion is preceded by an investigation of the properties of noncircular tokamaks, representing a second method of achieving higher values of β_t.

6.6. NONCIRCULAR TOKAMAKS

The second modification to the ohmically heated tokamak that leads to increased values of β_t is the use of noncircular cross sections. By introducing this additional degree of freedom into the equilibrium, one expects the possibility of further optimization.

There are, indeed, several improvements in tokamak operation that are associated with noncircularity. First, gains in β_t occur in both the ohmic regime $\beta_t \sim \epsilon^2$, and the high-β regime $\beta_t \sim \epsilon$. Second, noncircularity also allows higher values of $\bar{J}_\phi$ and I_0. This leads to increased ohmic heating and to improved confinement in the high-β regime where experimental measurements have shown that $\tau_E \propto I_0$.

There are a number of ways to demonstrate the improvements associated with noncircularity. However, the resulting scaling laws are not unique. The confusion arises because, depending upon application, different quantities must be held fixed as the cross section is varied. For instance, in comparing configurations should one hold the aspect ratio, the major radius, or the plasma volume fixed? Even more subtle is the question of the safety factor. If stability is an important limitation should q_0, q_a, or q_* be held fixed? Furthermore, if q_* is the appropriate quantity, is the correct geometric scaling built into the definition, or does the critical q_* itself depend upon cross section? The point is that noncircularity represents an additional degree of freedom and consequently any optimization subject to a set of meaningful physical constraints leads to improved performance. In the discussion that follows a set of scaling laws relating β_t, $\bar{J}_\phi$, and I_0 to the cross section is presented which is particularly relevant to the high-β tokamak regime.

Solution to the Grad–Shafranov Equation. Solutions to the Grad–Shafranov equation for noncircular plasmas usually require numerical calculations. However, a simple analytical solution can be obtained by generalizing the special model for the high-β tokamak discussed in Section 6.5. Recall that this model has the two free functions $B'_{\phi 2}(\psi_0)$ and $p'_1(\psi_0)$ taken as constants: $R_0^2 B_0 B'_{\phi 2}(\psi_0) = -A$ and $2\mu_0 R_0 p'_1(\psi_0) = -C$. The solution for ψ_0 given by Eq. (6.99) can be generalized to include noncircular cases by adding homogeneous solutions to Eq. (6.98). Hence, a general solution is given by

$$\psi_0 = \frac{A}{4} r^2 + \frac{C}{8} r^3 \cos\theta + \sum_{m=0}^{\infty} H_m r^m \cos m\theta \tag{6.116}$$

where the H_m are arbitrary coefficients.

The flux surfaces correspond to the contours of constant ψ_0. A wide variety of cross sections is possible by varying the H_m. For instance, the coefficient H_1 has the effect of shifting the outer flux surfaces with respect to the magnetic axis. The coefficient H_2 elongates the cross section in the shape of an ellipse. H_3 produces a triangularity in the flux surfaces, an important property related to the magnetic well and pressure-gradient-driven instabilities. A sample of flux-surface plots is illustrated in Fig. 6.18 demonstrating a variety of cross sections.

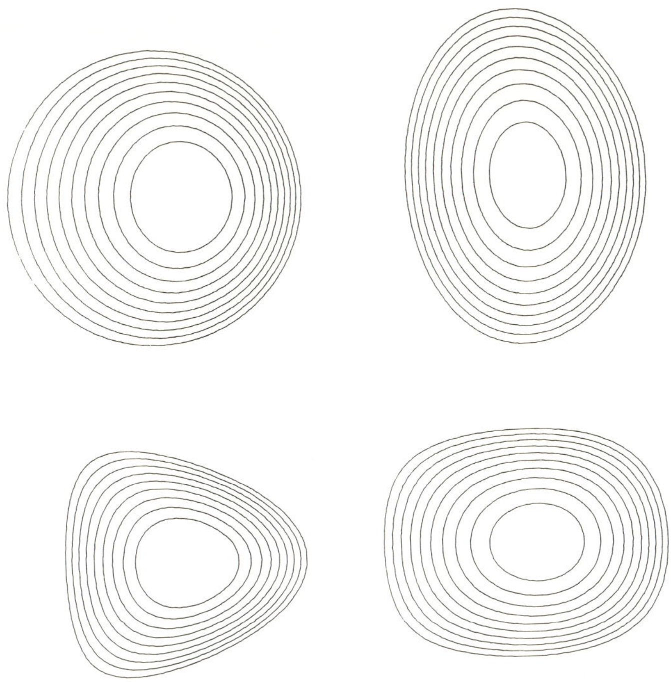

Figure 6.18. Sample of flux-surface plots from the analytic noncircular high-β tokamak equilibrium equations.

It is of particular interest to investigate the case of the elliptical cross section as it is possible to calculate analytically the important plasma parameters.

A straightforward calculation shows that the following expression satisfies the high-β Grad–Shafranov equation [Eq. (6.98)]

$$\psi_0 = K_1\left(\frac{x^2}{a^2} + \frac{y^2}{\kappa^2 a^2} - 1\right)(1 + K_2 x) \tag{6.117}$$

where $x = r\cos\theta$, $y = r\sin\theta$ are rectangular coordinates centered about $R = R_0$ and $Z = 0$, κ is the plasma elongation, and

$$\begin{aligned} K_1 &= \frac{a^2}{2}\left(\frac{\kappa^2}{1 + \kappa^2}\right)A \\ K_2 &= \left(\frac{1 + \kappa^2}{1 + 3\kappa^2}\right)\frac{C}{A} \end{aligned} \tag{6.118}$$

Note that the plasma surface $\psi_0 = 0$ is an ellipse of width a, height κa, and cross-sectional area $S = \pi a^2 \kappa$.

As in the circular case, it is useful to express K_1 and K_2 in terms of the physical parameters q_* and β_t. A simple calculation yields

$$K_1 = \frac{a^2 B_0}{q_*}\left(\frac{\kappa^2}{1+\kappa^2}\right)$$
$$K_2 = \frac{\beta_t q_*^2}{a\epsilon}\left[\frac{(1+\kappa^2)^2}{\kappa^2(1+3\kappa^2)}\right] \tag{6.119}$$

where in noncircular cases ϵ is still given by $\epsilon \equiv a/R_0$.

Plasma Parameters. Using these results it is now possible to evaluate the important plasma parameters. For the noncircular case it is instructive to first calculate q_a. A straightforward calculation gives

$$q_a = \frac{q_*}{(1-\nu^2)^{1/2}}\left(\frac{1+\kappa^2}{2\kappa}\right)$$
$$\nu = \frac{\beta_t q_*^2}{\epsilon}\left[\frac{(1+\kappa^2)^2}{\kappa^2(1+3\kappa^2)}\right] \tag{6.120}$$

Equation (6.120) demonstrates that the elliptical tokamak also has an equilibrium β_t limit when $\nu = 1$. This suggests that it is appropriate to fix $\nu = 1$ when comparing configurations with different elongations to determine which is capable of supporting the highest β_t.

In trying to maximize the equilibrium β_t limit it is also appropriate to consider ϵ and q_* as fixed. The aspect ratio $1/\epsilon$ should correspond to as tight a torus as is technologically feasible. The kink safety factor q_* should be at the minimum value possible to maintain stability against current-driven kink modes. Both of these conditions maximize the value of β_t.

The usefulness of holding ϵ fixed when comparing different elongations follows from the intuition that the tightest aspect ratio as set by technology is essentially independent of κ. The situation is not as clear regarding q_*. It is by no means obvious that the minimum q_* for stability is independent of cross section. In fact, this currently remains as one of the important unanswered questions in tokamak stability theory. Partial analysis of the problem indicates that the minimum q_* does indeed depend on elongation and shape. For the cases studied q_* is a slowly increasing function of elongation. Nevertheless, for the sake of simplicity the quantity q_* will be assumed fixed when comparing different cross sections.

Another point to discuss concerns the evaluation of I_0 and $\bar{J}_\phi$. In these calculations an additional degree of geometric freedom appears, since as κ varies both a and R_0 are not individually constrained but only their ratio $\epsilon = a/R_0$. This is resolved by assuming the total volume of the plasma $V_0 = 2\pi^2 R_0 a^2 \kappa$ is held constant, corresponding to the assumption of fixed total cost.

Plasma Parameters; High-β Regime. To compare the performance of high-β tokamaks as a function of cross section, one thus calculates the important plasma parameters, β_t, β_p, q_0, q_a, I_0, and $\bar{J}_\phi$ in terms of the elongation κ and the fixed quantities ϵ, q_*, and V_0. These results easily follow from the solution for ψ_0 and the definitions of the quantities of interest. The results are given by

β_t [Eq. (6.120)]:

$$\beta_t = \frac{\nu\epsilon}{q_*^2}\left[\frac{\kappa^2(1 + 3\kappa^2)}{(1 + \kappa^2)^2}\right] \tag{6.121}$$

β_p [Eqs. (6.32)–(6.34)]:

$$\beta_p = \frac{\nu}{\epsilon}\left[\frac{\kappa^2(1 + 3\kappa^2)}{(1 + \kappa^2)^2}\right] \tag{6.122}$$

q_0 [Eq. (6.106)]:

$$\begin{aligned} q_0 &= q_*\left[\frac{3}{\eta(2 + \eta)}\right]^{1/2}\left(\frac{1 + \kappa^2}{2\kappa}\right) \\ \eta &= (1 + 3\nu^2)^{1/2} \end{aligned} \tag{6.123}$$

q_a [Eq. (6.120)]:

$$q_a = \frac{q_*}{(1 - \nu^2)^{1/2}}\left(\frac{1 + \kappa^2}{2\kappa}\right) \tag{6.124}$$

I_0 [Eq. (6.36)]:

$$\begin{aligned} I_0 &= I_c \kappa^{2/3} \\ I_c &= \frac{(4\pi V_0 \epsilon^4)^{1/3} B_0}{\mu_0 q_*} \end{aligned} \tag{6.125}$$

$\bar{J}_\phi \equiv I_0/S$:

$$\begin{aligned} \bar{J}_\phi &= \bar{J}_c \kappa^{1/3} \\ \bar{J}_c &= \left(\frac{16\pi^2\epsilon^2}{V_0}\right)^{1/3} \frac{B_0}{\mu_0 q_*} \end{aligned} \tag{6.126}$$

Note the following points. As κ increases, β_t increases monotonically from its circular value eventually saturating as $\kappa \to \infty$. At the equilibrium limit $\nu = 1$, and for large κ, the value of β_t is given by

$$\beta_t = 3\epsilon/q_*^2 \tag{6.127}$$

which is three times larger than its circular value.

The quantity β_p has the same scaling with κ as β_t. The maximum of β_p occurs for $\nu = 1$ and is given by

$$\epsilon\beta_p = 3 \tag{6.128}$$

Both q_0 and q_a increase faster with κ than q_*; at large κ, $q_0/q_* \sim \kappa$ and $q_a/q_* \sim \kappa$. Also both q_0 and q_a exhibit a minimum about $\kappa = 1$. However, their ratio is independent of κ. The higher values of q_0 and q_a at fixed q_* for elongated plasmas may be advantageous for stabilizing certain classes of ideal and resistive MHD instabilities.

The current I_0 and average current density $\bar{J}_\phi$ each increases with elongation, thus implying improved ohmic heating and energy confinement time.

Plasma Parameters: Ohmic Regime. By making a minor modification to the analysis it is possible to use the results just derived to investigate the influence of noncircularity on ohmically heated tokamaks. As an example consider a series of elliptically shaped ohmic tokamaks all characterized by the same diamagnetism. For simplicity assume this diamagnetism is zero; the free function $F(\psi) = \text{const}$, implying that there exists a constraint relationship between the functions $p_1(\psi)$ and $B_{\phi 2}(\psi)$ introduced in Eq. (6.96)

$$\frac{2\mu_0}{B_0^2} p'(\psi) = \frac{2B'_{\phi 2}(\psi)}{B_0} \tag{6.129}$$

For the simple model under investigation in which $R_0^2 B_0 B'_{\phi 2} = -A$ and $2\mu_0 R_0 p'_1 = -C$, Eq. (6.129) indicates that the two constants A and C are no longer independent, but instead must satisfy

$$C = \frac{2A}{R_0} \tag{6.130}$$

The relationship between C and A serves to fix the quantity ν appearing in the high-β solutions at a low value, of order ϵ, consistent with ohmic operation. Simple algebraic elimination among Eqs. (6.118)–(6.120) shows that ν is given by

$$\nu = 2\epsilon\left(\frac{1 + \kappa^2}{1 + 3\kappa^2}\right) \tag{6.131}$$

Substituting this result into Eqs. (6.121)–(6.126) leads to the following scaling laws as a function of κ for the ohmically heated tokamak:

$$\begin{aligned}
\beta_t &= \frac{\epsilon^2}{q_*^2}\left(\frac{2\kappa^2}{1+\kappa^2}\right) \\
\beta_p &= \frac{2\kappa^2}{1+\kappa^2} \\
q_0 &\approx q_a \approx q_*\left(\frac{1+\kappa^2}{2\kappa}\right) \\
I_0 &= I_c\kappa^{2/3} \\
\bar{J}_\phi &= \bar{J}_c\kappa^{1/3}
\end{aligned} \tag{6.132}$$

These results are qualitatively similar to those of the high-β tokamak. β_t increases with κ and saturates as $\kappa \to \infty$. The gain over the circular case is a factor of 2. A similar gain is found for β_p. The q profile for the simple model remains flat in the ohmic regime, even with an elliptical cross section. Note that at low $\beta_t \sim \epsilon^2$, q_a is inversely proportional to I_0. The scalings of I_0 and $\bar{J}_\phi$ with κ remain unchanged from their high-β relationships.

Experimental Discussion/Numerical Results. Since all the important physical quantities of interest appear to improve with κ, one might expect tokamak experiments to be designed with very highly elongated cross sections. In practice, another problem arises that prevents this strategy. Elongated tokamaks have a strong susceptibility to axisymmetric instabilities. Specifically, a pure rigid vertical displacement of the plasma becomes unstable at sufficiently large elongations. Either passive or active feedback is required to maintain equilibrium in such configurations. In practice, elongations are usually restricted to the range $\kappa \lesssim 2$ to prevent these modes from occurring. Vertical instabilities are discussed in more detail in Chapter 10.

A combination of analytical and numerical studies has shown that the overall stability of noncircular tokamaks is further improved by adding an outward-pointing triangularity to the elongated cross section. This is particularly useful in suppressing pressure-driven modes and is also discussed in Chapter 10. Many of the latest generation of large tokamaks are designed to have this cross section. One example is the DIII-D experiment located at GA Technologies. A numerically computed equilibrium of this device is shown in Fig. 6.19.

Summary. To summarize, a noncircular cross section represents an additional degree of freedom in the design of tokamaks, allowing a

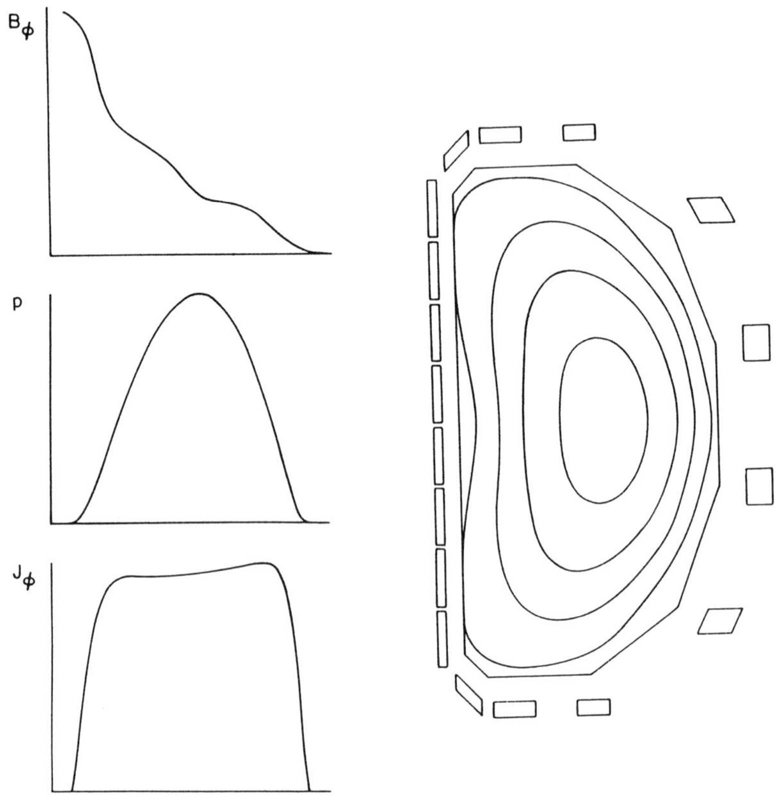

Figure 6.19. Numerically computed equilibrium of the noncircular, high-β tokamak, DIII-D located at GA Technologies. Shown are flux surface plots and midplane profiles. Courtesy J. Helton, GA Technologies.

further optimization in performance. A comparison of tokamaks with fixed ν, ϵ, q_*, and V_0 shows that as elongation increases, the maximum attainable β_t, I_0, and $\bar{J}_\phi$ also increase, both in the high-β and ohmic regimes of operation. Many current-generation tokamaks have elongated cross sections with outward-pointing triangularity. Very large elongations are not possible practically because of the appearance of vertical instabilities.

6.7. THE FLUX-CONSERVING TOKAMAK

The existence of an equilibrium β_t limit for high-β tokamaks is strongly coupled to the constraint that requires that $I_0 \propto 1/q_*$ be held fixed as β_t increases. While this constraint is convenient for numerical calculations, it does not in general represent the actual evolution of a given plasma discharge. Shafranov, in fact, first noted that a tokamak surrounded by a perfectly conducting shell could not evolve in such a way as to form a separatrix if one did not originally exist because of the topological constraints imposed by ideal MHD; thus, it would appear that such systems do not have an equilibrium β_t limit (Shafranov, 1971; Mukhovatov and Shafranov, 1971).

Flux-Conserving Evolution. This issue has been resolved with the introduction of the flux-conserving tokamak concept, a special subclass of high-β tokamak equilibria (Clarke and Sigmar, 1977; Dory and Peng, 1977). Here the evolution to higher β is governed by a different set of constraints and leads to the conclusion that no equilibrium β_t limit exists. The basic idea is to model the actual evolution of a single plasma discharge by devising a realistic sequence of functions for $p(\psi)$ and $F(\psi)$. This is done as follows. Consider a system that is initially operated as an ohmically heated tokamak, $\beta_t \sim \epsilon^2$, $\beta_p \sim 1$. The plasma is then heated by a high-power external source, for instance neutral beams or rf waves. It is assumed that the time scale for heating is long compared to the characteristic MHD time:

$$\tau_H \gg \tau_M \tag{6.133}$$

where $\tau_M = a/V_{T_i}$ is the MHD time and $\tau_H = T/(\partial T/\partial t)$ is the heating time. Under this assumption MHD inertial effects are negligible (i.e., the $\rho\, d\mathbf{v}/dt$ term in the MHD momentum equation can be neglected). The plasma evolution can thus be viewed as a series of quasistatic equilibria, each one satisfying the Grad–Shafranov equation.

On the other hand, the time scale for heating is assumed to be short compared to the magnetic diffusion time, whether it is classical or anomalous:

$$\tau_H \ll \tau_D \tag{6.134}$$

where $\tau_D = \mu_0 a^2/\eta$ is the diffusion time and η is the effective resistivity. In this situation the electrical properties of the plasma during the heating process are the same as those of a perfect conductor.

Keeping these assumptions in mind, consider the following flux-conserving tokamak prescription for determining $p(\psi)$ and $F(\psi)$. The function p is treated as a known function of ψ, slowly varying with t: $p = p(\psi, t)$. Initially, $p(\psi, 0) = p\Omega(\psi)$, where $p_\Omega(\psi)$ is the ohmic pressure function. As time progresses $p(\psi, t)$ increases due to the heating. Since the theory and computation of heat deposition in a plasma is reasonably well established, it is not very difficult to generate reasonably accurate models for $p(\psi, t)$. Often the simple assumption is made that $p(\psi, t) = K(t)p_\Omega(\psi)$ with $K(0) = 1$. This modeling directly specifies the evolution of p in the Grad–Shafranov equation and thus eliminates the need for solving the particle and energy transport equations. Note that in the function $p(\psi, t)$, the time t enters the calculation only as a parameter, consistent with the quasistatic assumption.

The interesting new feature of the flux-conserving tokamak is the specification of $F(\psi, t)$. Since the plasma acts like a perfect conductor during the heating process both the toroidal and poloidal fluxes are conserved, and it is this condition that determines $F(\psi, t)$. To find the relationship for $F(\psi, t)$, first imagine that in each quasistatic equilibrium the flux surfaces are labeled by $\psi = \text{const}$. Since the poloidal flux $\psi_p = 2\pi\psi$, then flux surfaces with the same ψ label in different equilibria, by definition, conserve poloidal flux. Now consider the toroidal flux contained within any of these surfaces

$$\psi_t(\psi) = \int B_\phi(r, \theta) r\, dr\, d\theta = \int_0^\psi d\psi' \int_0^{2\pi} \left(\frac{rB_\phi}{RB_\theta}\right)_{\psi', \theta'} d\theta' \quad (6.135)$$

where in the second integral a transformation of coordinates has been made, $\psi' = \psi'(r, \theta)$, $\theta' = \theta$, and the integrand is assumed to be a function of ψ' and θ'. A comparison of Eq. (6.135) with Eq. (6.25) indicates that

$$\frac{d\psi_t(\psi)}{d\psi} = \int_0^{2\pi} \left(\frac{rB_\phi}{RB_\theta}\right)_{\psi, \theta'} d\theta = 2\pi q(\psi) \quad (6.136)$$

It follows from Eq. (6.136) that in order for the toroidal flux to be conserved as the system evolves (i.e., the toroidal flux contained within a given ψ surface must be the same for each equilibrium) the function $F(\psi, t)$ must be chosen so that $q(\psi)$ is identical in each quasistatic state. Since this applies in particular to the initial state, the actual functional dependence of $q(\psi)$ corresponds to ohmically heated tokamak operation. Using Eq. (6.25) one can easily show that the Grad–Shafranov equation for ψ is given by

$$\Delta^*\psi = -\mu_0 R^2 \frac{dp}{d\psi} - \frac{1}{2}\frac{d}{d\psi}\left[q_\Omega^2\left(\frac{1}{2\pi}\oint \frac{dl_p}{R\,|\nabla\psi|}\right)^{-2}\right] \quad (6.137)$$

where $q_\Omega(\psi)$ is the initial ohmic MHD safety factor and dl_p is the poloidal arc length.

Simple Flux-Conserving Model. Equation (6.137) is a nonlinear partial integrodifferential equation which must in general be solved numerically. Nevertheless, one can get some insight into the flux-conserving tokamak without having to explicitly solve Eq. (6.137). One idea is as follows. As a simple approximation to flux conservation, assume that as the plasma evolves, only q_0 and q_a are exactly conserved. Since the $q(\psi)$ profile is a simple monotonically increasing function, fixing the end points, q_0 and q_a, should be sufficient to ascertain the main features of the evolution, even if the exact profiles are not conserved.

One can attempt to implement this idea using the simple circular high-β tokamak equilibrium discussed in Section 6.5. Observe, however, that this equilibrium is characterized by only two free parameters, A and C. One represents the plasma heating and is proportional to β_t. The other can be related to an appropriate form of the safety factor. Consequently, in this simple model there is freedom to fix one, but not both, of the quantities q_0 and q_a since β_t must be left free to vary. Nevertheless, even making the rather gross approximation that flux conservation requires only q_a conserved, it follows that the model no longer exhibits an equilibrium limit; that is, well-behaved solutions exist for arbitrarily large β_t.

This can be seen explicitly by introducing the heating parameter defined by

$$H \equiv \beta_t q_a^2/\epsilon \tag{6.138}$$

H is a useful measure of the heating if one assumes that the flux-conserving evolution occurs for fixed ϵ, B_0, and (by definition) q_a, the usual experimental situation. Under these conditions $H \propto \langle nT \rangle$, indicating that increases in plasma energy due to auxiliary heating correspond to increases in the value of H. For the high-β tokamak with fixed q_* rather than fixed q_a, the equivalent heating parameter is defined as

$$H = \beta_t q_*^2/\epsilon \tag{6.139}$$

and equilibria exist only in the range $0 < H < 1$.

The next step is the evaluation of the important plasma parameters β_p, I_0, q_*, B_v, and ρ_s (the location of the separatrix) for the flux-conserving evolution. The equation for β_p follows from Eqs. (6.102), (6.103), and (6.108) and is given by

$$(\epsilon\beta_p)^2 + \epsilon\beta_p/H - 1 = 0 \tag{6.140}$$

The solution can be written as

$$\epsilon\beta_p = \frac{2H}{(1 + 4H^2)^{1/2} + 1} \tag{6.141}$$

The quantities q_* and I_0 follow from substituting Eq. (6.141) into Eqs. (6.108) and (6.101)

$$q_* = q_a\left[\frac{2}{(1 + 4H^2)^{1/2} + 1}\right]^{1/2} \tag{6.142}$$

$$I_0 = I_\Omega\left[\frac{(1 + 4H^2)^{1/2} + 1}{2}\right]^{1/2} \tag{6.143}$$

where $I_\Omega = 2\pi a^2 B_0/\mu_0 R_0 q_a$ is the initial ohmic current. The remaining two quantities, B_v and ρ_s, are obtained by substituting Eqs. (6.141) and (6.143) into Eqs. (6.115) and (6.114). One finds

$$B_v = \frac{\epsilon B_0}{2q_a}\left[\frac{2H^2}{(1 + 4H^2)^{1/2} + 1}\right]^{1/2} \tag{6.144}$$

$$\rho_s = \left[\frac{(1 + 4H^2)^{1/2} + 1}{2H}\right]\left\{1 + \left[\frac{2}{(1 + 4H^2)^{1/2} + 1}\right]^{1/2}\right\} \tag{6.145}$$

These results are illustrated in Fig. 6.20. Also illustrated are the results for the high-β tokamak with fixed q_*. (Keep in mind that $H = \beta_t q_a^2/\epsilon$ for the flux-conserving tokamak while $H = \beta_t q_*^2/\epsilon$ for the high-β tokamak.)

Observe the following points. Each of the important plasma parameters is finite and well behaved for arbitrarily large β_t (i.e., H) when q_a is held fixed; there is no equilibrium β limit for the flux-conserving tokamak. However, when q_* is held fixed there does indeed exist a limit, $\beta_t = \epsilon/q_*^2$ corresponding to $H = 1$. The explanation is that in the flux-conserving case, the current I_0 must increase as β_t increases in order to keep q_a constant. A larger I_0 requires a smaller increase in the vertical field B_v to hold a given plasma β_t in toroidal force balance. Thus, the ratio of B_v/I_0 is smaller in a flux-conserving tokamak than in a high-β tokamak with fixed q_*. From Fig. 6.16 one can see intuitively that this behavior inhibits the motion of the separatrix toward the plasma surface.

Another important aspect of flux-conserving operation involves q_*. Since I_0 increases with H, the value of $q_* \sim 1/I_0$ decreases. Consequently, a system that satisfies the kink stability criterion $q_* > q_{*\min}$ in the ohmic regime may violate the condition as β is increased. This is a potential danger that must be avoided by insuring that the plasma current at the beginning of a flux-conserving evolution has an adequate safety margin from the critical value.

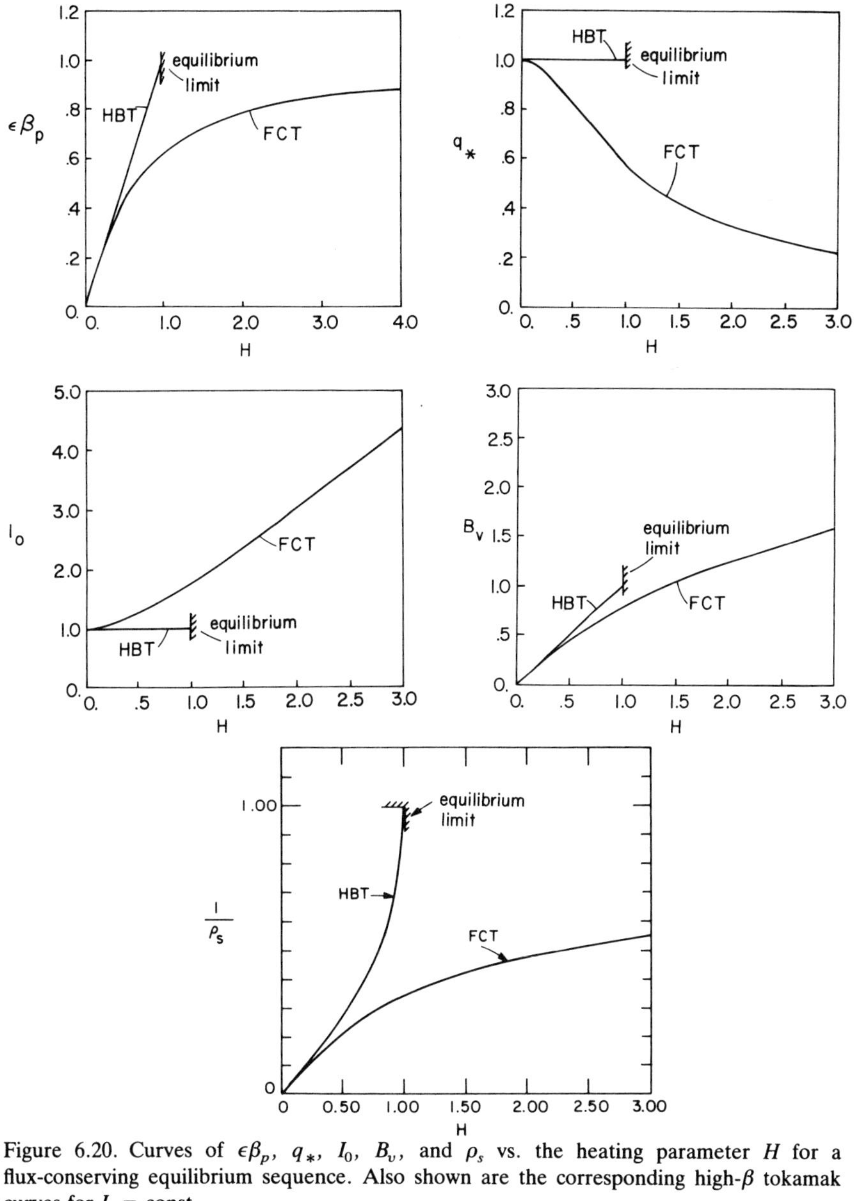

Figure 6.20. Curves of $\epsilon\beta_p$, q_*, I_0, B_v, and ρ_s vs. the heating parameter H for a flux-conserving equilibrium sequence. Also shown are the corresponding high-β tokamak curves for I_0 = const.

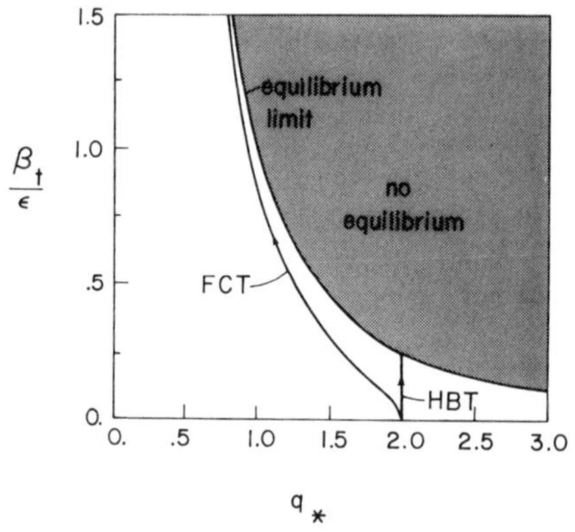

Figure 6.21. Flux-conserving trajectory in β_t vs. q_* space. Also shown is the trajectory corresponding to the high-β tokamak at fixed current.

These results can be conveniently summarized by plotting the trajectory of a flux-conserving evolution in a β_t vs. q_* space as illustrated in Fig. 6.21. Shown here is the equilibrium limit $\beta_t/\epsilon = 1/q_*^2$. No high-$\beta$ tokamak equilibrium can exist above this boundary. The flux-conserving evolution is obtained by eliminating H between Eqs. (6.138) and (6.142). The result is

$$\frac{\beta_t}{\epsilon} = \left[\frac{1}{q_*^2}\left(\frac{1}{q_*^2} - \frac{1}{q_a^2}\right)\right]^{1/2} \tag{6.146}$$

One can see that as β_t/ϵ increases, the value of q_* decreases in such a way that the equilibrium limit is never crossed. In contrast, the trajectory of a high-β tokamak with fixed q_* represents a vertical line in β_t/ϵ vs. q_* space. For sufficiently high β_t/ϵ the equilibrium boundary is always intersected.

Numerical Results. The basic features of the flux-conserving tokamak discussed above were first found in a more accurate and sophisticated analytic theory by Clarke and Sigmar (1977) and in a series of numerical computations by Dory and Peng (1977). The simple model presented here is in qualitative agreement with the analytic theory, although the scalings of I_0 and β_p with β_t are slightly different, presumably because of the gross approximations that have been made. An example of a series of numerically computed flux-conserving equilibria is illustrated in Fig. 6.22. Note the large shift and elongation of the flux surfaces as β_t increases. This is accompanied by a strong peaking of the current on the outside of the torus. Also plotted for comparison is the behavior of $\epsilon\beta_p$ and I_0 as a function of β_t which is proportional to the

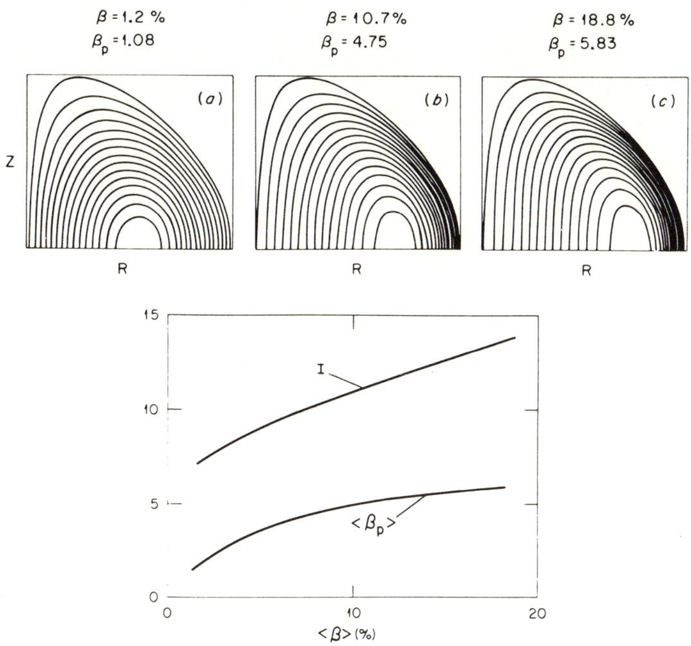

Figure 6.22. Sequence of numerically computed flux conserving equilibria for increasing values of β_t. Also shown are plots of $\epsilon\beta_p$ and I_0 vs. $\beta_t \sim H$. [From Dory and Peng (1977).]

heating parameter H. This should be compared with Fig. 6.20. In all cases, substantial increases in the toroidal current and vertical field are required to maintain equilibria as the plasma evolves.

Summary. A flux-conserving tokamak is an externally heated high-β tokamak in which the heating rate is slow compared to the MHD time and fast compared to the magnetic diffusion time. In this regime the plasma behaves as a perfect conductor, passing through a series of quasistatic equilibria as the heating phase evolves. The requirements of flux conservation, in addition to a simple modeling of the heating, serve to determine the functions $p(\psi)$ and $F(\psi)$ for each quasistatic state. The most interesting feature of the flux-conserving tokamak is that no equilibrium β_t limit exists, although when $\epsilon\beta_p \sim 1$, the profiles undergo a qualitative change, becoming quite peaked on the outside of the torus. The avoidance of the equilibrium limit is a result of the reduction in the ratio of B_v/I_0, implied by the requirements of flux conservation. In practice one cannot achieve arbitrarily high β_t by flux conservation because as β_t increases, q_* decreases, indicating the eventual onset of current-driven instabilities. Furthermore, at high β_t, modes driven by large pressure gradients will also be excited. Nevertheless, the fact that no equilibrium β_t limit is set by ideal MHD must be viewed as an optimistic result for tokamaks.

6.8. THE SPHERICAL TOKAMAK

The spherical tokamak is a recently proposed concept (Peng *et al.*, 1985) that attempts to exploit the aspect ratio dependence of the β limit inherent in the tokamak configuration; specifically the ohmic and high-β scalings which predict that for fixed q_*, $\beta_t \lesssim \epsilon^2/q_*^2$ and $\beta_t \lesssim \epsilon/q_*^2$, respectively. Consequently, tokamaks designed with ultratight aspect ratio (e.g., $R_0/a \sim 1.6$) realize two important benefits. First, tight aspect ratio implies a higher β_t limit, a definite advantage in achieving the densities and temperatures required for ignition. Second, tight aspect ratio by its very nature implies a compact design almost spherical in appearance. When viewed in the context of fusion reactor economics compactness is advantageous, as it corresponds to a more efficient mass utilization of the nuclear island: more kilowatts of electricity output per tonne of nuclear island structure. This lowers the initial capital investment to build the plant, thereby addressing one general criticism that has been made of fusion power: the relatively high anticipated capital cost.

Not surprisingly, there are also several problems associated with ultratight aspect ratio. First, there are limits to how tight an aspect ratio can be practically achieved since a certain minimum space is required in the center of the torus to accommodate the toroidal field coils, the shielding, and perhaps some blanket material. Second, with a very compact design the technology associated with the construction and maintenance of the device may be more difficult than for a "normal" tokamak. Third, large currents will have to be driven noninductively, a costly and physically difficult requirement.

The exact tradeoffs are difficult to assess at this time. Even so, the spherical tokamak is an interesting new physics concept worthy of further study. From the point of view of MHD equilibrium, the spherical tokamak, despite its compactness, still maintains toroidal axisymmetry and hence its equilibrium is described by the Grad–Shafranov equation.

Solution to the Grad–Shafranov Equation. As its name implies the spherical tokamak is an ultracompact device, spherical in appearance, that cannot be accurately approximated by means of an aspect ratio expansion. Radial pressure balance, toroidal force balance, and finite toroidicity all enter simultaneously in leading order. What is required is an exact solution of the unexpanded Grad–Shafranov equation, repeated here for convenience:

$$R\frac{\partial}{\partial R}\left(\frac{1}{R}\frac{\partial\psi}{\partial R}\right) + \frac{\partial^2\psi}{\partial Z^2} = -\mu_0 R^2\frac{dp}{d\psi} - F\frac{dF}{d\psi} \tag{6.147}$$

As in previous sections, one can develop intuition about a given concept by the use of special models which invoke simple choices for the free functions $p(\psi)$ and $F(\psi)$. The simplest choice that one can make that still retains the crucial physics of the spherical tokamak is $\mu_0 p' = -C$, $FF' = A$ with A and C constants. This class of equilibria was first introduced and investigated by Solev'ev (1968). The corresponding solution to the Grad–Shafranov equation is given by

$$\psi = -\frac{A}{2}Z^2 + \frac{C}{8}R^4 + c_1 + c_2R^2 + c_3(R^4 - 4R^2Z^2) \tag{6.148}$$

The first two terms are particular solutions while the last three are homogeneous solutions. The constants A and C are related to the plasma parameters β_t and q_*. The constants c_1, c_2, and c_3 are related to the geometry of the cross section.

Equation (6.148) can be cast in a more convenient form for identifying c_1, c_2, and c_3 as follows:

$$\psi = \frac{C\gamma}{8}[(R^2 - R_a^2)^2 - R_b^4] + \frac{C}{2}\left[(1-\gamma)R^2 - \frac{A}{C}\right]Z^2 \tag{6.149}$$

In this expression the constants γ, R_a, and R_b replace c_1, c_2, and c_3 and the flux function has been normalized so that $\psi(R, Z) = 0$ on the plasma surface.

The quantities R_a and R_b are identified by examining the dependence of ψ along the midplane $Z = 0$ as illustrated in Fig. 6.23. The point $R = R_a$, $Z = 0$ corresponds to the magnetic axis. The outer edge of the plasma is located at $R = (R_a^2 + R_b^2)^{1/2}$, $Z = 0$, which in terms of the standard tokamak notation corresponds to $(R_a^2 + R_b^2)^{1/2} = R_0 + a$. Similarly, the inner edge is located at $Z = 0$, $R = (R_a^2 - R_b^2)^{1/2} = R_0 - a$. These two relations can be solved simultaneously for R_a and R_b in terms of R_0 and a. The result is

$$\begin{aligned} R_a^2 &= R_0^2(1+\epsilon^2) \\ R_b^2 &= 2R_0^2\epsilon \end{aligned} \tag{6.150}$$

where $\epsilon = a/R_0$.

The remaining homogeneous constant γ is related to the plasma

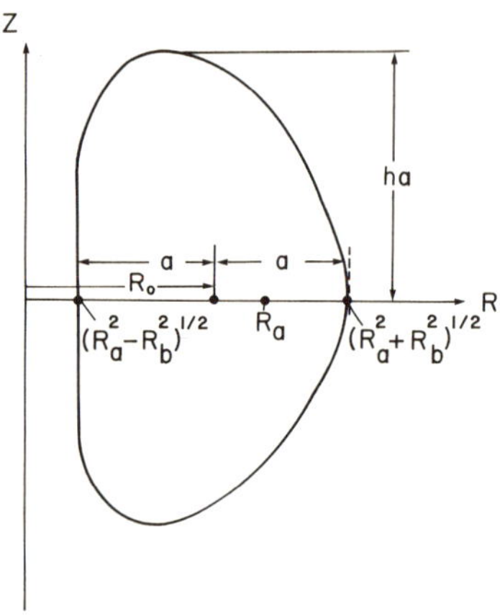

Figure 6.23. Geometry of the spherical torus.

elongation κ defined with Eq. (4.29). (Note that κ is closely related but not exactly equal to the height h shown in Fig. 6.23.) The relationship between γ and κ can be found analytically, but even with the simple model under consideration, the algebra quickly becomes cumbersome. Thus, strictly for the sake of analytic simplicity, the constant A is set to zero. This represents a pure vacuum toroidal field $B_\phi = B_0(R_0/R)$ and corresponds to "pure ohmic" operation. In the large aspect ratio limit ohmic scaling predicts $\beta_t \sim \epsilon^2$. When $\epsilon \sim 1$ the distinction between the ohmic and high-β regimes is not critical.

Even with the assumption $A = 0$, the evaluation of the physical quantities leads to complicated elliptic integrals which are functions of ϵ. As a further analytic simplification, all such integrals are expanded as a series in ϵ^2 keeping leading and first-order terms. This is clearly not very accurate when $\epsilon \to 1$, but it is sufficient to qualitatively and semiquantitatively predict the correct trends.

Returning to the evaluation of κ, a straightforward calculation making use of assumptions just described shows that

$$\gamma \approx \frac{\kappa^2}{1+\kappa^2} + O(\epsilon^4) \tag{6.151}$$

Having identified R_a, R_b, and γ, one can write a simplified expression for the flux function by introducing normalized coordinates x and y as follows:

$$\begin{aligned} R^2 &= R_a^2 + R_b^2 x \\ Z &= Z_0 y \end{aligned} \tag{6.152}$$

where the range of x is $-1 \leq x \leq 1$ and Z_0 is defined as

$$Z_0 = \frac{\kappa a}{(1 + \epsilon^2)^{1/2}} \tag{6.153}$$

The flux function is then given by

$$\psi = \psi_0[x^2 - 1 + (1 + \lambda x)y^2] \tag{6.154}$$

Here, $\lambda = 2\epsilon/(1 + \epsilon^2)$ and $\psi_0 = \gamma\epsilon^2 R_0^4 C/2$. The equation for the plasma surface, $\psi = 0$, has the form

$$y = \pm\left(\frac{1 - x^2}{1 + \lambda x}\right)^{1/2} \tag{6.155}$$

A set of surfaces is illustrated in Fig. 6.24 for $\epsilon = 0.6$.

Equilibrium β_t Limit. By making use of Eq. (6.155) one can calculate the maximum value of β_t possible for a given q_* in the regime of ohmic operation. It is of interest to focus on the effects of finite toroidicity. Specifically, one wants to learn how reliable the ohmic scaling result $\beta_t \sim \epsilon^2$ is as $\epsilon \rightarrow 1$; and if there are differences do they lead to higher or lower values of the β_t limit?

The relevant relations are obtained by evaluating β_t, q_*, and q_a as defined in Chapter 4. Straightforward algebra yields

$$\begin{aligned} \frac{\beta_t}{\epsilon^2} &\approx \frac{\hat{C}^2}{2}\left(\frac{\kappa^2}{1 + \kappa^2}\right)\left(1 - \frac{\epsilon^2}{8}\right) \\ q_* &\approx \frac{2}{\hat{C}}\left(1 + \frac{\epsilon^2}{8}\right) \\ q_a &\approx \frac{1}{\hat{C}}\left(\frac{1 + \kappa^2}{\kappa}\right)\left(1 + \frac{9}{4}\epsilon^2\right) \end{aligned} \tag{6.156}$$

where $\hat{C} = R_0^2 C/B_0$.

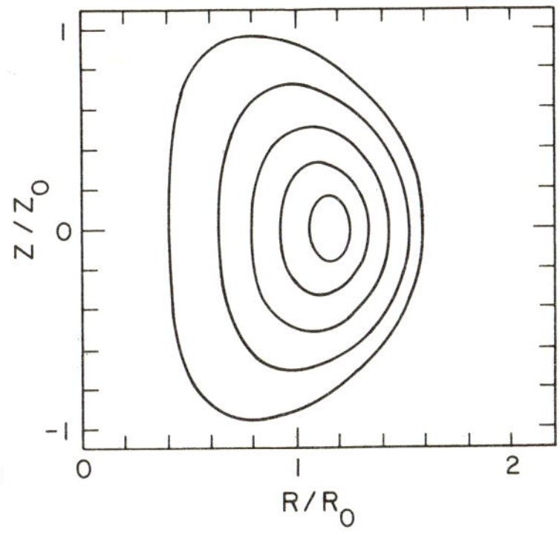

Figure 6.24. Flux surface plot for the analytic model of a spherical tokamak for $\epsilon = 0.6$.

Under the assumption that stability against current-driven kink modes is possible only above a critical value of q_*, then the maximum β_t occurs for the minimum q_* and is given by

$$\frac{\beta_t q_*^2}{\epsilon^2} \approx \frac{2\kappa^2}{1+\kappa^2}\left(1+\frac{\epsilon^2}{8}\right) \tag{6.157}$$

Observe that the large aspect ratio scaling is quite accurate even for finite values of ϵ. When ϵ is of order unity there is a slight additional gain (i.e., the $+\epsilon^2/8$ term) over and above the dominant ϵ^2 scaling of β_t. Elongation also helps, and for large κ there is a gain of a factor of 2 over the circular case.

A more optimistic result can be obtained by noting that

$$q_a \approx q_*\left(\frac{1+\kappa^2}{2\kappa}\right)\left(1+\frac{17}{8}\epsilon^2\right) \tag{6.158}$$

In the large aspect ratio limit $q_a \approx f(\kappa)q_*$. If it is now assumed that stability against kink modes is determined by q_a rather than q_*, then the β_t limit corresponding to the minimum q_a is given by

$$\frac{\beta_t q_a^2}{\epsilon^2} \approx \left(\frac{1+\kappa^2}{2}\right)\left(1+\frac{35}{8}\epsilon^2\right) \tag{6.159}$$

Equation (6.159) indicates that the gains in β_t associated with finite aspect ratio are substantial; for $\epsilon = 1/2$ the ratio β_t/ϵ^2 is more than double that predicted by the large aspect ratio expansion. Furthermore, the β_t limit increases continuously as κ increases.

One sees from the analysis just presented that qualitatively different results are possible for the spherical tokamak depending upon exactly which quantities are held fixed. In general, q_* is the appropriate parameter which describes the kink stability limit. However, in general one can expect the critical q_* to be a function of both κ and ϵ. This dependence, which has not as yet been established either theoretically or experimentally could lead to important changes in the behavior of the equilibrium β_t limit.

Recently, a spherical tokamak experiment has been proposed by the Oak Ridge National Laboratory. A numerically computed equilibrium for this experiment is illustrated in Fig. 6.25. The qualitative similarities between the numerical results and the simple analytic model are apparent.

Summary. In summary, the spherical tokamak offers the promise of higher β_t and compact construction in a tokamak system by designing up

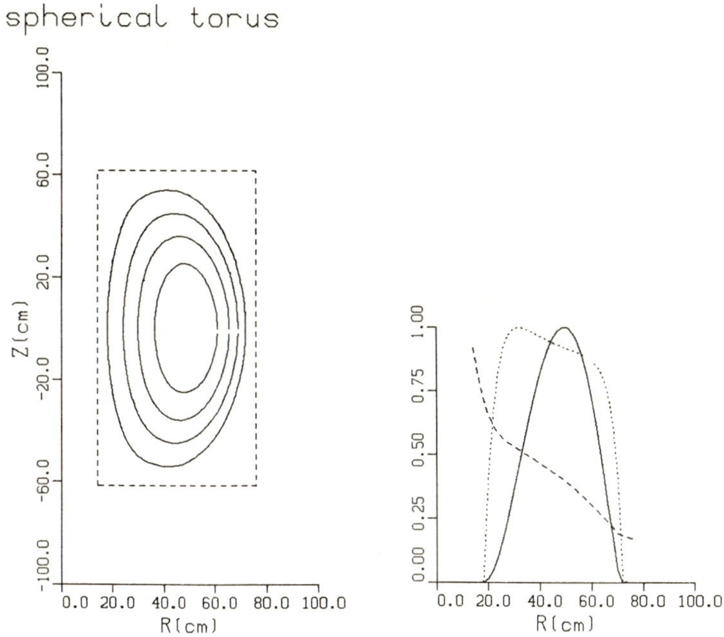

Figure 6.25. Numerically computed equilibrium for a spherical tokamak experiment. Courtesy M. Peng, Oak Ridge National Laboratory.

to the technological limits associated with tight aspect ratio. High β_t and compact design have important economic advantages for fusion power. These must be evaluated against the technological difficulties associated with compact construction.

6.9. TOKAMAKS WITH DIVERTORS

This section discusses the application of the Grad–Shafranov equation to the practical problem of the interface between a tokamak plasma and its external surroundings. In particular, it is shown how the various boundary conditions applicable to ideal MHD can be used to model actual experimental situations.

In typical experiments, the first wall surrounding the plasma is the vacuum chamber, usually a thin metallic shell of moderate resistivity. This construction allows the magnetic fields, especially the equilibrium vertical fields, to penetrate on a reasonably rapid time scale (on the order of milliseconds). The rapid penetration is important to achieve a satisfactory response time in the vertical field feedback control system and to keep the transient currents in the vacuum chamber at a tolerable level.

Any metallic wall in contact with the plasma forces the edge temperature to become very low, less than 100 eV. Consequently, as energy and particles are transported across the plasma they must be absorbed by the wall. If the vacuum chamber was indeed allowed to be in contact with the plasma, the wall, because of its fragile structure, would rapidly become damaged, perhaps developing burn spots and leaks. Repair or replacement of the vacuum chamber is a lengthy and costly process. Thus, actual experiments require an alternate type of interface between the plasma and its external surroundings.

The Mechanical Limiter. There are two main types of interfaces currently in use in tokamak experiments: the mechanical limiter and the magnetic divertor. A mechanical limiter is a robust piece of material, often made of tungsten, molybdenum, or graphite placed just inside the vacuum chamber. It protects the vacuum chamber from plasma bombardment and serves to define the edge of the plasma. Limiters come in various shapes. Two common examples are the poloidal ring limiter and the toroidal hoop limiter shown in Fig. 6.26.

The basic idea of limiter operation is as follows. To leading order, plasma ions spiral along magnetic field lines with a typical velocity equal to that of the ion thermal speed: $v_{\|} \sim V_{T_i}$. They slowly diffuse across the magnetic field on a collisional transport time: $v_r \ll V_{T_i}$. Hence, when a

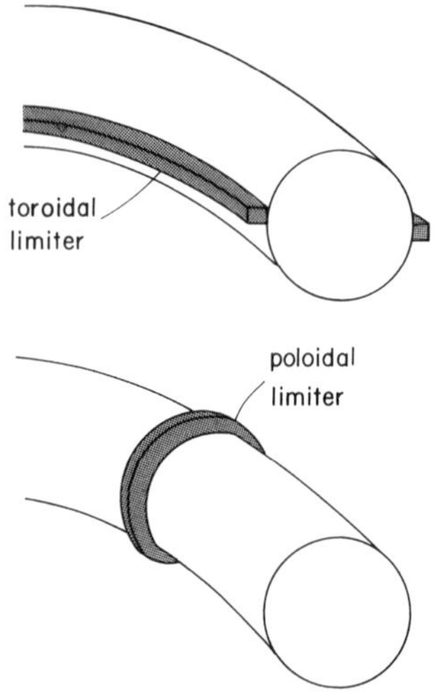

Figure 6.26. Schematic diagram of a toroidal and poloidal limiter.

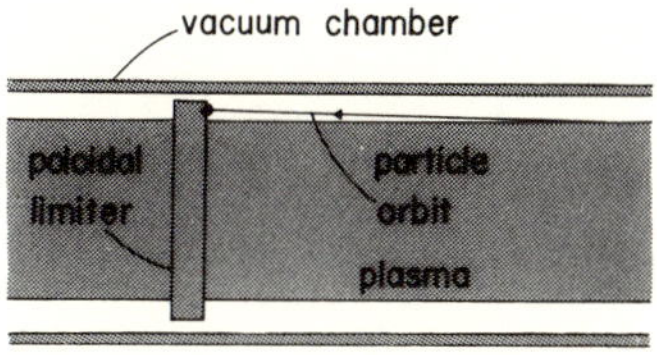

Figure 6.27. Orbit of a particle striking the limiter as it slowly tries to escape from the plasma surface.

particle crosses the edge of the plasma and enters the shadow region of the limiter, its large parallel velocity ensures that it will make contact with the limiter before undergoing sufficient collisions to allow it to diffuse into the vacuum chamber. See Fig. 6.27. In this way, the limiter protects the vacuum chamber and defines the edge of the plasma.

From the ideal MHD point of view, the boundary conditions appropriate to a system with a limiter correspond to those of a plasma isolated from a conducting wall by a vacuum region. Here, the vacuum region corresponds to the thin shadow region behind the limiter and the "resistive" vacuum chamber acts as a perfect conductor on the fast ideal MHD time scale.

Ideally one wants a limiter of sturdy and durable construction that can absorb a large flux of particles and energy without producing any adverse effects on the plasma. In practice, the plasma bombardment does cause damage. However, limiters are designed for much easier replacement than the vacuum chamber itself. Even so, there remains an important problem. The constant bombardment by the plasma causes some of the particles of the limiter surface to escape. Especially troublesome are neutral particles which can penetrate some distance into the plasma before becoming ionized. The high-Z impurities can lead to significant additional energy loss in the plasma through radiation. It is clearly important to minimize these losses.

A combination of clever design, clever material, and strong pumping is reasonably effective in achieving this goal in current experiments. It is not clear, however, how satisfactory limiters will be in ignition experiments and fusion reactors. In these applications the bombardment is more intense and extends over longer periods of time. Also, if the impurity level is too high, it may not be possible to achieve a high enough temperature to ignite.

The Magnetic Divertor. The issues just discussed have led to the use of magnetic divertors, a second alternative for the plasma interface. The idea of the divertor is to guide a narrower layer of magnetic lines away from the edge of the plasma as shown in Fig. 6.28. This is accomplished for a poloidal divertor by placing several coils near the surface to

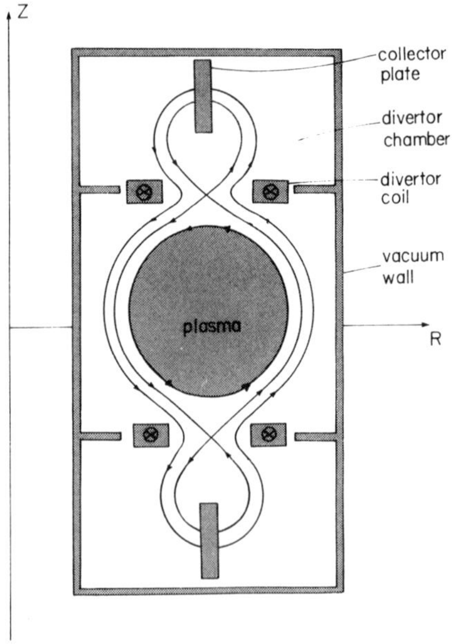

Figure 6.28. Schematic diagram of a double null poloidal divertor.

generate a local magnetic field which opposes the poloidal field in the plasma: the result is the formation of a separatrix. The magnetic lines just outside the separatrix are guided some distance into the divertor chamber where they emerge on a robust collector plate.

As particles diffuse across the magnetic field in the plasma they eventually enter the divertor layer. Once across the separatrix, the particles continue to spiral along the field line, ultimately being absorbed by the collector plate.

Although both alternatives protect the vacuum chamber, the divertor has several advantages over the limiter. Because the collector plate is remote from the plasma there is space available to spread out the magnetic lines. This produces a lower intensity of particles and energy bombarding the collector plate leading to a longer replacement time. Equally important, the remoteness of the collector plate makes it more difficult for impurities to migrate into the plasma. There are longer distances to travel and if, for instance, a neutral particle becomes ionized before or during the time it crosses the divertor layer on its way toward the plasma, its parallel motion then carries it back to the collector plate. Also, the large divertor chamber provides more access to pump out impurities. A further advantage of the divertor is that the plasma edge is not in direct contact with a solid material such as a limiter. Consequently, higher edge temperatures are possible, a circumstance that can lead to improved confinement times.

The disadvantages of having a divertor are that such experiments are larger and more complex in order to accommodate the divertor chamber. This translates into higher costs which must be balanced against the anticipated improvements in performance.

In terms of ideal MHD, the boundary conditions corresponding to a tokamak with a divertor are accurately modeled by solving the free boundary problem assuming the plasma is surrounded by external coils. Poloidal divertor systems maintain toroidal axisymmetry and therefore are described by the Grad–Shafranov equation. Numerically obtained solutions are of great importance in the design and operation of tokamak experiments. Examples of tokamaks with divertors are the ASDEX experiment in the Max Planck Laboratory in West Germany and the PDX experiment at Princeton. A diagram of the flux surfaces corresponding to the PDX experiment is illustrated in Fig. 6.29. By clever use of the divertor, the ASDEX group has discovered the "H mode" of operation in auxilliary heated tokamaks, a regime where the energy confinement time is enhanced substantially over that of limiter operation.

Applications such as this demonstrate one reason why MHD

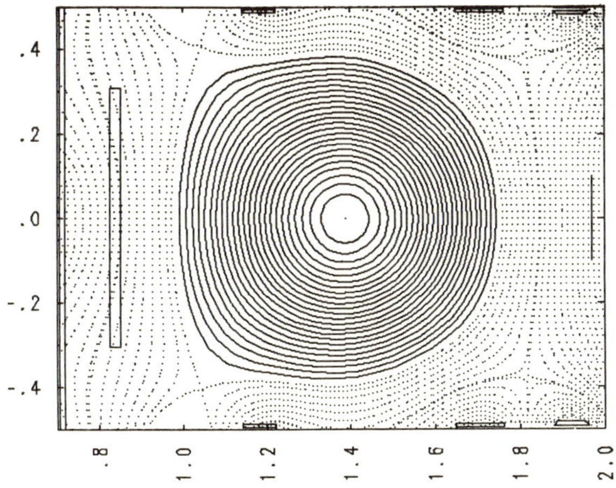

Figure 6.29. Numerically computed equilibrium for the poloidal divertor experiment (PDX) at the Princeton Plasma Physics Laboratory. Courtesy J. Manickam, Princeton Plasma Physics Laboratory.

equilibrium studies are so widely used and are such a valuable tool in the magnetic fusion program.

6.10. SUMMARY

This chapter demonstrates how the ideal MHD model can be used to investigate the equilibrium properties of two-dimensional, axisymmetric toroidal configurations. For such systems the equilibrium is described by a single nonlinear partial differential equation known as the Grad–Shafranov equation. This equation provides a complete description of radial pressure balance and toroidal force balance, and can be used to derive basic information pertaining to equilibrium β limits and the figures of merit for stability. A number of concepts are discussed, starting with large aspect ratio circular devices and ultimately evolving to compact noncircular configurations. A summary of these systems follows.

The Reversed Field Pinch. The RFP is a large aspect ratio circular configuration with $\beta_p \sim \beta_t \sim 1$, $q \sim \epsilon \ll 1$. This system provides radial pressure balance primarily by the toroidal current, similar to a Z pinch. Toroidal force balance is achieved by flux compression against a conducting wall. The combination of flat central pressure profile and B_ϕ reversal at the edge of the plasma provides stability against pressure-driven modes. The low value of $q_* \sim \epsilon$, however, is potentially dangerous for current-driven kink modes. These are suppressed by requiring a conducting wall at the edge of the plasma.

The Ohmically Heated Tokamak. The ohmic tokamak is a large aspect ratio circular device with $\beta_t \sim \epsilon^2$, $\beta_p \sim 1$, $q \sim 1$. Here, radial pressure balance, toroidal force balance, and plasma heating are provided by the toroidal current. The primary MHD function of the large and costly toroidal field is to provide stability. Because of the long pulse duration of tokamak experiments, toroidal force balance is accomplished by means of a vertical field rather than a conducting wall. Stability is achieved by a combination of shear, favorable magnetic well, and high q_*. The requirement of high q_* eliminates the need for a conducting wall.

High-β Tokamak. The high-β tokamak is a large aspect ratio configuration that attempts to achieve higher values of β_t than are possible in an ohmically heated tokamak. It accomplishes this by confining plasma in a small diamagnetic depression in the toroidal field, formed by the application of a large amount of auxiliary power; radial pressure balance is similar to that of a θ pinch. In the high-β regime $\beta_t \sim \epsilon$, $\beta_p \sim 1/\epsilon$, and $q \sim 1$. Thus, the toroidal field is required for both radial pressure balance and stability, while the poloidal field provides toroidal force balance and a small amount of ohmic heating. Stability is again provided by a combination of shear, favorable magnetic well, and high q_*. High-β tokamaks have the interesting property of exhibiting an equilibrium β_t limit at fixed q_* (i.e., fixed I) corresponding to a separatrix moving onto the plasma surface. The equilibrium limit scales as $\beta_t \lesssim \epsilon/q_*^2$.

Noncircular Tokamaks. The achievable β_t values in both ohmically heated and high-β tokamaks can be further raised by allowing the possibility of a noncircular cross section. In particular, elongated cross sections allow higher values of I and J_ϕ for the same q_*. This leads to a higher equilibrium β_t limit, larger ohmic heating, and improved confinement. However, there is a limit to the elongation of the cross section, typically $\kappa \lesssim 2$, set by the onset of axisymmetric vertical instabilities.

The Flux-Conserving Tokamak. The flux-conserving tokamak is a special subclass of high-β tokamak equilibria which attempts to model the actual evolution of a given tokamak discharge. When the auxiliary heating is applied at a rate slow compared to ideal MHD inertial effects but fast compared to resistive diffusion, the plasma evolves as a sequence of quasistatic perfectly conducting equilibria. Perfect conductivity implies that the $q(\psi)$ profile is preserved during the evolution. Since q_a in particular remains unchanged, a separatrix cannot form on the plasma surface. Thus, the flux-conserving tokamak does not exhibit an equilibrium β_t limit. Even so, in order to preserve the $q(\psi)$ profile, q_* decreases as β_t is raised. One must then be careful to avoid the onset of current-driven kink instabilities.

The Spherical Tokamak. The spherical tokamak attempts to achieve the highest β_t possible in a tokamak system by means of a compact, ultratight aspect ratio design, $R_0/a \sim 1.6$. The idea is valid in both the high-β and ohmic regimes of operation, leading to higher values of β_t since $\epsilon \rightarrow 1$. Depending upon exactly which parameters are held fixed, the gains can be even greater than those predicted by the simple large aspect ratio scalings $\beta_t \sim \epsilon$ or $\beta_t \sim \epsilon^2$.

Tokamaks with Divertors. An important practical problem in the operation of tokamak experiments is the interface between the edge of the plasma and its external surroundings. In most cases the plasma boundary is defined by either a mechanical limiter or a magnetic divertor. The divertor offers a more flexible means of removing particles from the plasma, protecting the vacuum chamber and preventing impurities from entering the plasma. However, divertors are more complicated and costly. Both limiter and divertor systems can be modeled by the Grad–Shafranov equation.

From the wide variety of realistic equilibria described by the Grad–Shafranov equation, one can appreciate why the ideal MHD model is so widely used in the fusion program.

REFERENCES

Bodin, H. A. B., and A. A. Newton (1980). *Nucl. Fusion* **20,** 1255.

Butt, E. P., R. Curruthers, J. T. D. Mitchell, R. S. Pease, P. C. Thonemann, M. A. Bird, J. Blears, and E. R. Hartill (1958). In *Proceedings of the Second United Nations International Conference on the Peaceful Uses of Atomic Energy,* United Nations, Geneva, Vol. 32, p. 42.

Clarke, J. F., and D. J. Sigmar (1977). *Phys. Rev. Lett.* **38,** 70.

Dory, R. A., and Y.-K. M. Peng (1977). *Nucl. Fusion* **17,** 21.

Grad, H., and H. Rubin (1958). In *Proceedings of the Second United Nations International Conference on the Peaceful Uses of Atomic Energy,* United Nations, Geneva, Vol. 31, p. 190.

Hass, F. A. (1972). *Phys. Fluids* **15,** 141.

Mukhovatov, V. S., and V. D. Shafranov (1971). *Nucl. Fusion* **11,** 605.

Peng, M., S. K. Boroski, B. R. Dalton, G. E. Gorker, W. R. Hamilton, M. H. Kunselman, V. D. Lee, R. L. Reid, E. C. Selcow, N. A. Uckan (1985). In *Proceedings of the Topical Conference on the Technology of Fusion Energy,* San Francisco, California, p. 19.

Robinson, D. C. (1971). *Plasma Phys.* **13,** 439.

Shafranov, V. D.(1960). *Sov. Phys.-JETP* **26,** 682.

Shafranov, V. D. (1966). In *Reviews of Plasma Physics,* edited by M. A. Leontovich, Consultants Bureau, New York, Vol. II.

Shafranov, V. D. (1971). *Plasma Phys.* **13,** 757.

Solev'ev, L. S. (1967). In *Reviews of Plasma Physics,* edited by M. A. Leontovich, Consultants Bureau, New York, Vol. III.

Solov'ev, L. S. (1968). *Sov. Phys.-JETP* **26,** 400.

Stratton, J. A. (1941). *Electromagnetic Theory,* McGraw-Hill, New York.

GENERAL REFERENCES

Reversed Field Pinch

Bodin, H. A. B., and A. A. Newton (1980). *Nucl. Fusion* **20,** 1255.

Robinson, D. C. (1971). *Plasma Phys.* **13,** 439.

Tokamaks (General)

Bateman, G. (1978). *MHD Instabilities,* MIT, Cambridge, Massachusetts.
Furth, H. P. (1975). *Nucl. Fusion* **15,** 487.
Shafranov, V. D.. (1966). In *Reviews of Plasma Physics,* edited by M. A. Leontovich, Consultants Bureau, New York, Vol. II.
Wesson, J. A. (1978). *Nucl. Fusion* **18,** 87.

High-β Tokamak

Callan, J. D., and R. A. Dory (1972). *Phys. Fluids* **15,** 1523.
Dory, R. A., and Y.-K. M. Peng (1977). *Nucl. Fusion* **17,** 21.
Freidberg, J. P., and F. A. Haas (1973). *Phys. Fluids* **16,** 1909.
Shafranov, V. D. (1971). *Plasma Phys.* **13,** 757.

Noncircular Tokamaks

Laval, G., and R. Pellat (1973). In *Controlled Fusion and Plasma Physics,* Proceedings of the Sixth European Conference, Moscow, Vol. II, p. 640.
Laval, G., R. Pellat, and J. L. Soule (1972). In *Controlled Fusion and Plasma Physics,* Proceedings of the Fifth European Conference, Euratom CEA, Grenoble, Vol. I, p. 25.
Laval, G., R. Pellat, and J. L. Soule (1974). *Phys. Fluids* **17,** 835.
Solov'ev, L. S., V. D. Shafranov, and E. I. Yurchenko (1969). In *Plasma Physics and Controlled Nuclear Fusion Research 1968,* IAEA, Vienna, Vol. I, p. 173.

Flux-Conserving Tokamak

Clarke, J. F., and D. J. Sigmar (1977). *Phys. Rev. Lett.* **38,** 70.
Dory, R. A., and Y.-K. M. Peng (1977). *Nucl. Fusion* **17,** 21.

PROBLEMS

6.1. a. Derive a two-dimensional Grad–Shafranov equation for a "straight stellarator." Assume all quantities are functions of the two independent variables r and $\alpha \equiv l\theta + hz$.
b. Show that in the limit where the helical field amplitude approaches zero [i.e., $\psi(r, \alpha) \to \psi(r)$], the general one-dimensional radial pressure balance relation for the screw pinch is obtained.

6.2. Using the ideal MHD model derive a Grad–Shafranov equation for stationary axisymmetric toroidal equilibrium in which there is a steady-state velocity $V_\phi(R, Z)$ flowing in the toroidal direction. For simplicity assume

$\rho = \rho(\psi)$. Hints: Show that

$$\mathbf{B} = B_\phi \mathbf{e}_\phi + \frac{1}{R}\nabla\psi \times \mathbf{e}_\phi$$

$$\mu_0 \mathbf{J} = -\frac{1}{R}\Delta^*\psi \mathbf{e}_\phi + \frac{1}{R}\nabla(RB_\phi) \times \mathbf{e}_\phi$$

$$V_\phi / R = \Omega(\psi)$$

$$p - \tfrac{1}{2}\rho\Omega^2 R^2 = \Pi(\psi)$$

$$RB_\phi = F(\psi)$$

$$\Delta^*\psi = -FF' - \mu_0 R^2 \Pi' - \frac{\mu_0 R^4}{2}(\rho\Omega^2)'$$

6.3. (Due to R. R. Parker, 1986.) Define $I_\phi(\psi)$ as the total ϕ-directed current within a flux surface labeled by ψ for an axisymmetric toroidal equilibrium. The Grad–Shafranov equation for this case can be put in the form $\Delta^*\psi = f(I'_\phi, p', B_p, R)$. Determine f.

6.4. Consider the transformation of coordinates (R, Z, ϕ') to (r, θ, ϕ) defined by

$$R = R_0 + r\cos\theta$$

$$Z = r\sin\theta$$

$$\phi' = -\phi$$

where (R, Z, ϕ') represents a standard cylindrical coordinate system. Show that the following relations hold in the (r, θ, ϕ) system:

$$\nabla\psi = \nabla_c\psi$$

$$\nabla \cdot \mathbf{B} = \nabla_c \cdot \mathbf{B} + \frac{\mathbf{e}_R \cdot \mathbf{B}}{R}$$

$$\nabla \times \mathbf{B} = \nabla_c \times \mathbf{B} - \frac{B_\phi}{R}\mathbf{e}_Z$$

$$\Delta^*\psi = \nabla_c^2\psi - \frac{\mathbf{e}_R \cdot \nabla\psi}{R}$$

Here

$$\nabla_c = \mathbf{e}_r \frac{\partial}{\partial r} + \frac{\mathbf{e}_\theta}{r}\frac{\partial}{\partial\theta} + \frac{\mathbf{e}_\phi}{R}\frac{\partial}{\partial\phi}$$

and

$$\mathbf{e}_R = \cos\theta\,\mathbf{e}_r - \sin\theta\,\mathbf{e}_\theta$$

$$\mathbf{e}_Z = \sin\theta\,\mathbf{e}_r + \cos\theta\,\mathbf{e}_\theta$$

$$\mathbf{e}_{\phi'} = -\mathbf{e}_\phi$$

6.5. Show that the value of the safety factor q evaluated on the magnetic axis, for an arbitrary axisymmetric torus (i.e., no small a/R expansion, no assumption of circular flux surfaces) is given by

$$q_0 = \Delta_0 B_\phi / (\psi_{rr}\psi_{\theta\theta})^{1/2}$$

where B_ϕ and ψ are to be evaluated on the magnetic axis, $r = \Delta_0$, $\theta = 0$. Hint: Taylor expand $\psi(r, \theta)$ about a coordinate system centered on the magnetic axis and consider the limit $r' \ll \Delta_0$. See the accompanying diagram.

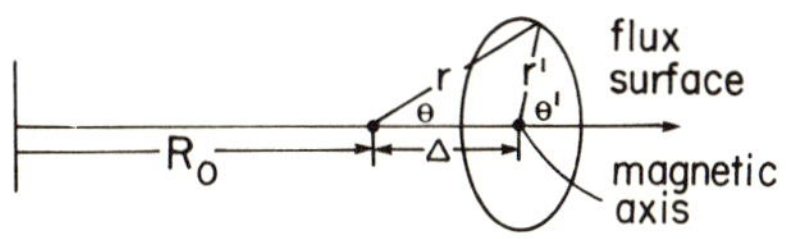

6.6. Consider a static cylindrically symmetric reversed field pinch characterized by $p(r)$, $B_\theta(r)$, $B_z(r)$. In realistic experimental situations Ohm's law should include an anisotropic resistivity: $\mathbf{E} = \eta_\parallel \mathbf{J}_\parallel + \eta_\perp \mathbf{J}_\perp$ with $\eta_\parallel$ and $\eta_\perp$ assumed constant for simplicity.

a. In steady state show that $E_\theta = 0$ and $E_z = E_0 = \text{const}$.

b. Show that B_z satisfies the differential equation

$$\frac{dB_z}{dr} = \frac{\eta_\parallel - \eta_\perp}{\eta_\parallel \eta_\perp} \frac{B_\theta B_z}{B_\theta^2 + B_z^2} \mu_0 E_0$$

c. Show that under the assumptions made, *no* solutions exist with B_z reversed for any positive values of $\eta_\parallel$, $\eta_\perp$. What does this suggest to you about the steady-state operation of an RFP?

6.7. Starting with the general one-dimensional radial pressure balance relation

$$\frac{d}{dr}\left(p + \frac{B_\phi^2}{2\mu_0}\right) + \frac{B_\theta}{\mu_0 r}\frac{d}{dr}(rB_\theta) = 0$$

and the equation for the perturbed flux

$$\nabla^2 \psi_1 = -\psi_1 \frac{d^2}{d\psi_0^2}\left(\mu_0 R_0^2 p + \frac{F^2}{2}\right) - \left(2\mu_0 R_0 r \frac{dp}{d\psi_0} - \frac{1}{R_0}\frac{d\psi_0}{dr}\right)\cos\theta$$

show that the solution for ψ_1 is given by $\psi_1(r, \theta) = \psi_1(r)\cos\theta$, where

$$\psi_1(r) = B_\theta(r)\int_r^b \frac{dx}{xB_\theta^2(x)}\int_0^x \left[2\mu_0 y^2 \frac{dp(y)}{dy} - yB_\theta^2(y)\right] dy$$

Assume there is a perfectly conducting circular wall located at $r = b$.

6.8. Show that in an ohmically heated tokamak the normal curvature, correct to order ϵ^2, is given by

$$\kappa_n(r, \theta) = -\frac{B_{\theta 1}^2}{rB_0^2} - \left(1 - \frac{r\cos\theta}{R_0}\right)\frac{\cos\theta}{R_0} + \frac{\Delta \sin^2\theta}{rR_0}$$

Here, $\Delta(r) = -\psi_1(r)/\psi_0'(r)$ is the shift of the flux surfaces.

6.9. Evaluate the magnetic well function for an ohmically heated tokamak as follows. Using the inverse aspect ratio expansion show that on a plasma flux surface $r = r_0 + \Delta(r_0)\cos\theta$,

$$B^2 \approx B_0^2\left[1 - 2\frac{r_0}{R_0}\cos\theta - \left(\frac{2\Delta}{R_0} - \frac{3r_0^2}{R_0^2}\right)\cos^2\theta + \frac{2B_{\phi 2}}{B_0} + \frac{B_{\theta 1}^2}{B_0^2}\right]$$

Here $\Delta(r_0) = -\psi_1(r_0)/\psi_0'(r_0)$ is the flux surface shift. From this result show that the magnetic well $W(r_0)$ is given by

$$W = -\frac{B_{\theta 1}^2}{B_0^2} - \frac{r_0}{2R_0}\left[\frac{(r_0^2\Delta)'}{r_0}\right] + \frac{r^2}{2R_0^2}$$

6.10. Show that Shafranov's formula for the vertical field required to hold a tokamak in toroidal force balance

$$B_v = \frac{\mu_0 I}{4\pi R_0}\left(\beta_p + \frac{l_i - 3}{2} + \ln\frac{8R_0}{a}\right)$$

can be written in the following more intuitive form:

$$2\pi R_0 I_0 B_v = \frac{I_0^2}{2}\frac{\partial}{\partial R_0}(L_e + L_i) + 4\pi^2\int_0^a\left(p - \frac{B_0 B_{\phi 2}}{\mu_0}\right)r\,dr$$

where L_e and L_i are the external and internal inductances associated with the plasma current. Hint: Make use of the fact that L_i is linearly dependent on R_0.

6.11. Consider an ohmically heated tokamak plasma held in equilibrium by a vertical field B_v. It is desired to measure β_p, l_i, and $\Delta(a)$ by placing various magnetic probes just outside the vacuum chamber, which for simplicity is assumed electrically insulating, and whose radius is c. See accompanying diagram.

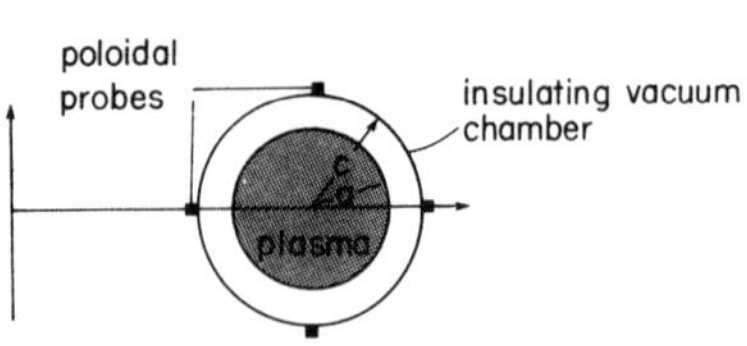

poloidal probe geometry

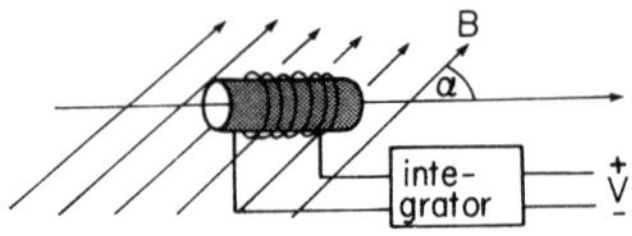

poloidal probe

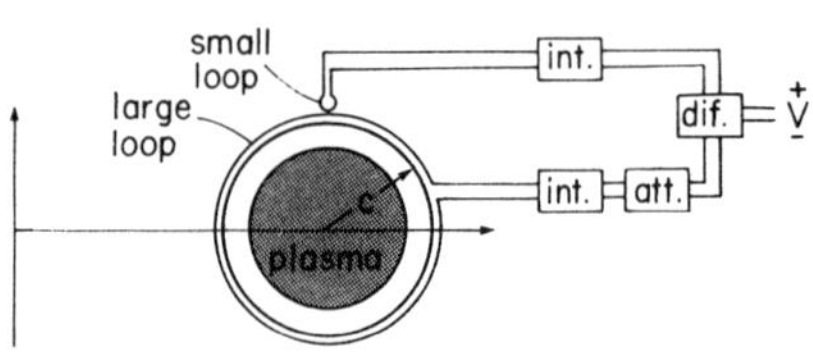

diamagnetic probe geometry

a. Consider an individual probe as illustrated. Assume the probe has N turns and cross-sectional area A. The **B** field makes an angle α with the axis of the probe. What signal appears on the output of the integrator?

b. Show that the toroidal correction to the flux function $\psi_1(r, \theta)$, evaluated at a radius r in the vicinity of the vacuum chamber, is given by

$$\psi_1(r, \theta) = \frac{I_0}{2\pi}\left[\frac{r^2 - a^2}{2r}\left(\beta_p + \frac{l_i - 1}{2}\right) + \frac{r}{2}\ln\frac{r}{a} - \Delta(a)\frac{R_0}{r}\right]\cos\theta$$

c. Calculate the difference in signals from two "B_θ" probes located at $r = c,\ \theta = 0$ and $r = c,\ \theta = \pi$.

d. Calculate the signal on a "B_r" probe located at $r = c,\ \theta = \pi/2$.

e. Assuming $c = a(1 + \delta)$ with $\delta \ll 1$, show how the "B_r" and "B_θ" signals can be combined to give separate measurements of $\Delta(a)$ and $\beta_p + l_i/2$.

f. Consider now a set of "diamagnetic"probes as shown in the diagram. This system consists of a small B_ϕ probe located at $r = c,\ \theta = \pi/2$ and a large single-turn loop of radius c, surrounding the vacuum chamber.
   1. By what factor must the large loop signal be attenuated in order to produce a null signal at the output of the "difference" circuit when no plasma is present?
   2. Calculate the difference signal V when plasma is present and show how it can be used to determine β_p.

g. Which of these measurements do you think might be difficult in a high-field experiment like Alcator C? Explain.

6.12. Consider a low β ohmically heated tokamak. Probe measurements on the outside of the discharge indicate that

$$\beta_p + \frac{l_i}{2} = 0.75$$

Profile measurements indicate peaked densities such that

$$n(r) = CT^2(r), \qquad C = \text{const}$$

In addition the plasma diamagnetism is nonzero and for simplicity is modeled as

$$\frac{B_{\phi 2}}{B_0} = \lambda\frac{\mu_0 p}{B_0^2}$$

with λ a constant, yet to be determined. The above data correspond to the "steady state" flattop portion of the current in an ohmic discharge. Assume Spitzer resistivity: $\eta = KT^{-3/2}$ with $T_e = T_i = T/2$ and $K = 8.4 \times 10^{-8}$, T in keV, η in Ω m.

a. Calculate the profile $B_\theta(r)/B_\theta(a)$.

b. If the discharge is operating just at the onset of sawtooth oscillations (i.e., $q_0 = 1$) calculate $q(a)$ assuming $J_\phi(a) = 0$.

c. Calculate β_p and the diamagnetic constant λ. Is the discharge diamagnetic or paramagnetic?

d. If $B_0 = 10\,\text{T}$, $R_0 = 0.64\,\text{m}$, and $a = 0.16\,\text{m}$, calculate the electron temperature if the toroidal loop voltage is 1.6 V. See the table at the end of the Problems section for helpful relations involving Bessel functions $J_0(x)$ and $J_1(x)$.

6.13. This problem demonstrates a practical application of MHD theory to the diagnosing of tokamak experiments. As a particular example, consider the geometry in the accompanying diagram. Assume that $R_0 = 0.64\,\text{m}$, $a = 0.16\,\text{m}$, $B_\phi = 8\,\text{T}$, and that the discharge is operating just at the onset of sawtooth oscillations (i.e., $q_0 = 1$). As a simple model of transport assume $n(r) \sim T(r) \sim J_\phi(r)$ and that $J_p(r) = 0$. Using the ohmically heated tokamak expansion calculate

a. $B_\theta(r)$;
b. $J_\phi(r)$;
c. $p(r)$;
d. I;
e. $q_a \approx q_*$;
f. β_t;
g. $B_t(0)$;
h. β_p;
i. l_i;
j. B_v (for zero shift of the outer surface);
k. Δ_0 (shift of the magnetic axis relative to the outer surface).

See the table at the end of the Problems section for helpful relations involving Bessel functions $J_0(x)$ and $J_1(x)$.

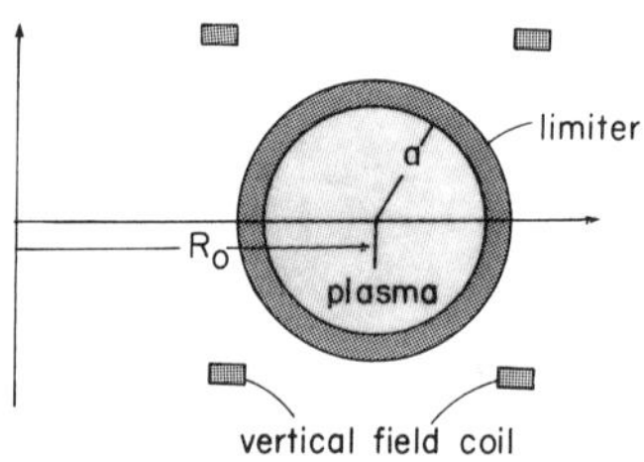

6.14. The MHD equations describing the equilibrium of a plasma with an external force derivable from a potential are given by

$$\mathbf{J} \times \mathbf{B} - \nabla p - \rho \nabla \phi = 0$$

$$\nabla \times \mathbf{B} = \mu_0 \mathbf{J} \qquad \nabla \cdot \mathbf{B} = 0$$

Assume (1) two-dimensional slab symmetry (all quantities are functions only of x and y), (2) $p = (T/m_i)\rho$ with T a constant, (3) all three components of $\mathbf{B}$ are nonzero, and (4) $\phi(x, y)$ is a known prescribed function.

a. Derive a two-dimensional Grad–Shafranov equation for this configuration: Hints: (1) Use A (the z component of vector potential) as the flux function; (2) show that $p \exp(m\phi/T) = \Pi(A)$, where $\Pi(A)$ is arbitrary; (3) show that $B_z = F(A)$, where $F(A)$ is arbitrary; (4) the final equation is $\nabla^2 A = -FF' - \Pi' \exp(m_i\phi/T)$.

b. Show that in the one-dimensional gravitational case ($\phi = -gx$ with $g = \text{const}$), the Grad–Shafranov equation reduces to the appropriate one-dimensional limit.

c. Consider now a two-dimensional case

$$m_i\phi/T = \epsilon f(x)y/L$$

with $\epsilon \ll 1$, $y \geq 0$, $-b \leq x \leq b$. Assume the currents and pressure are confined to the region $-a < x < a$ and that p and B_z are even about $x = 0$ while B_y is odd. See the accompanying figure. The region between a and b is a vacuum region and at $x = \pm b$ there is a perfectly conducting boundary. Assume $\epsilon \ll 1$ and expand $A(x, y) = A_0(x) + A_1(x, y) \ldots$ Find a solution for A_1 expressed in terms of integrals of the zero-order quantities.

d. Derive an expression for the shift of the last flux surface carrying current for the case $f(x) = -x/a$. Sketch the contour describing this surface, including the correction and give a physical description of the surface distortion resulting from $f(x)$.

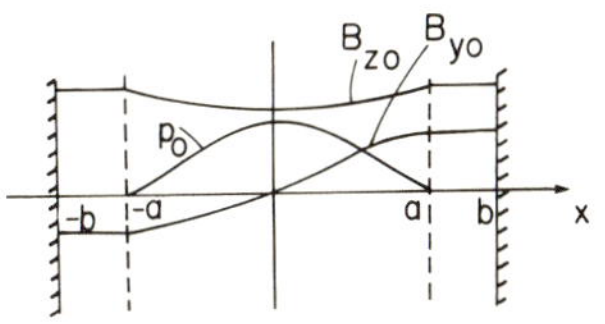

6.15. It has been known since the time of Newton that the Earth's gravitational field causes objects to "fall" towards the center of the Earth. Thus, if unsupported, a tokamak plasma in a 200-ms discharge would fall almost 0.2 m. during its lifetime. The purpose of this problem is to calculate the equilibrium of an axisymmetric toroidal tokamak under the influence of a downward gravitational force and to determine how effective the magnetic forces are in supporting the mass of the plasma against gravity.

a. Consider the axisymmetric geometry shown here. Assume there is an external potential $\phi(R, Z)$ and calculate the Grad–Shafranov equation for this system. For simplicity, assume $p = (T/m_i)\rho$ with $T = \text{const}$.

b. What choice of $\phi(R, Z)$ corresponds to a downward gravitational force?

c. Use this choice and treat ϕ as a small quantity; that is, assume gravitational effects enter in the same order as toroidal corrections. Expand the flux function as $\psi = \psi_0(r) + \psi_1(r, \theta) \ldots$ and find the contribution to $\psi_1(r, \theta)$ due to gravity. For simplicity, assume there is a conducting wall at $r = b$.

d. Calculate the influence of gravity on the shape and position of the last flux surface that carries current. Estimate the size of these corrections for typical fusion plasmas. Is gravity an important effect?

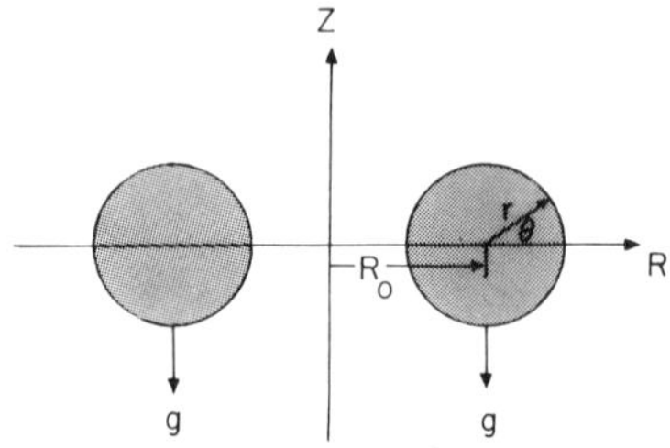

6.16. Consider the simple model of the high-β tokamak discussed in Section 6.5. Assume that as the plasma evolves it is constrained by stability considerations to operate at $q_0 = 1$, rather than at fixed I or in a flux-conserving mode. Show that when $q_0 = 1$, the maximum achievable plasma β_t does not occur at the equilibrium limit $\nu = 1$, but at some lower value. Calculate the value of ν and the corresponding value of $\beta_t q_0^2/\epsilon$ at the optimum condition. See accompanying figure.

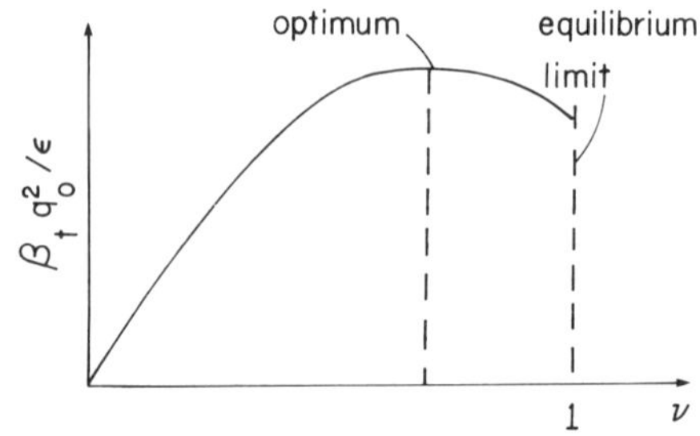

6.17. In Section 6.6 it was shown, using a simple elliptic model of a tokamak, that the highest equilibrium β_t occurs when $\nu = 1$ and $\kappa \to \infty$ assuming a fixed ϵ, q_*, and plasma volume V_0. At the limit, $\beta_t = 3\epsilon/q_*^2$ representing a gain of 300% over the circular case. Repeat the analyses assuming that the major radius is held fixed rather than the aspect ratio. Show that in this case there is an optimum elongation corresponding to the maximum β_t. Find the optimum κ and the corresponding values of β_t, J_ϕ, and I_0 and compare them to the circular case. Repeat the calculation for the choice of ν corresponding to ohmic operation.

6.18. Consider a simple model of a superconducting tokamak fusion reactor as illustrated in the accompanying figure.

a. Derive an expression for the total thermal power P_T associated with the 14-MeV neutrons.
b. Assume the wall loading $P_W \equiv P_T/S$ is fixed at the largest value consistent with the properties of the first wall material. Here, S is the plasma surface area. Also, assume the toroidal field coils operate at the maximum permissible value B_c which maintains superconductivity. Note that $B_\phi = B_c$ at $R = R_c$. Derive an expression for the required value of β_t/ϵ in terms of P_T, P_W, T_i, ϵ, ϵ_T, and B_c. Here, $\epsilon = a/R_0$, $\epsilon_T = (a + b)/R_0$, $T_i = T_e = T/2$, and β_t is defined as $\beta_t = 4\mu_0 n T_i/B_0^2$, with B_0 the field at the center of the plasma, $R = R_0$.

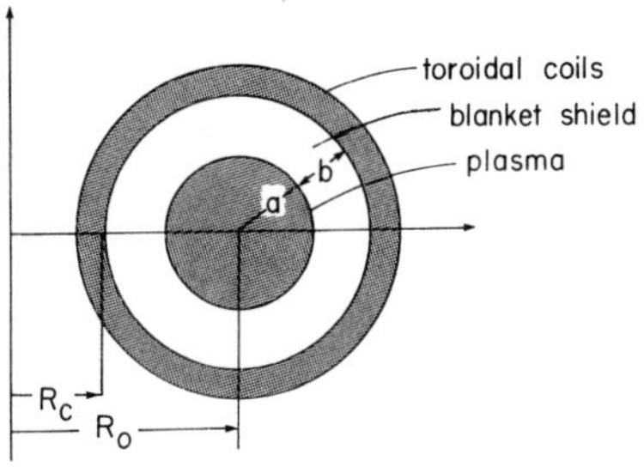

c. Assume P_T, P_W, T_i, B_c, and b are fixed. Show that the required β_t/ϵ exhibits a minimum as a function of ϵ. Derive an expression for the optimum aspect ratio.
d. At the optimum point calculate the values of ϵ, β_t/ϵ, β_t, a, and R_0 if $T_i = 15\,\text{keV}$, $P_T = 3500\,\text{MW}$, $P_W = 4\,\text{MW/m}^2$, $b = 1\,\text{m}$, and $B_c = 11$ T. Note: at $T_i = 15\,\text{keV}$, $\overline{\sigma v} = 3 \times 10^{-22}\,\text{m}^{-3}/\text{s}$ for D-T fusion reactions.

Helpful Relations for Bessel Functions $J_0(x)$, $J_1(x)$

$$J_0'' + \frac{J_0'}{x} + J_0 = 0$$

$$J_1'' + \frac{J_1'}{x} - \left(\frac{1}{x^2} - 1\right)J_1 = 0$$

$$J_1 = -J_0'$$

$$J_0 \approx 1 \qquad \text{as} \quad x \to 0$$

$$J_1 \approx \frac{x}{2} \qquad \text{as} \quad x \to 0$$

$$\int xJ_0^2\,dx = \frac{x^2}{2}(J_0^2 + J_1^2)$$

$$\int xJ_1^2\,dx = \frac{x^2}{2}(J_0^2 + J_1^2) - xJ_0J_1$$

$$\int \frac{J_0}{J_1}\,dx = \ln(xJ_1)$$

$$\int \frac{xJ_0^2}{J_1^2}\,dx = -\frac{xJ_0}{J_1} + 2\ln(xJ_1) - \frac{x^2}{2}$$

$$J_0(x_0) = 0 \to x_0 = 2.4 \text{ (first zero)}$$

$$J_1(x_0) = 0.52$$

7

Equilibrium
Three-Dimensional Configurations

7.1. INTRODUCTION

Chapter 6 has shown that toroidal MHD equilibria possessing axisymmetry can be constructed from the ideal MHD model. Specifically, the RFP and each of the different types of tokamaks are of this class. While radial pressure balance can be achieved by either θ-pinch or Z-pinch confinement, all axisymmetric configurations require sufficient toroidal current to produce toroidal force balance.

This chapter considers toroidal MHD configurations that are inherently three-dimensional. The motivation for investigating such obviously more complicated equilibria is that they possess the potential for providing toroidal confinement without the need of a net toroidal current. This leads to a number of advantages with respect to fusion power generation. First, the attractive feature of steady-state operation becomes more feasible since such configurations no longer require a toroidal current transformer. Second, with zero net current, one potentially dangerous class of MHD instabilities, the current-driven kink modes, is eliminated. Finally, three-dimensional configurations possess nondegenerate flux surfaces even in the absence of plasma pressure and plasma current. (Note that this is not true in an axisymmetric torus.) Consequently, there is a reasonable degree of external experimental control in establishing favorable profiles of shear and magnetic well for stability without relying on the cooperation of the plasma.

Force Balance Considerations. Although there is an enormous range of possible three-dimensional equilibria, the configurations of interest are accurately described as axisymmetric tori with superimposed helical fields; furthermore, they possess no net toroidal current. This last feature has interesting implications regarding the equilibrium force balance. First, as one might expect, the only option for radial pressure balance with $I_0 = 0$ is that corresponding to θ-pinch confinement. Second, the question of toroidal force balance is more subtle. In current-free systems there is no possibility of generating a $2\pi R_0 B_v I_0$ vertical field restoring force. Instead, two different, and not so obvious, restoring forces are developed: the helical sideband force and the toroidal dipole current force. Each is discussed in detail in Chapter 7.

Consequences of Three-Dimensionality. In the investigation of the properties of three-dimensional systems, the lack of symmetry adds major complications to the analysis. Recall that in two-dimensional systems it is possible to derive an exact, single, self-contained, equation describing the equilibria: the Grad–Shafranov equation. One then introduces various asymptotic expansions into this equation to analyze the different configurations of interest.

In systems without symmetry, no correspondingly general single equation can be derived. Consequently, one is forced to introduce asymptotic expansions before, rather than after, the formulation of the equilibrium problem. This requires more physical intuition into systems that are inherently more complicated, a situation that not unexpectedly can lead to considerable amounts of trial and error analysis.

As ludicrous as it might seem, it is often not clear in three-dimensional asymptotic expansions when the calculation "ends." For instance, in certain cases it may be necessary to expand the different vector components of $\mathbf{B}$ to different orders so that the system can be closed. Even more subtle, in many cases it is necessary to introduce homogeneous solutions at early stages of the analysis which are only determined several orders later by means of naturally arising periodicity constraints.

Fortunately, through years of experience, many of these difficulties have been resolved.

Solution Procedure. For the systems of interest, there now exist relatively well-established procedures for the reduction of the MHD equilibrium equations to tractable form. Qualitatively, the solution procedure consists of two steps. First, on the assumption that the flux contours $\psi(\mathbf{r}) = \text{const}$ are given, one calculates the magnetic fields and currents in terms of ψ to whatever order is desired. This is usually quite straightforward. Second, the flux function itself is determined in higher order by naturally arising periodicity constraints. Experience has shown that this last step is greatly simplified by making use of the so-called parallel current constraint. This constraint is a special form of the relation $\nabla \cdot \mathbf{J} = 0$, which has the virtue of requiring a relatively small amount of analysis to arrive at the final equation for ψ. A detailed discussion of the parallel current constraint, including its physical significance, is given in Section 7.2.

Applications. The configurations of interest naturally subdivide into two qualitatively different classes of three-dimensional equilibria: helical sideband equilibria and toroidal dipole current equilibria. The configurations are distinguished by the dominant toroidal-force-balance restoring force.

A superficial examination of these equilibria demonstrates many similarities. Both classes have θ-pinch radial pressure balance, large aspect ratio, helical fields, and zero net current. Even so, it is shown that the mathematical formulation describing each class is qualitatively different.

A simple guideline for distinguishing the two classes of equilibria is as follows. If the parameter

$$\beta q^2/\epsilon \gg 1 \tag{7.1}$$

the system corresponds to a helical sideband equilibrium. These systems are essentially linear systems, gently bent into a torus solely for the purpose of eliminating endloss. Toroidicity plays only a minor role in stability and has no significant effect on the establishment of equilibrium β limits. The Elmo bumpy torus and the Scyllac High-β Stellarator are configurations of this type. A general analysis of helical sideband equilibria, along with a detailed description of the Elmo bumpy torus, is presented in Sections 7.3 and 7.4, respectively.

If, on the other hand, the parameter

$$\beta q^2/\epsilon \lesssim 1 \tag{7.2}$$

then the configuration corresponds to a toroidal dipole-current equilibrium. In these systems toroidicity plays a major role in determining the overall behavior of the plasma; specifically, toroidicity is crucial in the creation of a favorable magnetic well and in the establishment of equilibrium β limits associated with toroidal force balance. Stellarators, heliotrons, and torsatrons are configurations of this class. A general description of toroidal dipole-current equilibria, including a detailed discussion of stellarators, heliotrons, and torsatrons, is given in Sections 7.5 and 7.6. Also, for historical accuracy these equilibria will hereafter be referred to as "stellarator equilibria" where "stellarator" now designates the general class of toroidal dipole configurations.

"Exact" Three-Dimensional Equilibria. A final point of interest concerns the question of whether or not exact three-dimensional equilibrium solutions exist to the ideal MHD equations. Not surprisingly, most practical calculations involve either numerical computations or asymptotic expansions. However, Grad (1967) has provided evidence that "exact" analytic three-dimensional equilibria do not exist; that is, infinitesimal changes in the boundary conditions can lead to finite changes in the topological structure of the flux surfaces, for example, the formation of fine-scale magnetic islands. The existence of such fine-scale structures adds considerable complexity to the mathematics, although such phenomena may be unimportant or dominated by other physics in an actual experiment. In this regard, asymptotic analysis may not only be mathematically simpler, but may represent the most useful and accurate procedure for practical applications. This is the point of view that is adopted here.

7.2. THE PARALLEL CURRENT CONSTRAINT

The first step in the analysis of three-dimensional MHD equilibria is the derivation of the parallel current constraint. This constraint is a general integral relationship obtained from the requirement that $\nabla \cdot \mathbf{J} = 0$ and is satisfied by all MHD equilibria. In systems with one or two degrees of symmetry the constraint is trivially satisfied. However, in three-dimensional systems the constraint yields nontrivial information. In particular, it corresponds to the simplest form of the solvability constraint that determines the leading-order flux surfaces.

Keep in mind that it is not necessary to use the constraint in the specific form discussed below. There are other straightforward, self-consistent expansion procedures which, when substituted into the MHD equations, ultimately yield the same constraint information. The virtues of the form given below are that it is general and that in all cases of practical interest it requires a minimum of algebra to arrive at the final equation for the flux surfaces.

The derivation of the parallel current constraint proceeds as follows. $\mathbf{J}$ is written as $\mathbf{J} = J_\parallel \mathbf{b} + \mathbf{J}_\perp$ and then substituted into the condition $\nabla \cdot \mathbf{J} = 0$, yielding

$$\mathbf{B} \cdot \nabla \frac{J_\parallel}{B} = -\nabla \cdot \mathbf{J}_\perp \tag{7.3}$$

$\mathbf{J}_\perp$ is obtained directly from the momentum equation: $\mathbf{J}_\perp = \mathbf{B} \times \nabla p / B^2$. This gives

$$\mathbf{B} \cdot \nabla \frac{J_\parallel}{B} = -\nabla p \cdot \nabla \times \frac{\mathbf{B}}{B^2} \tag{7.4}$$

A simple calculation shows that

$$\left(\nabla \times \frac{\mathbf{B}}{B^2}\right)_\perp = -\frac{\nabla p \times \mathbf{B}}{B^4} - \frac{2\boldsymbol{\kappa} \times \mathbf{B}}{B^2} \tag{7.5}$$

One then obtains

$$\frac{\partial}{\partial l}\left(\frac{J_\parallel}{B}\right) = \frac{2}{B^2}(\mathbf{b} \cdot \nabla p \times \boldsymbol{\kappa}) \tag{7.6}$$

where $\partial/\partial l \equiv \mathbf{b} \cdot \nabla$ is the derivative along the field line. This is the desired form of the $\nabla \cdot \mathbf{J} = 0$ relation from which the parallel current constraint is derived.

Note that there are a number of ways to rewrite the right-hand side of Eq. (7.6) using the equilibrium relations. The specific form shown has been chosen so that the first nonvanishing contribution is already of higher order in any of the standard expansions. It is for this reason that only a minimum of algebra is required to evaluate the constraint condition.

A practical form of the parallel current constraint is obtained by expanding the operator $\partial/\partial l$ in Eq. (7.6):

$$\frac{\partial}{\partial \phi}\left(\frac{J_\parallel}{B}\right) = \frac{R}{B_\phi}\left[\frac{2}{B}(\mathbf{b} \cdot \nabla p \times \boldsymbol{\kappa}) - \mathbf{B}_p \cdot \nabla\left(\frac{J_\parallel}{B}\right)\right] \tag{7.7}$$

The constraint, which is valid for closed line and ergodic systems, follows from the observation that the left-hand side of Eq. (7.7) averages to zero upon integrating once around the torus. Thus, any self-consistent MHD equilibrium must satisfy

$$C \equiv \int_0^{2\pi} \frac{R\, d\phi}{B_\phi}\left[\frac{2}{B}(\mathbf{b} \cdot \nabla p \times \boldsymbol{\kappa}) - \mathbf{B}_p \cdot \nabla\left(\frac{J_\parallel}{B}\right)\right] = 0 \tag{7.8}$$

This form is useful because all the quantities in the integrand are evaluated in an Eulerian coordinate system and not along the field line trajectories.

The parallel current constraint forces the flux surfaces to assume a shape that allows a physically acceptable $J_\parallel$ to flow, making the total current divergence-free. If the flux surfaces did not satisfy the constraint, the $J_\parallel$ required for $\nabla \cdot \mathbf{J} = 0$ would not lie in the surfaces.

7.3. HELICAL SIDEBAND EQUILIBRIA

A helical sideband equilibrium is a high-β large aspect ratio configuration in which the magnetic field consists primarily of (1) a large toroidal field, (2) moderate-size helical fields, and (3) no net toroidal current. Here, a helical field is one whose geometric dependence is characterized by $l\theta - N\phi$, where l represents the poloidal multipolarity of the field and $L_h = 2\pi R_0/N$ is the helical period.

Sideband Expansion. The basic physics issues associated with helical sideband equilibria are addressed by the "sideband expansion" (Blank *et al.*, 1969; Rosenbluth *et al.*, 1969). The appropriate ordering parameter is the ratio of the helical field amplitude to the main toroidal field

$$\delta = B_{\text{hel}}/B_\phi \ll 1 \tag{7.9}$$

In terms of δ the sideband expansion is given by

$$\begin{aligned} \beta_t &\sim 1 \\ ha &\sim 1 \\ a/R_0 &\sim \delta^2 \\ q &\sim 1 \end{aligned} \tag{7.10}$$

where $h = 2\pi/L_h = N/R_0$ is the pitch number of the helical field. Qualitatively, one sees that the equilibria of interest have the following properties. The β is very high: $\beta_t \sim 1$. The helical wavelength is comparable to the minor dimension of the plasma: $ha \sim 1$. The aspect ratio is very large: $R_0/a \sim 1/\delta^2$. There are many helical periods along the length of the torus: $N = hR_0 \sim 1/\delta^2$. The MHD safety factor is of order unity, but does not play a major role in the equilibrium: $q \sim 1$.

In the sideband expansion the fields are written as

$$\begin{aligned} \mathbf{B} &= B_{\phi 0}(r, \theta)\mathbf{e}_\phi + \mathbf{B}_1(r, \theta, \phi) + \cdots \\ p &= p_0(r, \theta) + p_1(r, \theta, \phi) + \cdots \end{aligned} \tag{7.11}$$

Here, p_0 and $B_{\phi 0}$ are the leading-order fields corresponding to θ-pinch radial pressure balance and $p_1 \sim \delta$. Also $\mathbf{B}_1 \sim \delta$ represents the applied and induced helical fields. It is crucial to assume that the leading ordering quantities are functions of r and θ indicating that the flux surfaces are in general not small deviations from a circle. The tempting hypothesis which assumes all zero-order quantities to be cylindrically symmetric (i.e., only functions of r) leads to an overdetermined system with no solutions.

The analysis proceeds as follows. First, assuming the leading-order flux surfaces are known [i.e., $p_0(r, \theta)$ is specified], it is then straightforward to calculate $B_{\phi 0}$, $\mathbf{B}_1$, and p_1 in terms of p_0. Since $a/R_0 \sim \delta^2$ toroidicity does not enter these calculations, which are identical to those in a straight helix. In the second part of the calculation, p_0 is determined by evaluation of the parallel current constraint and it is here that toroidicity enters. Remarkably, however, at no point does one require an explicit knowledge of the second-order, toroidally corrected, magnetic fields.

Calculation of the Fields. One now substitutes the sideband expansion into the MHD equilibrium equations. The zeroth-order equations involving $B_{\phi 0}$ and p_0 are easily shown to reduce to the familiar θ-pinch pressure balance relation

$$\nabla_\perp\left(p_0 + \frac{B_{\phi 0}^2}{2\mu_0}\right) = 0 \tag{7.12}$$

where $\nabla_\perp$ is the poloidal gradient

$$\nabla_\perp \equiv \mathbf{e}_r \frac{\partial}{\partial r} + \frac{\mathbf{e}_\theta}{r}\frac{\partial}{\partial \theta} \tag{7.13}$$

For convenience, a normalized form of p_0 is introduced:

$$\beta(r, \theta) \equiv \frac{2\mu_0 p_0(r, \theta)}{B_0^2} \sim 1 \tag{7.14}$$

where B_0 is the applied toroidal field in the midplane $R = R_0$. $B_{\phi 0}$ is related to β by

$$B_{\phi 0}(r, \theta) = B_0(1 - \beta)^{1/2} \tag{7.15}$$

The first-order fields are assumed to contain an arbitrary number of helical fields with different l values, but all with the same number of helical periods N. Multiple l values are essential because in practical applications the applied field often consists of a main helical field with significant, and sometimes deliberately generated, harmonics. Even more important, when a pure, single-l, helical field is applied, toroidal effects automatically couple $l \pm 1$ harmonics. Thus, the first-order fields can be expressed as

$$\begin{aligned} \mathbf{B}_1(r, \theta, \phi) &= \hat{\mathbf{B}}_1(r, \theta)e^{-iN\phi} + \hat{\mathbf{B}}_1^*(r, \theta)e^{iN\phi} \\ p_1(r, \theta, \phi) &= \hat{p}_1(r, \theta)e^{-iN\phi} + \hat{p}_1^*(r, \theta)e^{iN\phi} \end{aligned} \tag{7.16}$$

The amplitudes $\hat{\mathbf{B}}_1$ and $\hat{p}_1$ are functions only of r and θ and are calculated as follows. The first-order condition $\nabla \cdot \mathbf{B}_1 = 0$ yields an expression for $\hat{B}_{\phi 1}(r, \theta)$ in terms of the poloidal field $\hat{\mathbf{B}}_{p1}(r, \theta)$:

$$\hat{B}_{\phi 1} = -\frac{i}{h}\nabla_\perp \cdot \hat{\mathbf{B}}_{p1} \tag{7.17}$$

Next, the θ-pinch nature of the radial pressure balance is exploited by substituting $\mu_0\mathbf{J} = \nabla \times \mathbf{B}$ directly into the momentum equation and then expanding. The result is

$$\mathbf{B}_1 \cdot \nabla(B_{\phi 0}\mathbf{e}_\phi) + B_{\phi 0}\mathbf{e}_\phi \cdot \nabla\mathbf{B}_1 - \nabla(p_1 + B_{\phi 0}B_{\phi 1}) = 0 \tag{7.18}$$

It is now convenient to decompose Eq. (7.18) into vector components by forming the operations $\mathbf{e}_\phi \cdot \nabla \times$ and $\mathbf{e}_\phi \times$. The operation $\mathbf{e}_\phi \cdot \nabla \times$ yields

$$\mathbf{e}_\phi \cdot \nabla \times (B_{\phi 0}\hat{\mathbf{B}}_{p1}) = 0 \tag{7.19}$$

which implies that $\hat{\mathbf{B}}_{p1}$ can be written as

$$\hat{\mathbf{B}}_{p1}(r, \theta) = \frac{B_0}{(1-\beta)^{1/2}}\nabla_\perp V \tag{7.20}$$

Here, $V(r, \theta)$ is a potential function which assumes the role of the primary dependent variable. Ultimately, all of the first-order fields are expressed in terms of V and β.

Forming the operation $\mathbf{e}_\phi \times$ on Eq. (7.18) and then substituting Eq. (7.20) leads to the relationship

$$\mathbf{e}_\phi \times \nabla_\perp(\mu_0\hat{p}_1 + B_{\phi 0}\hat{B}_{\phi 1} + ihB_0^2 V) = 0 \tag{7.21}$$

The poloidal components of pressure balance are redundant, yielding only the single piece of information

$$\mu_0\hat{p}_1 + B_{\phi 0}\hat{B}_{\phi 1} = -ihB_0^2 V \tag{7.22}$$

The third and remaining piece of independent information contained in Eq. (7.18) is obtained by forming the dot product with $\mathbf{e}_\phi$. This gives

$$\hat{\mathbf{B}}_{p1} \cdot \nabla B_{\phi 0} - ihB_{\phi 0}\hat{B}_{\phi 1} + ih(\mu_0\hat{p}_1 + B_{\phi 0}\hat{B}_{\phi 1}) = 0 \tag{7.23}$$

Substituting for $B_{\phi 0}$, $\hat{\mathbf{B}}_{p1}$, $\hat{B}_{\phi 1}$, and $(\mu_0\hat{p}_1 + B_{\phi 0}\hat{B}_{\phi 1})$ leads to a single partial differential equation for V in terms of β:

$$\nabla_\perp \cdot \left(\frac{\nabla_\perp V}{1-\beta}\right) - \frac{h^2 V}{1-\beta} = 0 \tag{7.24}$$

Summarizing, for a prescribed $\beta(r, \theta)$, the first-order fields are expressed in terms of V and β as follows:

$$\begin{aligned}\frac{\hat{\mathbf{B}}_{p1}}{B_0} &= \frac{\nabla_\perp V}{(1-\beta)^{1/2}} \\ \frac{\hat{B}_{\phi 1}}{B_0} &= \frac{i}{h}\left[\frac{1}{2}\frac{\nabla_\perp V\cdot\nabla_\perp\beta}{(1-\beta)^{3/2}} - \frac{h^2 V}{(1-\beta)^{1/2}}\right] \\ \frac{\mu_0\hat{p}_1}{B_0^2} &= -\frac{i}{2h}\frac{\nabla_\perp V\cdot\nabla_\perp\beta}{1-\beta}\end{aligned} \tag{7.25}$$

with V determined by Eq. (7.24).

Evaluation of $\beta(r,\theta)$. The equation governing the behavior of $\beta(r,\theta)$ follows from substitution into the parallel current constraint. There are several points in the derivation worth noting. First, the leading-order contribution to the constraint is of order δ but vanishes automatically because of helical symmetry. Thus, the constraint must be evaluated to order δ^2. However, in carrying out this calculation only a knowledge of the zeroth- and first-order fields is explicitly required. Second, the only direct appearance of toroidicity occurs as a simple contribution to the curvature. Finally, despite the substantial simplifications that arise by the use of the parallel current constraint, the analysis is still unexpectedly tedious.

The calculation begins with the order δ contribution to the parallel current constraint. Straightforward substitution into Eq. (7.7) leads to the following expression for $J_\parallel/B \sim \delta$:

$$\frac{\partial}{\partial\phi}\left(\frac{\mu_0 J_\parallel}{B}\right)_1 = \frac{1}{1-\beta}\left(\mathbf{e}_\phi\cdot\nabla_\perp\beta\times\frac{\partial\mathbf{b}_{p1}}{\partial\phi}\right) \tag{7.26}$$

where

$$\mathbf{b}_{p1} = \frac{\mathbf{B}_p}{B} = \frac{1}{1-\beta}(\nabla_\perp V e^{-iN\phi} + \nabla_\perp V^* e^{iN\phi}) \tag{7.27}$$

The right-hand side of Eq. (7.26) is a perfect derivative in ϕ. Hence, if one integrates around one circuit of the torus the parallel current constraint is trivially satisfied. Equation (7.26) also yields the useful information

$$\left(\frac{\mu_0 J_\parallel}{B}\right)_1 = \frac{1}{1-\beta}(\mathbf{e}_\phi\cdot\nabla_\perp\beta\times\mathbf{b}_{p1}) \tag{7.28}$$

which is required for the order δ^2 calculation.

The next step is to expand all quantities appearing in the parallel current constraint [Eq. (7.8)] one order higher in δ. After some simple algebra one obtains

$$C_2 \equiv \frac{R_0}{\mu_0}\int_0^{2\pi} d\phi\left\{\frac{1}{1-\beta}[\mathbf{e}_\phi \cdot \nabla_\perp \beta \times (\boldsymbol{\kappa}_2 - 2b_{\phi 1}\boldsymbol{\kappa}_1) + \mathbf{e}_\phi \cdot \nabla_\perp \beta_1 \times \boldsymbol{\kappa}_1] - \mathbf{b}_{p1} \cdot \nabla\left(\frac{\mu_0 J_\parallel}{B}\right)_1\right\} \tag{7.29}$$

where $b_{\phi 1} = B_{\phi 1}/B_0$, $\beta_1 = 2\mu_0 p_1/B_0^2$ and

$$\boldsymbol{\kappa} = \boldsymbol{\kappa}_1 + \boldsymbol{\kappa}_2 = \frac{1}{R_0}\frac{\partial \mathbf{b}_{p1}}{\partial \phi} + \left[\mathbf{b}_{p1} \cdot \nabla_\perp \mathbf{b}_{p1} + \frac{1}{R_0}\frac{\partial}{\partial \phi}(\mathbf{b}_{p2} - b_{\phi 1}\mathbf{b}_{p1}) - \frac{\mathbf{e}_R}{R_0}\right] \tag{7.30}$$

Although Eqs. (7.29) and (7.30) are somewhat complicated, there are nevertheless two important points that can be made. First, the only explicit appearance of toroidicity arises from the $\mathbf{e}_\phi \cdot \nabla \mathbf{e}_\phi$ contribution to $\boldsymbol{\kappa}$. Second, the quantity $\boldsymbol{\kappa}_2$ appears to require a knowledge of $\mathbf{B}_{p2}$, a second-order field which has not been calculated. However, this term appears as a perfect ϕ derivative which averages to zero. Hence, $\mathbf{B}_{p2}$ is not explicitly required.

Upon substituting into the constraint integral and setting $C_2 = 0$, one obtains, after substantial algebra (Freidberg *et al.*, 1979)

$$\frac{\mu_0 C_2}{R_0} = -\pi\left[\frac{e_\phi \cdot \nabla_\perp \beta \times \nabla_\perp G}{(1-\beta)^3}\right] = 0 \tag{7.31}$$

$$G \equiv |\nabla_\perp V|^2 + h^2|V|^2 + (1-\beta)^2\frac{r}{R_0}\cos\theta \tag{7.32}$$

The general solution to this equation is $G = G(\beta)$, where $G(\beta)$ is now considered as a free function. Once $G(\beta)$ is specified then Eqs. (7.24) and (7.32) form a set of coupled equations for V and β. These equations thus constitute a complete mathematical description of helical sideband equilibria.

Discussion. To gain some insight into helical sideband equilibria consider the basic set of governing equations, repeated here for convenience:

$$\nabla_\perp \cdot \left(\frac{\nabla_\perp V}{1-\beta}\right) - \frac{h^2 V}{1-\beta} = 0 \tag{7.33}$$

$$|\nabla_\perp V|^2 + h^2|V|^2 + (1-\beta)^2\frac{r}{R_0}\cos\theta = G(\beta) \tag{7.34}$$

These are a pair of coupled, nonlinear, partial differential equations which in general must be solved numerically. The situation is further complicated by the fact that β is related to $\nabla_\perp V$ via Eq. (7.34). This implies that the highest-order V derivative terms in Eq. (7.33) are much more complicated than a simple Laplacian operator. In the next section an approximate analytic solution to these equations is given corresponding to the Elmo bumpy torus equilibrium.

The next point to note is that the system of equilibrium equations is characterized by a single free function $G(\beta)$. The second free function is related to the net current flowing on each surface. This function has been automatically set to zero through order δ by virtue of the form assumed for the magnetic fields. If one continues the calculation to order δ^2, the second free function explicitly appears and can be used to adjust $I(\beta)$ as desired.

The most interesting feature of the equilibrium equations is the mechanism for generating a toroidal restoring force. This is slightly involved but can be understood as follows. First, consider the limit of an infinitely long helically symmetric system: $R_0 \to \infty$ and $V(r, \theta) = V_l(r)\exp(il\theta)$. In this limit the $(r/R_0)\cos\theta$ term is negligible and $\beta(r, \theta) \to \hat{\beta}(r)$. One solves the system of equations by assuming $\hat{\beta}(r)$ to be the free function. Equation (7.33) determines $V(r)$. Substituting $\beta(r)$ and $V(r)$ into Eq. (7.34) then gives $G[\hat{\beta}(r)]$ as an *a posteriori* calculation.

When toroidicity is present the situation changes substantially. β is no longer a function only of r (even with a single applied helical field); instead, it develops a toroidal correction. If r/R_0 is very small, but still finite, this correction assumes the form of a shift to balance the $(r/R_0)\cos\theta$ term in Eq. (7.34): $\beta(r, \theta) \approx \hat{\beta}(r) + \tilde{\beta}(r)\cos\theta$. This shift, when substituted into Eq. (7.33), generates helical sideband fields about the applied helical field: $V(r, \theta) = V_l(r)\exp(il\theta) + V_{l\pm1}(r)\exp[i(l \pm 1)\theta]$. The system of equations must then be solved simultaneously for $V_l(r)$, $V_{l\pm1}(r)$, and $\tilde{\beta}(r)$, assuming $\hat{\beta}(r)$ is given as a free function.

This leads to the following physical picture of toroidal force balance. A straight helically symmetric system has a well-defined geometric center about which the plasma is confined. In a large aspect ratio torus the plasma is shifted along R, away from this center, because of toroidal forces; that is, the flux surfaces develop a shift $\beta(r, \theta) = \hat{\beta}(r) + \tilde{\beta}(r)\cos\theta$. Because of the plasma pressure a helical diamagnetic current is induced. In particular, one component of current associated with $\tilde{\beta}$ is given by

$$2\mu_0\mathbf{J}_{SB} \approx \mathbf{B}_l \times \nabla(\tilde{\beta}\cos\theta) \tag{7.35}$$

where $\mathbf{B}_l$ is the applied helical field. The geometric dependence of this current corresponds to $l \pm 1$ sideband fields caused by the plasma shift. The force generated by this current, $\mathbf{J}_{SB} \times \mathbf{B}_l$, consists of two geometrically different types of contributions, one proportional to $\cos[(2l + 1)\theta - 2N\phi]$ and the other to $\cos\theta$. The latter term, when integrated over the plasma volume, gives rise to a net, ϕ-averaged, body force along $\mathbf{e}_R$:

$$F_R \approx \int \mathbf{J}_{SB} \times \mathbf{B}_l \cdot \mathbf{e}_r \cos\theta \, d\mathbf{r} \tag{7.36}$$

This force can partially compensate the outward toroidal drift force. Since $F_R \sim \tilde{\beta}$, the plasma continues to shift until the forces exactly balance. A more detailed calculation (Miller, 1977) shows that there are additional sideband forces of comparable magnitude that enter the calculation, for instance, those due to $\mathbf{J}_l \times \mathbf{B}_{SB}$ and $\Delta \cos\theta \, \partial(\boldsymbol{J}_l \times \mathbf{B}_l)/\partial r$ with $\Delta = -\tilde{\beta}/\hat{\beta}'$ the toroidal shift. To obtain quantitative agreement with Eqs. (7.33) and (7.34) all of these terms must be included in the evaluation of F_R.

The point is that each of these terms is generically the same. Toroidal effects induce sideband fields and it is the interaction of the sideband fields with the basic applied helical field that produces the toroidal restoring force. Hence the name "helical sideband equilibria."

7.4. THE ELMO BUMPY TORUS (EBT)

One interesting application of the helical sideband equilibrium theory is to the Elmo bumpy torus (EBT), a device developed and built at the Oak Ridge National Laboratory.

Experimental Description. The Elmo bumpy torus is a large aspect ratio, closed line, toroidal configuration consisting of a series of linked mirrors as shown schematically in Fig. 7.1. The mirror coils produce a magnetic field with periodic variation in the toroidal direction, thus producing localized mirror cells or "bumps." There is no net toroidal current in the EBT configuration. Midway between each mirror coil, in the bulge region, is a hot electron ring generated by a high-power rf source resonant with the electron cyclotron frequency. The resonant radius where the ring is formed determines the radial boundary of the main toroidal plasma. The electron rings have several crucial roles in the EBT concept. First, they must ultimately provide a large fraction of the heating in the main toroidal plasma, since they themselves are typically at several hundred keV owing to their resonant formation. Second, they are

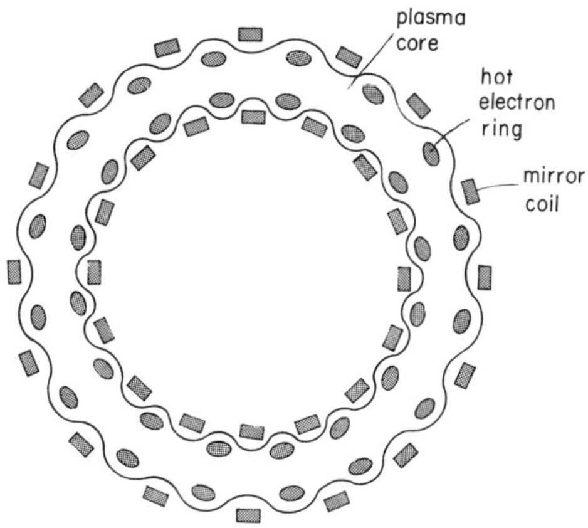

Figure 7.1. Schematic diagram of EBT.

vital for the stability of the main plasma core. Without the rings, the system would be unstable to both gross and localized MHD modes. Interestingly enough, the rings are not required to produce the equilibrium itself, only to stabilize it. For this reason the treatment of the rings can be greatly simplified for equilibrium calculations by assuming the ring pressure is isotropic. Consequently, the pressure p appearing in the MHD equilibrium equations consists of two contributions: $p = p_c + p_e$, where p_c is the core plasma pressure and p_e is the electron ring pressure.

The basic MHD issues associated with the EBT experiment are reliably approximated by the helical sideband expansion. Note first that the closed line, bumpy magnetic field configuration corresponds to a particularly simple "helical" field, one whose poloidal multipolarity is characterized by $l = 0$. Second, the basic ordering parameter defining the expansion is proportional to the amplitude of the bumpy field component. Specifically, δ is defined by

$$\delta \equiv \left.\frac{B_{\max} - B_{\min}}{B_{\max} + B_{\min}}\right|_{r=0} \ll 1 \tag{7.37}$$

Typically, $\delta \approx 1/3$ in actual experiments, so that while the expansion may be qualitatively correct, it should not be used for quantitative predictions.

EBT Solution. With this as background, one can now proceed to solve the sideband expansion equations for EBT equilibria. Because of the inherent nonlinearity, additional assumptions must be made in order to obtain an analytic solution. This is accomplished by introducing a subsidiary expansion within the sideband expansion; that is, a new ordering parameter ν is introduced such that

$$\delta \ll \nu \ll 1 \tag{7.38}$$

Equation (7.38) is more a mathematical convenience than an accurate representation of some aspect of the actual physical situation. Its purpose is to separate in a mathematical sense the various physical phenomena taking place. This leads to an order-by-order (in ν) analytic solution for the equilibrium.

The subsidiary expansion occurs in two stages. In the first, only the inverse aspect ratio is ordered smaller. This assumption leads to an expansion type solution for V and β which transforms the nonlinear partial differential equations into a truncated set of linear ordinary differential equations. These equations, while easy to solve numerically, are still too difficult to treat analytically. The difficulty is resolved with the second stage of the subsidiary expansion in which the order of β is also reduced. This leads to a complete analytic solution for the equilibrium.

The analysis begins with the introduction of the subsidiary expansion. It is assumed that the inverse aspect ratio is sufficiently small that the configuration becomes a perturbation about a linear system. The first stage of the subsidiary expansion is defined below and compared with the original sideband expansion:

Subsidiary expansion	Original expansion
$a/R_0 \sim \delta^2\nu$	$a/R_0 \sim \delta^2$
$\beta \sim 1$	$\beta \sim 1$
$V \sim \delta$	$V \sim \delta$
$ha \sim 1$	$ha \sim 1$

The virtue of introducing ν as a small parameter is that it allows both $V(r, \theta)$ and $\beta(r, \theta)$ to be expanded as follows:

$$\begin{aligned} V(r, \theta) &= V_0(r) + V_1(r)\cos\theta + \cdots \\ \beta(r, \theta) &= \beta_0(r) + \beta_1(r)\cos\theta + \cdots \end{aligned} \tag{7.39}$$

where

$$\beta_0 \sim 1, \qquad \beta_1 \sim \nu, \qquad V_0 \sim \delta, \qquad V_1 \sim \delta\nu \tag{7.40}$$

Note that the flux surfaces assume the shape of slightly shifted circles.

Substitution of the subsidiary expansion into the sideband expansion [Eqs. (7.33) and (7.34)] leads to a set of linear ordinary differential equations

$$\frac{1}{r}\left(\frac{rV_0'}{1-\beta_0}\right)' - \frac{h^2V_0}{1-\beta_0} = 0 \tag{7.41}$$

$$\frac{1}{r}\left(\frac{rV_1'}{1-\beta_0}\right)' - \left(h^2 + \frac{1}{r^2}\right)\frac{V_1}{1-\beta_0} = -\frac{V_0'}{1-\beta_0}\left(\frac{\beta_1}{1-\beta_0}\right)' \tag{7.42}$$

$$G(\beta_0) = V_0'^2 + h^2V_0^2 \tag{7.43}$$

$$\beta_1 = \frac{1}{dG/d\beta_0}\left[(1-\beta_0)\frac{r}{R_0} + 2(V_0'V_1' + h^2V_0V_1)\right] \tag{7.44}$$

In these equations prime denotes differentiation with respect to r, and $\beta_0(r)$ should be viewed as the free function. Equation (7.43) thus defines $G(r) \equiv G[\beta_0(r)]$ in terms of $V_0(r)$. While the subsidiary equations are straightforward to solve numerically, they are still too complicated to treat analytically.

An analytic solution can, however, be obtained if one further assumes that β_0 is small. Specifically, in the second stage of the expansion it is assumed that

Subsidiary expansion (second stage)
$a/R_0 \sim \delta^2 \nu$
$\beta_0 \sim \nu$
$\beta_1 \sim \nu^2$
$ha \sim 1$

A straightforward comparison of terms then indicates that

$$\begin{aligned} V_0 &\sim \delta \\ V_1 &\sim \delta \nu^2 \end{aligned} \tag{7.45}$$

Under these conditions the equilibrium equations can be solved sequentially as follows. Equation (7.41) reduces to

$$\frac{1}{r}(rV_0')' - h^2V_0 = 0 \tag{7.46}$$

which has as its solution

$$V_0 = c_1 I_0(hr) \tag{7.47}$$

where I_0 is the modified Bessel function. The constant c_1 can be related to the applied mirror ratio along the axis $r = 0$. Using the definition given by Eq. (7.37) one obtains

$$V_0 = \frac{(M-1)}{4h} I_0(hr) \tag{7.48}$$

where $M = 1 + 2\delta$ is the applied mirror ratio, assumed to be a known experimental parameter.

The quantity β_1 is obtained from Eq. (7.44)

$$\beta_1 = \frac{1}{dG/d\beta_0}\left(\frac{r}{R_0}\right) \tag{7.49}$$

where $dG/d\beta_0$ follows from Eq. (7.43):

$$\frac{dG}{d\beta_0} = \frac{G'}{\beta_0'} = \frac{1}{\beta_0'}\left(4h^2 V_0 V_0' - 2\frac{V_0'^2}{r}\right) \tag{7.50}$$

The final step is to calculate V_1 from Eq. (7.42):

$$\frac{1}{r}(rV_1')' - \left(h^2 + \frac{1}{r^2}\right)V_1 = -V_0'\beta_1' \tag{7.51}$$

Assuming V_0 and β_1 have been calculated, one finds

$$V_1 = I_1 \int_r^\infty \frac{dx}{xI_1^2} \int_0^x yV_0'\beta_1'\,dy \tag{7.52}$$

where $I_1(hr)$ is the modified Bessel function.

Equations (7.48), (7.49), and (7.52) constitute the analytic equilibrium solution for the EBT configuration.

Discussion. The basic physics properties of the Elmo bumpy torus follow from an examination of the analytic equilibrium solutions. Radial pressure balance corresponds to that of a θ pinch. A typical pressure profile of the combined plasma core–electron ring is illustrated in Fig. 7.2. Observe that the total pressure is hollow because of the presence of

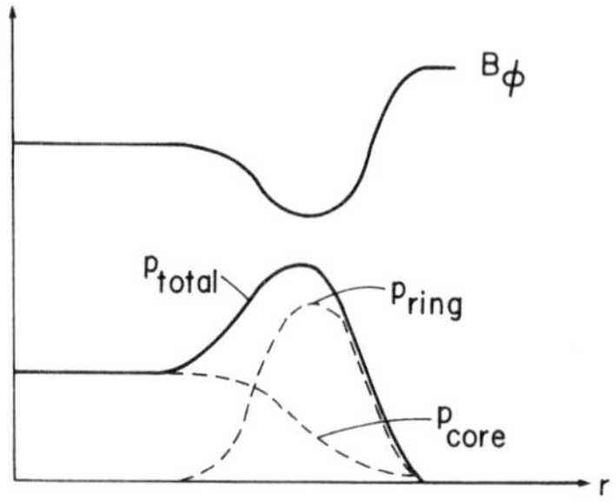

Figure 7.2. Typical radial pressure-balance profiles for a high-β EBT.

the electron ring and that the central core profile is quite flat. The electron ring also causes a corresponding depression in the toroidal magnetic field.

Since EBT is a closed-line configuration, there is no transform and no shear in the magnetic field. Also, since there is zero net toroidal current on each flux surface, current-driven kink modes do not present a problem: $q_* = \infty$.

The stability of the EBT configuration arises from the formation of a favorable magnetic well. The existence of a favorable well is, however, strongly coupled to the non-MHD behavior of the hot electron rings. This is a somewhat involved situation, and the question of the magnetic well and EBT stability is deferred until Chapter 10.

An important property of the EBT configuration is that it does not possess an equilibrium β limit. Although the analytic solution presented here requires small $\beta \sim \nu$, this was assumed solely for mathematical convenience. Numerical calculations involving the sideband expansion equations as well as the full three-dimensional model have been carried out and no equilibrium β limit has been found.

The most interesting feature of the equilibrium is the toroidal shift of the flux surfaces. If the surfaces are defined as $r = r_0 + \Delta(r_0)\cos\theta$, then from Eq. (7.39) one finds $\Delta = -\beta_1/\beta_0'$. Note that Δ measures the shift of the surfaces with respect to the center of the "helical" field, in this case corresponding to the center of the mirror coils. A simple calculation gives

$$\Delta(r) = -\frac{8}{h^2R_0(M-1)^2}\left(\frac{1}{D}\right) \tag{7.53}$$

where

$$D(r) = \frac{I_1}{hr}\left(2I_0 - \frac{I_1}{hr}\right) \tag{7.54}$$

Since $D(r)$ is always positive one obtains the surprising result that $\Delta(r) < 0$; in particular, on axis

$$\Delta_0 = -\frac{32}{3}\left[\frac{1}{h^2R_0(M-1)^2}\right] \tag{7.55}$$

Remarkably, the plasma is shifted inward despite the fact that the toroidal forces attempt to move the plasma outward.

A physical picture describing this behavior in terms of a simple mechanical model has been given by Rosenbluth *et al.* (1969). Consider first as a reference case the intuitive situation where the plasma is shifted outward. Assume that in the limit of an infinitely long straight helix the plasma is confined in a stable potential well, as illustrated by the ball in Fig. 7.3a. When bent into a torus, toroidal forces tend to push the plasma outward, corresponding to a decreasing potential as shown in Fig. 7.3b. If one now sums the two contributions, the combined potential has a minimum shifted slightly outward from its original position, as seen in Fig. 7.3c. As "expected," the introduction of toroidicity leads to an outward shift of the plasma.

In contrast, assume now that the infinitely long straight helix provides equilibrium in an unstable potential well (Fig. 7.4a). Toroidal effects still tend to push the plasma outward, giving rise to the decreasing potential shown in Fig. 7.4b. Summing these contributions, one now obtains a combined potential where the peak is shifted inward (Fig. 7.4c). This is the situation in EBT, and for that matter in all helical sideband equilibria. (See Problem 7.3.)

To the extent that this picture is valid, the implication is that high-β toroidal EBT equilibria exist but they are very unstable. If the hot electron rings were to behave as an MHD plasma this would indeed be the case. It is precisely the fact that the rings behave in a very non-MHD manner that gives rise to the stability of the system. This question is discussed in detail in Chapter 10.

Finally, it is useful to illustrate the asymptotic analytic solution. An

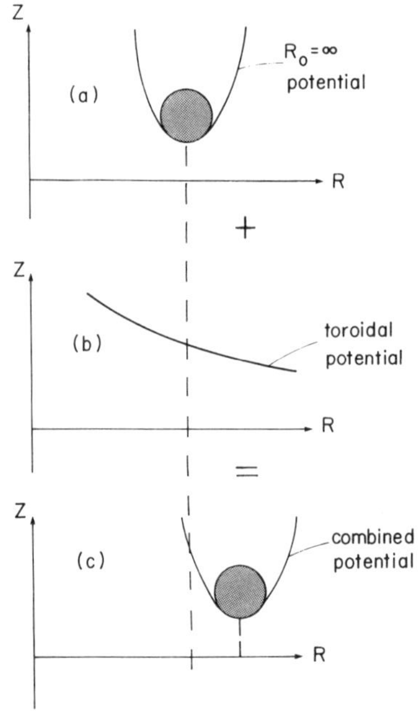

Figure 7.3. Simple mechanical model illustrating toroidal force balance in a stable system: (a) $R_0 = \infty$ potential; (b) $1/R$ potential; (c) combined $\infty + 1/R$ potentials. Note that toroidicity produces an outward shift.

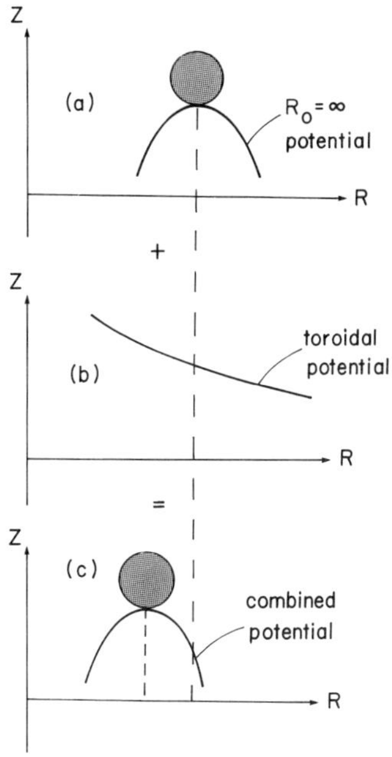

Figure 7.4. Simple mechanical model illustrating toroidal force balance in an unstable system: (a) $R_0 = \infty$ potential; (b) $1/R$ potential; (c) combined $\infty + 1/R$ potentials. Note that toroidicity produces an inward shift.

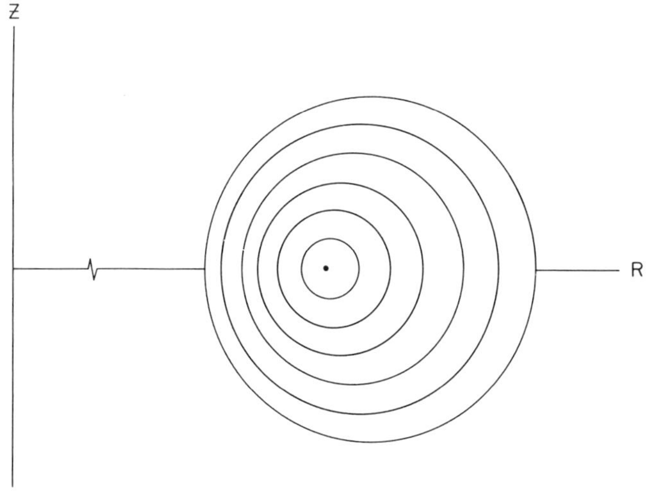

Figure 7.5. Numerically computed three-dimensional equilibrium of the EBT configuration. Shown are constant-pressure contours.

example of such a computation, corresponding to the Oak Ridge EBT experiment, is given in Fig. 7.5. Observe as the general shape of the flux surfaces including the inward toroidal shift.

7.5. STELLARATOR EQUILIBRIA

Stellarator equilibria represent the second basic class of three-dimensional toroidal systems to be investigated. These configurations are low-β, large aspect ratio devices in which the magnetic field consists primarily of (1) a large *vacuum* toroidal field, (2) a moderate size *vacuum* helical field, and (3) a small induced poloidal dipole field. The systems of interest normally contain no net toroidal current, although the theory presented allows such a current to flow.

Stellarator Expansion. The basic equilibrium issues are addressed by the "stellarator expansion" introduced by Greene and Johnson (1961). As in the sideband expansion the appropriate ordering parameter is the ratio of the helical field to the main toroidal field

$$\delta = B_{\text{hel}}/B_0 \ll 1 \tag{7.56}$$

In terms of δ the stellarator expansion is given by

$$\begin{aligned} \beta_t &\sim \delta^2 \\ ha &\sim 1 \\ a/R_0 &\sim \delta^2 \\ q &\sim 1 \end{aligned} \tag{7.57}$$

where, as before, $h = N/R_0$ is the pitch number of the helical field.

Qualitatively one observes that the equilibria of interest have the following properties. The helical wavelength is comparable to the plasma radius: $ha \sim 1$. The aspect ratio is large: $R_0/a \sim 1/\delta^2$. There are many helical periods along the length of the torus: $N = hR_0 \sim 1/\delta^2$. These features are similar to those of helical sideband equilibria. The safety factor is also of order unity, $q \sim 1$, but here it plays a major role in the quality of the equilibria. The main difference between the two classes of equilibria is the ordering of β. Stellarator equilibria have $\beta_t \sim \delta^2$, which is low compared to sideband systems where $\beta_t \sim 1$ but comparable to "high-β" tokamaks where $\beta_t \sim a/R_0$.

It is important to point out that the behavior of stellarator equilibria is not simply given by the low-β limit of the sideband equations. While there are many terms that overlap, new effects enter the stellarator model in a competitive order, which are negligible in sideband systems.

In the stellarator expansion the fields are written as

$$\begin{aligned} \mathbf{B} &= B_0\mathbf{e}_\phi + \mathbf{B}_1(r, \theta, \phi) + \mathbf{B}_2(r, \theta) + \cdots \\ p &= p_2(r, \theta) + p_3(r, \theta, \phi) + \cdots \end{aligned} \tag{7.58}$$

Here, $B_0 = \text{const}$ is the leading-order vacuum toroidal field and $\mathbf{B}_1$ is the first-order helical field. Since β_t is low, $\mathbf{B}_1$ is also a vacuum field. The pressure first enters in second order and p_2 and $\mathbf{B}_{\phi 2}$ give rise to a θ-pinch radial pressure balance. $\mathbf{B}_2$ also contains a poloidal magnetic field and it is this term that generates the toroidal "dipole current." The quantity p_3 represents the helical modification of the pressure contours. The variables listed above are essentially the minimum set required to carry out the calculation and to test stability.

The analysis is slightly involved and an outline is given as follows. The first part of the calculation is straightforward. The fields described by Eq. (7.58) are expressed in terms of two functions $p_2(r, \theta)$ and $A_2(r, \theta)$. The pressure p_2 represents the leading-order flux contours and A_2 is the second-order vector potential describing the ϕ independent component of the poloidal magnetic field. A_2 enters the calculation as a homogeneous solution to the second-order equations. It is crucial to keep this term or else the system does not close properly in high order. Once p_2 and A_2 are known, all the important equilibrium and stability properties of stellarator equilibria can be readily calculated.

The second part of the analysis involves the calculation of the two unknowns, p_2 and A_2. The quantity p_2 is determined from the parallel current constraint. The quantity A_2 is also determined by a high-order periodicity constraint. While there are several ways to extract the A_2 constraint information, the most efficient method makes use of the parallel component of the momentum equation $\mathbf{B} \cdot \nabla p = 0$, which can be written as

$$\frac{\partial p}{\partial \phi} = -\frac{R}{B_\phi}(\mathbf{B}_p \cdot \nabla p) \tag{7.59}$$

Since p is periodic in ϕ it follows that

$$\int_0^{2\pi} \frac{R\, d\phi}{B_\phi}(\mathbf{B}_p \cdot \nabla p) = 0 \tag{7.60}$$

where, as with the parallel current constraint, the integrand is evaluated in Eulerian variables. Equation (7.60) is in the desired form to evaluate

A_2. Note that in the sideband expansion there is an analogous calculation for the poloidal fields, but it appears in higher order and is not explicitly required.

Calculation of the Fields. The stellarator expansion is now substituted into the MHD equilibrium equations. The resulting set of equations is solved order by order until the fields defined in Eq. (7.58) are determined.

The zeroth-order information is trivial, requiring only that $B_0 =$ const, corresponding to a vacuum toroidal field. In first order, the quantity $\mathbf{B}_1$ is easy to evaluate since it represents a vacuum helical field. In analogy with the sideband expansion, $\mathbf{B}_1$ is assumed to contain an arbitrary number of helical harmonics with different l values, but all with the same number of helical periods. Thus, $\mathbf{B}_1$ can be expressed as

$$\mathbf{B}_1 = B_0\nabla[V(r, \theta)e^{-iN\phi} + V^*(r, \theta)e^{iN\phi}] \tag{7.61}$$

where $V(r, \theta)$ is a potential function. Since $\mathbf{B}_1$ is a vacuum field, V must satisfy

$$\nabla_\perp^2 V - h^2 V = 0 \tag{7.62}$$

where $hR_0 = N$. The solution to this equation is readily expressible in terms of modified Bessel functions

$$V(r, \theta) = \sum_l V_l I_l(hr)e^{il\theta} \tag{7.63}$$

Here, the V_l are known coefficients related to the amplitudes of the applied helical harmonics. As a consequence, in the stellarator expansion the potential $V(r, \theta)$ is assumed known.

The second-order fields are ϕ independent and describe the first influence of the plasma. The equation $\nabla \cdot \mathbf{B} = 0$ expanded to second order can be written as

$$\nabla_\perp \cdot \boldsymbol{B}_{p2}(r, \theta) = 0 \tag{7.64}$$

which implies that

$$\mathbf{B}_{p2} = \mathbf{e}_\phi \times \nabla_\perp A_2 \tag{7.65}$$

where $A_2(r, \theta)$ is the ϕ component of the vector potential. A_2 enters the calculation as a homogeneous solution to Eq. (7.64) and is completely arbitrary in second order. It is ultimately determined by one of the higher-order periodicity constraints.

The next step in the analysis is to substitute $\mu_0\mathbf{J} = \nabla \times \mathbf{B}$ into the momentum equation and then expand to second order. One finds

$$\mathbf{B}_2 \cdot \nabla \mathbf{B}_0 + \mathbf{B}_0 \cdot \nabla \mathbf{B}_2 + \mathbf{B}_1 \cdot \nabla \mathbf{B}_1 + (\mathbf{B}_0 \cdot \nabla \mathbf{B}_0)_2 = \nabla(\mu_0 p_2 + B_0 B_{\phi 2} + B_1^2/2) \quad (7.66)$$

The first term vanishes because B_0 = const. The second term vanishes because $\mathbf{B}_2$ is not a function of ϕ. The combination $\mathbf{B}_1 \cdot \nabla \mathbf{B}_1 - \nabla B_1^2/2 = -\mathbf{B}_1 \times \nabla \times \mathbf{B}_1$ vanishes because $\mathbf{B}_1$ is a vacuum field: $\nabla \times \mathbf{B}_1 = 0$. The term $(\mathbf{B}_0 \cdot \nabla \mathbf{B}_0)_2$ represents a second-order contribution arising from the curvature of the $\mathbf{e}_\phi$ vector: $\mathbf{B}_0 \cdot \nabla \mathbf{B}_0 = B_0^2(\mathbf{e}_\phi \cdot \nabla \mathbf{e}_\phi) \approx (-B_0^2/R_0)\mathbf{e}_R \sim \delta^2$. The nonzero terms in Eq. (7.66) can be combined and rewritten as follows:

$$\nabla\left(\mu_0 p_2 + B_0 B_{\phi 2} + B_0^2 \frac{r}{R_0} \cos\theta\right) = 0 \quad (7.67)$$

The solution is given by

$$\frac{B_{\phi 2}}{B_0} = -\frac{\beta}{2} - \frac{r}{R_0} \cos\theta \quad (7.68)$$

where for convenience a normalized form of p_2 has been introduced:

$$\beta(r, \theta) \equiv \frac{2\mu_0 p_2(r, \theta)}{B_0^2} \sim \delta^2 \quad (7.69)$$

Equation (7.67) represents a θ-pinch radial pressure balance. The β term is the plasma diamagnetism and $\hat{B}_{\phi 2} \equiv B_{\phi 2} + B_0(r/R_0)\cos\theta$ is the net toroidal field once the $(r/R_0)\cos\theta$ vacuum correction has been subtracted out.

The last step in the calculation of the fields is the evaluation of p_3, which is required for the constraint relations. This quantity is easily evaluated without solving the full set of third-order equations. If one substitutes into the parallel pressure gradient relation given by Eq. (7.59) then p_3 is found to satisfy

$$\frac{\partial p_3}{\partial \phi} = -\frac{R_0}{B_0}(\mathbf{B}_{p1} \cdot \nabla p_2) \quad (7.70)$$

which has as its solution

$$\frac{2\mu_0 p_3}{B_0^2} = -\frac{i}{h} \nabla\beta \cdot [e^{-iN\phi}\nabla V - e^{iN\phi}\nabla V^*] \quad (7.71)$$

Summarizing, for a given $\beta(r, \theta)$ and $A(r, \theta) \equiv A_2(r, \theta)/B_0$ the fields that are specified in Eq. (7.58) are given by

$$\begin{aligned} B_\phi &= B_0\left\{1 - [ih(Ve^{-iN\phi} - V^*e^{iN\phi})] - \left[\frac{\beta}{2} + \frac{r}{R_0}\cos\theta\right]\right\} \\ \mathbf{B}_p &= B_0[(e^{-iN\phi}\nabla_\perp V + e^{iN\phi}\nabla_\perp V^*) + \mathbf{e}_\phi \times \nabla_\perp A] \\ p &= \frac{B_0^2}{2\mu_0}\left\{\beta - \left[\frac{i}{h}\nabla_\perp\beta\cdot(e^{-iN\phi}\nabla_\perp V - e^{iN\phi}\nabla_\perp V^*)\right]\right\} \end{aligned} \tag{7.72}$$

with V given by Eq. (7.63).

Evaluation of A. The quantity A is determined by substitution into the parallel pressure gradient constraint given by Eq. (7.60). The first contribution that does not vanish trivially occurs in fourth order. This relation is given by

$$\frac{R_0}{B_0}\int_0^{2\pi} d\phi[\mathbf{B}_{p1}\cdot\nabla_\perp p_3 + \mathbf{B}_{p2}\cdot\nabla_\perp p_2] = 0 \tag{7.73}$$

After a short calculation Eq. (7.73) reduces to

$$\mathbf{e}_\phi\cdot\nabla_\perp\beta\times\nabla_\perp\left(\frac{i}{h}\mathbf{e}_\phi\cdot\nabla_\perp V^*\times\nabla_\perp V - A\right) = 0 \tag{7.74}$$

The general solution is given by

$$A = \frac{i}{h}\mathbf{e}_\phi\cdot\nabla_\perp V^*\times\nabla_\perp V + H(\beta) \tag{7.75}$$

where $H(\beta)$ is an arbitrary free function.

$H(\beta)$ can be interpreted as follows. Consider the ϕ-averaged poloidal flux ψ_p between the magnetic axis and a given pressure contour:

$$\psi_p = \int \mathbf{B}_p\cdot d\mathbf{S} \tag{7.76}$$

A straightforward calculation shows that (see Problem 7.4)

$$\psi_p = 2\pi R_0 B_0\left(A - \frac{i}{h}\mathbf{e}_\phi\cdot\nabla_\perp V^*\times\nabla_\perp V\right) \tag{7.77}$$

Consequently, one can identify

$$H(\beta) = \psi \tag{7.78}$$

where $\psi(r, \theta) = \psi_p/2\pi R_0 B_0$ is the normalized poloidal flux. As is customary in the stellarator expansion ψ is chosen in place of β as the flux surface label; that is, ψ becomes the primary dependent variable in the problem and the flux surfaces correspond to the ψ = const contours. In terms of ψ, A is given by

$$A = \psi + \frac{i}{h}\mathbf{e}_\phi \cdot \nabla_\perp V^* \times \nabla_\perp V \tag{7.79}$$

and

$$\beta = \beta(\psi) \tag{7.80}$$

plays the role of a free function.

Evaluation of ψ. The final step in the analysis is the evaluation of ψ from the parallel current constraint. The calculation is quite similar to that of the helical sideband expansion, including the tedious algebra required. All of the quantities needed for the calculation have already been evaluated, except $J_\parallel$. This quantity has nonzero second- and third-order contributions which enter the constraint relation. The second-order term follows directly from taking the curl of Eq. (7.65):

$$\left(\frac{\mu_0 J_\parallel}{B}\right)_2 = \nabla_\perp^2 A \tag{7.81}$$

The third-order contribution is most easily obtainable from the parallel current relation itself [Eq. (7.7)]:

$$\frac{\partial}{\partial\phi}\left(\frac{J_\parallel}{B}\right)_3 = \frac{R_0}{B_0^2}\left[2\mathbf{e}_\phi \cdot \nabla_\perp p_2 \times \boldsymbol{\kappa}_1 - \mathbf{b}_{p1}\cdot\nabla_\perp\left(\frac{J_\parallel}{B}\right)_2\right] \tag{7.82}$$

where $\mathbf{b}_{p1} = \mathbf{B}_{p1}/B_0$ and $\boldsymbol{\kappa}_1 = (1/R_0)(\partial\mathbf{b}_{p1}/\partial\phi)$. This expression is easily integrated, yielding

$$\left(\frac{\mu_0 J_\parallel}{B}\right)_3 = \left[\mathbf{e}_\phi \cdot \nabla_\perp \beta \times \nabla_\perp V - \frac{i}{h}(\nabla_\perp V \cdot \nabla_\perp)\nabla_\perp^2 A\right]e^{-iN\phi} + \text{c.c.} \tag{7.83}$$

Next, one substitutes into the parallel current constraint [Eq. (7.8)]. The first contribution that does not trivially vanish is of fourth order and is given by

$$C_4 \equiv \frac{R_0}{\mu_0}\int_0^{2\pi} d\phi\left\{\mathbf{e}_\phi \cdot \nabla_\perp\beta \times (\boldsymbol{\kappa}_2 - 2b_{\phi 1}\boldsymbol{\kappa}_1) + \mathbf{e}_\phi\cdot\nabla_\perp\beta_3 \times \boldsymbol{\kappa}_1 \right.$$
$$\left. - \mathbf{b}_{p1}\cdot\nabla_\perp\left(\frac{J_\parallel}{B}\right)_3 - (\mathbf{b}_{p2} - b_{\phi 1}\mathbf{b}_{p1})\cdot\nabla_\perp\left(\frac{J_\parallel}{B}\right)_2\right\} \tag{7.84}$$

with $b_{\phi 1} = B_{\phi 1}/B_0$, $\mathbf{b}_{p2} = \mathbf{B}_{p2}/B_0$, $\beta_3 = 2\mu_0 p_3/B_0^2$ and

$$\boldsymbol{\kappa} = \boldsymbol{\kappa}_1 + \boldsymbol{\kappa}_2 = \frac{1}{R_0}\frac{\partial\mathbf{b}_{p1}}{\partial\phi} + \left[\mathbf{b}_{p1}\cdot\nabla\mathbf{b}_{p1} - \frac{1}{R_0}\frac{\partial}{\partial\phi}(b_{\phi 1}\mathbf{b}_{p1}) - \frac{\mathbf{e}_R}{R_0}\right] \tag{7.85}$$

As in the sideband expansion toroidicity only appears explicitly in the $\mathbf{e}_\phi \cdot \nabla \mathbf{e}_\phi$ contribution to the curvature. After substantial algebra, one can evaluate C_4 and set it to zero. One obtains

$$e_\phi \cdot \nabla_\perp \psi \times \nabla_\perp \left[\nabla_\perp^2 A + \frac{d\beta}{d\psi} \left(\frac{r}{R_0} \cos\theta + |\nabla_\perp V|^2 + h^2 |V|^2 \right) \right] = 0 \quad (7.86)$$

The general solution is

$$\nabla_\perp^2 A = -\frac{d\beta}{d\psi} \left(\frac{r}{R_0} \cos\theta + |\nabla_\perp V|^2 + h^2 |V|^2 \right) - G(\psi) \quad (7.87)$$

where $G(\psi)$ is a second arbitrary free function related to the net current flowing on each flux surface.

Once $G(\psi)$ and $\beta(\psi)$ are specified, Eqs. (7.62), (7.79), and (7.87) constitute a complete mathematical description of stellarator equilibria.

Discussion. The basic structure of the stellarator equilibrium equations can be ascertained by eliminating A from Eq. (7.87) by means of Eq. (7.79). One finds

$$\nabla_\perp^2 \psi = -G(\psi) - \frac{d\beta}{d\psi} \left(\frac{r}{R_0} \cos\theta + |\nabla_\perp V|^2 + h^2 |V|^2 \right)$$
$$- \frac{i}{h} \nabla_\perp^2 (\mathbf{e}_\phi \cdot \nabla_\perp V^* \times \nabla_\perp V) \quad (7.88)$$

where V is a known function given by Eq. (7.63) and repeated here for convenience:

$$V(r, \theta) = \sum_l V_l I_l(hr) e^{il\theta} \quad (7.89)$$

The V_l are the prescribed amplitudes of the applied helical fields.

Equation (7.88) is a nonlinear, two-dimensional, elliptic partial differential equation for the flux function $\psi(r, \theta)$. It is similar in form to the Grad–Shafranov equation for axisymmetric tori. In fact, if one sets the helical fields to zero (i.e., $V = 0$) the resulting equation is identical to that of the high-β tokamak Eq. (6.97).

It is of interest to understand the mechanism by which a toroidal restoring force is generated in stellarator equilibria. In this connection it is useful to first identify those forces that do *not* play a major role. For simplicity assume the applied helical field consists of a single harmonic: $V(r, \theta) = V_l I_l(hr) \exp(il\theta)$. The terms involving V in Eq. (7.88) are then only functions of r and consequently cannot balance the $(r/R_0)\cos\theta$ toroidal term; that is, the helical sideband force does not play a major role in toroidal force balance in single helicity systems. The sideband force is actually of order δ^6 and as a result does not appear in the calculation.

Next, assume that $G(\psi)$ is chosen so that no net toroidal current flows on any flux surface. In this situation an applied vertical field does not produce a net body force on the plasma. While the vertical field plays an important role in defining the vacuum flux surfaces it must be emphasized that such a field is not required for the establishment of toroidal equilibrium. This is in contrast to the tokamak configuration.

In stellarator equilibria, the toroidal restoring force is generated by the interaction of the induced toroidal dipole current with the effective average value of the vacuum helical field on the plasma flux surface. The dipole current J_{DP}, sometimes known as the Pfirsch–Schluter current, is induced by toroidal effects and is required to maintain **J** divergence free. Setting $\mu_0 J_{DP} = \nabla_\perp^2 A_{DP}$ one finds from Eq. (7.87) that J_{DP} is given by

$$\frac{\mu_0 J_{DP}}{B_0} = -\frac{d\beta}{d\psi}\left(\frac{r}{R_0}\cos\theta\right) \tag{7.90}$$

On a flux surface the helical magnetic field has a small nonzero ϕ averaged component due to the helical distortion of the surface. In particular, if the poloidal magnetic field is of the form $B_\theta \approx \hat{B}_\theta(r)\cos(l\theta - N\phi)$, and the flux surface is given by $r = r_0 + r_1(\theta, \phi)$ with $r_1 \approx (\hat{B}_r/hB_0)\cos(l\theta - iN\phi)$, then the poloidally averaged value of B_θ is $\langle B_\theta \rangle \approx \hat{B}_r\hat{B}_\theta'/2hB_0$. The force density $\mathbf{J}_{DP} \times \langle \mathbf{B}_\theta \rangle \cdot \mathbf{e}_R$ when integrated over the plasma volume gives rise to the dipole current restoring force.

7.6. STELLARATORS, HELIOTRONS, AND TORSATRONS

One of the main applications of the stellarator theory just described is to the stellarator configuration first suggested by the Princeton Plasma Physics group (Spitzer, 1958).

Experimental Description. The stellarator configuration, one of the earliest in the fusion program, is a large aspect ratio, low-to-moderate β toroidal device whose main field components are a large toroidal field and a somewhat smaller single helicity helical field, usually $l = 2$ or $l = 3$. It can in principle operate with zero net toroidal current. Start-up and plasma heating are accomplished by a combination of rf heating and neutral beam injection. Some stellarators also operate with ohmic heating, at least for part of the discharge, in order to provide preliminary heating and to establish a good target for the auxiliary heating power.

Typical time scales of stellarator experiments are tens to hundreds of milliseconds. Recent experiments have demonstrated confinement times and temperatures quite similar to those of comparably sized tokamaks. These are indeed optimistic results. Since current-free discharges in a stellarator are possible both in principle and practice, future extrapolations are thought of in terms of steady-state operation.

The helical fields in a stellarator are generated by means of external coils, and three different magnet configurations have been suggested: the stellarator, the heliotron, and the torsatron, illustrated schematically in Fig. 7.6. The stellarator consists of $2l$ helical wires with alternating direction of current flow, situated in a large toroidal field. This is perhaps the most flexible configuration from the point of view of physics, but there exist very large forces, alternating inward and outward on the helical coils, making extrapolations to higher fields and larger experi-

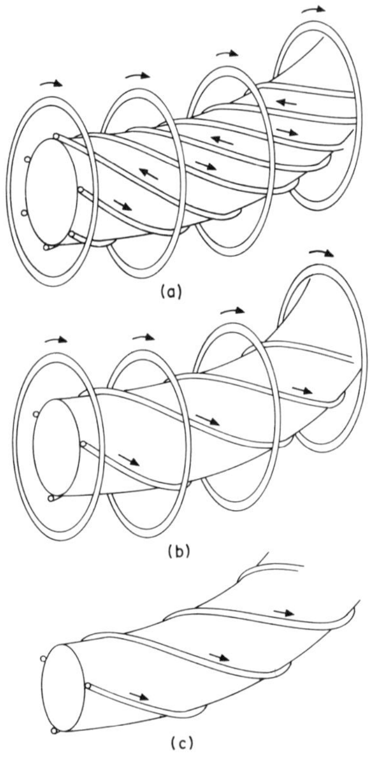

Figure 7.6. Schematic diagram of (a) stellarator, (b) heliotron, and (c) torsatron windings.

ments very unattractive if not impossible. The constraints imposed by the radially inward forces are more severe since the helical coils are usually mounted on a relatively fragile vacuum chamber. By eliminating the appropriate set of alternate helical coils, the remaining coil set experiences only outward forces. This is the heliotron. Although not quite as efficient as a stellarator in generating helical fields it has a considerable structural advantage. Note that the helical coil system itself produces a net toroidal field since all the currents flow in one direction. If in fact the external toroidal field is eliminated the heliotron is transformed into a torsatron. This is the least flexible method of generating helical fields in that the ratio of the helical field amplitude to the toroidal field amplitude can no longer be varied independently. However, by appropriately choosing the pitch of the windings the torsatron configuration can be made essentially force-free, thereby dramatically reducing the structural support problem. Although quite different in physical appearance, the stellarator, heliotron, and torsatron each produce virtually the same fields in the region occupied by plasma. Consequently, in the theoretical analysis these configurations are indistinguishable.

It is also worth pointing out that there has been considerable interest recently in designing modular stellarator coils (as opposed to the continuous types just discussed) in order to alleviate the access and maintainability problems inherent in reactor-size devices.

Rotational Transform. The analysis of the stellarator configuration begins with the evaluation of the rotational transform. This calculation is important as it demonstrates how the vacuum helical field establishes the basic structure of the flux surfaces. Also, the helical transform represents a convenient means of measuring the amplitude of the helical fields.

Calculating the transform in three-dimensional systems often involves a numerical integration along field-line trajectories many circuits around the torus. However, within the context of the stellarator expansion, the calculation can be carried out essentially analytically. The main difficulty is that the field-line trajectories are relatively complicated. In particular, a field line moves a finite distance over the poloidal cross section during one full circuit of the torus. Hence, no simple approximations are valid during the entire trajectory.

The difficulty is resolved by a two-step procedure which exploits the fact that there are many helical periods around the torus. Since each helical period is short, the transform per helical period is small and can be calculated by perturbation theory. Once determined, this transform can then be summed over many helical periods to evaluate the full transform.

The analysis proceeds as follows. The equations for a field line trajectory are given by

$$\begin{aligned} \frac{dr}{d\phi} &= \frac{RB_r}{B_\phi} \\ \frac{d\theta}{d\phi} &= \frac{RB_\theta}{rB_\phi} \end{aligned} \tag{7.91}$$

Consider a magnetic line with initial coordinates $r = r_0$, $\theta = \theta_0$, $\phi = \phi_0$. Since the helical fields are small in amplitude, of order δ, it follows that r and θ do not change much over the short distance corresponding to one helical period. Consequently, it is appropriate to expand $r = r_0 + r_1 + r_2 + \cdots$, $\theta = \theta_0 + \theta_1 + \theta_2 + \cdots$. The first-order equations are given by

$$\begin{aligned} \frac{dr_1}{d\phi} &= \left(\frac{R_0 B_{r1}}{B_0}\right)_{r_0, \theta_0} \\ \frac{d\theta_1}{d\phi} &= \left(\frac{R_0 B_{\theta 1}}{r_0 B_0}\right)_{r_0, \theta_0} \end{aligned} \tag{7.92}$$

Substituting $\mathbf{B}_1$ from Eq. (7.72) one finds

$$\begin{aligned} r_1 &= \frac{i}{h}\left(\frac{\partial V}{\partial r_0} e^{-iN\phi} - \frac{\partial V^*}{\partial r_0} e^{iN\phi}\right) \\ \theta_1 &= \frac{i}{hr_0^2}\left(\frac{\partial V}{\partial \theta_0} e^{-iN\phi} - \frac{\partial V^*}{\partial \theta_0} e^{iN\phi}\right) \end{aligned} \tag{7.93}$$

Note that over one helical period θ_1 returns to its original value; there is no contribution to the transform of order δ because of the helical symmetry.

There is, however, a second-order contribution obtained by solving

$$\frac{d\theta_2}{d\phi} = R_0\left[\left(r_1 \frac{\partial}{\partial r_0} + \theta_1 \frac{\partial}{\partial \theta_0}\right)\left(\frac{B_{\theta 1}}{r_0 B_0}\right) - \frac{B_{\theta 1} B_{\phi 1}}{r_0 B_0^2} + \frac{B_{\theta 2}}{r_0 B_0}\right]_{r_0, \theta_0} \tag{7.94}$$

The value of $\theta = \theta_0 + \theta_2(\phi_0 + 2\pi/N)$ represents the new starting value of θ_0 for the next helical period; that is, the change in θ_0 over one helical period $\delta\phi = 2\pi/N$ is given by $\delta\theta_0 = \theta_2(\phi_0 + 2\pi/N)$. After a straightforward calculation, making use of Eqs. (7.72) and (7.79), one obtains

$$\frac{\delta\theta_0}{\delta\phi} = \frac{R_0}{r}\frac{\partial\psi}{\partial r} \tag{7.95}$$

Equation (7.95) represents the rotational transform per helical period.

The total transform is calculated by summing over helical periods as follows. The angle $\Delta\phi$ transversed by a magnetic line as it makes one complete poloidal circuit is given by

$$\Delta\phi = \int_0^{\Delta\phi} \delta\phi = \int_0^{2\pi} \left(\frac{\delta\phi}{\delta\theta}\right) d\theta \tag{7.96}$$

where for convenience the zero subscript has been dropped from θ. The rotational transform is the average value of $\Delta\theta$ corresponding to one complete toroidal circuit $\delta\phi = 2\pi$, and is obtained by taking simple proportions:

$$\frac{\iota}{2\pi} = \frac{2\pi}{\Delta\phi} \tag{7.97}$$

or

$$\hat{\iota} = 2\pi R_0 \left(\int_0^{2\pi} \frac{r\,d\theta}{\partial\psi/\partial r}\right)^{-1} \tag{7.98}$$

where $\hat{\iota} = \iota/2\pi$ is the transform normalized to 2π. The integrand in Eq. (7.98) is to be evaluated on the flux surface $r = r(\theta)$ corresponding to the contour $\psi(r, \theta) = \text{const}$.

In general, the flux surfaces in a stellarator can be described as a set of toroidally averaged noncircular nested contours $\psi(r, \theta) = \text{const}$, with a superimposed helical distortion $\psi_1 = -p_3/(dp_2/d\psi)$ which rotates around the cross section as one progresses along ϕ. Both the flux surfaces and the transform simplify considerably for a single helicity vacuum field: $V(r, \theta) = V_l I_l(hr)\exp(il\theta)$. For the vacuum case the vector potential $A = 0$ and $\psi = (i/h)(\mathbf{e}_\phi \cdot \nabla V^* \times \nabla V)$ or

$$\psi = 2hlV_l^2\left(\frac{I_l I_l'}{u}\right) \tag{7.99}$$

where $I_l = I_l(u)$ and $u = hr$.

For a single helicity system with no applied vertical field, $\psi(r, \theta) \to \psi(r)$. The toroidally averaged flux surfaces are circles. Consequently, it follows that

$$\hat{\iota} = 2lh^3 R_0 V_l^2 \left[\frac{1}{u}\left(\frac{I_l I_l'}{u}\right)'\right] \tag{7.100}$$

There are several useful limits of Eq. (7.100) to investigate. For a relatively loosely wound helix corresponding to $hr \ll 1$, $\hat{\iota}$ reduces to

$$\hat{\iota} = \hat{\iota}_H \left(\frac{r}{a}\right)^{2l-4} \tag{7.101}$$

where

$$\hat{\iota}_H = \frac{4h^3R_0V_l^2(ha)^{2l-4}l^2(l-1)}{(2^l l!)^2} \tag{7.102}$$

is the vacuum helical transform on the plasma surface $r = a$.

Note that for $l \geq 3$ systems, such as the CLEO stellarator, the magnetic field possesses shear everywhere, but has zero transform on axis. For $l = 2$, on the other hand, there is a nonzero constant transform everywhere but the shear is zero, as in the Wendelstein stellarator. Also, stellarators with $l \geq 2$ helical fields have a circular magnetic axis while $l = 1$ systems have a helically rotating magnetic axis.

One can design configurations with both nonzero transform and nonzero shear by using a tightly wound $l = 2$ helical field: $hr \gtrsim 1$. This is the case for the Heliotron and ATF torsatron experiments. When $hr \sim 1$, the full Bessel functions must be maintained. A reasonable approximation to the transform is obtained by expanding the Bessel functions one order higher in u. For an $l = 2$ system one obtains

$$\hat{\iota} = \hat{\iota}_H\left[\frac{1 + h^2r^2/2}{1 + h^2a^2/2}\right] \tag{7.103}$$

Observe that $\hat{\iota}$ increases quadratically with r. A rotational transform profile that monotonically increases with radius is a general property of the stellarator configuration. This represents one major difference from the tokamak where the transform is normally a decreasing function.

A comparably complicated set of calculations can be carried out for the magnetic well. The stability issues are, however, somewhat involved and a full discussion is deferred until Chapter 10. At the present point it is sufficient to note that a straight helix has an unfavorable vacuum magnetic well. (See Problem 7.6.) To obtain favorable curvature one needs to include the effects of toroidicity and the vertical field.

Stellarator Equilibria in the Limit $\beta/\epsilon \ll 1$. Having established the important properties of the vacuum surfaces and the rotational transform one can now proceed to solve the stellarator equations including the effect of plasma pressure and current. Because of the complexity of the equations it is useful to begin by considering single helicity systems with low β, specifically $\beta/\epsilon \ll 1$. This is the regime analogous to the ohmically heated tokamak.

For low β it is appropriate to expand $\psi(r, \theta)$ and the free function $G(\psi)$ as follows:

$$\psi(r, \theta) = \psi_0(r) + \psi_1(r, \theta) + \cdots \tag{7.104}$$

$$G(\psi) = G_1(\psi_0) + \cdots \tag{7.105}$$

with $\psi_1/\psi_0 \sim \beta/\epsilon$. The leading-order contribution to the stellarator flux equation [Eq. (7.88)] reduces to

$$\begin{aligned}\nabla_\perp^2 \psi_0 &= -\frac{i}{h}\nabla_\perp^2(\mathbf{e}_\phi \cdot \nabla_\perp V^* \times \nabla_\perp V) \\ V &= V_l I_l(hr)e^{il\theta}\end{aligned} \tag{7.106}$$

which has as its solution

$$\psi_0(r) = 2hlV_l^2\left(\frac{I_l I_l'}{u}\right) \tag{7.107}$$

where, as before, $I_l = I_l(u)$ and $u = hr$. Equation (7.107) indicates that in leading order the flux surfaces and transform coincide with those of the vacuum helical fields.

In first order, the equation for ψ_1 can be written as

$$\nabla_\perp^2 \psi_1 = -G_1(\psi_0) - \frac{d\beta}{d\psi_0}\left(\frac{r}{R_0}\cos\theta + |\nabla_\perp V|^2 + h^2|V|^2\right) \tag{7.108}$$

Consider now the determination of $G_1(\psi_0)$ for the interesting case where there is zero net toroidal current flowing on every flux surface. Recalling that the right-hand side of Eq. (7.108) is proportional to J_ϕ, and that the leading-order flux surfaces are circles, it follows that

$$G_1[\psi_0(r)] = -\frac{d\beta}{d\psi_0}(|\nabla_\perp V|^2 + h^2|V|^2) \tag{7.109}$$

With this choice, ψ_1 satisfies

$$\nabla_\perp^2 \psi_1 = -\frac{d\beta}{d\psi_0}\frac{r}{R_0}\cos\theta \tag{7.110}$$

If one now treats $\beta(r)$ rather than $\beta(\psi_0)$ as the free function, and makes use of Eq. (7.98), then ψ_1 can be easily evaluated:

$$\psi_1(r, \theta) = \left[\int_r^\infty \frac{dx}{x^3}\int_0^x \frac{y^2\beta'(y)\,dy}{\iota(y)} + \frac{B_v}{B_0}\right] r\cos\theta \tag{7.111}$$

In obtaining ψ_1 it has been assumed that far from the plasma a small vertical field of magnitude B_v has been applied.

Equation (7.111) demonstrates that low-β stellarator equilibria exist for a wide variety of helical fields and pressure profiles. Furthermore, while a vertical field influences the form of ψ_1, such a field is not required for equilibria; well-behaved solutions exist for $B_v = 0$. Equation (7.111) also shows that for $l \geq 3$, the pressure profile must be very flat near the origin for the integrals to converge, since $\iota(r) \propto r^{2l-4}$

near $r = 0$. This is intuitively clear since high l values produce no magnetic fields near the origin and hence cannot support any pressure gradient.

Since $\psi_1(r, \theta) = \hat{\psi}_1(r) \cos\theta$, the flux surfaces are shifted circles. The toroidal shift defined by $\Delta(r) = -\hat{\psi}_1(r)/\psi_0'(r)$ is easily obtained and is given by

$$\frac{\Delta}{R_0} = -\frac{1}{\iota(r)}\left[\int_r^\infty \frac{dx}{x^3}\int_0^x \frac{y^2\beta'(y)\,dy}{\iota(y)} + \frac{B_v}{B_0}\right] \tag{7.112}$$

There are two points to be noted. First, since $\beta' < 0$, the introduction of plasma pressure produces an outward shift of the flux surfaces. In contrast, helical sideband equilibria are characterized by an inward shift. Recall that an outward shift is in the direction corresponding to stability, at least with respect to rigid horizontal displacements. Second, observe that the vertical field produces a shift of the vacuum flux surfaces. In fact, for $l \geq 3$, the shift of the magnetic axis is infinite. This is again associated with the vanishing of the helical field near the origin for $l \geq 3$. Thus, for $l \geq 3$ the slightly shifted circle assumption breaks down and the full two-dimensional stellarator equations must be solved.

In summary, a simple physical picture of the low-β stellarator is given as follows. Vacuum flux surfaces are established by means of a helical field and perhaps a vertical field. Plasma is then "loaded" onto these flux surfaces. The toroidal forces tend to push the plasma outward. When the pressure is low, a small toroidal dipole current is induced which produces a restoring force. The plasma is shifted slightly outward until toroidal force balance is achieved. The vertical field contributes to the toroidal shift but is not required for equilibria to exist.

Stellarator Equilibria with $\beta/\epsilon \sim 1$; Zero Current Case. The next system of interest is the moderate-β stellarator with zero net toroidal current. This system is analogous in many ways to the high-β tokamak. In particular, the flux surfaces become noncircular, the shifts become finite, and the system exhibits an equilibrium β limit.

Since the equilibrium is two-dimensional, a number of simplifying assumptions must be made in order to obtain an analytic solution. First, it is assumed that the vacuum helical field corresponds to a loosely wound ($hr \ll 1$), low shear, $l = 2$ configuration. (Systems with shear and other l numbers are discussed in Problem 7.7.) Second, the functions G and $d\beta/d\psi$ are chosen as follows: $G = -K$, $d\beta/d\psi = -C$ with K and C constants. Note that with G a constant there is freedom to set the net current to zero on only one flux surface. In the analysis this surface is chosen to coincide with the boundary of the plasma, thereby implying that no net toroidal current flows in the plasma. Finally, the toroidally averaged plasma boundary is assumed to be circular. Hence, the boundary condition on ψ is $\psi(a, \theta) = 0$.

Under these conditions the equation for ψ reduces to

$$\nabla_\perp^2 \psi = K + C\left(\frac{r}{R_0}\right)\cos\theta + \nabla_\perp^2\left(\frac{\iota_H r^2}{2R_0}\right) \tag{7.113}$$

where ι_H is the vacuum transform associated with the helical field. The solution for ψ satisfying the boundary conditions is given by

$$\psi = \frac{a^2}{2R_0}(\hat{K} + \iota_H + \hat{C}\rho\cos\theta)(\rho^2 - 1) \tag{7.114}$$

with $\rho = r/a$, $\hat{K} = KR_0/2$, and $\hat{C} = Ca/4$.

One now expresses $\hat{K}$ and $\hat{C}$ in terms of physical parameters. $\hat{K}$ can be related to the net toroidal current flowing in the plasma. If one recalls that

$$\frac{\mu_0 J_\phi}{B_0} = K + C\left(\frac{r}{R_0}\right)\cos\theta \tag{7.115}$$

then $\hat{K}$ is given by

$$\hat{K} = \iota_I \equiv \frac{\mu_0 R_0 I_0}{2\pi a^2 B_0} \tag{7.116}$$

The value of $\hat{C}$ is related to the toroidal β_t defined as

$$\beta_t = \frac{1}{\pi a^2}\int \beta r\, dr\, d\theta \tag{7.117}$$

A straightforward calculation which makes use of the relation $\beta = -C\psi$ yields

$$\hat{C} = \frac{\beta_t}{\epsilon(\iota_I + \iota_H)} \tag{7.118}$$

By using Eqs. (7.116) and (7.118) one can calculate the MHD transform on the plasma surface ι_a, and the magnetic axis shift Δ_0. From Eq. (7.98) it can easily be shown that

$$\iota_a = (\iota_I + \iota_H)(1 - \nu^2)^{1/2} \tag{7.119}$$

where

$$\nu = \frac{\hat{C}}{\iota_I + \iota_H} = \frac{\beta_t}{\epsilon(\iota_I + \iota_H)^2} \tag{7.120}$$

The final quantity is the magnetic axis shift, defined as the radius where

$\partial\psi(r, \theta = 0)/\partial r = 0$. A simple calculation gives

$$\frac{\Delta_0}{a} = \frac{\nu}{1 + (1 + 3\nu^2)^{1/2}} \tag{7.121}$$

Consider the application of these relations to the stellarator with zero net toroidal current: $\hat{\iota}_I = 0$. Equation (7.119) shows that at low β (i.e., $\nu \to 0$) $\hat{\iota}_a = \hat{\iota}_H$. As expected, the MHD transform is equal to the applied helical transform. As ν increases at fixed helical field $\hat{\iota}_H$, $\hat{\iota}_a$ decreases. Ultimately, when $\nu \to 1$ then $\hat{\iota}_a \to 0$: the rotational transform vanishes on the plasma surface. This is equivalent to the equilibrium limit in a high-β tokamak at fixed q_*. At the equilibrium limit β_t is given by

$$\beta_t = \epsilon \hat{\iota}_H^2 \tag{7.122}$$

which is identical in form to the high-β tokamak limit given by $\beta_t = \epsilon/q_*^2$. Observe also that as ν increases, the magnetic axis shift increases and at the limit $\nu = 1$

$$\frac{\Delta_a}{a} = \frac{1}{3} \tag{7.123}$$

In comparing tokamaks and stellarators it is worth making the following point regarding the equilibrium β_t limit. Although both criteria are of the same form, the actual tokamak limit is set by the lowest value of q_* allowed by current-driven kink instabilities. Since a stellarator has zero net current, kink modes are not an important issue. Instead, the practical β limit is set largely by technological considerations: how high can $\hat{\iota}_a$ be made with realistic coils? Whereas $q_{*\min} \sim 2$–3 for a tokamak, values of $\hat{\iota}_H \approx 1$ are achievable in stellarator experiments. Thus, in practice ideal MHD equilibrium β limits appear to be less serious in a stellarator than in a tokamak.

Stellarator Equilibria with $\beta/\epsilon \sim 1$: Flux-Conserving Operation. The final configuration to be analyzed is the moderate-β stellarator operating in the flux-conserving mode. As might be expected, this system is quite analogous to the high-β flux-conserving tokamak. In particular, equilibria exist for arbitrarily high β/ϵ. One interesting feature of this configuration is that a net toroidal current is induced as the plasma is heated in order to conserve the flux. Therefore, a stellarator that is initially current-free does not remain so when heated in the flux-conserving mode. The magnitude and consequences of this current are evaluated in the analysis that follows.

By using the same equilibrium model as for the moderate-β, current-free stellarator one can easily derive the relevant relations for flux-conserving operation. The critical step in the analysis is to hold the MHD transform on the surface $\hat{\iota}_a$ fixed during the evolution. This represents the simplest model of flux conservation and replaces the condition $\hat{\iota}_I = 0$ used for the current-free stellarator.

For the systems under consideration the important relations to calculate are the functional dependences of the net toroidal current and the magnetic axis shift on the heating parameter, proportional to β_t. These relations are calculated as follows. The solution given by Eq. (7.114) is valid for the flux-conserving stellarator. What is required is a proper interpretation of the equilibrium relations given by Eqs. (7.119)–(7.121).

In this connection assume that prior to the heating phase the plasma is in a low-β, current-free state in which the MHD transform on the surface coincides with the vacuum helical transform: $\hat{\iota}_a = \hat{\iota}_H$. During the entire flux-conserving evolution $\hat{\iota}_a$ remains fixed at its initial value $\hat{\iota}_H$. Setting $\hat{\iota}_a = \hat{\iota}_H$ in Eq. (7.119) gives

$$\nu^2 = 1 - \left(\frac{\hat{\iota}_H}{\hat{\iota}_I + \hat{\iota}_H}\right)^2 \tag{7.124}$$

The parameter ν is related to β_t through Eq. (7.120). Upon substituting, one can then solve for $\hat{\iota}_I$ as a function of the heating parameter $H \equiv \beta_t/\epsilon\hat{\iota}_H^2$. The result is

$$\frac{\hat{\iota}_I}{\hat{\iota}_H} = \left\{\frac{1}{2}[1 + (1 + 4H^2)^{1/2}]\right\}^{1/2} - 1 \tag{7.125}$$

Similarly, elimination of ν from Eq. (7.121) leads to the following expression for the magnetic axis shift

$$\begin{aligned} \frac{\Delta_0}{a} &= \frac{\nu}{1 + (1 + 3\nu^2)^{1/2}} \\ \nu &= \frac{2H}{1 + (1 + 4H^2)^{1/2}} \end{aligned} \tag{7.126}$$

Equations (7.125) and (7.126) are illustrated in Fig. 7.7. As in the high-β flux-conserving tokamak, well-behaved equilibrium solutions exist for arbitrarily high β_t/ϵ. Consider now the induced net toroidal current. In the low- and high-β_t/ϵ limits

$$\begin{aligned} \frac{\hat{\iota}_I}{\hat{\iota}_H} &\approx \frac{H^2}{2}, \qquad H \ll 1 \\ \frac{\hat{\iota}_I}{\hat{\iota}_H} &\approx H^{1/2}, \qquad H \gg 1 \end{aligned} \tag{7.127}$$

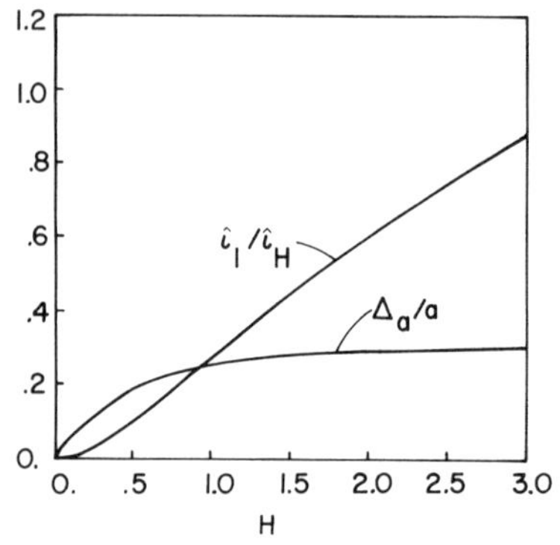

Figure 7.7. Toroidal current (in units of $\hat{\iota}_I/\hat{\iota}_H$) and magnetic axis shift Δ_a/a versus the heating parameter $H = \beta_t/\epsilon\hat{\iota}_H^2$ in a flux-conserving stellarator.

Thus, if β_t becomes large enough a substantial net current can be induced during flux-conserving operation. This current ultimately decays on the resistive diffusion time scale (i.e., the time scale for which flux conservation no longer is valid). However, this time is much longer than the characteristic MHD time. Hence, care must be taken, either by limiting the final β_t or starting with a reversed $\hat{\iota}_I$, to ensure that current-driven kink modes are not excited during the flux-conserving evolution.

Numerical Equilibria. The results described above and in previous sections are in good qualitative agreement with multidimensional numerical calculations. In recent years, codes have been developed which solve the two-dimensional stellarator expansion equations. A large effort has also been devoted to the development of full three-dimensional codes. Several different numerical procedures have been proposed and the progress has been quite remarkable.

One of the fastest and most efficient three-dimensional codes has been developed recently by Hirshman (1983) of the Oak Ridge National Laboratory. A series of poloidal flux plots at various toroidal angles is illustrated in Fig. 7.8 from the Hirshman code. The configuration corresponds to the ATF experiment currently under construction at Oak Ridge. This device is a high-shear $l = 2$ torsatron with the vacuum helical transform varying between 0.35 on axis to 0.9 at the plasma edge. It should be capable of achieving β_t values in excess of 5%. Observe that the flux contours vary considerably from one toroidal angle to another. Also in no poloidal cross section are the flux contours nearly circular. In this regard, the analytic theory might at best be expected to provide good qualitative insight into the behavior of stellarators but not to be very accurate quantitatively. In practice, the analytic theory is often more

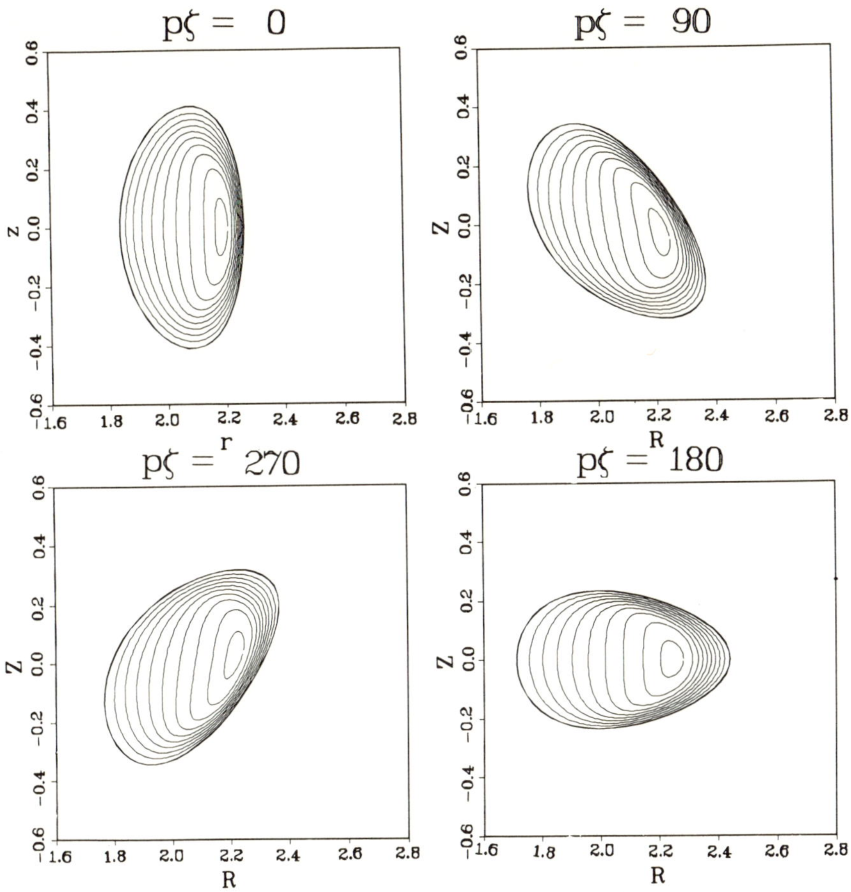

Figure 7.8. Poloidal flux plots at various toroidal angles for the ATF stellarator at Oak Ridge National Laboratory as calculated from a three-dimensional numerical code. Courtesy S. Hirshman, Oak Ridge National Laboratory.

accurate than one has a right to expect, although still not accurate enough for the purposes of experimental design and data interpretation.

Summary. One sees from the analysis in this section that it is possible to operate a stellarator in a variety of ways: low β, current free, or flux conserving with each regime having an analog in the tokamak concept. In general, stellarators and tokamaks of comparable size produce comparable performance in terms of densities, temperatures, and confinement times. Furthermore, in absolute units, this performance is quite attractive. The stellarator has the advantages of potential steady-state operation with current-free operation, and immunity from kink instabilities. It has the disadvantages of a more complicated magnet technology and a stronger reliance on auxiliary power for start-up and heating. Future experiments will decide which of these concepts offers the most promise for fusion.

7.7. SUMMARY

In Chapter 7 the MHD model has been used to calculate the equilibrium properties of three-dimensional toroidal configurations. These systems, which are inherently more complicated than two-dimensional axisymmetric tori, are of interest because of their potential for steady-state operation with zero net toroidal current. As no relatively simple, exact Grad–Shafranov equation exists for such configurations, one is forced to introduce asymptotic expansions at the outset, rather than the completion, of the analysis. This is a difficult task because of the complex geometry and the corresponding lack of intuition. Even so, efficient methods of analysis have been developed, leading to the identification of two basically different types of three-dimensional systems. A summary is given below.

The Parallel Current Constraint. The parallel current constraint is a convenient form of the relationship $\nabla \cdot \mathbf{J} = 0$. it represents perhaps the most efficient method (i.e. minimum algebra) of closing the system of asymptotic equations resulting from any of the standard expansions. The constraint effectively determines the shape of the leading-order flux surfaces by ensuring that the $J_{\|}$ required to make $\nabla \cdot \mathbf{J} = 0$ lies on the surface.

Helical Sideband Equilibria. The first basic class of toroidal equilibria corresponds to helical sideband configurations. These systems have high β and are described by the following expansion: $\beta \sim \iota \sim ha \sim 1$, $\epsilon \sim \delta^2$. In configurations with zero net current, toroidal force balance is achieved by the interaction of the applied helical fields with induced helical sideband currents generated by toroidicity. The equations describing helical sideband equilibria have the form of a pair of coupled, nonlinear, two-dimensional, partial differential equations.

The Elmo Bumpy Torus. One interesting example of helical sideband equilibria is the Elmo bumpy torus. This configuration is an $l = 0$ toroidal helix consisting of a series of linked mirror cells. A hot electron ring, generated by electron cyclotron resonance heating, is located in the bulge region of each cell. The equilibrium analysis leads to the unexpected result that the toroidal shift of the flux surfaces is inward even though the toroidal forces tend to push the plasma outward. This is a property of all single helicity helical sideband equilibria. The physical explanation of the inward shift is associated with the fact that helical sideband equilibria are essentially confined in an unstable MHD potential well, i.e., a hill. If it were not for the distinctly non-MHD behavior of the hot electron rings, EBT would indeed be very unstable.

Stellarator Equilibria. Stellarator configurations represent the second basic class of toroidal equilibria. These systems have lower values of β than sideband systems. Specifically, the expansion describing stellarator equilibria is given by $\hat{\imath} \sim ha \sim 1$, $\beta \sim \epsilon \sim \delta^2$. These configurations are also capable of operation with zero net toroidal current. In such cases, the toroidal restoring force is generated by the interaction of an induced toroidal dipole current with the small nonzero average value of the vacuum helical field on any given flux surface. The equation describing these equilibria has the form of a nonlinear, second-order, two-dimensional, partial differential equation of the Grad–Shafranov type. When the helical fields are set to zero, the equation reduces to that of the high-β tokamak.

Stellarators, Heliotrons, and Torsatrons. An important application of the theory is to stellarators, heliotrons, and torsatrons. These are actually three different coil configurations which, at least in the context of the theory, generate the same toroidal–helical magnetic field. The stellarator is most difficult to build but is the most flexible in operation. The opposite is true for the torsatron, with the heliotron being intermediate. Stellarators can operate in a variety of ways: low β, zero net toroidal current, or flux conserving. In each of these systems the toroidal shift is outward, indicating confinement in an MHD potential well that is stable against rigid horizontal displacements. Most stellarators have either $l = 2$ or $l = 3$ helical fields. In general, their performance is comparable to that of tokamaks of the same size. This is an optimistic result. It represents one major motivation for the interest in this concept. Compared to tokamaks, stellarators have the advantage of potential steady-state operation without the presence of kink modes, but have the disadvantage of a more complicated magnet technology.

In conclusion, Chapters 4–7 have shown that the ideal MHD model provides a very reliable description of the equilibria of magnetically confined fusion plasmas.

REFERENCES

Blank, A. A., H. Grad, and H. Weitzner (1969). In *Plasma Physics and Controlled Nuclear Fusion Research 1968,* IAEA, Vienna, Vol. II, p. 607.
Freidberg, J. P., R. Y. Dagazian, and D. C. Barnes (1979). *Phys. Fluids* **22,** 926.
Grad, H. (1967). *Phys. Fluids* **10,** 137.
Greene, J. M., and J. L. Johnson (1961). *Phys. Fluids* **4,** 875.
Hirshman, S. P., and J. C. Whitson (1983). *Phys. Fluids* **26,** 3553.
Miller, G. (1977). *Phys. Fluids* **20,** 928.
Rosenbluth, M. N., J. L. Johnson, J. M. Greene, and K. E. Weimer (1969). *Phys. Fluids* **12,** 726.
Spitzer, L., Jr. (1958). *Phys. Fluids* **1,** 253.

GENERAL REFERENCES

EBT

Dandl, R. A., F. W. Baity, Jr., K. H. Carpenter, J. A. Cobble, H. O. Eason, J. C. Glowienka, *et al.* (1979). In *Plasma Physics and Controlled Nuclear Fusion Research,* IAEA, Vienna, Vol. II, p. 365.
Freidberg, J. P. (1980). *Nucl. Fusion* **20,** 673.
Grad, H. (1967). *Phys. Fluids* **10,** 137.
Hedrick, C. L., R. A. Dandl, J. A. Cobble, R. A. Dory, H. O. Eason, *et al.* (1977). In *Plasma Physics and Controlled Nuclear Fusion Research,* IAEA, Vienna, Vol. II, p. 145.
McAlees, D. G., N. Uckan, L. M. Lidsky, *et al.* (1976). Oak Ridge National Laboratory Report ORNL/TM-5669 (unpublished).
Uckan, N. A. (1980). Editor, Proceedings of the EBT Ring Physics Workshop CONF-791228, Oak Ridge National Laboratory.

Stellarators

Miyamoto, K. (1978). *Nucl. Fusion* **18,** 243.
Shafranov, V. D. (1980). *Nucl. Fusion* **20,** 1075.
Shafranov, V. D. (1983). *Phys. Fluids* **26,** 357.
Shohet, J. L. (1977). *Comments Plasma Phys. Controlled Fusion* **3,** 25.
Solov'ev, L. S., and V. D. Shafranov (1979). In *Reviews of Plasma Physics,* edited by M. A. Leontovich, Consultants Bureau, New York, Vol. V.

Three-Dimensional MHD Computations

Anania, G., and J. L. Johnson (1983). *Phys. Fluids* **26,** 3070.
Bauer, F., O. Betancourt, and P. Garabedian (1980). *J. Comput. Phys.* **39,** 341.
Carraras, B. A., J. L. Cantrell, L. A. Charlton, L. Garcia, J. H. Harris, T. C. Hender, *et al.* (1984). In *Proceedings of the International Conference on Plasma Physics and Controlled Nuclear Fusion Research,* London, UK.
Hirshman, S. P., and J. C. Whitson (1983). *Phys. Fluids* **26,** 3553.
Strauss, H. R., and D. A. Monticello (1981). *Phys. Fluids* **24,** 1148.

PROBLEMS

7.1. Show that the parallel current constraint given by Eq. (7.8) is automatically satisfied for an arbitrary one-dimensional screw pinch equilibrium. Repeat the calculation for an arbitrary axisymmetric torus as described by the Grad–Shafranov equation.

7.2. In Section 7.4 it was shown that in the EBT configuration the flux surfaces are shifted inward as a result of toroidicity. This problem demonstrates how the inward shift can be eliminated by the application of a small, external $l = \pm 1$ sideband field. To carry out the calculation consider the expanded equilibrium model given by Eqs. (7.41)–(7.44). To make the algebra tractable, assume $ha \ll 1$. In fact, the entire problem is greatly simplified by introducing the following low-β, large aspect ratio expansion:

$$\beta \sim \delta^2$$

$$\frac{\partial V_0}{\partial r} \sim \delta^2$$

$$ha \sim \delta$$

$$\frac{\partial V_1}{\partial r} \sim \delta^3$$

$$a/R_0 \sim \delta^5$$

Using this expansion derive a formula for the shift of the flux surfaces. Express your answer in terms of ha, a/R_0, and the two parameters $B_{l=0}/B_0$ and $B_{l=1}/B_0$. The last two quantities are measures of the $l = 0$ and $l = 1$ field amplitudes defined as follows:

$$B_{l=0} \equiv |B_{\phi 1}|_{r=0}$$

$$B_{l=1} \equiv |B_{\theta 1}|_{r=0}$$

Show that to leading order, the shift $\Delta(r)$ is independent of r. Show that the shift vanishes when the $l = 0$ and $l = 1$ amplitudes are chosen such that

$$\frac{B_{l=0}B_{l=1}}{B_0^2} = -\frac{4}{3hR_0}$$

Hint: When calculating $V_0(r)$ it is necessary to carry out the δ expansion to second order: $V_0(r) = c_1(1 + c_2h^2r^2 + \cdots)$. The c_2 contribution must be kept in order to evaluate V_0'.

7.3. Following the procedure described in Section 7.4 derive the flux surface shift of an arbitrary l helical sideband stellarator. As in the EBT calculation assume β is very small. Your final expression should be similar in form to Eq. (7.53). Express your results in terms of ha, a/R_0, and the helical field amplitude B_l/B_0 defined as

$$B_l = |B_{\theta 1}|_{r=a}$$

7.4. Show that for stellarator equilibria, the average poloidal flux is related to $A(r, \theta)$ and $V(r, \theta)$ as follows [Eq. (7.77)]:

$$\psi_p(r, \theta) = 2\pi R_0 B_0\left(A - \frac{i}{h}\mathbf{e}_\phi \cdot \nabla_\perp V^* \times \nabla_\perp V\right)$$

7.5. Consider a stellarator equilibrium where the applied fields consist of a toroidal field, a helical field, and a vertical field. For simplicity assume $ha \ll 1$. Using Eqs. (7.77), (7.100), and (7.101) show that the shift of the magnetic axis for the vacuum fields is given by

$$\frac{\Delta_0}{a} = \left(\frac{RB_v}{aB_0\hat{\iota}_H}\right)^{1/(2l-3)}$$

for $l \geq 2$.

7.6. This problem describes the procedure for determining the magnetic well in a stellarator. The calculation closely follows that presented in Section 7.6 for the evaluation of the rotational transform.

a. Using the stellarator expansion write the equations for a flux surface as $r = r_0(\theta_0) + r_1(r_0, \theta_0, \phi)$, $\theta = \theta_0 + \theta_1(r_0, \theta_0, \phi)$. Evaluate B_ϕ and B on the flux surface. Keep terms up to and including order δ^2.

b. Show that as ϕ changes by $2\pi/N$ (i.e., one helical period) the starting angle θ_0 changes by

$$\delta\theta_0 = \frac{2\pi R_0}{N}\left(\frac{1}{r_0}\frac{\partial\psi}{\partial r_0}\right)$$

Next, show that $\langle B^2 + 2\mu_0 p\rangle$ over the given helical period is given by

$$\delta\langle B^2 + 2\mu_0 p\rangle = \frac{B_0^2}{N}\left(1 - 2\frac{r_0}{R_0}\cos\theta - 2|\nabla_\perp V|^2 - 2h^2|V|^2\right)$$

c. Calculate the change in ϕ and $\langle B^2 + 2\mu_0 p\rangle$ after the field line makes one complete poloidal circuit ($\Delta\theta_0 = 2\pi$). Show that

$$\Delta\phi = \frac{N}{R_0}\int_0^{2\pi} \frac{r_0\, d\theta_0}{\partial\psi/\partial r_0}$$

$$\Delta\langle B^2 + 2\mu_0 p\rangle = \frac{B_0^2}{2\pi R_0} \times \int_0^{2\pi} \frac{[1 - 2(r_0/R_0)\cos\theta_0 - 2|\nabla_\perp V|^2 - 2h^2|V|^2]r_0\, d\theta_0}{\partial\psi_0/\partial r_0}$$

d. By taking simple proportions show that

$$\frac{2V}{\langle B^2\rangle}\frac{d\langle B^2/2\rangle}{dV} = -r\frac{d}{dr}\langle (r_0/R)\cos\theta_0 + |\nabla_\perp V|^2 + h^2|V|^2\rangle$$

where

$$\langle Q\rangle \equiv \int_0^{2\pi} \frac{Qr_0\, d\theta_0}{\partial\psi/\partial r_0} \bigg/ \int_0^{2\pi} \frac{r_0\, d\theta_0}{\partial\psi/\partial r_0}$$

and r is the flux label: $r_0 = r_0(\theta_0, r)$.

e. Evaluate W for a low-β single helicity stellarator with no vertical field. Show that when $V(r, \theta) = V_l I_l(hr)\exp(il\theta)$,

$$W = -h^2V_l^2 u(I_l'^2 + I_l^2/u^2 + I_l^2)'$$

where $u = hr$. Note that $W < 0$, indicating that a single helicity stellarator with zero vertical field has an unfavorable magnetic well.

7.7. By following the analysis presented in Section 7.6 one can calculate the equilibrium of a single helicity, $\beta \sim \epsilon$ stellarator for arbitrary l. Assume the free functions G and β are chosen as $G = -K$, $d\beta/d\psi = -C$ with K and C constants. Also, use the small argument expansion ($ha \ll 1$) of the Bessel functions to simplify the evaluation of $V(r, \theta)$.

a. Show that the solution for ψ corresponding to a current-free stellarator is given by

$$\psi = \frac{a^2}{2R_0}\left[\frac{l\beta_t}{2\iota_H\epsilon}(\rho^3 - \rho)\cos\theta + \frac{\iota_H}{l-1}(\rho^{2l-2} - 1)\right]$$

where $\rho = r/a$ and ι_H is the vacuum transform/2π at the edge of the plasma $r = a$.

b. Calculate the MHD safety factor at the edge of the plasma. Show that there is an equilibrium β_t limit given by

$$\beta_t < \frac{2\iota_H^2\epsilon}{l}$$

8

Stability

General Considerations

8.1. INTRODUCTION

The remainder of the textbook is concerned with determining the stability properties of ideal MHD equilibria. The problem of stability can be stated qualitatively as follows. The existence of an MHD equilibrium state implies a situation where the sum of the forces acting on the plasma is zero. If the plasma is perturbed from this state, the resulting perturbed forces either restore the plasma to its original equilibrium (stability) or cause a further enhancement of the initial disturbance (instability). The question of ideal MHD stability is crucial since, as a general rule, plasmas suffer catastrophic termination as a consequence of such instabilities. Thus, there is consensus in the fusion community that ideal MHD stability is a necessary requirement for a fusion reactor. To a large extent it is the avoidance of ideal MHD instabilities that has led to the somewhat unusual, ingenious, and sometimes technologically complicated magnetic configurations in the international fusion program.

The goal of Chapter 8 is to develop a basic understanding of the various mechanisms that drive instabilities and to discuss possible ways to avoid them. There are a number of techniques available for investigating ideal MHD stability. By far, the greatest effort is devoted to linear stability analysis. This is often adequate for practical situations. Nonlinear analysis of ideal MHD modes essentially provides the details of plasma self-destruction; that is, these modes are so virulent that it is more important in practice to avoid them (a linear stability issue) rather than ascertain their nonlinear evolution. Because of this and the fact that physical intuition is most easily acquired when analytic treatment is possible, the discussion presented in Chapter 8 is concerned primarily with analytic linear theory. However, in trying to assess the merits of actual experimental concepts one must rely on the results of numerical studies of linear stability; the expansions used in the analytic theories are not accurate enough, or general enough, for realistic designs. The numerical results are discussed in conjunction with the analytic theory in Chapters 9–11, which are concerned with applications.

Chapter 8 begins with a discussion of a mathematical definition of stability particularly applicable to ideal MHD. Next, a short derivation is presented of the dispersion relation for waves in an infinite homogeneous MHD plasma. These waves play an important role in providing intuition into the dynamical behavior of an MHD plasma. Following this is a rather extensive discussion of the formulation of the general linearized stability problem, starting with the equations of motion and culminating with the development of the Energy Principle, a powerful method for testing stability. Having established the Energy Principle, it is then possible to address the role of plasma compressibility on ideal MHD stability in a quite general manner. This demonstrates the relationship between the ideal and collisionless MHD stability boundaries. Next, a rather subtle point is investigated concerning whether the region outside the main plasma core is more accurately described by a vacuum or by a relatively cold, but still perfectly conducting, force-free plasma. Finally, a brief discussion is presented of the general classification of ideal MHD instabilities.

8.2. DEFINITION OF STABILITY

A useful starting point for the analysis is the definition of stability. Actually there are a number of different definitions that might be appropriate depending upon the particular physical properties of the system under investigation. Several possible mechanical analogs are illustrated in Fig. 8.1.

In Fig. 8.1a, if the ball is moved a small distance from its equilibrium position, it simply oscillates indefinitely about this point, assuming an ideal frictionless system. Even though the ball never returns to rest at its equilibrium position, the system can be considered stable; that is, the ball always remains close to its original position. This is the best that one can expect in a dissipationless system. Figure 8.1b indicates instability. A small perturbation off the top of the hill sets the ball rolling continually further away from its equilibrium position. In Fig. 8.1c the ball is neutrally or marginally stable. If the ball is placed a short distance away from its initial position, it neither continues to move further away, nor

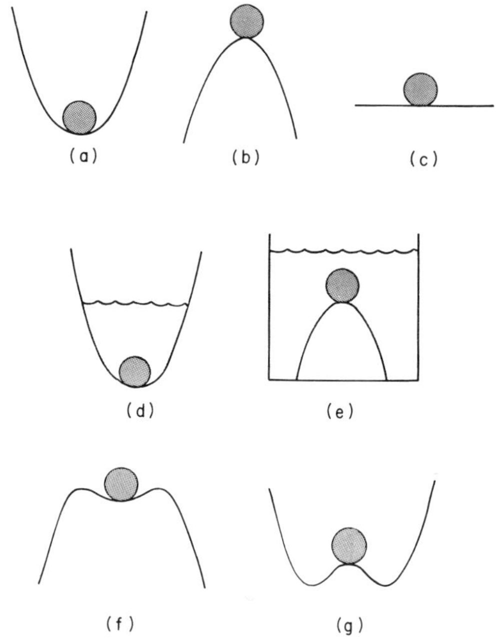

Figure 8.1. Mechanical analogs of various types of stability.

does it return and oscillate about its original position. Oftentimes in marginally stable plasmas, a small change in some parameter, e.g., β or I, transforms the system into one of type Fig. 8.1a or Fig. 8.1b. Thus, the condition for neutral stability defines the boundary between stability and instability.

Some of the effects of dissipation can be ascertained by imagining that the ball is immersed in a viscous fluid such as oil. In Fig. 8.1d, if the ball is moved from its equilibrium position, it will eventually return and come to rest at this position: the dissipation damps out the oscillations. Figure 8.1e shows that while dissipation can slow down the motion of the ball, it cannot in general convert an unstable system into a stable one. Figures 8.1f and 8.1g show systems that are linearly stable but nonlinearly unstable and vice versa, respectively.

To a good approximation ideal MHD is closely analogous to the set of situations illustrated in Figs. 8.1a–8.1c. There is no dissipation and the instabilities are so virulent that nonlinearities do not dramatically change the behavior. With this in mind one can introduce the following definition of ideal MHD instability. Assume that all quantities of interest are linearized about their equilibrium values

$$Q(\mathbf{r}, t) = Q_0(\mathbf{r}) + \tilde{Q}_1(\mathbf{r}, t) \tag{8.1}$$

Here, $Q_0(\mathbf{r})$ is the equilibrium value and $\tilde{Q}_1/|Q_0| \ll 1$ is a small first-order perturbation. Since the equilibrium, by definition, is time independent, a general form of $\tilde{Q}_1$ can be written as

$$\tilde{Q}_1(\mathbf{r}, t) = Q_1(\mathbf{r})\epsilon^{-i\omega t} \tag{8.2}$$

With very few exceptions the simplest and most reliable definition of stability corresponds to exponential stability; that is, if any of the eigenfrequencies ω corresponds to exponential growth the system is said to be exponentially unstable. If not it is considered exponentially stable:

$$\begin{aligned} &\operatorname{Im} \omega > 0 \qquad \text{exponential instability} \\ &\operatorname{Im} \omega \le 0 \qquad \text{exponential stability} \end{aligned} \tag{8.3}$$

Implicit in this definition is the assumption that the modes are discrete with distinguishable eigenfrequencies. This is not true in general. However, most of the difficulties occur in the stable part of the spectrum and are associated with existence of continua. These questions, and the frequency spectrum of ideal MHD, are discussed in more detail in Section 8.5.4. Consequently, if one is primarily concerned with the existence of ideal MHD instabilities and the conditions for avoiding them, it is the unstable part of the complex ω plane that is of interest. The definition given by Eq. (8.3) then provides a simple and reliable test for stability.

8.3. WAVES IN AN INFINITE HOMOGENEOUS PLASMA

As a first investigation into the dynamical behavior of an ideal MHD plasma consider a configuration in which the magnetic field is infinite in extent, homogeneous, and unidirectional. This system, which is not toroidal and does not confine plasma, is nonetheless very useful to investigate, for its stability corresponds to a determination of the basic waves which propagate in an MHD plasma. It thus forms a foundation upon which one can develop intuition.

The equilibrium of the homogeneous system is given by

$$\begin{aligned} \mathbf{B} &= B_0\mathbf{e}_z \\ \mathbf{J} &= 0 \\ p &= p_0 \\ \rho &= \rho_0 \\ \mathbf{v} &= 0 \end{aligned} \tag{8.4}$$

where B_0, ρ_0, and p_0 are constants and it has been assumed that $\mathbf{B}$ points in the $\mathbf{e}_z$ direction. Note that this "equilibrium" has no gradients so that $\nabla p = \nabla \rho = \mathbf{J} = 0$, thus trivially satisfying the MHD equilibrium equations.

The linear stability of this configuration is determined by linearizing all quantities as follows: $Q(\mathbf{r}, t) = Q_0 + \tilde{Q}_1(\mathbf{r}, t)$ with $\tilde{Q}_1$ being a small first-order perturbation. Since the equilibrium is both time and spatially independent, the most general form of perturbation can be written as

$$\begin{aligned} \tilde{Q}_1(\mathbf{r}, t) &= Q_1 \exp[-i(\omega t - \mathbf{k} \cdot \mathbf{r})] \\ \mathbf{k} &= k_\perp \mathbf{e}_y + k_\| \mathbf{e}_z \\ \mathbf{k} \cdot \mathbf{r} &= k_\perp y + k_\| z \end{aligned} \tag{8.5}$$

Here, without loss in generality it has been assumed that the wave vector $\mathbf{k}$ lies in the (y, z) plane. Also, the subscripts $\perp$ and $\|$ refer to perpendicular and parallel to the *equilibrium* magnetic field, respectively.

One substitutes Eq. (8.5) into the linearized MHD equations. After some straightforward algebra (not using the momentum equation) all quantities can be expressed in terms of the perturbed velocity $\mathbf{v}_1$. The results are as follows:

$$\begin{aligned} &\text{Conservation of mass:} & \omega \rho_1 &= \rho_0(\mathbf{k} \cdot \mathbf{v}_1) \\ &\text{Conservation of energy:} & \omega p_1 &= \gamma p_0(\mathbf{k} \cdot \mathbf{v}_1) \\ &\text{Faraday's law:} & \omega \mathbf{B}_1 &= -\mathbf{k} \times (\mathbf{v}_1 \times \mathbf{B}_0) \\ &\text{Ampere's law:} & \mu_0 \omega \mathbf{J}_1 &= -i\mathbf{k} \times [\mathbf{k} \times (\mathbf{v}_1 \times \mathbf{B}_0)] \end{aligned} \tag{8.6}$$

Note that the $\nabla \cdot \mathbf{B} = 0$ relation reduces to $\mathbf{k} \cdot \mathbf{B}_1 = 0$ and is a trivial consequence of Faraday's Law. Substituting Eq. (8.6) into the linearized momentum equation yields the following three vector component relations:

$$(\omega^2 - k_\|^2 V_a^2) v_{1x} = 0 \tag{8.7}$$

$$(\omega^2 - k_\perp^2 V_s^2 - k^2 V_a^2) v_{1y} - k_\perp k_\| V_s^2 v_{1z} = 0 \tag{8.8}$$

$$-k_\perp k_\| V_s^2 v_{1y} + (\omega^2 - k_\|^2 V_s^2) v_{1z} = 0 \tag{8.9}$$

where $k^2 \doteq k_\perp^2 + k_\|^2$, $V_a = (B_0^2/\mu_0 \rho_0)^{1/2}$ is the Alfvén speed and $V_s = (\gamma p_0/\rho_0)^{1/2}$ is the adiabatic sound speed. Setting the determinant of this system to zero yields the dispersion relation

$$\omega^2 = k_\|^2 V_a^2 \tag{8.10}$$

$$\omega^2 = \tfrac{1}{2} k^2 (V_a^2 + V_s^2)[1 \pm (1 - \alpha^2)^{1/2}] \tag{8.11}$$

where

$$\alpha^2 = 4\frac{k_\parallel^2}{k^2}\frac{V_s^2V_a^2}{(V_s^2 + V_a^2)^2} \tag{8.12}$$

Note that there are three branches to the dispersion relation. Since $0 \leq \alpha^2 \leq 1$, each corresponds to a purely oscillatory solution: $\text{Im}\,\omega = 0$. Consequently, the homogeneous magnetic field configuration is exponentially stable. This is not surprising since the system is in thermodynamic equilibrium and there are no sources of free energy available to drive instabilities.

Consider each branch of the dispersion relation. The first branch, $\omega_a^2 = k_\parallel^2 V_a^2$, is known as the shear Alfvén wave and is independent of $k_\perp$ even when $k_\perp^2 \gg k_\parallel^2$. It is polarized so that the perturbed magnetic field B_{1x} and velocity v_{1x} are always perpendicular to $\mathbf{B}_0$ and $\mathbf{k}$; the wave is purely transverse. This causes the magnetic lines to bend. Plasma is carried with the magnetic perturbation by the $\mathbf{E} \times \mathbf{B}/B^2$ velocity. See Fig. 8.2. Furthermore, the quantities v_{1y}, v_{1z}, ρ_1, p_1, and $\nabla \cdot \mathbf{v}_1$ are each zero for the shear Alfvén wave; the mode is incompressible, producing no density or pressure fluctuations. The shear Alfvén wave describes a basic oscillation between perpendicular plasma kinetic energy and perpendicular "line bending" magnetic energy; i.e., a balance between inertia and field line tension.

The second branch of the dispersion relation corresponding to the + sign in Eq. (8.11) describes the fast magnetoacoustic wave, ω_f^2. A simple calculation shows that $\omega_f^2 \geq \omega_a^2$. This is a wave in which both the magnetic field and the plasma pressure are compressed so that $\nabla \cdot \mathbf{v}_1$ and

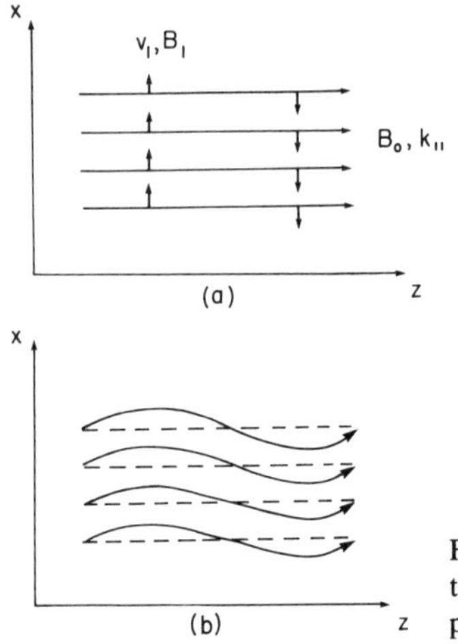

Figure 8.2. Magnetic and velocity perturbation for the shear Alfvén wave shown in (a) vector component form and (b) combined form.

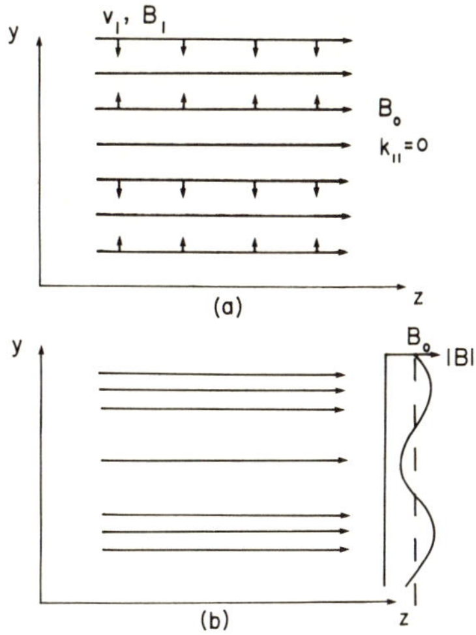

Figure 8.3. Magnetic and velocity perturbations for the compressional Alfvén wave shown in (a) vector component form and (b) combined form.

p_1 are nonzero and $\mathbf{B}_1$ has both y and z components. See Fig. 8.3. In the interesting limit where $\beta \sim V_s^2/V_a^2 \ll 1$, the fast magnetoacoustic wave reduces to the compressional Alfvén wave

$$\omega_f^2 \approx (k_\perp^2 + k_\parallel^2)V_a^2 \tag{8.13}$$

In the low-β limit, it can easily be shown that $\mu_0 p_1/B_0 B_{1z} \sim \beta \ll 1$, indicating that most of the compression involves the magnetic field and not the plasma. The compressional Alfvén wave describes a basic oscillation between perpendicular plasma kinetic energy (plasma inertia) and compressional (field line pressure) plus line bending (field line tension) magnetic energy. Also, since $v_{1z}/v_{1y} \sim \beta \ll 1$ the plasma motion is nearly transverse.

The third branch of the dispersion relation corresponds to the slow magnetoacoustic wave, ω_s^2. This wave always satisfies $\omega_s^2 \leq \omega_a^2$. As in the fast branch the wave is polarized so that both the plasma pressure and the magnetic field are compressed. In the low-β limit, $V_s^2/V_a^2 \ll 1$, the slow magnetoacoustic wave reduces to a sound wave

$$\omega_s^2 \approx k_\parallel^2 V_s^2 \tag{8.14}$$

In this limit the mode is nearly longitudinal since $V_{1y}/V_{1z} \sim \beta \ll 1$. See Fig. 8.4. Hence, the sound wave describes a basic oscillation between parallel plasma kinetic energy (inertia) and plasma internal energy (plasma pressure). A subtle point concerning the sound wave is that the dispersion relation is identical to that of the ion acoustic wave of two-fluid theory. What appears as a paradox is that the ion acoustic wave

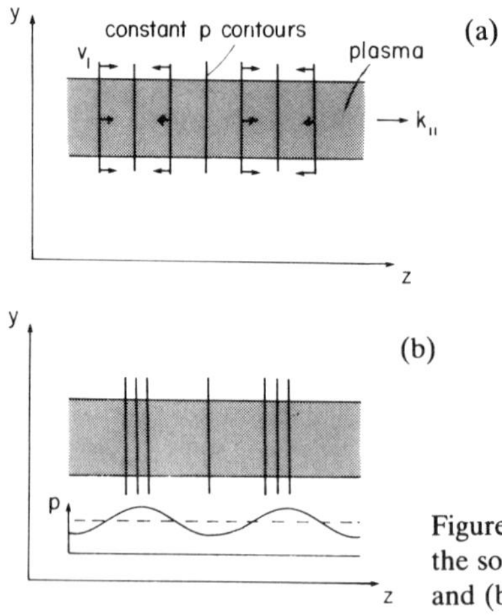

Figure 8.4. Velocity and pressure perturbations for the sound wave shown in (a) vector component form and (b) combined form.

is essentially a longitudinal electrostatic oscillation with $E_\parallel = -ik_\parallel\phi$, while in ideal MHD, $E_\parallel = 0$. Both modes are in fact the same mode and the paradox is resolved in Problem 8.3.

It is interesting to note that if one repeats this calculation for the collisionless MHD model, the resulting dispersion relation consists of two branches given by

$$\begin{aligned} \omega^2 &= k_\parallel^2 V_a^2 \\ \omega^2 &= (k_\perp^2 + k_\parallel^2) V_a^2 \end{aligned} \tag{8.15}$$

These correspond to the shear Alfvén and compressional Alfvén waves, respectively. As expected, the sound wave is suppressed. This result can be obtained from the ideal MHD dispersion relation by setting $V_s^2 = 0$ and suppressing the slow magnetoacoustic wave.

The three branches of the dispersion relation describe the basic wave propagation characteristics of an ideal MHD plasma. In the homogeneous geometry, all are stable. In more physically interesting inhomogeneous geometries each of these waves is modified and can couple to one another. An important result concerning stability is that, for reasons to be discussed later, the most unstable perturbations almost always involve the shear Alfvén wave.

8.4. GENERAL LINEARIZED STABILITY EQUATIONS

There exists an elegant and powerful procedure, known as the Energy Principle, for testing ideal MHD stability in arbitrary three-dimensional geometry. To derive this principle several steps are required. First, the stability problem is formulated as an initial value problem using the general linearized equations of motion. Second, these equations are cast in the form of a normal-mode eigenvalue problem. Next, the eigenmode formulation is transformed into a variational principle. Finally, the variational principle is reduced to the Energy Principle. Each of these steps is discussed in detail in the sections that follow. The present section is concerned with the initial value formulation of the general linearized equations of motion.

8.4.1. Initial Value Formulation

To begin assume that a static ideal MHD equilibrium, satisfying

$$\begin{aligned} \mathbf{J}_0 \times \mathbf{B}_0 &= \nabla p_0 \\ \mu_0 \mathbf{J}_0 &= \nabla \times \mathbf{B}_0 \\ \nabla \cdot \mathbf{B}_0 &= 0 \\ \mathbf{v}_0 &= 0 \end{aligned} \tag{8.16}$$

is given. All quantities are linearized about this background state: $Q(\mathbf{r}, t) = Q_0(\mathbf{r}) + \tilde{Q}_1(\mathbf{r}, t)$ with $\tilde{Q}_1/Q_0 \ll 1$. When substituting into the MHD equations it is convenient to express all perturbed quantities in terms of a vector $\boldsymbol{\xi}$ defined by

$$\tilde{\mathbf{v}}_1 = \frac{\partial \boldsymbol{\xi}}{\partial t} \tag{8.17}$$

The vector $\boldsymbol{\xi}$ represents the displacement of the plasma away from its equilibrium position. The aim of the analysis that follows is to express all perturbed quantities in terms of $\boldsymbol{\xi}$ and then obtain a single equation describing the time evolution of $\boldsymbol{\xi}$.

In an initial value formulation one needs to specify appropriate initial data. A very convenient choice of initial data for stability problems is as follows:

$$\begin{aligned} &\boldsymbol{\xi}(\mathbf{r}, 0) = \tilde{\mathbf{B}}_1(\mathbf{r}, 0) = \tilde{\rho}_1(\mathbf{r}, 0) = \tilde{p}_1(\mathbf{r}, 0) = 0 \\ &\frac{\partial \boldsymbol{\xi}(\mathbf{r}, 0)}{\partial t} \equiv \tilde{\mathbf{v}}_1(\mathbf{r}, 0) \neq 0 \end{aligned} \tag{8.18}$$

This corresponds to the situation where at $t = 0$, the plasma is in its exact equilibrium position but is moving away with a small velocity $\tilde{\mathbf{v}}_1(\mathbf{r}, 0)$. Under such conditions the linearized form of the mass conservation equation, energy relation, and Faraday's Law can be integrated with respect to time, yielding

$$\begin{aligned} \tilde{\rho}_1 &= -\nabla \cdot (\rho_0 \tilde{\xi}) \\ \tilde{p}_1 &= -\tilde{\xi} \cdot \nabla p_0 - \gamma p_0 \nabla \cdot \tilde{\xi} \\ \tilde{\mathbf{B}}_1 &= \nabla \times (\tilde{\xi} \times \mathbf{B}_0) \end{aligned} \tag{8.19}$$

Note that the relation $\nabla \cdot \tilde{\mathbf{B}}_1 = 0$ is trivially satisfied by virtue of Faraday's Law.

These quantities can be substituted into the momentum equation leading to a single vector equation for the displacement $\tilde{\xi}$:

$$\rho \frac{\partial^2 \tilde{\xi}}{\partial t^2} = \mathbf{F}(\tilde{\xi}) \tag{8.20}$$

where the force operator

$$\mathbf{F}(\tilde{\xi}) = \mathbf{J} \times \tilde{\mathbf{B}}_1 + \tilde{\mathbf{J}}_1 \times \mathbf{B} - \nabla \tilde{p}_1 \tag{8.21}$$

can be written as

$$\mathbf{F}(\tilde{\xi}) = \frac{1}{\mu_0}(\nabla \times \mathbf{B}) \times \tilde{\mathbf{Q}} + \frac{1}{\mu_0}(\nabla \times \tilde{\mathbf{Q}}) \times \mathbf{B} + \nabla(\tilde{\xi} \cdot \nabla p + \gamma p \nabla \cdot \tilde{\xi}) \tag{8.22}$$

with $\tilde{\mathbf{Q}} \equiv \tilde{\mathbf{B}}_1 = \nabla \times (\tilde{\xi} \times \mathbf{B})$. In Eqs. (8.20)–(8.22) and hereafter, the zero subscript has been dropped from all equilibrium quantities. Equation (8.20), subject to $\tilde{\xi}(\mathbf{r}, 0) = 0$, $\partial \tilde{\xi}(\mathbf{r}, 0)/\partial t = \tilde{\mathbf{v}}_1(\mathbf{r}, 0)$ plus appropriate boundary conditions (as discussed in Section 3.2) constitute the formulation of the general linearized stability equations as an initial value problem.

The initial value approach has the advantage of directly determining the actual time evolution of a given initial perturbation. It is also useful in the numerical formulation of the full nonlinear problem. The drawback is that it often contains much more information than is required to determine stability, one consequence of which is that numerical calculations are usually very time-consuming.

8.4.2. Normal-Mode Formulation

A more efficient way to investigate linear stability is to reformulate Eq. (8.20) as a normal mode problem. This can easily be done by letting all perturbed quantities vary as follows: $\tilde{Q}_1(\mathbf{r}, t) = Q_1(\mathbf{r}) \exp(-i\omega t)$. Observe that the right-hand sides of Eqs. (8.19) and (8.20) contain no explicit time derivatives. Hence, the conservation of mass, conservation of energy, and Faraday's Law reduce to

$$\rho_1 = -\nabla \cdot (\rho \boldsymbol{\xi}) \tag{8.23}$$

$$p_1 = -\boldsymbol{\xi} \cdot \nabla p - \gamma p \nabla \cdot \boldsymbol{\xi} \tag{8.24}$$

$$\mathbf{Q} \equiv \mathbf{B}_1 = \nabla \times (\boldsymbol{\xi} \times \mathbf{B}) \tag{8.25}$$

Upon substituting these relations into the momentum equation, one finds

$$-\omega^2 \rho \boldsymbol{\xi} = \mathbf{F}(\boldsymbol{\xi})$$

$$\mathbf{F}(\boldsymbol{\xi}) = \frac{1}{\mu_0}(\nabla \times \mathbf{Q}) \times \mathbf{B} + \frac{1}{\mu_0}(\nabla \times \mathbf{B}) \times \mathbf{Q} + \nabla(\boldsymbol{\xi} \cdot \nabla p + \gamma p \nabla \cdot \boldsymbol{\xi}) \tag{8.26}$$

Equation (8.26) represents the normal-mode formulation of the linearized MHD stability problem for general three-dimensional equilibria. In this approach only appropriate boundary conditions on $\boldsymbol{\xi}$ are required. Equation (8.26) can then be solved as an eigenvalue problem for the eigenvalue ω^2. As compared to the initial value approach, the normal mode method is more amenable to analysis and more efficient with respect to numerical computations. However, there is no obvious way to exploit these advantages in a full nonlinear formulation.

It is worth pointing out that there is a subtlety involved with the normal-mode approach: its usefulness is strongly coupled to the assumption that for the problems of interest the eigenvalues are discrete and distinguishable so that the concept of exponential stability is valid. This is indeed true for the unstable part of the spectrum, although the full situation is more complicated. To obtain a more complete understanding, additional detailed knowledge of the properties of the force operator **F** is required. This question is discussed in the next section.

8.5. PROPERTIES OF THE FORCE OPERATOR F

The force operator **F** possesses an important mathematical property that greatly aids in the analysis of linearized MHD stability. In particular, **F** is a self-adjoint operator. This implies that any eigenvalue ω^2 is purely real. Hence, stability transitions always occur when ω^2 crosses zero, rather than at some general point on the real axis where $\mathrm{Re}\,\omega \neq 0$. This fact ultimately leads to an elegant and efficient formulation for testing linear stability known as the Energy Principle.

A further examination of F shows that the frequency spectrum consists not only of discrete modes, but of continua as well. Fortunately, in all cases so far investigated the continua lie on the stable side of the ω plane or at most reach the origin, $\omega^2 = 0$. Thus, when attention is restricted to exponential instabilities, the difficulties associated with the continuous spectra are avoided.

Each of these issues is now discussed separately.

8.5.1. Self-Adjointness of F

As stated, the self-adjointness of **F** has a major impact on both the analytic and the numerical formulation of linearized MHD stability. To demonstrate this property it is necessary to show that for any two arbitrary vectors $\boldsymbol{\xi}$ and $\boldsymbol{\eta}$ satisfying appropriate boundary conditions, such as those discussed in Section 3.2, the following relation holds:

$$\int \boldsymbol{\eta} \cdot \mathbf{F}(\boldsymbol{\xi})\, d\mathbf{r} = \int \boldsymbol{\xi} \cdot \mathbf{F}(\boldsymbol{\eta})\, d\mathbf{r} \tag{8.27}$$

Equation (8.27) can be demonstrated by a direct but lengthy calculation requiring several integrations by parts and substantial amounts of apparently mindless algebra. The result of such a calculation is a form of the integral that is automatically self-adjoint by construction. In fact several such forms have been derived in the literature. One particular example derived in Appendix A is given by (Berge and Freidberg, 1975)

$$\begin{aligned} \int \boldsymbol{\eta} \cdot \mathbf{F}(\boldsymbol{\xi})\, d\mathbf{r} = & -\int d\mathbf{r}\bigg[\frac{1}{\mu_0}(\mathbf{B}\cdot\nabla\boldsymbol{\xi}_\perp)\cdot(\mathbf{B}\cdot\nabla\boldsymbol{\eta}_\perp) + \gamma p(\nabla\cdot\boldsymbol{\xi})(\nabla\cdot\boldsymbol{\eta}) \\ & + \frac{B^2}{\mu_0}(\nabla\cdot\boldsymbol{\xi}_\perp + 2\boldsymbol{\xi}_\perp\cdot\boldsymbol{\kappa})(\nabla\cdot\boldsymbol{\eta}_\perp + 2\boldsymbol{\eta}_\perp\cdot\boldsymbol{\kappa}) \\ & - \frac{4B^2}{\mu_0}(\boldsymbol{\xi}_\perp\cdot\boldsymbol{\kappa})(\boldsymbol{\eta}_\perp\cdot\boldsymbol{\kappa}) + (\boldsymbol{\eta}_\perp\boldsymbol{\xi}_\perp : \nabla\nabla)\left(p + \frac{B^2}{2\mu_0}\right)\bigg] \end{aligned} \tag{8.28}$$

In obtaining Eq. (8.28) $\boldsymbol{\xi}$ and $\boldsymbol{\eta}$ have been written as $\boldsymbol{\xi} = \boldsymbol{\xi}_\perp + \xi_\parallel \mathbf{b}$, $\boldsymbol{\eta} = \boldsymbol{\eta}_\perp + \eta_\parallel \mathbf{b}$, where $\perp$ and $\parallel$ refer to the equilibrium magnetic field. Also, for simplicity it has been assumed that the plasma is surrounded by a perfectly conducting wall so that $\mathbf{n}\cdot\boldsymbol{\xi} = \mathbf{n}\cdot\boldsymbol{\eta} = 0$ on the boundary. However, the proof can easily be generalized to include boundary conditions which allow a vacuum region.

Equation (8.28) demonstrates the self-adjointness of **F**: if $\boldsymbol{\xi}$ and $\boldsymbol{\eta}$ are interchanged, then by inspection the integral remains unchanged.

8.5.2. Real ω^2

By making use of the self-adjointness of **F** it is straightforward to show that for any discrete normal mode, the corresponding eigenvalue ω^2

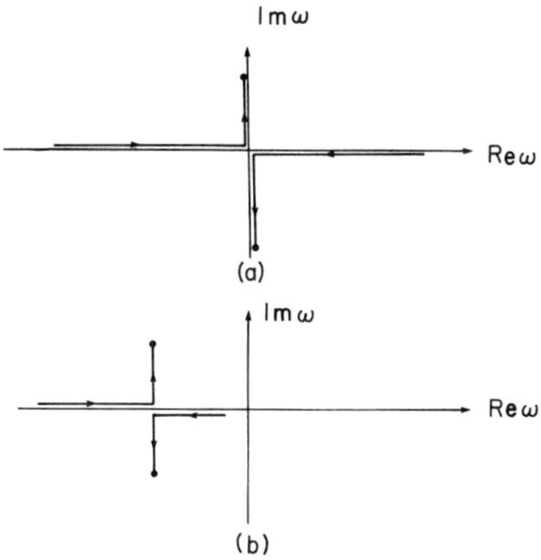

Figure 8.5. Stability transition (a) through $\omega^2 = 0$ in a self-adjoint system and (b) through $\text{Re}\,\omega \neq 0$ in a non-self-adjoint system.

is purely real. The proof is obtained by forming the dot product of Eq. (8.26) with $\boldsymbol{\xi}^*(\mathbf{r})$ and integrating over the plasma volume. The result is

$$\omega^2 \int \rho\, |\boldsymbol{\xi}|^2 \, d\mathbf{r} = -\int \boldsymbol{\xi}^* \cdot \mathbf{F}(\boldsymbol{\xi})\, d\mathbf{r} \tag{8.29}$$

The procedure is then repeated with the vector $\boldsymbol{\xi}(\mathbf{r})$ and the complex conjugate of Eq. (8.26). Using the self-adjoint property of $\mathbf{F}$ it follows that

$$(\omega^2 - \omega^{*2}) \int \rho\, |\boldsymbol{\xi}|^2 \, d\mathbf{r} = 0 \tag{8.30}$$

or $\omega^2 = \omega^{*2}$; that is, ω^2 is purely real. Furthermore, the fact that the differential operators in Eq. (8.26) are real implies that $\boldsymbol{\xi}(\mathbf{r})$ itself is real.

In terms of the definition of exponential stability a normal mode with $\omega^2 > 0$ corresponds to a pure oscillation and hence would be considered stable. Conversely, for a mode with $\omega^2 < 0$, one branch must grow exponentially and thus would correspond to instability. Clearly, the transition from stability to instability occurs when $\omega^2 = 0$. See Fig. 8.5. It is worthwhile emphasizing the importance of this result. In general a stability transition occurs when $\text{Im}\,\omega = 0$ with $\text{Re}\,\omega \neq 0$. This makes the determination of marginal stability boundaries considerably more complicated since $\text{Re}\,\omega$ must also be calculated. However, in ideal MHD the self-adjointness of $\mathbf{F}$ guarantees that at any stability boundary $\text{Im}\,\omega = 0$, the $\text{Re}\,\omega$ must also be zero simultaneously.

8.5.3. Orthogonality of the Normal Modes

A further consequence of the self-adjointness of $\mathbf{F}$ is that the discrete normal modes are orthogonal. To show this consider two modes characterized by eigenfunctions and eigenvalues (ξ_m, ω_m^2) and (ξ_n, ω_n^2). Each mode satisfies

$$-\omega_m^2 \rho \xi_m = \mathbf{F}(\xi_m) \tag{8.31}$$

$$-\omega_n^2 \rho \xi_n = \mathbf{F}(\xi_n) \tag{8.32}$$

One now forms the dot product of Eq. (8.31) with ξ_n and Eq. (8.32) with ξ_m. Upon subtracting the equations and making use of the self-adjointness of $\mathbf{F}$, one obtains

$$(\omega_n^2 - \omega_m^2) \int \rho \xi_m \cdot \xi_n \, d\mathbf{r} = 0 \tag{8.33}$$

Equation (8.33) demonstrates that the nondegenerate discrete normal modes are orthogonal; that is, for two distinct modes with $\omega_m^2 \neq \omega_n^2$ it follows that

$$\int \rho \xi_m \cdot \xi_n \, d\mathbf{r} = 0 \tag{8.34}$$

The modes are orthogonal with weight function ρ.

8.5.4. Spectrum of F

Because of the self-adjointness of $\mathbf{F}$ one is strongly motivated to choose the normal-mode approach rather than the initial-value approach when considering the linearized stability of ideal MHD plasmas. In fact, if the operator $\mathbf{F}$ allowed only discrete eigenvalues, the concept of exponential stability could easily be extended to include both oscillatory and damped motions of the plasma. However, this is not the case. To show this in detail would require a complete spectral analysis of the force operator $\mathbf{F}$, a task beyond the scope of the present work and one that has indeed only been carried out for special MHD equilibria. Nevertheless, based on these special results, it is worthwhile indicating typical spectral properties that can occur and their influence on stability (Grad, 1973; Goedbloed, 1975, 1979).

The spectral properties of $\mathbf{F}$ follow from an examination of the operator $(\mathbf{F}/\rho - \lambda)^{-1}$ for all complex λ. If this operator exists and is bounded for a given λ, the linearized inhomogeneous MHD equation $(\mathbf{F}/\rho - \lambda)\xi = \mathbf{a}$ (such as occurs when solving an initial value problem by Laplace transforms) can be inverted yielding $\xi = (\mathbf{F}/\rho - \lambda)^{-1}\mathbf{a}$. The spectrum of $\mathbf{F}$ consists of those values of λ for which the operator

$(\mathbf{F}/\rho - \lambda)^{-1}$ cannot be inverted. There are two important cases. First, the familiar situation when λ is such that $(\mathbf{F}/\rho - \lambda)\boldsymbol{\xi} = 0$ possesses a nontrivial solution. These values of λ correspond to the point or discrete spectra of $\mathbf{F}$ and represent the normal mode eigenvalues to be examined for exponential stability. In this case it is clear that the operator $(\mathbf{F}/\rho - \lambda)^{-1}$ does not exist.

In the second case of interest λ is such that $(\mathbf{F}/\rho - \lambda)^{-1}$ exists, but is unbounded on some surface in the plasma. For example, in a cylindrical system one might find $\mathbf{F}/\rho - \lambda = k_{\parallel}^2 V_a^2(r) - \lambda$, where $V_a(r)$ is the local Alfvén velocity. In these situations there is a continuous range of λ [determined by setting $\lambda = k_{\parallel}^2 V_a^2(r)$ and varying r over the range $0 < r < a$] over which the operator is ill-behaved. These values of λ constitute the continuous spectra.

A typical ideal MHD spectrum is illustrated schematically in Fig. 8.6. Notice that the continuous spectra could significantly increase the complexity of determining plasma stability if the corresponding frequencies were located in that part of the complex ω plane where $\text{Im}\,\omega > 0$. Fortunately, this appears not to be the case. In all the problems thus far investigated containing continua the frequencies are located on the real axis (i.e., $\omega^2 > 0$). Grad has conjectured that this should be true for any arbitrary static ideal MHD equilibrium. A recent calculation by Tataronis and Salat (1981) supports this conjecture. Their calculation implies that under a wide range of conditions, unstable continua do not exist for general three-dimensional geometries.

One consequence of the existence of a continuum is that discrete modes can accumulate at either edge of its boundaries. If one boundary extends to the origin $\omega^2 = 0$, as it often does, it is possible to have an accumulation of unstable modes with growth rates approaching zero. (See Fig. 8.6.) Although such behavior is perfectly acceptable mathematically, if not accounted for properly it can obscure the concept of marginal stability. On the other hand, if it is known that a continuum

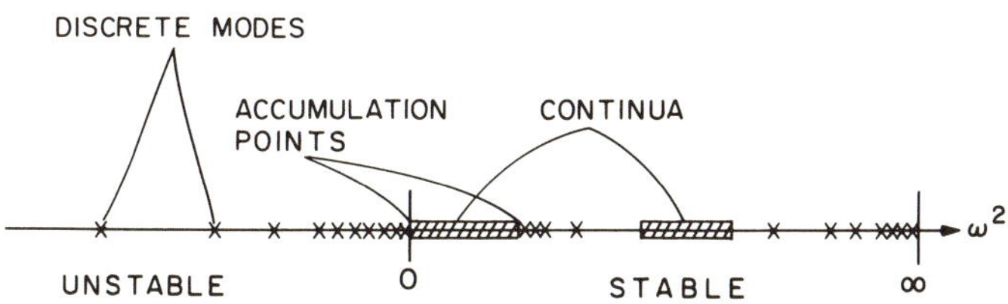

Figure 8.6. Typical ideal MHD spectrum.

extends to the origin it is often possible to derive an analytic criterion, depending only upon the equilibrium variables, which determines whether or not there is an accumulation of unstable modes. This procedure is far simpler than computing normal modes and gives rise to a sufficient condition for instability.

In summary, the spectrum of the force operator $\mathbf{F}(\boldsymbol{\xi})$ contains both discrete eigenvalues and continua. However, the continua exist only for $\omega^2 \geq 0$. This result provides further motivation for examining plasma stability by the normal mode approach, restricting attention only to the question of whether or not exponentially growing modes exist. Even so, the situation can become somewhat complicated as $\omega^2 \to 0$ since unstable modes can either accumulate or make smooth transitions from instability to stability.

8.6. ELEMENTS OF VARIATIONAL CALCULUS

The next step in the analysis is the transformation of the normal mode differential equations into an integral representation having the form of a variational principle. This section presents a brief review of the elements of variational calculus for those unfamiliar with the subject. The review is not intended to be complete or comprehensive, and more detailed information can be obtained from a standard reference textbook such as listed at the end of the chapter. However, the discussion presented here should be sufficient for application to the MHD stability problem.

Construction of the Variational Principle. To begin, consider the classic eigenvalue problem given by

$$\frac{d}{dx}\left(f\frac{dy}{dx}\right) + (\lambda - g)y = 0 \tag{8.35}$$

$$y(0) = y(1) = 0 \tag{8.36}$$

where $f(x)$ and $g(x)$ are two known functions and λ is the eigenvalue. Equations (8.35) and (8.36) correspond to the differential equation representation of the eigenvalue problem. In special cases where f and g are sufficiently simple the problem can be solved analytically. For the general case, one can sometimes make use of power series expansions, asymptotic expansions, or transform methods. Mostly, one is forced to numerical calculations for an accurate evaluation of λ.

By the use of variational calculus, Eq. (8.35) can be cast in the form of an integral relation, exactly equivalent to the original problem. One advantage of the variational formulation is that one can substitute analytic "guesses," known as trial functions, in place of the true eigenfunction in order to estimate the eigenvalue. Furthermore, if the variational principle is properly formulated, the estimated eigenvalue is more accurate than the trial function itself.

These features are demonstrated as follows. To obtain the variational principle Eq. (8.35) is multiplied by y and integrated over the region $0 \leq x \leq 1$. A simple calculation gives

$$\lambda = \frac{\int (fy'^2 + gy^2)\, dx}{\int y^2\, dx} \tag{8.37}$$

In this integral and all remaining ones in Section 8.6 the limits of integration are $x = 0$ to $x = 1$. Equation (8.37) is exactly equivalent to Eq. (8.35); if all allowable trial functions $y(x)$ are substituted into Eq. (8.37) the resulting values of λ will exhibit an extremum only for those trial functions that happen to coincide with actual eigenfunctions. An allowable trial function is any arbitrary function that is bounded and satisfies the boundary conditions while an extremum corresponds to a maximum, minimum, or saddle point in λ.

To prove this, assume an allowable trial function $y_0(x)$ is substituted into Eq. (8.37) yielding a corresponding value, λ_0. Now modify the trial function by a small but arbitrary perturbation: $y(x) = y_0(x) + \delta y(x)$ with $\delta y(0) = \delta y(1) = 0$. This produces a change $\lambda = \lambda_0 + \delta\lambda$ given by

$$\delta\lambda = \frac{\int [f(y_0 + \delta y)'^2 + g(y_0 + \delta y)^2]\, dx}{\int (y_0 + \delta y)^2\, dx} - \frac{\int (fy_0'^2 + gy_0'^2)\, dx}{\int y_0^2\, dx} \tag{8.38}$$

or, for small δy,

$$\delta\lambda \approx \frac{2 \int [\delta y' f y_0' + \delta y\, y_0(g - \lambda_0)]\, dx}{\int y_0^2\, dx} \tag{8.39}$$

The first term can be integrated by parts, yielding

$$\delta\lambda = -\frac{2 \int \delta y[(fy_0')' + (\lambda_0 - g)y_0]\, dx}{\int y_0^2\, dx} \tag{8.40}$$

For λ_0 to correspond to an extremum then $\delta\lambda = 0$ for *arbitrary* δy.

This can only occur if

$$(fy_0')' + (\lambda_0 - g)y_0 = 0 \tag{8.41}$$

which is equivalent to the original eigenvalue equation.

Thus, Eq. (8.37) is the variational principle for the eigenvalue λ. An important point to emphasize is that there exists an infinite number of integral relations for λ that are correct mathematical expressions but that are not variational. For instance, if Eq. (8.35) is multiplied by $h(x)y(x)$ rather than $y(x)$ then a simple calculation gives

$$\lambda = \frac{\int \{hfy'^2 + [hg - (fh')'/2]y^2\}\, dx}{\int hy^2\, dx} \tag{8.42}$$

Although this is a correct mathematical expression, it is not variational; the function $y(x)$ that makes $\delta\lambda = 0$ does not satisfy the original equation unless $h(x) = 1$.

The fact that $\delta\lambda = 0$ when $y(x)$ coincides with a true eigenfunction implies that an estimate for λ obtained from a trial function approximation is more accurate than the trial function itself. This follows by writing the trial function as $y = y_0 + \delta y$, where y_0 is a true eigenfunction and δy is the error, assumed to be of order ϵ. Upon substituting into Eq. (8.37) one finds

$$\lambda = \lambda_0 + \frac{\int [f(\delta y')^2 + g(\delta y)^2]\, dx}{\int y_0^2\, dx} + O[(\delta y)^3] \tag{8.43}$$

where λ_0 is the true eigenvalue. Equation (8.43) implies that λ differs from λ_0 by a correction of order ϵ^2, which is one order more accurate than the trial function itself.

Generalized Boundary Conditions. There is a generalization of Eq. (8.37) that proves to be quite useful when applied to MHD systems containing a vacuum region. The generalization involves the boundary conditions. Consider the eigenvalue problem defined by Eq. (8.35) but subject to the boundary conditions

$$\begin{aligned} y(0) &= 0 \\ y'(1) &= Ay(1) \end{aligned} \tag{8.44}$$

where A is a constant.

One can again obtain a variational principle by multiplying Eq. (8.35) by $y(x)$ and integrating. In this case a boundary term appears in the integral relation for λ

$$\lambda = \frac{\int (fy'^2 + gy^2)\, dx - fy'y|_{x=1}}{\int y^2\, dx} \tag{8.45}$$

Setting $y = y_0 + \delta y$ and substituting into Eq. (8.45) leads to the following expression for $\delta\lambda$:

$$\delta\lambda = -\frac{2\int \delta y[(fy_0')' + (\lambda_0 - g)y_0]\,dx + f(y_0'\delta y - y_0\delta y')|_{x=1}}{\int y_0^2\,dx} \tag{8.46}$$

$\delta\lambda$ vanishes if (1) y_0 satisfies the original differential equation and (2) both y_0 and δy satisfy the boundary conditions $y_0'(1) = Ay_0(1)$, $\delta y'(1) = A\delta y(1)$.

Thus, Eq. (8.45) is a proper variational principle equivalent to the original eigenvalue problem. However, there is an important practical drawback to its usage. If one does not know the true eigenfunction, but is instead substituting a sequence of trial functions in an attempt to extremize λ, then each trial function must satisfy the boundary condition $y'(1) = Ay(1)$. Although this is not an unexpected requirement, it is often difficult to implement practically.

An alternate variational principle, which is more elegant and more convenient, is obtained by replacing $y'(1)$ in the boundary term with $Ay(1)$. The variational principle becomes

$$\lambda = \frac{\int [fy'^2 + gy^2]\,dx - Afy^2|_{x=1}}{\int y^2\,dx} \tag{8.47}$$

Again setting $y = y_0 + \delta y$ yields an expression for $\delta\lambda$ given by

$$\delta\lambda = \frac{2\int \delta y[(fy_0')' + (\lambda_0 - g)y_0]\,dx + 2f\delta y(y_0' - Ay_0)|_{x=1}}{\int y_0^2\,dx} \tag{8.48}$$

Consider the evaluation of $\delta\lambda$ for the class of trial functions that satisfy $y(0) = 0$, but not $y'(1) = Ay(1)$. Instead, $y(1)$ is allowed to float freely: $\delta y(1) \neq 0$. Equation (8.48) implies that in order to set $\delta\lambda = 0$ for all $\delta y(x)$ with $\delta y(1) \neq 0$, y_0 must satisfy (1) the original differential equation and (2) $y_0'(1) = Ay_0(1)$; the variational principle forces the trial function to satisfy the correct boundary condition.

The variational principle given by Eq. (8.47) is said to contain the "natural boundary condition" $y'(1) = Ay(1)$. It has the important advantage of allowing trial functions to be substituted that do not automatically satisfy this condition.

Summary. The variational principle is an alternate but equivalent integral representation of the original differential equation eigenvalue problem. It can be used in two ways. First, if a function $y_0(x)$ can be found that extremizes λ over all allowable $y(x)$, then $y_0(x)$ and the corresponding value of λ_0 are a correct eigenfunction and eigenvalue of the system. Second, if a function $y_0(x)$ can be found that extremizes λ

over a special, and presumably simple, subclass of trial functions then this $y_0(x)$ is the best approximation within the subclass to an actual eigenfunction. If the error in the trial function is of order ϵ then the error between the estimated and actual eigenvalue is of order ϵ^2.

For those unfamiliar with variational techniques there is often a gap between the basic theory, which is relatively straightforward, and the process of extracting useful information in actual applications. In the chapters that follow detailed examples are presented that demonstrate the power and utility of variational techniques.

8.7. VARIATIONAL FORMULATION

Equation (8.26) represents the normal mode formulation of the general three-dimensional linearized MHD stability problem. These equations have the form of a set of three coupled homogeneous partial differential equations for the components of $\boldsymbol{\xi}$ with eigenvalue ω^2.

Because of the self-adjointness of **F** the problem can be easily cast in the form of a variational principle (Bernstein *et al.*, 1958). The dot product of Eq. (8.26) with $\boldsymbol{\xi}^*$ is formed and then integrated over the plasma volume, yielding

$$\omega^2 = \frac{\delta W(\boldsymbol{\xi}^*, \boldsymbol{\xi})}{K(\boldsymbol{\xi}^*, \boldsymbol{\xi})} \tag{8.49}$$

where

$$\begin{aligned}
\delta W(\boldsymbol{\xi}^*, \boldsymbol{\xi}) &= -\frac{1}{2}\int \boldsymbol{\xi}^* \cdot \mathbf{F}(\boldsymbol{\xi})\, d\mathbf{r} \\
&= -\frac{1}{2}\int \boldsymbol{\xi}^* \cdot \left[\frac{1}{\mu_0}(\nabla \times \mathbf{Q}) \times \mathbf{B} + \frac{1}{\mu_0}(\nabla \times \mathbf{B})\right. \\
&\qquad \left. \times\, \mathbf{Q} + \nabla(\gamma p \nabla \cdot \boldsymbol{\xi} + \boldsymbol{\xi} \cdot \nabla p)\right] d\mathbf{r} \\
K(\boldsymbol{\xi}^*, \boldsymbol{\xi}) &= \frac{1}{2}\int \rho\, |\boldsymbol{\xi}|^2\, d\mathbf{r}
\end{aligned} \tag{8.50}$$

Although $\boldsymbol{\xi}(\mathbf{r})$ and ω^2 have been shown to be real, $\boldsymbol{\xi}(\mathbf{r})$ is treated as complex in Eq. (8.49) in anticipation of cases where, because of symmetry, several spatial coordinates can be Fourier analyzed. The variational principle states that any allowable function $\boldsymbol{\xi}$ for which ω^2 becomes an extremum is an eigenfunction of the ideal MHD normal-mode equations with eigenvalue ω^2. The proof follows by letting

$\boldsymbol{\xi} \to \boldsymbol{\xi} + \delta\boldsymbol{\xi}$, $\omega^2 \to \omega^2 + \delta\omega^2$, and setting the resulting $\delta\omega^2 = 0$ (corresponding to ω^2 being an extremum). One finds

$$\omega^2 + \delta\omega^2 = \frac{\delta W(\boldsymbol{\xi}^*, \boldsymbol{\xi}) + \delta W(\delta\boldsymbol{\xi}^*, \boldsymbol{\xi}) + \delta W(\boldsymbol{\xi}^*, \delta\boldsymbol{\xi}) + \delta W(\delta\boldsymbol{\xi})}{K(\boldsymbol{\xi}^*, \boldsymbol{\xi}) + K(\delta\boldsymbol{\xi}^*, \boldsymbol{\xi}) + K(\boldsymbol{\xi}^*, \delta\boldsymbol{\xi}) + K(\delta\boldsymbol{\xi}^*, \delta\boldsymbol{\xi})} \tag{8.51}$$

For small $\delta\boldsymbol{\xi}$ and $\delta\omega^2$, Eq. (8.51) leads to the following expression for $\delta\omega^2$:

$$\delta\omega^2 = \frac{\delta W(\delta\boldsymbol{\xi}^*, \boldsymbol{\xi}) + \delta W(\boldsymbol{\xi}^*, \delta\boldsymbol{\xi}) - \omega^2[K(\delta\boldsymbol{\xi}^*, \boldsymbol{\xi}) + K(\boldsymbol{\xi}^*, \delta\boldsymbol{\xi})]}{K(\boldsymbol{\xi}^*, \boldsymbol{\xi})} \tag{8.52}$$

Using the self-adjoint property of $\mathbf{F}$, and setting $\delta\omega^2 = 0$, one can write Eq. (8.52) as

$$\int d\mathbf{r}\{\delta\boldsymbol{\xi}^* \cdot [\mathbf{F}(\boldsymbol{\xi}) + \omega^2\rho\boldsymbol{\xi}] + \delta\boldsymbol{\xi} \cdot [\mathbf{F}(\boldsymbol{\xi}^*) + \omega^2\rho\boldsymbol{\xi}^*]\} = 0 \tag{8.53}$$

Since $\delta\boldsymbol{\xi}$ is arbitrary, Eq. (8.53) implies that

$$-\omega^2\rho\boldsymbol{\xi} = \mathbf{F}(\boldsymbol{\xi}) \tag{8.54}$$

This completes the proof and demonstrates that the normal-mode eigenvalue equation [Eq. (8.26)] and the variational principle [Eq. (8.49)] are equivalent formulations of the linearized ideal MHD stability problem.

The quantity δW represents the change in potential energy associated with the perturbation and is equal to the work done against the force $\mathbf{F}(\boldsymbol{\xi})$ in displacing the plasma by an amount $\boldsymbol{\xi}$. The quantity K is proportional to the kinetic energy.

8.8. THE ENERGY PRINCIPLE

It is often of primary interest to determine whether a given system is stable or unstable without being particularly concerned about the specific values of growth rates. The reason is that the growth times ($\tau \lesssim 50\,\mu$s) are typically much shorter than experimental times ($\tau \gtrsim 1$ ms). Thus, it is far more important to determine the conditions for avoiding instabilities than to calculate precise growth rates (which can easily be estimated). In such cases, the variational formulation just derived can be further simplified, leading to a powerful minimizing principle which exactly determines stability boundaries but only estimates growth rates.

This formulation is known as the Energy Principle and represents the most efficient and often the most intuitive method of determining plasma stability (Bernstein *et al.*, 1958). It receives wide application in the literature and will be used extensively throughout the remainder of the textbook.

The Energy Principle can be applied in a direct manner to systems in which the plasma is surrounded by a conducting wall. It also applies, in principle, to the situation in which a vacuum region is present. However, in this case the actual calculations can become somewhat cumbersome because of the complicated boundary conditions across the plasma–vacuum interface, specifically the linearized pressure-balance condition. This difficulty is resolved by introducing a natural boundary condition into the Extended Energy Principle. The vacuum contribution is explicitly isolated, thereby eliminating the need for the pressure-balance jump condition. With the establishment of the Extended Energy Principle it then becomes useful to derive an intuitive form of the energy integral which is used in carrying out specific applications.

8.8.1. Statement and Proof of the Energy Principle

The physical basis for the Energy Principle is the fact that energy is exactly conserved in the ideal MHD model. As a consequence the extremum corresponding to the most negative eigenvalue of ω^2 actually represents a minimum in potential energy δW. This in turn implies that the question of stability or instability can be determined by analyzing only the sign of $\delta W(\boldsymbol{\xi}, \boldsymbol{\xi})$ and not the full variational problem. Specifically, the Energy Principle states that an equilibrium is stable if and only if

$$\delta W(\boldsymbol{\xi}^*, \boldsymbol{\xi}) \geq 0 \tag{8.55}$$

for all allowable displacements (i.e., $\boldsymbol{\xi}$ bounded in energy and satisfying appropriate boundary conditions); that is, if the minimum value of potential energy is positive for all displacements, the system is stable. If it is negative for any displacement, the system is unstable.

The proof of the Energy Principle would be straightforward if $\mathbf{F}$ allowed only discrete normal modes, constituting a complete set of basis functions, $\boldsymbol{\xi}_n$, each satisfying

$$-\omega_n^2 \rho \boldsymbol{\xi}_n = \mathbf{F}(\boldsymbol{\xi}_n) \tag{8.56}$$

In this case, any arbitrary trial function $\boldsymbol{\xi}(\mathbf{r})$ could be expanded as

$$\boldsymbol{\xi}(\mathbf{r}) = \sum a_n \boldsymbol{\xi}_n(r) \tag{8.57}$$

The $\boldsymbol{\xi}_n$ have been shown to be orthogonal (Section 8.5.3). If they are normalized to be orthonormal with respect to the weight function ρ,

$$\int \rho \boldsymbol{\xi}_n^* \cdot \boldsymbol{\xi}_m \, d\mathbf{r} = \delta_{nm} \tag{8.58}$$

then δW can be written as

$$\delta W(\boldsymbol{\xi}^*, \boldsymbol{\xi}) = \tfrac{1}{2} \sum_n |a_m|^2 \omega_m^2 \tag{8.59}$$

Thus, if a $\boldsymbol{\xi}(\mathbf{r})$ could be found for which $\delta W < 0$, at least one $\omega_m^2 < 0$, indicating instability. Conversely, if $\delta W \geq 0$ for all $\boldsymbol{\xi}(\mathbf{r})$ then each $\omega_m^2 \geq 0$, indicating exponential stability. Unfortunately, this proof, first given by Bernstein *et al.* (1958), is not completely valid for the general case because of the existence of continua.

An elegant proof of the Energy Principle, outlined below, has been given by Laval *et al.* (1965), which does not invoke the assumption of a complete set of discrete normal modes. The proof is based on the conservation of energy and is carried out in the real time domain. Consider the energy $H(t)$ given by

$$\begin{aligned} H &= K\left(\frac{\partial \boldsymbol{\xi}}{\partial t}, \frac{\partial \boldsymbol{\xi}}{\partial t}\right) + \delta W(\boldsymbol{\xi}, \boldsymbol{\xi})) \\ &= \frac{1}{2} \int d\mathbf{r} \left[\rho \left(\frac{\partial \boldsymbol{\xi}}{\partial t}\right)^2 - \boldsymbol{\xi} \cdot \mathbf{F}(\boldsymbol{\xi})\right] \end{aligned} \tag{8.60}$$

where $\boldsymbol{\xi} = \boldsymbol{\xi}(\mathbf{r}, t)$. A simple calculation that makes use of the self-adjoint property of $\mathbf{F}$ yields

$$\frac{dH}{dt} = \int d\mathbf{r} \frac{\partial \boldsymbol{\xi}}{\partial t} \cdot \left[\rho \frac{\partial^2 \boldsymbol{\xi}}{\partial t^2} - \mathbf{F}(\boldsymbol{\xi})\right] = 0 \tag{8.61}$$

Equation (8.61) shows that $H(t) = H_0 = \text{const}$ and corresponds to energy conservation.

To show sufficiency of the Energy Principle, assume that δW is positive for all allowable $\boldsymbol{\xi}(\mathbf{r}, t)$. Energy conservation implies that

$$\delta W = H_0 - K \tag{8.62}$$

Hence, unbounded growth of the kinetic energy K (e.g., exponential instability) would violate energy conservation: $\delta W > 0$ for all allowable $\boldsymbol{\xi}$ is sufficient for stability.

To show necessity of the Energy Principle assume a perturbation $\boldsymbol{\eta}(\mathbf{r})$ exists such that $\delta W(\boldsymbol{\eta}, \boldsymbol{\eta}) < 0$. Consider a displacement $\boldsymbol{\xi}(\mathbf{r}, t)$

satisfying initial conditions

$$\begin{aligned} \boldsymbol{\xi}(\mathbf{r},0) &= \boldsymbol{\eta}(\mathbf{r}) \\ \frac{\partial \boldsymbol{\xi}}{\partial t}(\mathbf{r},0) &= 0 \end{aligned} \tag{8.63}$$

Energy conservation implies that

$$H_0 = (\delta W + K)_{t=0} = \delta W(\boldsymbol{\eta}, \boldsymbol{\eta}) < 0 \tag{8.64}$$

One now calculates d^2I/dt^2, where

$$I(t) = K(\boldsymbol{\xi}, \boldsymbol{\xi}) = \frac{1}{2}\int d\mathbf{r}\, \rho(\boldsymbol{\xi})^2 \tag{8.65}$$

A simple calculation gives

$$\begin{aligned} \frac{d^2I}{dt^2} &= 2\left[K\left(\frac{\partial \boldsymbol{\xi}}{\partial t}, \frac{\partial \boldsymbol{\xi}}{\partial t}\right) - \delta W(\boldsymbol{\xi}, \boldsymbol{\xi})\right] \\ &= \int d\mathbf{r}\left[\rho\left(\frac{\partial \boldsymbol{\xi}}{\partial t}\right)^2 + \boldsymbol{\xi}\cdot\mathbf{F}(\boldsymbol{\xi})\right] \end{aligned} \tag{8.66}$$

Setting $\delta W(\boldsymbol{\xi}, \boldsymbol{\xi}) = H_0 - K(\partial\boldsymbol{\xi}/\partial t, \partial\boldsymbol{\xi}/\partial t)$ leads to the result

$$\frac{d^2I}{dt^2} = 2\left[2K\left(\frac{\partial \boldsymbol{\xi}}{\partial t}, \frac{\partial \boldsymbol{\xi}}{\partial t}\right) - H_0\right] > -2H_0 > 0 \tag{8.67}$$

Equation (8.67) implies that I grows without bound as $t \to \infty$, indicating that $\boldsymbol{\xi}$ increases as least as fast as t. Laval *et al.* (1965) showed by a somewhat more complicated analysis that if $\delta W(\boldsymbol{\eta}, \boldsymbol{\eta}) < 0$ there exists a $\boldsymbol{\xi}$ that grows exponentially [i.e., $\boldsymbol{\xi} \sim \exp(\lambda t)$] with a growth rate at least as fast as $\lambda = [-\delta W(\boldsymbol{\eta}, \boldsymbol{\eta})/K(\boldsymbol{\eta}, \boldsymbol{\eta})]^{1/2}$. The implication of this analysis is that if any perturbation $\boldsymbol{\eta}$ exists that makes $\delta W < 0$, then the system is unstable; the displacement grows without bound. Consequently, $\delta W > 0$ for all allowable trial functions is a necessary condition for stability. This completes the proof of the Energy Principle.

Consider now the advantages of using the Energy Principle. First, if one has some intuition about the form of an unstable perturbation, this form can be used as a trial function to minimize δW. If the minimum $\delta W < 0$, the Energy Principle guarantees that the actual eigenvalue ω^2 must be smaller (i.e., more unstable) than the value $\omega^2 = \delta W/K < 0$ calculated by using the trial function; that is, the existence of an allowable trial function which makes $\delta W < 0$ is sufficient for instability. Second, the Euler–Lagrange differential equation following from the variation of δW corresponds to the neutral stability form (i.e., $\omega^2 = 0$) of the normal mode equations and, as will be shown in future chapters, is far simpler to analyze. Third, in the integral form of δW the various stabilizing and destabilizing mechanisms appear quite transparently, thus helping in the development of physical intuition.

Finally, there exists a very practical method for numerically testing MHD stability based on the Energy Principle. In this procedure one chooses a suitable set of simple but complete basis functions $\boldsymbol{\xi}_n$ and writes $\boldsymbol{\xi}$ as an arbitrary sum of these functions: $\boldsymbol{\xi} = \sum a_n \boldsymbol{\xi}_n$. Substituting into the expression for δW then yields $\delta W = \sum A_{mn} a_m a_n$ where the $A_{mn} = \delta W(\boldsymbol{\xi}_m, \boldsymbol{\xi}_n)$ are well-defined, computable matrix elements. Once the A_{mn} are known, the expression for δW can be minimized with respect to the coefficients a_m, given some suitable normalization. Clearly a clever choice and/or a sufficient number of basis functions gives an increasingly accurate indication of whether or not δW can be made negative. Note that the minimization with respect to the a_m can be carried out for any convenient choice of normalization, not necessarily $K(\boldsymbol{\xi}, \boldsymbol{\xi}) = \text{const}$. This procedure for determining stability is very effective in multidimensional geometries where numerical shooting methods are not possible.

8.8.2. The Extended Energy Principle

The Energy Principle, calculated with $\mathbf{F}(\boldsymbol{\xi})$ given by Eq. (8.26), is valid if the plasma is either directly surrounded by a conducting wall or isolated from the wall by a vacuum region. In the first case application of the Energy Principle is straightforward since the only boundary condition that need be satisfied is

$$\mathbf{n} \cdot \boldsymbol{\xi}_\perp|_{\mathbf{r}_w} = 0 \tag{8.68}$$

where $\mathbf{r}_w$ is the wall radius.

When a vacuum region is present the situation is more complicated because the vacuum fields do not appear explicitly in δW but enter only through the boundary conditions at the plasma surface. To see this, consider the set of linearized boundary conditions obtained by expanding Eqs. (3.6), (3.7), and (3.9) about the perturbed surface $\mathbf{r}_s = \mathbf{r}_p + \boldsymbol{\xi}$, where $\mathbf{r}_p$ is the unperturbed surface and $\boldsymbol{\xi} = \boldsymbol{\xi}(\mathbf{r}_p)$:

$$\mathbf{n} \cdot \hat{\mathbf{B}}_1|_{\mathbf{r}_w} = 0 \tag{8.69}$$

$$\mathbf{n} \cdot \hat{\mathbf{B}}_1|_{\mathbf{r}_p} = \mathbf{n} \cdot \nabla \times (\boldsymbol{\xi}_\perp \times \hat{\mathbf{B}})|_{\mathbf{r}_p} \tag{8.70}$$

$$\mathbf{n} \cdot \mathbf{B}_1|_{\mathbf{r}_p} = \mathbf{n} \cdot \nabla \times (\boldsymbol{\xi}_\perp \times \mathbf{B})|_{\mathbf{r}_p} \tag{8.71}$$

$$\left[p_1 + \frac{\mathbf{B} \cdot \mathbf{B}_1}{\mu_0} + \boldsymbol{\xi} \cdot \nabla\left(p + \frac{B^2}{2\mu_0}\right)\right]_{\mathbf{r}_p} = \left[\frac{\hat{\mathbf{B}} \cdot \hat{\mathbf{B}}_1}{\mu_0} + \boldsymbol{\xi} \cdot \nabla \frac{\hat{B}^2}{2\mu_0}\right]_{\mathbf{r}_p} \tag{8.72}$$

Here, the caret represents the vacuum region. As intuitively obvious as Eq. (8.70) may seem, its derivation is nontrivial since $\boldsymbol{\xi}$ is not defined in the vacuum. The result is derived by evaluating $\mathbf{B}$ on the perturbed surface and expanding the normal vector as $\mathbf{n} = \mathbf{n}_0 + \mathbf{n}_1$ with $\mathbf{n}_1 = -(\nabla\boldsymbol{\xi})\cdot\mathbf{n}_0 + \mathbf{n}_0[\mathbf{n}_0\cdot(\nabla\boldsymbol{\xi})\cdot\mathbf{n}_0]$. Note that for the plasma Eq. (8.71) is automatically satisfied by virtue of Eq. (8.25). Equation (8.70) couples the vacuum field $\hat{\mathbf{B}}_1$ to the plasma displacement $\boldsymbol{\xi}$. The pressure balance jump condition is straightforward to derive but has a rather complicated form. As a consequence, direct application of the Energy Principle to a system with a vacuum region can be cumbersome, first because of the difficulty of generating trial functions satisfying Eq. (8.72) and second because the form of δW with $\mathbf{F}$ given by Eq. (8.26) is not very intuitive (since the vacuum contribution does not appear explicitly).

These issues are resolved by a reformulation of δW known as the Extended Energy Principle (Bernstein *et al.*, 1958, and also Goedbloed, 1979) in which the vacuum contribution explicitly appears and Eq. (8.72) occurs as a natural boundary condition.

The first step in the reformulation is to perform an integration by parts of the $\boldsymbol{\xi}^*\cdot\nabla(\gamma p\nabla\cdot\boldsymbol{\xi})$ and $\boldsymbol{\xi}^*\cdot(\nabla\times\mathbf{Q})\times\mathbf{B}$ terms in δW. See Eq. (8.50). One obtains

$$\begin{aligned}\delta W &= \frac{1}{2}\int d\mathbf{r}\left\{\frac{|\mathbf{Q}|^2}{\mu_0} + \gamma p\,|\nabla\cdot\boldsymbol{\xi}|^2 - \boldsymbol{\xi}^*\cdot[\mathbf{J}\times\mathbf{Q} + \nabla(\boldsymbol{\xi}\cdot\nabla p)]\right\}\\ &\quad - \frac{1}{2}\int dS(\mathbf{n}\cdot\boldsymbol{\xi}^*)\left(\gamma p\nabla\cdot\boldsymbol{\xi} - \frac{\mathbf{B}\cdot\mathbf{B}_1}{\mu_0}\right)\end{aligned} \tag{8.73}$$

Next, writing $\boldsymbol{\xi} = \boldsymbol{\xi}_\perp + \xi_\parallel\mathbf{b}$, it can easily be shown that $\xi_\parallel^*\mathbf{b}\cdot[\mathbf{J}\times\mathbf{Q} + \nabla(\boldsymbol{\xi}\cdot\nabla p)] = 0$ [see Eq. (A.4) in Appendix A]. The term $\boldsymbol{\xi}_\perp^*\cdot\nabla(\boldsymbol{\xi}\cdot\nabla p)$ is then integrated by parts. The result is

$$\delta W = \delta W_F + \text{B.T.} \tag{8.74}$$

where δW_F is given by

$$\delta W_F = \frac{1}{2}\int_P d\mathbf{r}\left[\frac{|\mathbf{Q}|^2}{\mu_0} - \boldsymbol{\xi}_\perp^*\cdot\mathbf{J}\times\mathbf{Q} + \gamma p\,|\nabla\cdot\boldsymbol{\xi}|^2 + (\boldsymbol{\xi}_\perp\cdot\nabla p)\nabla\cdot\boldsymbol{\xi}_\perp^*\right] \tag{8.75}$$

and is known as the standard form of the fluid energy. The boundary term B.T. has the form

$$\text{B.T.} = \frac{1}{2}\int_S dS(\mathbf{n}\cdot\boldsymbol{\xi}_\perp^*)\left(\frac{\mathbf{B}\cdot\mathbf{B}_1}{\mu_0} - \gamma p\nabla\cdot\boldsymbol{\xi} - \boldsymbol{\xi}_\perp\cdot\nabla p\right) \tag{8.76}$$

The labels P, S, and V appearing on the integrals correspond to the unperturbed plasma volume, plasma surface, and vacuum volume, respectively.

The complicated pressure balance jump relation given by Eq. (8.72) can be introduced into the Energy Principle as a natural boundary condition by substituting Eq. (8.72) into Eq. (8.76). This leads to the following expression for the boundary terms:

$$\text{B.T.} = \delta W_S + \frac{1}{2}\int_S dS(\mathbf{n}\cdot\boldsymbol{\xi}_\perp^*)\left(\frac{\hat{\mathbf{B}}\cdot\hat{\mathbf{B}}_1}{\mu_0}\right) \tag{8.77}$$

with

$$\delta W_S = \frac{1}{2}\int_S dS\,|\mathbf{n}\cdot\boldsymbol{\xi}_\perp|^2\mathbf{n}\cdot\left[\!\left[\nabla\left(p + \frac{B^2}{2\mu_0}\right)\right]\!\right] \tag{8.78}$$

and $[\![T]\!]$ denoting the jump in T from vacuum to plasma. In deriving Eq. (8.78), use has been made of the equilibrium relation $[\![p + B^2/2\mu_0]\!] = 0$ to infer that the tangential jump in $[\![\mathbf{n}\times\nabla(p + B^2/2\mu_0)]\!]$ vanishes. The quantity δW_S is usually referred to as the surface contribution and vanishes unless surface currents flow on the plasma–vacuum boundary.

The remaining term in Eq. (8.77) is identical to the perturbed magnetic energy in the vacuum region

$$\delta W_V = \frac{1}{2}\int_V d\mathbf{r}\frac{|\hat{\mathbf{B}}_1|^2}{\mu_0} \tag{8.79}$$

To show this $\hat{\mathbf{B}}_1$ is written as $\hat{\mathbf{B}}_1 = \nabla\times\hat{\mathbf{A}}_1$ with $\nabla\times\nabla\times\hat{\mathbf{A}}_1 = 0$. Integrating by parts then converts δW_V into a surface integral:

$$\delta W_V = -\frac{1}{2}\int_S dS\frac{\mathbf{n}\cdot(\hat{\mathbf{A}}_1^*\times\hat{\mathbf{B}}_1)}{\mu_0} \tag{8.80}$$

Note that the contribution on the outer conductor vanishes by virtue of Eq. (8.69). On the surface $\hat{\mathbf{A}}_1$ can be expressed in terms of $\mathbf{n}\cdot\boldsymbol{\xi}$ by using the relation $\mathbf{n}\cdot\hat{\mathbf{B}}_1 = \mathbf{n}\cdot\nabla\times\hat{\mathbf{A}}_1 = \mathbf{n}\cdot\nabla\times(\boldsymbol{\xi}_\perp\times\hat{\mathbf{B}})$; that is, $\hat{\mathbf{A}}_1 = \boldsymbol{\xi}_\perp\times\mathbf{B} + \nabla\phi$. If ϕ is chosen so that the gauge condition is $\hat{\mathbf{B}}_1\cdot(\mathbf{n}\times\nabla\phi) = 0$ then Eq. (8.80) reduces to

$$\delta W_V = \frac{1}{2}\int_S dS(\mathbf{n}\cdot\boldsymbol{\xi}_\perp^*)\left(\frac{\hat{\mathbf{B}}\cdot\hat{\mathbf{B}}_1}{\mu_0}\right) \tag{8.81}$$

which is identical to the remaining term in Eq. (8.77).

This completes the derivation. To summarize, in the Extended Energy Principle a system is exponentially stable if and only if $\delta W \geq 0$ for all allowable perturbations with

$$\delta W = \delta W_F + \delta W_S + \delta W_V \tag{8.82}$$

where the fluid, surface, and vacuum contributions are given by

$$\delta W_F = \frac{1}{2}\int_P d\mathbf{r}\left[\frac{|\mathbf{Q}|^2}{\mu_0} - \boldsymbol{\xi}_\perp^* \cdot (\mathbf{J} \times \mathbf{Q}) + \gamma p\,|\nabla \cdot \boldsymbol{\xi}|^2 + (\boldsymbol{\xi}_\perp \cdot \nabla p)\nabla \cdot \boldsymbol{\xi}_\perp^*\right] \tag{8.83a}$$

$$\delta W_S = \frac{1}{2}\int_S dS\,|\mathbf{n} \cdot \boldsymbol{\xi}_\perp|^2 \mathbf{n} \cdot \left[\!\left[\nabla\left(p + \frac{B^2}{2\mu_0}\right)\right]\!\right] \tag{8.83b}$$

$$\delta W_V = \frac{1}{2}\int_V d\mathbf{r}\,\frac{|\hat{\mathbf{B}}_1|^2}{\mu_0} \tag{8.83c}$$

The vacuum contribution appears explicitly and is coupled to the plasma displacement by solving $\nabla \times \hat{\mathbf{B}}_1 = \nabla \cdot \hat{\mathbf{B}}_1 = 0$ subject to the conditions

$$\mathbf{n} \cdot \hat{\mathbf{B}}_1|_{\mathbf{r}_w} = 0 \tag{8.84}$$

$$\mathbf{n} \cdot \hat{\mathbf{B}}_1|_{\mathbf{r}_p} = \mathbf{n} \cdot \nabla \times (\boldsymbol{\xi}_\perp \times \hat{\mathbf{B}})|_{\mathbf{r}_p} \tag{8.85}$$

Note that the complicated pressure-balance jump condition is no longer explicitly required as it enters the calculation as a natural boundary condition. Furthermore, after a somewhat lengthy calculation (see Problem 8.10) it can be shown that Eq. (8.85) can be expressed solely in terms of $\mathbf{n} \cdot \boldsymbol{\xi}_\perp$ on the boundary by the relation

$$\mathbf{n} \cdot \hat{\mathbf{B}}_1|_{\mathbf{r}_p} = \hat{\mathbf{B}}_1 \cdot \nabla(\mathbf{n} \cdot \boldsymbol{\xi}_\perp) - (\mathbf{n} \cdot \boldsymbol{\xi}_\perp)[\mathbf{n} \cdot (\mathbf{n} \cdot \nabla)\hat{\mathbf{B}}_1]|_{\mathbf{r}_p} \tag{8.86}$$

Thus, in a plasma–vacuum system one can specify an arbitrary surface distortion $\mathbf{n} \cdot \boldsymbol{\xi}_\perp$, which is then sufficient to determine both the surface and vacuum contributions to δW. For the special case where there is no vacuum (i.e., the plasma extends to the wall) then $\mathbf{n} \cdot \boldsymbol{\xi}_\perp$ on the boundary must vanish and $\delta W = \delta W_F$. Equation (8.82) is the standard form of the Extended Energy Principle and is used widely throughout the literature.

8.8.3. The Intuitive Form of δW_F

The Extended Energy Principle, described by Eqs. (8.82) and (8.83), is completely adequate for application to any magnetic geometry. Even so, it is useful to derive an alternate form of δW_F that provides additional physical insight into the behavior of MHD instabilities, and is used for most of the applications in the remainder of the textbook. Once the intuitive form is derived, a brief discussion is given, comparing the Energy and Variational Principles for the ideal and collisionless MHD models.

The starting point for the derivation is the "standard" form of δW_F

given by Eq. (8.75) and repeated here for convenience:

$$\delta W_F = \frac{1}{2}\int_P d\mathbf{r}\left[\frac{|\mathbf{Q}|^2}{\mu_0} - \boldsymbol{\xi}_\perp^* \cdot \mathbf{J} \times \mathbf{Q} + \gamma p\,|\nabla \cdot \boldsymbol{\xi}|^2 + (\boldsymbol{\xi}_\perp \cdot \nabla p)\nabla \cdot \boldsymbol{\xi}_\perp^*\right] \quad (8.87)$$

The terms in the integrand are rearranged as follows. The first two terms are separated into components

$$\begin{aligned} |\mathbf{Q}|^2 &= |\mathbf{Q}_\perp|^2 + |Q_\|}|^2 \\ \boldsymbol{\xi}_\perp^* \cdot \mathbf{J} \times \mathbf{Q} &= J_\|(\boldsymbol{\xi}_\perp^* \times \mathbf{b}) \cdot \mathbf{Q}_\perp + Q_\|\boldsymbol{\xi}_\perp^* \cdot \mathbf{J}_\perp \times \mathbf{b} \end{aligned} \quad (8.88)$$

where $\mathbf{J}_\perp$ and $Q_\|$ can be written as

$$\begin{aligned} \mathbf{J}_\perp &= \frac{\mathbf{b} \times \nabla p}{B} \\ Q_\| &= \mathbf{b} \cdot \nabla \times (\boldsymbol{\xi}_\perp \times \mathbf{B}) \\ &= \mathbf{b} \cdot (\mathbf{B} \cdot \nabla\boldsymbol{\xi}_\perp - \boldsymbol{\xi}_\perp \cdot \nabla\mathbf{B} - \mathbf{B}\nabla \cdot \boldsymbol{\xi}_\perp) \\ &= -B(\nabla \cdot \boldsymbol{\xi}_\perp + 2\boldsymbol{\xi}_\perp \cdot \boldsymbol{\kappa}) + \frac{\mu_0}{B}\boldsymbol{\xi}_\perp \cdot \nabla p \end{aligned} \quad (8.89)$$

Equation (8.89) is substituted into Eq. (8.88), which is then substituted back into Eq. (8.87). After a short calculation one obtains

$$\begin{aligned} \delta W_F = \frac{1}{2}\int_P d\mathbf{r}\bigg[&\frac{|\mathbf{Q}_\perp|^2}{\mu_0} + \frac{B^2}{\mu_0}|\nabla \cdot \boldsymbol{\xi}_\perp + 2\boldsymbol{\xi}_\perp \cdot \boldsymbol{\kappa}|^2 + \gamma p\,|\nabla \cdot \boldsymbol{\xi}|^2 \\ &- 2(\boldsymbol{\xi}_\perp \cdot \nabla p)(\boldsymbol{\kappa} \cdot \boldsymbol{\xi}_\perp^*) - J_\|(\boldsymbol{\xi}_\perp^* \times \mathbf{b}) \cdot \mathbf{Q}_\perp\bigg] \end{aligned} \quad (8.90)$$

This is the intuitive form of δW_F originally suggested by Furth *et al.* (1965) and Greene and Johnson (1968). The terms in Eq. (8.90) have the following simple physical interpretation. The $|\mathbf{Q}_\perp|^2$ term represents the energy required to bend magnetic field lines. It is the dominant potential energy contribution to the shear Alfvén wave. The second term corresponds to the energy necessary to compress the magnetic field and describes the major potential energy contribution to the compressional Alfvén wave. The $\gamma p\,|\nabla \cdot \boldsymbol{\xi}|^2$ term represents the energy required to compress the plasma. It is the main source of potential energy for the sound wave. Each of the contributions just described is stabilizing. The remaining two terms can be positive or negative and thus can drive instabilities. The first of these is proportional to $\nabla p \sim \mathbf{J}_\perp \times \mathbf{B}$ while the second is proportional to $J_\|$. Thus, while a vacuum field is MHD stable, either perpendicular or parallel currents represent potential sources of instability. The former type are sometimes referred to as pressure-driven modes and the latter as current-driven modes. A more detailed discussion of the classification of MHD instabilities is given in Section 8.11.

As a final point it is useful to note that δW_F can be written as

$$\delta W_F(\boldsymbol{\xi}^*, \boldsymbol{\xi}) = \delta W_\perp(\boldsymbol{\xi}_\perp^*, \boldsymbol{\xi}_\perp) + \frac{1}{2}\int \gamma p \, |\nabla \cdot \boldsymbol{\xi}|^2 \, d\mathbf{r} \tag{8.91}$$

Similarly, the ideal MHD normalization K can be expressed as

$$K(\boldsymbol{\xi}^*, \boldsymbol{\xi}) = K_\perp(\boldsymbol{\xi}_\perp^*, \boldsymbol{\xi}_\perp) + \frac{1}{2}\int \rho \, |\xi_\parallel|^2 \, d\mathbf{r} \tag{8.92}$$

Both $\delta W_\perp$ and $K_\perp$ are independent of $\xi_\parallel$. It is now possible to make a direct comparison of the energy and variational principles corresponding to the ideal and collisionless MHD models. For the Energy Principle, a necessary and sufficient condition for exponential stability is

$$\begin{aligned} &\delta W_F + \delta W_S + \delta W_V \geq 0 \qquad \text{(ideal MHD)} \\ &\delta W_\perp + \delta W_S + \delta W_V \geq 0 \qquad \text{(collisionless MHD)} \end{aligned} \tag{8.93}$$

The corresponding variational principles are given by

$$\begin{aligned} \omega^2 &= (\delta W_F + \delta W_S + \delta W_V)/K \qquad \text{(ideal MHD)} \\ \omega^2 &= (\delta W_\perp + \delta W_S + \delta W_V)/K_\perp \qquad \text{(collisionless MHD)} \end{aligned} \tag{8.94}$$

As might be expected, the quantity $\xi_\parallel$ never appears in the collisionless MHD formulation. The only difference between these models is the plasma compressibility contribution, which is almost always very small if not identically zero. A more detailed comparison is made in the next section where the significance of incompressibility ($\nabla \cdot \boldsymbol{\xi} = 0$) is discussed.

8.8.4. Summary

In Section 8.8 a powerful minimizing principle, known as the Energy Principle, has been derived which will be extensively applied throughout the remainder of the textbook to determine the stability boundaries of different MHD equilibria. The Energy Principle states that a necessary and sufficient condition for MHD stability is that

$$\delta W \geq 0 \tag{8.95}$$

for all allowable plasma displacements. Here,

$$\delta W = \delta W_F + \delta W_S + \delta W_V \tag{8.96}$$

with

$$\delta W_F = \frac{1}{2}\int_P d\mathbf{r}\left[\frac{|\mathbf{Q}_\perp|^2}{\mu_0} + \frac{B^2}{\mu_0}|\nabla\cdot\boldsymbol{\xi}_\perp + 2\boldsymbol{\xi}_\perp\cdot\boldsymbol{\kappa}|^2 + \gamma p\,|\nabla\cdot\boldsymbol{\xi}|^2 - 2(\boldsymbol{\xi}_\perp\cdot\nabla p)(\boldsymbol{\kappa}\cdot\boldsymbol{\xi}_\perp^*) - J_\parallel(\boldsymbol{\xi}_\perp^*\times\mathbf{b})\cdot\mathbf{Q}_\perp\right] \tag{8.97a}$$

$$\delta W_S = \frac{1}{2}\int_S dS\,|\mathbf{n}\cdot\boldsymbol{\xi}_\perp|^2\mathbf{n}\cdot\left[\!\left[\nabla\left(p + \frac{B^2}{2\mu_0}\right)\right]\!\right] \tag{8.97b}$$

$$\delta W_V = \frac{1}{2}\int_V d\mathbf{r}\,\frac{|\hat{\mathbf{B}}_1|^2}{\mu_0} \tag{8.97c}$$

The vacuum magnetic field $\hat{\mathbf{B}}_1$ satisfies

$$\nabla\times\hat{\mathbf{B}}_1 = \nabla\cdot\hat{\mathbf{B}}_1 = 0 \tag{8.98}$$

and is related to the normal component of the plasma displacement evaluated on the plasma surface by the boundary conditions

$$\begin{aligned} \mathbf{n}\cdot\hat{\mathbf{B}}_1|_{S_w} &= 0 \\ \mathbf{n}\cdot\hat{\mathbf{B}}_1|_{S_p} &= \hat{\mathbf{B}}\cdot\nabla(\mathbf{n}\cdot\boldsymbol{\xi}_\perp) - (\mathbf{n}\cdot\boldsymbol{\xi}_\perp)[\mathbf{n}\cdot(\mathbf{n}\cdot\nabla)\hat{\mathbf{B}}]|_{S_p} \end{aligned} \tag{8.99}$$

where S_w and S_p refer to the conducting wall and the plasma surface, respectively. Equations (8.95)–(8.99) constitute a complete formulation of the ideal MHD stability problem.

8.9. INCOMPRESSIBILITY

Because of the simple way in which $\xi_\parallel$ appears in δW it is possible to minimize the potential energy once and for all with respect to $\xi_\parallel$ for arbitrary geometry. The resulting form of δW is only a function of $\boldsymbol{\xi}_\perp$ and thus, the number of unknown dependent variables is reduced from three to two.

The minimizing condition follows from letting $\xi_\parallel \to \xi_\parallel + \delta\xi_\parallel$ in Eq. (8.97) and then setting the corresponding variation to zero. Since $\xi_\parallel$ appears only in the $\gamma p\,|\nabla\cdot\boldsymbol{\xi}|^2$ term, one finds

$$\delta(\delta W)_{\xi_\parallel} = \frac{1}{2}\int_P d\mathbf{r}\,\gamma p\left[(\nabla\cdot\boldsymbol{\xi}^*)\nabla\cdot\left(\frac{\delta\xi_\parallel\mathbf{B}}{B}\right) + (\nabla\cdot\boldsymbol{\xi})\nabla\cdot\left(\frac{\delta\xi_\parallel^*}{B}\mathbf{B}\right)\right] \tag{8.100}$$

where $\delta(\delta W)_{\xi_\parallel}$ is the variation of δW with respect to $\xi_\parallel$. Integrating the

$\delta\xi_\parallel$ terms by parts yields

$$\delta(\delta W)_{\xi_\parallel} = -\frac{1}{2}\int_P d\mathbf{r}\left[\frac{\delta\xi_\parallel}{B}\mathbf{B}\cdot\nabla(\gamma p\nabla\cdot\boldsymbol{\xi}^*) + \frac{\delta\xi_\parallel^*}{B}\mathbf{B}\cdot\nabla(\gamma p\nabla\cdot\boldsymbol{\xi})\right] \quad (8.101)$$

The boundary term vanishes since $\mathbf{n}\cdot\mathbf{b} = 0$ on the plasma surface. If the variation $\delta(\delta W)_{\xi_\parallel}$ is set to zero for arbitrary $\delta\xi_\parallel$ and use is made of the fact that $\mathbf{B}\cdot\nabla p = 0$, then one obtains the general minimizing condition

$$\mathbf{B}\cdot\nabla(\nabla\cdot\boldsymbol{\xi}) = 0 \quad (8.102)$$

For most configurations the operator $\mathbf{B}\cdot\nabla$ is nonsingular and consequently Eq. (8.102) implies that the general minimizing condition is

$$\nabla\cdot\boldsymbol{\xi} = 0 \quad (8.103)$$

That is, the most unstable perturbations (in the context of minimizing δW) are incompressible.

In a sense this is obvious from Eq. (8.97) since the only term containing $\xi_\parallel$ is positive. Thus, its smallest value is zero and is obtained by setting $\nabla\cdot\boldsymbol{\xi} = 0$. However, it is the components of $\boldsymbol{\xi}$ which are the physical quantities and not $\nabla\cdot\boldsymbol{\xi}$ itself. Therefore, in order to set $\nabla\cdot\boldsymbol{\xi} = 0$, one must be able to choose a $\xi_\parallel$ such that

$$\mathbf{B}\cdot\nabla\left(\frac{\xi_\parallel}{B}\right) = -\nabla\cdot\boldsymbol{\xi}_\perp \quad (8.104)$$

As with Eqs. (8.102) and (8.103) physically acceptable solutions for $\xi_\parallel$ are possible only if the operator $\mathbf{B}\cdot\nabla$ is nonsingular as is the case in most configurations of interest. Even if $\mathbf{B}\cdot\nabla$ vanishes on *isolated* magnetic surfaces, a $\xi_\parallel$ can be constructed that is bounded in the vicinity of these surfaces but that makes a vanishingly small contribution to the plasma compressional energy.

It is worth noting that there are two special cases in which the operator $\mathbf{B}\cdot\nabla$ cannot be inverted. In the first case if there exists sufficient equilibrium symmetry the operator $\mathbf{B}\cdot\nabla$ identically vanishes for certain special modes. As an example, consider a pure Z pinch with equilibrium magnetic field $\mathbf{B} = B_\theta(r)\mathbf{e}_\theta$. If the perturbations are Fourier analyzed with respect to θ and z so that $\boldsymbol{\xi} \sim \boldsymbol{\xi}(r)\exp[i(m\theta + kz)]$ then for the $m = 0$ mode the operator $\mathbf{B}\cdot\nabla$ acting on any scalar identically vanishes. In this situation $\nabla\cdot\boldsymbol{\xi} = \nabla\cdot\boldsymbol{\xi}_\perp + \mathbf{B}\cdot\nabla(\xi_\parallel/B) = \nabla_\perp\cdot\boldsymbol{\xi}_\perp$; that is, $\xi_\parallel$ does not appear in δW and the compressibility term $\gamma p\,|\nabla_\perp\cdot\boldsymbol{\xi}_\perp|^2$ must be maintained and included in the minimization with respect to $\boldsymbol{\xi}_\perp$.

The second case corresponds to the situation of closed-line symmetry. Here the operator $\mathbf{B}\cdot\nabla$ is not in general singular but there is a periodicity constraint requiring $\xi_\parallel(l) = \xi_\parallel(l + L)$ on every magnetic line since the lines are closed. (l is arc length and L is the total length of a given field line.) In this case it is necessary to add a homogeneous solution to Eq. (8.103) so that

$$\nabla \cdot \boldsymbol{\xi} = F(p) \tag{8.105}$$

where F is an arbitrary function of the pressure p which satisfies the homogeneous equation $\mathbf{B} \cdot \nabla p = 0$. Solving Eq. (8.105) for $\xi_\parallel$ yields

$$\frac{\xi_\parallel}{B} = -\int_0^l \frac{dl}{B} \nabla \cdot \boldsymbol{\xi}_\perp + F(p) \int_0^l \frac{dl}{B} \tag{8.106}$$

Note that $F(p)$ can be taken outside the integration since p is constant along a field line. The periodicity constraint requires that $F(p)$ be chosen as follows:

$$F(p) = \langle \nabla \cdot \boldsymbol{\xi}_\perp \rangle \tag{8.107}$$

$$\langle \nabla \cdot \boldsymbol{\xi}_\perp \rangle = \oint \frac{dl}{B} \nabla \cdot \boldsymbol{\xi}_\perp \bigg/ \oint \frac{dl}{B}$$

The end result is that in closed-line systems the minimum value of the plasma compressibility term cannot be made zero, but has the value

$$\int_P \gamma p \, |\nabla \cdot \boldsymbol{\xi}|^2 \, d\mathbf{r} = \int_P \gamma p \, |\langle \nabla \cdot \boldsymbol{\xi}_\perp \rangle^2| \, d\mathbf{r} \tag{8.108}$$

On the basis of this discussion it is now possible to ascertain the relationship between the stability predictions of the ideal and collisionless MHD models. In the general case where $\nabla \cdot \boldsymbol{\xi} = 0$, δW (ideal MHD) = δW (collisionless MHD), indicating that both models predict the identical stability boundaries. Also, the resulting form of δW is independent of the energy equation of state since γ, the ratio of specific heats, does not appear.

For either of the special cases discussed, there is an additional stabilizing term in ideal MHD, proportional to γ. Considering the lack of validity of the ideal MHD energy equation in most cases of physical interest, the contribution of this term is likely to be unreliable. Similarly, in the collisionless MHD model the assumptions made restricting the class of allowable motions are also violated for the two special cases, indicating that this model is no more reliable than ideal MHD. In summary, for cases where $\nabla \cdot \boldsymbol{\xi} = 0$, both models give the same predictions; for either special case, neither model is particularly reliable.

Finally, it is worth pointing out the following mathematical result. For the full variational problem, applied to the general or either special case, it can be shown that for unstable modes

$$\omega^2(\text{coll. MHD}) \le \omega^2(\text{ideal MHD}) \tag{8.109}$$

Thus, the collisionless MHD model is more pessimistic with regard to growth rates. See Problem 8.11.

8.10. VACUUM VERSUS FORCE-FREE PLASMA

An important but somewhat subtle issue is the general stability relationship between configurations that are surrounded by either a vacuum or a force-free plasma. [A detailed discussion of this topic, including examples, has been given by Goedbloed (1979).] This problem arises physically as follows. Many confinement configurations consist of a hot core of plasma carrying nearly all of the plasma current. This core is surrounded by a cold, low-density, current-free plasma which ultimately makes contact with the first wall or limiter. Since the electrical conductivity of the outer plasma is much less than that of the core it might appear reasonable to treat this region as a perfect insulator (i.e., a vacuum).

However, it is more often the case that the conductivity of the outer region is still sufficiently high that its resistive diffusion time is long compared to characteristic MHD time. It would thus seem more realistic to treat this region not as a vacuum, but as an ideal MHD plasma with low pressure (i.e., force free) carrying no equilibrium current.

The replacement of a perfectly insulating region with a perfectly conducting region might be expected to have a large effect on the overall stability. While this is often true, there are many cases where there is no effect at all. To understand this situation, compare the outer region contribution to δW for each case: (a) the vacuum, given by Eq. (8.97c) and (b) the force-free plasma with zero current, given by δW_F [Eq. (8.87)] with $\mathbf{J} \to 0$, $p \to 0$:

$$\begin{aligned} \delta W_V &= \frac{1}{2}\int_V \frac{|\hat{\mathbf{B}}_1|^2}{\mu_0}\, d\mathbf{r} \qquad \text{(vacuum)} \\ \delta W_{\hat{F}} &= \frac{1}{2}\int_V \frac{|\mathbf{Q}|^2}{\mu_0}\, d\mathbf{r} \qquad \text{(force-free plasma)} \end{aligned} \tag{8.110}$$

Since $\mathbf{Q} \equiv \mathbf{B}_1 = \nabla \times (\boldsymbol{\xi}_\perp \times \mathbf{B})$ both energy contributions have the identical form. One might then be led to conclude that the magnetic perturbation $\mathbf{B}_1$, which minimizes the vacuum energy, also minimizes the force-free plasma energy and thus, both contributions should always be identical. This would be true if it were not for an additional constraint on the force-free plasma. In particular, the plasma displacement resulting from the minimizing magnetic perturbation must correspond to a physically allowable motion [i.e., $K(\boldsymbol{\xi}_\perp^*, \boldsymbol{\xi}_\perp)$ must be bounded]. No such constraint exists in the vacuum, since $\boldsymbol{\xi}_\perp$ is not defined in this region.

For a well-defined, bounded magnetic perturbation $\mathbf{B}_1$ the force-free plasma topological constraint arises when attempting to invert the relationship $\mathbf{B}_1 = \nabla \times (\boldsymbol{\xi}_\perp \times \mathbf{B})$ to determine $\boldsymbol{\xi}_\perp$. Whether or not the inversion is singular is closely related to the properties of the $\mathbf{B} \cdot \nabla$ operator. For example, in a general one-dimensional screw pinch the relationship between B_{1r} and ξ_r is given by $B_{1r} = \mathbf{B} \cdot \nabla \xi_r$. After Fourier analyzing with respect to θ and z one can invert this relationship yielding $\xi_r = -iB_{1r}/F$, $F = kB_z(r) + mB_\theta(r)/r$. Hence, if $F(r)$ vanishes anywhere in the outer region ξ_r is singular and $K(\boldsymbol{\xi}_\perp^*, \boldsymbol{\xi}_\perp)$ is unbounded. When this situation occurs, the stability of the force-free region must be recomputed with $\mathbf{n} \cdot \mathbf{B}_1$ set to zero, not on the outer conductor, but on the singular surface in order that ξ_r remain bounded. The force-free plasma is more stable than the vacuum since the new boundary condition is equivalent to moving the conducting boundary inward, a more constrained situation.

Even with the above distinction between vacuum and force-free plasma there still remains one crucial physics issue. Each of the two cases discussed represents a limiting description of finite resistivity. If such an effect is included in the Ohm's law, the plasma can diffuse through the magnetic field and the ideal MHD topological constraint associated with perfect conductivity is no longer required. Under this situation, the marginal stability boundaries are identical with the vacuum case (since no additional constraints are necessary) but the growth rates are much lower and depend upon the value of the resistivity in the region of the singular surface.

In summary, the stability behavior of the outer region is as follows. The most pessimistic description corresponds to the vacuum case. Here, no additional constraints need be imposed and the characteristic growth times are those of ideal MHD. If the vacuum region is replaced by an ideal force-free plasma the stability may or may not change depending upon whether the particular mode under consideration has a singular surface in the outer region. With no singular surface the situation is identical to the vacuum case. If there is a singular surface the force-free plasma is considerably more stable because of the topological constraint: the conducting wall is effectively moved inward to the singular surface. Finally, when the outer region is replaced with a resistive force-free plasma the topological constraint is eliminated. The marginal stability boundaries are again identical to the vacuum case, but the growth rates are much lower when a singular surface exists.

The resistive force-free plasma provides the most realistic description of the outer region. Therefore, since the remainder of the textbook is primarily concerned with the determination of *stability boundaries,* whenever the outer region plays an important role in the stability of a given mode it suffices to treat this region as a vacuum.

8.11. CLASSIFICATION OF MHD INSTABILITIES

This section describes a method of classifying MHD instabilities; the general structure of the modes, the main driving sources of the instabilities, and those properties of the equilibrium magnetic field that are effective in reducing or eliminating specific types of modes. The classification system naturally divides into two basic categories. First, one must distinguish whether a given instability is predominantly an internal or external mode. Second, each instability can be characterized by its dominant driving source. Thus, a given mode can be either pressure driven or current driven. A detailed description of the classification system is given below.

8.11.1. Internal/Fixed Boundary Modes

Consider a magnetic configuration in which the plasma is surrounded by a vacuum. Instabilities whose mode structure does not require any motion of the plasma–vacuum interface away from its equilibrium position are called internal or fixed boundary modes. Oftentimes, an internal mode will have a singular surface (i.e., a surface where $\mathbf{B} \cdot \nabla$ vanishes) inside the plasma. The boundary condition applying to such modes is $\mathbf{n} \cdot \boldsymbol{\xi}|_{S_p} = 0$ and is equivalent to moving the conducting wall onto the plasma surface. For internal modes it is only necessary to minimize δW_F since $\delta W_S = \delta W_V = 0$.

8.11.2. External/Free-Boundary Modes

If the plasma–vacuum interface moves from its equilibrium position during an unstable MHD perturbation, the corresponding modes are known as external or free-boundary modes. These perturbations often possess a singular surface in the vacuum region. For such instabilities δW_S and δW_V must be evaluated as well as δW_F since $\mathbf{n} \cdot \boldsymbol{\xi}|_{S_p} \neq 0$. Although an internal mode can be viewed as the special case of an external mode with

$\mathbf{n} \cdot \boldsymbol{\xi}|_{S_p} = 0$, the definition assumed here makes each mode mutually exclusive; that is, if $\mathbf{n} \cdot \boldsymbol{\xi}|_{S_p} = 0$, the mode is an internal mode. If not, it is an external mode.

8.11.3. Pressure-Driven Modes

It was shown in Section 8.3 that a plasma immersed in an infinite, homogeneous, unidirectional magnetic field is always MHD stable. It was later shown by Eq. (8.90) that there are two possible sources of MHD instability, one proportional to ∇p and the other to $J_{\|}$. MHD instabilities in which the dominant destablizing term is the one proportional to ∇p are known as pressure-driven modes. They are driven by perpendicular currents and can exist even if no parallel currents are present in the plasma. The most unstable pressure-driven instabilities are internal modes with very short wavelengths perpendicular to the magnetic field but long wavelengths parallel to the field. Pressure-driven modes are usually subdivided into two categories: interchanges and ballooning modes.

Interchange Instabilities. Interchange instabilities are very similar in nature to the Rayleigh–Taylor instability (Chandraseker, 1961). Actually, except in special asymptotic limits the interchange perturbation is not a true mode of the system but represents a special choice of trial function which attempts to minimize the line bending contribution in δW_F (i.e., for an interchange, $\mathbf{B}_{1\perp} \approx 0$). The interchange instability is important in the RFP, the Elmo bumpy torus, the stellarator, and sometimes in tokamaks.

The interchange perturbation can lead to instability depending upon the relative sign of the magnetic field line curvature with respect to the pressure gradient. If the field lines are concave toward the plasma their tension tends to make them shorten and collapse inward. The plasma pressure, on the other hand, has a natural tendency to expand outward. In such cases a perturbation that "interchanges" two flux tubes at different radii leads to a system with lower potential energy and hence instability. See Fig. 8.7. Because of the way the surfaces are fluted, the perturbations are also sometimes known as "flute modes." When the field lines are convex to the plasma, the system is stable to interchange perturbations.

From this description it follows that interchange instabilities represent plasma perturbations which are nearly constant along a field line (i.e., no line bending). Perpendicular to the field the most unstable perturbations have very rapid variations (i.e., $k_{\|}/k_{\perp} \ll 1$, $k_{\perp} a \gg 1$).

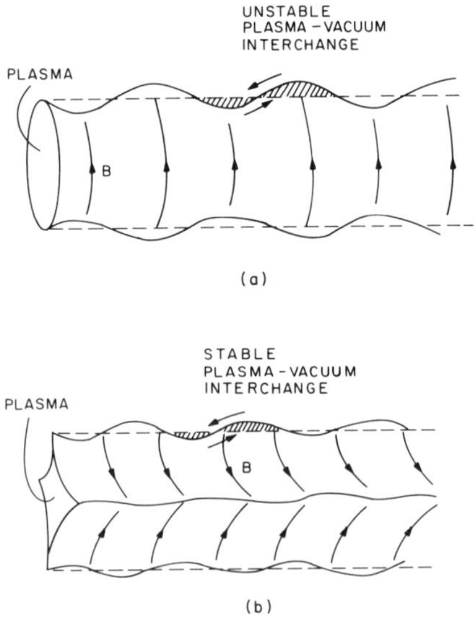

Figure 8.7. Illustration of (a) unstable and (b) stable interchange perturbations.

These are known as localized interchanges and are very amenable to analysis. In general, they lead to necessary conditions for stability which can be expressed solely in terms of local values of the equilibrium quantities. Examples of such conditions are the Suydam criterion (one-dimensional) and the Mercier criterion (two-dimensional).

There are several features of the magnetic geometry that can be effective in stabilizing interchanges. First, if there is shear in the magnetic field from one flux surface to another, it is not possible to interchange two neighboring flux tubes without some line bending occurring. If the pressure gradient is sufficiently small compared to the shear, the interchange can be stabilized. A second method makes use of the fact that on the inside of a torus the local toroidal field line curvature is favorable. Thus, as a given magnetic line encircles the torus it in general passes through alternating regions of favorable and unfavorable curvature. By carefully designing the configuration the "average" curvature can be made favorable thus stabilizing interchanges. Such configurations are said to possess a "magnetic well" or "average-favorable curvature."

Ballooning Modes. Ballooning modes are internal pressure-driven instabilities that occur in tori or other multidimensional configurations. These instabilities are important because they determine one set of criteria which limits the maximum achievable value of β.

The designation "ballooning" refers to the fact that in multidimen-

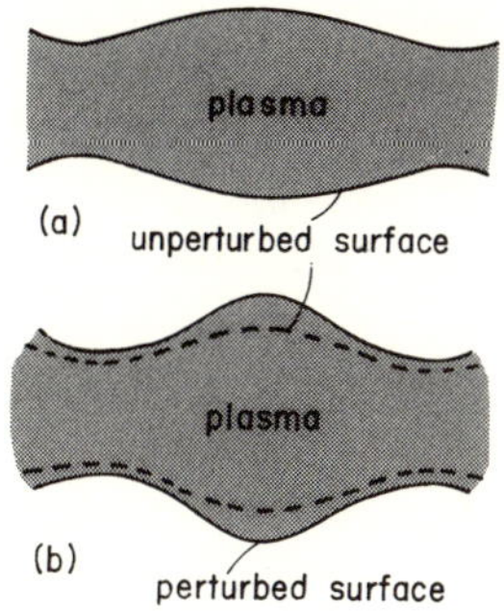

Figure 8.8. (a) Plasma with local regions of favorable and unfavorable curvature. (b) Plasma surface showing a ballooning perturbation.

sional geometries the curvature of the magnetic field line often alternates between regions of favorable and unfavorable curvature. Thus, a perturbation that is not constant, but varies slowly along a field line in such a way that the mode is concentrated in the unfavorable curvature region, can lead to more unstable situations than the simple interchange perturbation. See Fig. 8.8. In effect, the "ballooning" nature of the perturbation in the unfavorable curvature region increases the pressure-driven destabilizing contribution to δW_F. If the localization is not too severe, the accompanying increase in stability from line bending cannot compensate this destabilizing effect.

Magnetic shear can be helpful in stabilizing ballooning modes. However, once in the regime where ballooning modes are important the most effective way to stabilize given magnetic field profiles is to keep β below some critical value.

Like interchanges, ballooning modes are quite amenable to analysis. The most unstable modes occur for $k_{\|}/k_{\perp} \ll 1$ and $k_{\perp}a \gg 1$. By exploiting the short perpendicular wavelength nature of the instabilities, the general multidimensional stability problem reduces to the solution of a one-dimensional differential equation on each flux surface. Substantial progress has been made in the study of ballooning modes in tokamaks using this procedure. One property of the ballooning mode equation is that in a special limit, it leads to the localized interchange criterion (i.e., the Mercier criterion).

8.11.4. Current-Driven Modes

A current-driven mode is one in which the dominant driving source of instability is proportional to the $J_{\|}$ term in Eq. (8.90). These modes are driven by parallel currents and can exist even in a zero-pressure force-free plasma. Current-driven instabilities are often known as "kink" modes. In general the most unstable perturbations have long parallel wavelengths $k_{\|}/k_{\perp} \ll 1$, and perpendicular wavelengths that are macroscopic in scale $k_{\perp}a \sim 1$. Current-driven modes can have the form of either internal or external perturbations depending upon the location of the singular surface. As such, these modes are usually subdivided into two categories: external kinks and internal kinks.

External Kink Modes. The external kink mode is theoretically the dominant instability in a tokamak, although the internal kink also plays an important role experimentally. At low β the main destabilizing term for higher m modes (m is the azimuthal wave number) is the radial gradient in the parallel current. For low m numbers, $m = 1$ in particular, the current profile is not as important as the total current itself. The basic form of the perturbation is such that the plasma surface "kinks" into a helix as illustrated in Fig. 8.9. The unstable modes occur for long parallel wavelength [i.e., $k_{\|}/k_{\perp} \ll 1$] in order to minimize line bending. Usually the most dangerous kinks correspond to low perpendicular wave numbers [i.e., $k_{\perp}a \sim m \sim 1$].

At higher values of β the modes develop a distinct ballooning structure. The harmonic content is concentrated at low m values, unlike the short wavelength internal ballooning modes. At sufficiently high β the "external ballooning-kink" mode becomes unstable. The corresponding β limit is often the most severe stability criterion in the system.

There are several ways in which the stability of a given configuration to external kink modes can be improved. First, for a prescribed geometry there is usually a critical parallel current, indicating the onset of

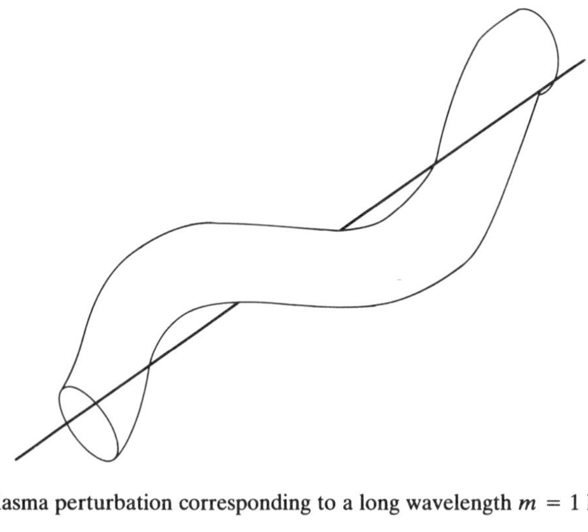

Figure 8.9. Plasma perturbation corresponding to a long wavelength $m = 1$ kink instability.

instability. Stability can be achieved by keeping the parallel current below this value. Likewise, for a fixed parallel current a toroidal device with sufficiently small circumference will prohibit the formation of long-wavelength kinks. Second, to the extent that one can experimentally control the field profiles, higher m kink modes can be stabilized by peaking the parallel current profile, keeping the total current fixed. Third, since low-m-number kink perturbations have a broad radial extent, they can be stabilized if a perfectly conducting shell is placed sufficiently close to the plasma surface. Finally, one must guarantee that β does not become too large, or else a ballooning-kink instability could develop.

Internal Kink Modes. The internal kink has many properties similar to those of the external kink, although it is in general a weaker instability. In fact for a tokamak it is usually only $m = 1$ that can become unstable. As before, a combination of low current and/or tight aspect ratio can help to stabilize internal kink modes particularly in a tokamak. Neither of these methods is effective in the RFP and there, interestingly enough, a broad profile helps to provide stability; that is, for internal modes, broadening the current profile at fixed total current effectively moves the conducting wall closer to the current channel, thereby providing stability. Although broad profiles would be detrimental for external kink stability, such modes are not expected to be important in the RFP since the plasma is in fact surrounded by a conducting shell.

8.12. SUMMARY

In Chapter 8 the general features of MHD stability theory have been described and are summarized below.

Exponential Stability. To determine whether or not a given configuration is MHD stable, the following definition of stability is adopted. A plasma is said to be exponentially stable if $\text{Im}\ \omega \leq 0$ for every discrete normal mode of the system.

Waves in a Homogeneous Plasma. As a simple test of stability, the infinite, homogeneous plasma in a uniform magnetic field is investigated. This system is shown to be stable. The dispersion relation predicts the existence of three distinct propagating waves: the shear Alfvén wave, the fast magnetoacoustic wave (compressional Alfvén wave), and the slow magnetoacoustic wave (sound wave). The shear Alfvén wave plays the dominant role in plasma instabilities.

General Linearized Stability Equations. The general three-dimensional MHD stability analysis is formulated first as an initial value problem and then transformed into a normal mode eigenvalue problem. All of the perturbed quantities are expressed in terms of the plasma displacement $\boldsymbol{\xi}$. The resulting system has the form of three coupled, homogeneous partial differential equations, with eigenvalue ω^2: $-\omega^2\rho\boldsymbol{\xi} = \mathbf{F}(\boldsymbol{\xi})$.

Properties of the Force Operator **F**: The force operator **F** appearing in the normal mode equation is shown to possess an important mathematical property, self-adjointness. This leads to the conclusion that the eigenvalue ω^2 is purely real. Consequently, normal modes either oscillate, or grow purely exponentially. The self-adjointness also leads to the conclusion that the normal modes are orthogonal. As a general rule, the unstable part of the spectrum associated with **F** contains only discrete modes. The stable part contains both discrete modes and continua.

Variational Formulation. By making use of the self-adjointness of **F**, the normal mode eigenvalue problem is recast into the form of a variational principle. This principle is an equivalent integral representation of the eigenvalue differential equation.

Energy Principle. A further analysis of the variational principle shows that there exists a powerful minimizing principle, known as the Energy Principle, for testing MHD stability. The physical basis for this principle is the exact, nonlinear conservation of energy in ideal MHD. The Energy Principle provides an exact test for the stability boundaries, although it provides only an estimate of the actual growth rates. Stability is tested by examining the sign of δW for all allowable displacements. The system is stable if and only if $\delta W \geq 0$.

Incompressibility. In attempting to find the most unstable trial functions that minimize δW, it is shown that the minimizing $\xi_\parallel$ can be calculated once and for all for arbitrary geometries. This reduces the number of unknowns to two, corresponding to the components of $\boldsymbol{\xi}_\perp$. For most systems, in which $\mathbf{B} \cdot \nabla$ is nonsingular, or at most singular on isolated surfaces, $\xi_\parallel$ must be chosen to make $\nabla \cdot \boldsymbol{\xi} = 0$. In these cases ideal and collisionless MHD give identical stability boundaries. For the special cases involving a high degree of symmetry or closed-line geometry there is an additional stabilizing term in δW due to the plasma compressibility. This contribution can be expressed explicitly in terms of $\boldsymbol{\xi}_\perp$.

Vacuum versus Force-Free Plasma. It is shown that it is important to distinguish whether the region surrounding the main plasma core is a vacuum, an ideal MHD plasma, or a resistive MHD plasma. The ideal MHD plasma often predicts considerably greater stability because the class of allowable perturbations is restricted as a result of the perfect conductivity topological constraint. The most realistic, but difficult, situation corresponds to the resistive plasma. This model gives rise to the identical stability boundaries as in the vacuum case, but much lower growth rates. Since the primary concern is the stability boundaries it is sufficient to treat the outer region as a vacuum in the applications that follow.

Classification of MHD Instabilities. A discussion is presented describing the various types of MHD instabilities that can occur. In general, instabilities can be driven by currents flowing parallel to the field (current-driven modes) or perpendicular to the field (pressure-driven modes). Modes are also distinguished by whether or not the unstable displacement perturbs the plasma surface: external or internal modes. These classifications were further categorized into internal kinks, external kinks, interchanges, and ballooning modes. Depending on the mode in question such equilibrium features as a conducting wall, maximum current, maximum β, shear, magnetic well, or tight aspect ratio might be used to improve stability. As a general feature essentially all MHD instabilities correspond to plasma displacements which are nearly constant along a field line: $k_{\|} \approx 0$ to minimize line bending. However, perpendicular to the field $\boldsymbol{\xi}$ can vary macroscopically (as for kinks) or very rapidly (as for interchange and ballooning modes).

REFERENCES

Berge, G., and J. P. Freidberg (1975). *Phys. Fluids* **18,** 1362.

Bernstein, I. B., E. A. Frieman, M. D. Kruskal, and R. M. Kulsrud (1958). *Proc. R. Soc. London, Ser. A* **244,** 17.

Chandraseker, S. (1961). *Hydrodynamic and Hydromagnetic Stability,* Clarendon, Oxford.

Furth, H. P., J. Killeen, M. N. Rosenbluth, and B. Coppi (1965). In *Plasma Physics and Controlled Nuclear Fusion Research 1964,* IAEA, Vienna, Vol. I, p. 103.

Goedbloed, J. P. (1975). *Phys. Fluids* **18,** 1258.

Goedbloed, J. P. (1979). *Lecture Notes on Ideal Magnetohydrodynamics,* Fom-Instituut voor Plasmafysica, Nieuwegein, Netherlands.

Grad, H. (1973). *Proc. Natl. Acad. Sci. USA* **70,** 3277.

Greene, J. M., and J. L. Johnson (1968). *Plasma Phys.* **10,** 729.

Laval, G., C. Mercier, and R. M. Pellat (1965). *Nucl. Fusion* **5,** 156.

Tataronis, J. A., and A. Salat (1981). Courant Institute of Mathematical Sciences Report No. DOE/ER/03077-171, MF-98.

GENERAL REFERENCES

General MHD Theory

Bateman, G. (1978). *MHD Instabilities*, MIT, Cambridge, Massachusetts.

Bernstein, I. B., E. A. Frieman, M. D. Kruskal, and R. M. Kulsrud (1958). *Proc. R. Soc. London, Ser. A* **244,** 17.

Goedbloed, J. P. (1979). *Lecture Notes on Ideal Magnetohydrodynamics,* Fom-Instituut voor Plasmafysica, Nieuwegein, Netherlands.

Kadomtsev, B. B. (1966). In *Reviews of Plasma Physics,* edited by M. A. Leontovich, Consultants Bureau, New York, Vol. II.

Variational Calculus

Morse, P. M., and H. Feshbach (1953). *Methods of Theoretical Physics,* McGraw-Hill, New York, Vol. II.

PROBLEMS

8.1. Following the procedure given in Section 8.3, derive the dispersion relation for waves propagating in an infinite homogeneous plasma described by the collisionless MHD model. Show that the roots to the dispersion relation are given by Eq. (8.15).

8.2. a. Calculate the dispersion relation for waves propagating in an infinite homogeneous MHD plasma with uniform magnetic field $\mathbf{B} = B_0\mathbf{e}_z$. Include the effect of a small resistivity in Ohm's law: $\mathbf{E} + \mathbf{v} \times \mathbf{B} = \eta\mathbf{J}$, $\eta = \text{const}$. Show that the shear Alfvén branch of the dispersion relation can be written as

$$\omega^2\left(1 + i\frac{\gamma_D}{\omega}\right) - k_\parallel^2 V_a^2 = 0$$

Similarly, show that the magnetoacoustic branch has the form

$$\omega^2(\omega^2 - k^2V_s^2)\left(1 + i\frac{\gamma_D}{\omega}\right) - (\omega^2 - k_\parallel^2 V_s^2)k^2V_a^2 = 0$$

where $\gamma_D = \eta k^2/\mu_0$.

b. Prove that each of the three basic MHD waves becomes slightly damped in the presence of resistivity.

8.3. This problem investigates the apparent paradox associated with the ion acoustic wave in a homogeneous plasma. Specifically, the fact that both ideal MHD and two-fluid theory yield the same dispersion relation although $E_\parallel = 0$ in ideal MHD while $E_\parallel \neq 0$ is crucial in two-fluid theory.

Consider the following simple two-fluid model: massless electrons, isothermal electrons and ions, quasineutrality, and negligible resistivity.

a. Assume a background state $\mathbf{B} = B_0\mathbf{e}_z$, $n = n_0$, $\mathbf{v}_e = \mathbf{v}_i = 0$. Show that the linearized equations describing parallel propagation are given by

$$m_i \frac{\partial v_{zi}}{\partial t} = eE_z - \frac{T_i}{n_0}\frac{\partial n_i}{\partial z}$$

$$0 = -eE_z - \frac{T_e}{n_0}\frac{\partial n_e}{\partial z}$$

$$\frac{\partial n_i}{\partial t} + \frac{\partial}{\partial z} n_0 v_{zi} = 0$$

$$n_e = n_i$$

$$\nabla \times (E_z \mathbf{e}_z) = 0$$

b. Calculate the dispersion relation and show that $\omega^2 = k_{\|}^2(T_e + T_i)/m_i$.

c. To obtain a single-fluid derivation, eliminate E_z from the momentum equations and set $n_i = n_e = n$. Note that the resulting mass and momentum equations are identical to those of ideal MHD for parallel propagation. Show that the mass and momentum equations, by themselves, yield the dispersion relation $\omega^2 = k_{\|}^2(T_e + T_i)/m_i$. One thus concludes that the effects of $E_{\|}$ *are included* in the ideal MHD momentum equation. It is in Ohm's law that $E_{\|}$ is neglected, but this relation is not required for the ion acoustic mode. Thus, the paradox is resolved.

8.4. Consider the integral

$$I = \int_0^1 \eta F(\xi)\, dx$$

where $\eta(x)$, $\xi(x)$ are two complex functions and F is a differential operator. Determine which of the following forms of F corresponds to a self-adjoint operator:

a. $F = 1$

b. $F = \dfrac{\partial}{\partial x}$, $\quad$ BC$_1$

c. $F = \dfrac{\partial^2}{\partial x^2}$, $\quad$ BC$_1$

d. $F = \dfrac{\partial}{\partial x}\left(h \dfrac{\partial}{\partial x}\right)$, $\quad$ BC$_1$

e. $F = \dfrac{\partial^2}{\partial x^2}$, $\quad$ BC$_2$

Here, $h(x)$ is a known function and BC$_1$, BC$_2$ represent the following boundary conditions:

$$\text{BC}_1\text{: } \xi(0) = \eta(0) = \xi(1) = \eta(1) = 0$$
$$\text{BC}_2\text{: } \xi(0) = \eta(0) = 0, \qquad \xi'(1) = A\xi(1), \qquad \eta'(1) = A^*\eta(1)$$

8.5. Consider the Grad–Shafranov equation

$$\Delta^*\psi = -\mu_0 R^2 \frac{dp}{d\psi} - F\frac{dF}{d\psi} \tag{1}$$

Derive a Lagrangian density, $\hat{L} = \hat{L}(\psi, \partial\psi/\partial R, \partial\psi/\partial Z, R, Z)$ such that the total Lagrangian

$$L = \int \hat{L}\, d\mathbf{r}$$

$$d\mathbf{r} = 2\pi R\, dR\, dZ$$

represents a variational principle for the Grad–Shafranov equation [i.e., letting $\psi \to \psi + \delta\psi$ in $\hat{L}$ and setting $\delta L = 0$ gives Eq. (1)]. State what boundary conditions you have assumed.

8.6. Consider the eigenvalue problem

$$y'' + \frac{y'}{x} + \lambda y = 0 \tag{1}$$

$$y'(0) = 0, \qquad y(1) = 0$$

a. Multiply Eq. (1) by y and then integrate over the region $0 \le x \le 1$. Note that the integral relation obtained is correct but not variational. Substitute the trial function $y = 1 - x^2$ and evaluate λ.
b. Derive a variational form for the eigenvalue problem.
c. Substitute the trial function $y = 1 - x^2$ into the variational form and evaluate λ.
d. Substitute the trial function $y = (1 - x^2)^\nu$ into the variational form and determine ν so that the Lagrangian is an extremum:

$$\partial L/\partial \nu = 0$$

e. Compare these results with the exact answer.

8.7. To demonstrate the advantage of using natural boundary conditions consider the following problem:

$$\frac{d^2y}{dx^2} + (\lambda - x^2)y = 0 \tag{1}$$

$$y'(0) = 0 \tag{2}$$

$$y'(1) = 3y(1) \tag{3}$$

a. Derive a Lagrangian with the property that Eq. (3) appears as a natural boundary condition.

b. Consider now the one-parameter trial function $y = 1 - \alpha x^2$. Choose α so that Eq. (3) is satisfied and evaluate the eigenvalue λ.

c. Recall that with natural boundary conditions one is not forced to choose a trial function such that Eq. (3) is automatically satisfied. The variational principle will do "as good a job as possible" to ensure that this condition is satisfied. Instead, treat α as a variational parameter determined by setting $\delta L = 0$. Use this value of α and calculate the eigenvalue λ.

d. Compare (b) and (c) with the exact result.

8.8. Assume the existence of an MHD equilibrium for which the operator $\mathbf{F}/\rho - \omega^2$ exhibits only a discrete spectrum; that is, there are no continua. Evaluate

$$\omega^2 = \delta W(\boldsymbol{\xi}, \boldsymbol{\xi})/K(\boldsymbol{\xi}, \boldsymbol{\xi}) \tag{1}$$

assuming

$$\boldsymbol{\xi} = \boldsymbol{\xi}_0 + \delta\boldsymbol{\xi} \tag{2}$$

where

$$\delta\boldsymbol{\xi} = \sum_{1}^{\infty} a_m \boldsymbol{\xi}_m \tag{3}$$

Here, $\boldsymbol{\xi}_0$ is the eigenfunction corresponding to the lowest (i.e., most unstable) eigenvalue ω_0^2. Show that

$$\omega^2 = \omega_0^2 + \frac{\sum_1^\infty a_m^2(\omega_m^2 - \omega_0^2)}{1 + \sum_1^\infty a_m^2} \tag{4}$$

Equation (4) explicitly demonstrates the fact that any trial function [as given by Eq. (3)] different from $\boldsymbol{\xi}_0$ always yields an estimate for ω^2 which is greater (more stable) than the lowest eigenvalue.

8.9. The boundary condition given by Eq. (8.85)

$$\mathbf{n} \cdot \hat{\mathbf{B}}_1|_{\mathbf{r}_p} = \mathbf{n} \cdot \nabla \times (\boldsymbol{\xi} \times \hat{\mathbf{B}})|_{\mathbf{r}_p} \tag{1}$$

is intuitively obvious although there are subtleties involved. In particular, the plasma displacement is not defined in the vacuum region. A more rigorous derivation starts from the basic condition $\mathbf{n} \cdot \hat{\mathbf{B}} = 0$. Linearize this relation about the equilibrium solution ($\mathbf{n} = \mathbf{n}_0 + \mathbf{n}_1$, $\hat{\mathbf{B}} = \hat{\mathbf{B}}_0 + \hat{\mathbf{B}}_1$, $\mathbf{r} = \mathbf{r}_p + \boldsymbol{\xi}$) and show that Eq. (1) is indeed the correct boundary condition. Hint: Show that $\mathbf{n}_1 = -(\nabla\boldsymbol{\xi}) \cdot \mathbf{n}_0 + \mathbf{n}_0[\mathbf{n}_0 \cdot (\nabla\boldsymbol{\xi}) \cdot \mathbf{n}_0]$.

8.10. The boundary condition

$$\mathbf{n} \cdot \mathbf{B}_1|_{\mathbf{r}_p} = \mathbf{n} \cdot \nabla \times (\boldsymbol{\xi} \times \mathbf{B})|_{\mathbf{r}_p}$$

can be expressed solely in terms of the normal component of the plasma displacement evaluated on the plasma surface: $\xi_n \equiv \mathbf{n} \cdot \boldsymbol{\xi}$. Specifically, show that [Eq. (8.86)]

$$\mathbf{n} \cdot \nabla \times (\boldsymbol{\xi} \times \mathbf{B})|_{\mathbf{r}_p} = \mathbf{B} \cdot \nabla \xi_n - \xi_n[\mathbf{n} \cdot (\mathbf{n} \cdot \nabla)\mathbf{B}]|_{\mathbf{r}_p}$$

8.11. Show that the eigenvalue corresponding to a given *unstable* mode is always lower (i.e., more unstable) for the collisionless MHD model than for ideal MHD. Hint: Use the unstable eigenfunction for ideal MHD as a trial function for collisionless MHD.

8.12. Show by the use of trial functions that the presence of a perfectly conducting wall is always stabilizing.

9

Stability

One-Dimensional Configurations

9.1. INTRODUCTION

This chapter describes the application of the Energy Principle to one-dimensional cylindrical configurations. First, the stability of the two basic configurations, the θ pinch and the Z pinch, is investigated. It is shown that the θ pinch has inherently favorable stability properties while the Z pinch is strongly unstable.

Next, the Energy Principle is applied to the general screw pinch. Even this relatively basic configuration exhibits a rather high level of complexity. A general second-order equation is derived for the plasma displacement with boundary conditions corresponding to a perfectly conducting wall, an isolating vacuum region and a resistive wall. Three important results are then discussed: Suydam's criterion for localized interchanges, Newcomb's general procedure for testing stability, and the oscillation theorem describing the eigenvalue behavior for the full linearized stability equations.

Based on these results a relatively complete description of the stability of the reversed field pinch is given. The ordering of the fields with respect to aspect ratio is such that the one-dimensional stability analysis is essentially all that is required for the reversed field pinch, the toroidal effects leading to only small corrections. Finally, a discussion is given of the "straight tokamak." Although a relatively complete stability picture is again obtained the results must be used cautiously since for the tokamak there are toroidal modifications that are of comparable importance.

9.2. THE θ PINCH

As a first application of the Energy Principle consider the θ pinch. Recall that the equilibrium is described by $p(r)$, $\mathbf{B} = B_z(r)\mathbf{e}_z$, where

$$p(r) + \frac{B_z^2}{2\mu_0} = \frac{B_0^2}{2\mu_0} \tag{9.1}$$

and B_0^2 is the externally applied field. Since the equilibrium is symmetric with respect to θ and z, the plasma displacement can be Fourier analyzed as follows:

$$\boldsymbol{\xi}(\mathbf{r}) = \boldsymbol{\xi}(r)\exp[i(m\theta + kz)] \tag{9.2}$$

Here, m and k correspond to the "poloidal" and "toroidal" wave numbers, respectively, and are interpreted graphically in Fig. 9.1. Observe that $m = 2$, $k = 0$ produces an elliptic distortion of the cross section while $m = 3$, $k = 0$ produces a triangular deformation. Similarly, $m = 0$, $k \neq 0$ gives rise to a sequence of bulges along the column. A combination of m and k both nonzero leads to a helical distortion as shown for $m = 1$.

The first step in the minimization of δW is to note that the incompressibility condition $\nabla \cdot \boldsymbol{\xi} = 0$ yields the following expression for $\xi_\parallel \equiv \xi_z$:

$$\xi_z = \frac{i}{kr}[(r\xi_r)' + im\xi_\theta] \tag{9.3}$$

Hence, as long as $k \neq 0$, the minimizing perturbations are incompressible: $\nabla \cdot \boldsymbol{\xi} = 0$.

The next step is the evaluation of δW_F [Eq. (8.97a)]. The individual

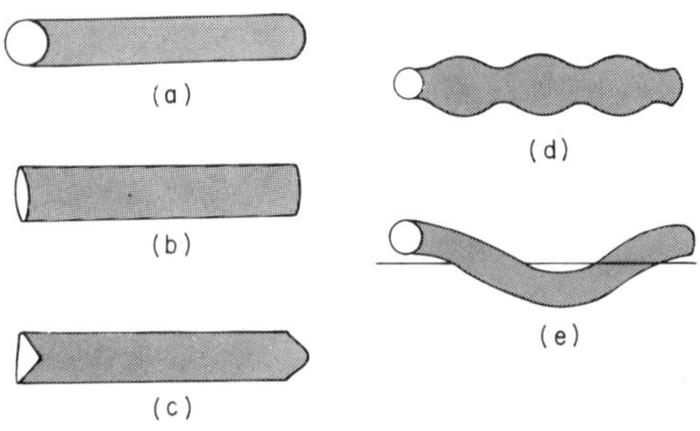

Figure 9.1. Graphical interpretation of the poloidal wave number m and toroidal wave number k: (a) unperturbed column; (b) $m = 2$, $k = 0$; (c) $m = 3$, $k = 0$; (d) $m = 0$, $k \neq 0$; (e) $m = 1$, $k \neq 0$.

terms are easily calculated:

$$
\begin{aligned}
\mathbf{Q}_\perp &= ikB_z\boldsymbol{\xi}_\perp = ikB_z(\xi_r\mathbf{e}_r + \xi_\theta\mathbf{e}_\theta)\\
\nabla\cdot\boldsymbol{\xi}_\perp &= \frac{1}{r}(r\xi_r)' + \frac{im}{r}\xi_\theta\\
\boldsymbol{\kappa} &= 0\\
J_\parallel &= 0
\end{aligned}
\tag{9.4}
$$

Substituting into δW_F one obtains

$$
\begin{aligned}
\frac{\delta W_F}{2\pi R_0} &= \frac{\pi}{\mu_0}\int_0^a W(r)r\,dr\\
W(r) &= B_z^2\left[k^2(|\xi_r|^2 + |\xi_\theta|^2) + \frac{1}{r^2}|(r\xi_r)'|^2 + \frac{m^2}{r^2}|\xi_\theta|^2\right.\\
&\quad\left. + \frac{im}{r^2}(r\xi_r^*)'\xi_\theta - \frac{im}{r^2}(r\xi_r)'\xi_\theta^*\right]
\end{aligned}
\tag{9.5}
$$

where a is the outer radius of the plasma and $2\pi R_0$ is the length of the column. Note that ξ_θ appears only algebraically in this expression. Consequently, the ξ_θ terms can be combined by completing the squares as follows:

$$
W(r) = B_z^2\left\{\left|k_0\xi_\theta - \frac{im}{k_0r^2}(r\xi_r)'\right|^2 + \frac{k^2}{k_0^2r^2}[|(r\xi)'|^2 + k_0^2r^2|\xi|^2]\right\} \tag{9.6}
$$

where $k_0^2 = k^2 + m^2/r^2$. Note that ξ_θ appears only in the first term. Since this term is positive, its minimum value is zero. Thus, δW_F is minimized by choosing

$$
\xi_\theta = \frac{im}{m^2 + k^2r^2}(r\xi_r)' \tag{9.7}
$$

δW_F then has the form

$$
\frac{\delta W_F}{2\pi R_0} = \frac{\pi}{\mu_0}\int_0^a r\,dr\frac{k^2B_z^2}{m^2 + k^2r^2}[|(r\xi)'|^2 + (m^2 + k^2r^2)\,|\xi|^2] \tag{9.8}
$$

where $\xi \equiv \xi_r$.

Observe that $\delta W_F > 0$ for any nonzero k^2 and $\delta W_F \to 0$ as $k^2 \to 0$; that is, at any value of β the θ pinch is positively stable for finite wavelengths and approaches marginal stability for very long wavelengths.

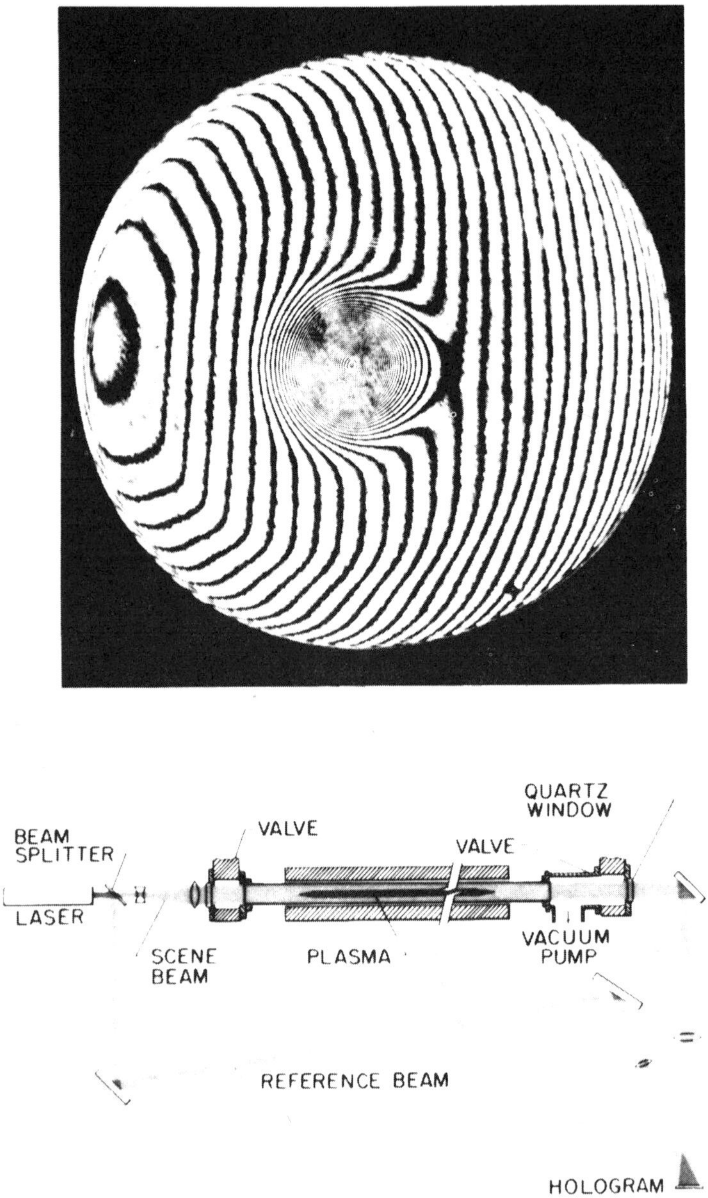

Figure 9.2. Experimental observation of the good radial confinement properties of an early Los Alamos θ pinch. Shown is an end-on hologram which measures the length-averaged density profile. Courtesy F. Jahoda, Los Alamos National Laboratory.

This represents the favorable stability properties previously mentioned. The inclusion of the vacuum term removes the restriction that $\xi(a) = 0$ in Eq. (9.8) but does not alter the stability; δW_F remains positive, δW_V is positive by definition, and $\delta W_S = 0$ for profiles that decay smoothly to zero at $r = a$.

An experimental demonstration of the stability of linear θ pinches is illustrated in Fig. 9.2. Shown here is an end-on holographic interferometer measurement of the density contours of a θ pinch, averaged along the length of the discharge. Note that the plasma cross section is circular and well confined, showing no indication of instability.

A simple physical explanation for the stability is as follows: A θ-pinch equilibrium has no parallel currents so that current-driven modes cannot be excited. Likewise, since the field lines are straight, their curvature is zero (i.e., radius of curvature is infinite) and pressure-driven modes cannot be excited. Any perturbation to the equilibrium either bends or compresses the magnetic field lines, and both are stabilizing influences.

9.3. THE Z PINCH

For the next application of the Energy Principle, consider the Z pinch described by $p(r)$, $\mathbf{B} = B_\theta(r)\mathbf{e}_\theta$, where

$$\frac{dp}{dr} + \frac{B_\theta}{\mu_0 r}\frac{d}{dr}(rB_\theta) = 0 \tag{9.9}$$

As before, the equilibrium symmetry implies that the perturbations can be Fourier analyzed with respect to θ and z: $\boldsymbol{\xi}(\mathbf{r}) = \boldsymbol{\xi}(r)\exp[i(m\theta + kz)]$.

The first step in the minimization procedure is to examine the incompressibility condition. For $m \neq 0$ it follows from $\nabla \cdot \boldsymbol{\xi} = 0$ that $\xi_\parallel \equiv \xi_\theta$ is given by

$$\xi_\theta = \frac{i}{m}[(r\xi_r)' + ikr\xi_z] \tag{9.10}$$

For $m = 0$ the general minimizing condition $\mathbf{B} \cdot \nabla(\nabla \cdot \boldsymbol{\xi}) = 0$ is automatically satisfied and ξ_θ never explicitly appears in the calculation. In this case, $\nabla \cdot \boldsymbol{\xi} = \nabla \cdot \boldsymbol{\xi}_\perp$ and the plasma compressibility must be maintained. Thus, the two cases $m \neq 0$ and $m = 0$ must be treated separately.

9.3.1. $m \neq 0$ Modes

Consider first internal modes with $m \neq 0$. The individual terms required for the evaluation of δW_F are easily calculated:

$$
\begin{aligned}
\mathbf{Q}_\perp &= \frac{imB_\theta}{r}(\xi_r\mathbf{e}_r + \xi_z\mathbf{e}_z) \\
\boldsymbol{\kappa} &= -\frac{\mathbf{e}_r}{r} \\
\nabla \cdot \boldsymbol{\xi}_\perp + 2\boldsymbol{\xi}_\perp \cdot \boldsymbol{\kappa} &= r\left(\frac{\xi_r}{r}\right)' + ik\xi_z \\
J_\parallel &= 0
\end{aligned}
\tag{9.11}
$$

This is substituted into Eq. (8.97a) leading to the following expression for δW_F

$$
\begin{aligned}
\frac{\delta W_F}{2\pi R_0} &= \frac{\pi}{\mu_0}\int_0^a W(r) r\, dr \\
W(r) &= \frac{m^2 B_\theta^2}{r^2}(|\xi_r|^2 + |\xi_z|^2) + B_\theta^2\left|r\left(\frac{\xi_r}{r}\right)' + ikr\xi_z\right|^2 + \frac{2\mu_0 p'}{r}|\xi_r|^2
\end{aligned}
\tag{9.12}
$$

In analogy with the θ pinch, the quantity ξ_z appears only algebraically. Thus, one can again complete the squares. A simple calculation shows that the value of ξ_z that minimizes δW is given by

$$\xi_z = \frac{ikr^3}{m^2 + k^2r^2}\left(\frac{\xi_r}{r}\right)' \tag{9.13}$$

Upon substituting Eq. (9.13) into the expression for δW_F one obtains

$$\frac{\delta W_F}{2\pi R_0} = \frac{\pi}{\mu_0}\int_0^a r\, dr\left[(2\mu_0 rp' + m^2B_\theta^2)\frac{|\xi|^2}{r^2} + \frac{m^2r^2B_\theta^2}{m^2 + k^2r^2}\left|\left(\frac{\xi}{r}\right)'\right|^2\right] \tag{9.14}$$

where $\xi \equiv \xi_r$.

Note that k^2 appears only in the denominator of a positive term. Consequently, for any arbitrary trial function $\xi(r)$, δW_F is minimized by letting $k^2 \to \infty$. In this limit δW_F reduces to

$$\frac{\delta W_F}{2\pi R_0} = \frac{\pi}{\mu_0}\int_0^a r\, dr(2\mu_0 rp' + m^2B_\theta^2)\frac{|\xi|^2}{r^2} \tag{9.15}$$

This is the desired form of δW_F.

Equation (9.15) implies that the necessary and sufficient condition

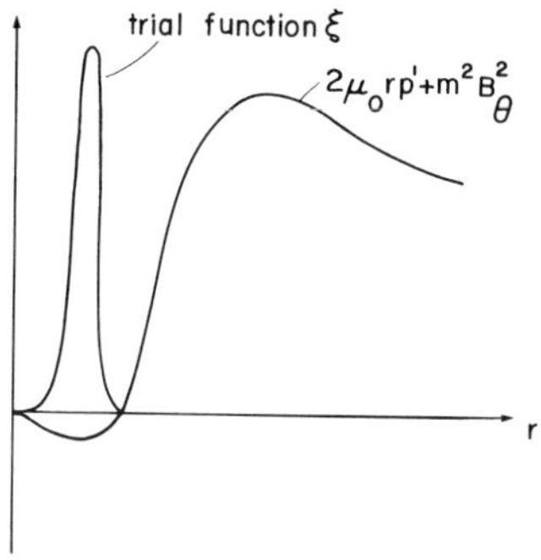

Figure 9.3. Trial function ξ making $\delta W_F < 0$ for $m \neq 0$ modes in a pure Z pinch. Note that $\xi \neq 0$ only in the region where $2\mu_0 rp' + m^2 B_\theta^2 < 0$.

for stability against $m \neq 0$ modes is given by (Kadomstev, 1966)

$$2r\frac{dp}{dr} + \frac{m^2 B_\theta^2}{\mu_0} > 0 \tag{9.16}$$

Were this quantity to be negative anywhere, a trial function localized within this region could be constructed that would make $\delta W_F < 0$, implying instability. See Fig. 9.3.

Using the equilibrium relation [Eq. (9.9)] the stability condition can be rewritten in the following two forms:

$$\frac{r^2}{B_\theta}\left(\frac{B_\theta}{r}\right)' < \frac{1}{2}(m^2 - 4) \tag{9.17a}$$

$$\frac{1}{B_\theta^2}(rB_\theta^2)' < m^2 - 1 \tag{9.17b}$$

For standard Z-pinch profiles B_θ/r is a decreasing function of radius: $(B_\theta/r)' < 0$. For such profiles Eq. (9.17a) predicts stability for $m \geq 2$. At large radii where the current is low, $B_\theta \sim 1/r$ corresponding to a vacuum field; in this region Eq. (9.17b) implies that $m = 1$ is also stable. However, near the origin $B_\theta \sim r$. Consequently, Eq. (9.17b) predicts that the inner core of a Z pinch is always unstable to an $m = 1$ mode.

The source of the instability can be seen in Fig. 9.4. As the plasma undergoes an $m = 1$ deformation the magnetic lines concentrate in the tighter portion of the column, raising the value of B_θ. The corresponding increased magnetic tension produces a force in the direction to further increase the $m = 1$ deformation; hence, instability.

Although the plasma distortion has the appearance of a helix, it does not correspond to a kink mode since $J_\parallel = 0$. The minimizing perturbation is best described as a competition between line bending and unfavorable curvature, with the magnetic compression making a negligibly small contribution.

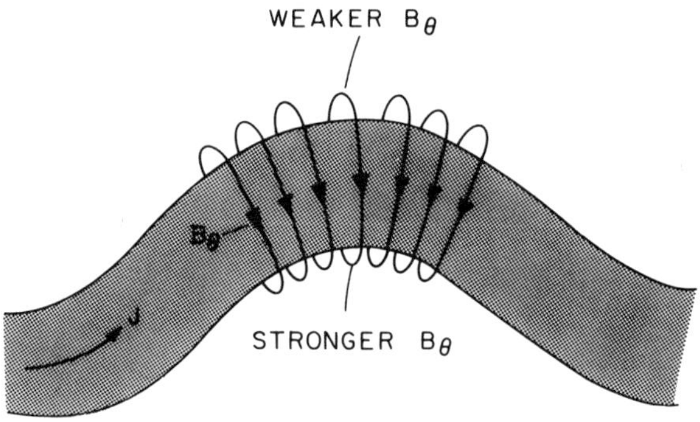

Figure 9.4. Physical mechanism of the $m = 1$ instability in a pure Z pinch.

9.3.2. $m = 0$ Mode

Consider now the $m = 0$ mode. Recall from the discussion following Eq. (9.10) that $\xi_\parallel \equiv \xi_\theta$ never appears in the calculation for $m = 0$. Hence, $\nabla \cdot \boldsymbol{\xi}$ cannot be set to zero and must be maintained in the minimization. The evaluation of δW_F is straightforward and again shows that ξ_z appears only algebraically:

$$\frac{\delta W_F}{2\pi R_0} = \pi \int_0^a W(r) r\, dr$$
$$W(r) = \frac{B_\theta^2}{\mu_0}\left| r\left(\frac{\xi_r}{r}\right)' + ikr\xi_z\right|^2 + \gamma p \left|\frac{(r\xi_r)'}{r} + ik\xi_z\right|^2 + \frac{2p'}{r}|\xi_r|^2 \tag{9.18}$$

One completes the squares with respect to ξ_z and finds that the perturbation that minimizes δW_F is given by

$$\xi_z = \frac{i}{\gamma p + B_\theta^2/\mu_0}\left[\frac{rB_\theta^2}{\mu_0}\left(\frac{\xi_r}{r}\right)' + \frac{\gamma p}{r}(r\xi_r)'\right] \tag{9.19}$$

Substituting back into δW_F yields

$$\frac{\delta W_F}{2\pi R_0} = \int_0^a r\, dr\left[\frac{4\gamma p B_\theta^2/\mu_0}{\gamma p + B_\theta^2/\mu_0} + 2rp'\right]\frac{|\xi|^2}{r^2} \tag{9.20}$$

The necessary and sufficient condition for stability when $m = 0$ can then be written as (Kadomstev, 1966)

$$-\frac{rp'}{p} < \frac{2\gamma B_\theta^2/\mu_0}{\gamma p + B_\theta^2/\mu_0} \tag{9.21}$$

This condition can be satisfied, at least theoretically, for sufficiently gradual pressure profiles. In practice it is difficult to meet this criterion experimentally. For example, if one substitutes the relatively gradual Bennett profile given by [Eq. (5.17)]

$$\begin{aligned} B_\theta &= \frac{\mu_0 I_0}{2\pi}\frac{r}{r^2 + r_0^2} \\ p &= \frac{\mu_0 I_0^2}{8\pi^2}\frac{r_0^2}{(r^2 + r_0^2)^2} \end{aligned} \tag{9.22}$$

then the stability condition given by Eq. (9.21) reduces to

$$\gamma > 2 \tag{9.23}$$

Since $\gamma = 5/3$ for ideal MHD the condition is violated.

One should note that Eq. (9.21) is a strong function of the ideal MHD energy equation, which provides a poor description of the physics in a collisionless plasma. The corresponding condition from collisionless MHD, equally unreliable since $\mathbf{B} \cdot \nabla \equiv 0$, is obtained by setting $\gamma = 0$. The resulting stability criterion, $-rp'/p < 0$ is always violated for a confined plasma. Thus, both theories predict instability, although important questions remain regarding the validity of each model. Perhaps the best guide is the experimental results which demonstrate catastrophic destruction of the plasma in several microseconds. This can be seen in Fig. 9.5, which shows a side-on streak photograph of an early linear Z pinch at Los Alamos.

The $m = 0$ instability is an interchange mode often known as the "sausage instability." The basic nature of the mode is illustrated in Fig. 9.6. When an $m = 0$ "sausage" perturbation is superimposed on the Z-pinch equilibrium, the magnetic field in the throat region increases since the plasma carries the same current in a smaller cross section. The increased magnetic tension produces a force which tends to further constrict the column.

The minimizing perturbation for the $m = 0$ interchange instability produces a competition between unfavorable curvature and compression of the plasma. The line bending is zero.

It is instructive to examine the interchange instability from the single particle point of view to see how the curvature enters. In this connection examine Fig. 9.7, which illustrates a linear Z pinch. The surface of the plasma has been fluted by interchanging tubes of plasma with flux tubes of vacuum magnetic field. Because of the curvature of the B_θ field, electrons and ions drift in opposite directions along the Z axis. The drifts shown correspond to the negative curvature characteristic of a Z pinch

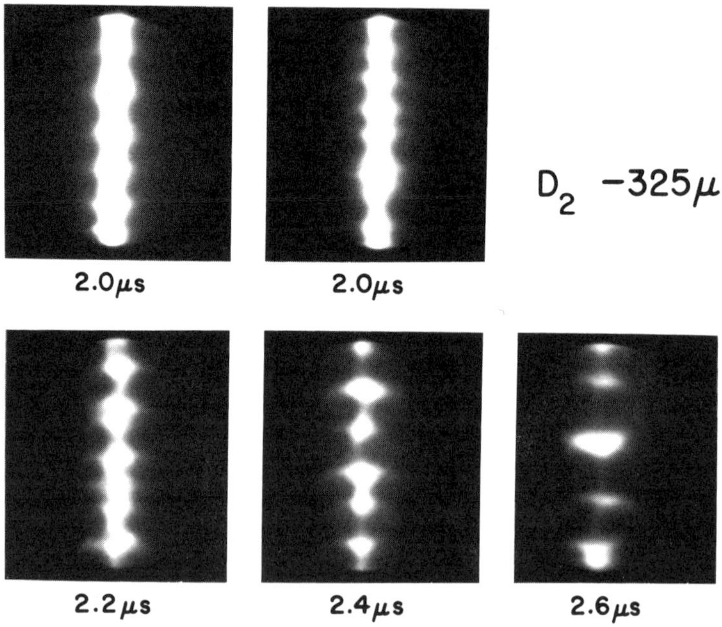

Figure 9.5. Side-on streak photographs of the Los Alamos Columba experiment demonstrating the development of the $m = 0$ sausage instability. Courtesy J. A. Phillips, Los Alamos National Laboratory.

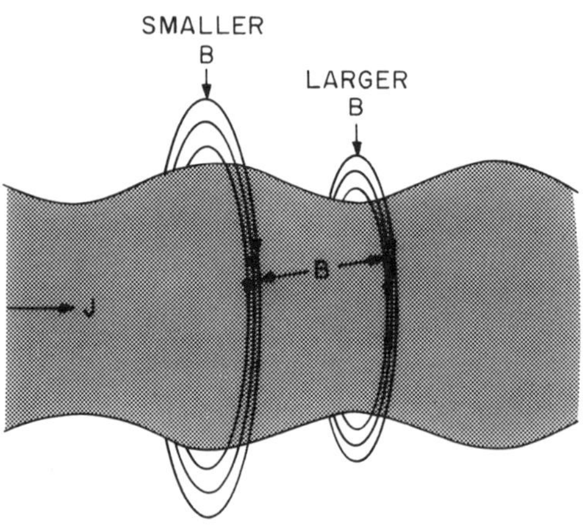

Figure 9.6. Physical mechanism of the $m = 0$ sausage instability in a pure Z pinch.

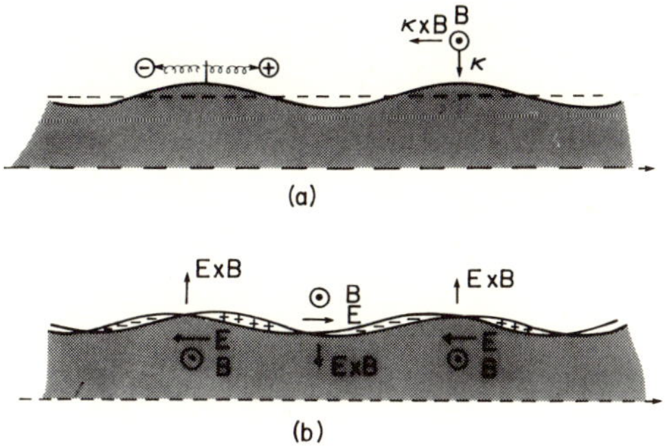

Figure 9.7. Single-particle picture of the $m = 0$ interchange mode in a pure Z pinch: (a) interchange perturbation flutes the plasma surface (b) opposing curvature drifts set up a charge separation leading to a destabilizing $\mathbf{E} \times \mathbf{B}$ drift.

field: $\boldsymbol{\kappa} = -\mathbf{e}_r/r$. When the surface is fluted, these opposing currents set up a charge separation that creates an electric field as shown. This electric field in turn produces an $\mathbf{E} \times \mathbf{B}$ drift of the plasma. When the curvature is negative the $\mathbf{E} \times \mathbf{B}$ drift reinforces the perturbation, leading to instability. When the curvature is positive, the $\mathbf{E} \times \mathbf{B}$ drift is in the opposite direction, and the perturbation is stabilized.

In summary, a Z pinch is always unstable to $m = 1$ perturbations and is likely to be unstable to $m = 0$ as well. The unstable modes are quite virulent and have the form of pressure-driven interchanges. These results form the basis of earlier statements attributing very poor stability characteristics to the pure Z pinch.

9.4. THE GENERAL SCREW PINCH

9.4.1. Evaluation of δW

As a first step in finding configurations that combine the favorable equilibrium properties of the Z pinch with the favorable stability properties of the θ pinch, consider the general screw pinch. The relevant equilibrium quantities are $p(r)$, $\mathbf{B} = B_\theta(r)\mathbf{e}_\theta + B_z(r)\mathbf{e}_z$ and satisfy

$$\left(p + \frac{B_\theta^2 + B_z^2}{2\mu_0}\right)' + \frac{B_\theta^2}{\mu_0 r} = 0 \tag{9.24}$$

Although two field components are present, the system still possesses (θ, z) symmetry so that the perturbations can be Fourier analyzed: $\boldsymbol{\xi}(\mathbf{r}) = \boldsymbol{\xi}(r)\exp[i(m\theta + kz)]$. In fact, it is this dual symmetry that is ultimately responsible for the algebraic elimination of two components of $\boldsymbol{\xi}$ in the minimization procedure.

For the screw pinch it is convenient to decompose the displacement vector as follows:

$$\boldsymbol{\xi} = \xi \mathbf{e}_r + \eta \mathbf{e}_\eta + \xi_\| \mathbf{b} \tag{9.25}$$

where

$$\begin{aligned} \xi_\| &= (\xi_\theta B_\theta + \xi_z B_z)/B \\ \eta &= (\xi_\theta B_z - \xi_z B_\theta)/B \\ \mathbf{b} &= (B_\theta \mathbf{e}_\theta + B_z \mathbf{e}_z)/B \\ \mathbf{e}_\eta &= (B_z \mathbf{e}_\theta - B_\theta \mathbf{e}_z)/B \end{aligned} \tag{9.26}$$

Incompressibility. The first step in the minimization procedure is to examine the incompressibility condition. For a given m and k, setting $\nabla \cdot \boldsymbol{\xi} = 0$ yields

$$\xi_\| = i\frac{B}{F}\nabla \cdot \boldsymbol{\xi}_\perp \tag{9.27}$$

where

$$F = \mathbf{k} \cdot \mathbf{B} = mB_\theta/r + kB_z \tag{9.28}$$

Excluding the very special case of zero shear [i.e., $(B_\theta/rB_z)' = 0$], F will in general be nonzero except perhaps at a finite number of discrete radii. If F is nonzero everywhere a well-behaved $\xi_\|$ can be chosen in accordance with Eq. (9.27), making the plasma compressibility term in δW_F vanish. Even when isolated $F = 0$ singular surfaces exist (as they almost always do) the compressibility term can be made negligibly small with a well-behaved $\xi_\|$.

The situation is illustrated in Fig. 9.8 for the case of a single resonant surface at $r = r_s$. The dashed curve represents the curve of $\xi_\|$ corresponding to Eq. (9.27). Note the singularity at $r = r_s$. The solid

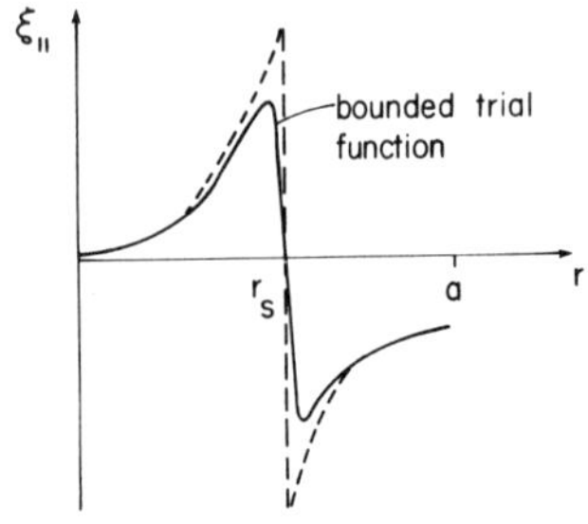

Figure 9.8. Bounded trial function for $\xi_\|$ which makes the plasma compressibility contribution to δW_F vanish.

curve represents a trial function for $\xi_\parallel$ of the form

$$\xi_\parallel = \frac{iBF}{F^2 + \sigma^2} \nabla \cdot \boldsymbol{\xi}_\perp \tag{9.29}$$

where $\sigma^2 > 0$ is a small but nonzero constant. This form of $\xi_\parallel$ is bounded, but no longer corresponds to $\nabla \cdot \boldsymbol{\xi} = 0$. However, for small σ^2, the deviation from $\nabla \cdot \boldsymbol{\xi} = 0$ is negligible everywhere except in the immediate vicinity of $r = r_s$; specifically, substituting Eq. (9.29) into $\nabla \cdot \boldsymbol{\xi}$ yields

$$\nabla \cdot \boldsymbol{\xi} = \frac{\sigma^2}{F^2 + \sigma^2} \nabla \cdot \boldsymbol{\xi}_\perp \tag{9.30}$$

The contribution to δW from the plasma compressibility, denoted by $\delta W_\parallel$, can now be easily evaluated by expanding the integrand about the singular surface: $r = r_s + x$, $F \approx F'(r_s)x$. One finds

$$\frac{\delta W_\parallel}{2\pi R_0} = \pi[\gamma pr \, |\nabla \cdot \boldsymbol{\xi}_\perp|^2]_{r_s} \int dx \frac{\sigma^4}{(F'^2 x^2 + \sigma^2)^2} \tag{9.31}$$

Since the integrand decays rapidly for large $|x|$, the limits of integration in Eq. (9.31) can be set to $-\infty < x < \infty$ with negligible error. Carrying out the integration gives

$$\frac{\delta W_\parallel}{2\pi R_0} = \frac{\pi^2 |\sigma|}{2} \left[\frac{\gamma pr \, |\nabla \cdot \boldsymbol{\xi}_\perp|^2}{|F'|} \right]_{r_s} \tag{9.32}$$

Thus, for arbitrarily small but nonzero σ,

$$\frac{\delta W_\parallel}{2\pi R_0} \to 0 \tag{9.33}$$

indicating that the plasma compressibility makes a vanishingly small contribution to δW.

Evaluation of δW_F. The remaining terms in δW_F are functions only of ξ and η. The minimization of δW_F with respect to ξ and η is conceptually quite analogous to that of the Z pinch. However, with two nonzero field components (B_θ, B_z) the algebra is considerably more cumbersome. Thus, the discussion presented below simply outlines the main steps and the results of the analysis. The details are given in Appendix B.

As in previous cases, substitution of $\boldsymbol{\xi}_\perp = \xi \mathbf{e}_r + \eta \mathbf{e}_\eta$ into δW_F shows that η appears only algebraically. The minimizing condition gives

$$\eta = \frac{i}{rk_0^2 B} [G(r\xi)' + 2kB_\theta \xi] \tag{9.34}$$

with $k_0^2 = k^2 + m^2/r^2$ and $G = mB_z/r - kB_\theta$. Equation (9.34) is substituted back into the expression for δW_F, which, after some algebra, can be cast in the following form (Newcomb, 1960):

$$\frac{\delta W_F}{2\pi R_0} = \frac{\pi}{\mu_0}\left\{\int_0^a (f\xi'^2 + g\xi^2)\,dr + \left[\frac{k^2r^2B_z^2 - m^2B_\theta^2}{k_0^2r^2}\right]_a \xi_a^2\right\} \tag{9.35}$$

where

$$\begin{aligned} f &= \frac{rF^2}{k_0^2} \\ g &= 2\frac{k^2}{k_0^2}(\mu_0 p)' + \left(\frac{k_0^2r^2 - 1}{k_0^2r^2}\right)rF^2 + 2\frac{k^2}{rk_0^4}\left(kB_z - \frac{mB_\theta}{r}\right)F \end{aligned} \tag{9.36}$$

$\xi_a = \xi(a)$, and $\xi(r)$ is a real quantity. The boundary term in Eq. (9.35) is a consequence of an integration by parts in δW_F. This term vanishes for internal modes but plays an important role in external modes. Equation (9.35) is the standard form of δW_F for the general screw pinch.

Evaluation of δW_V and δW_S. To obtain the complete δW, Eq. (9.35) must be combined with the surface and vacuum contributions to the potential energy. If the assumption is made that no surface currents flow on the plasma then $\delta W_S = 0$.

The vacuum term is calculated as follows. In the vacuum $\hat{\mathbf{B}}_1 = \nabla\hat{V}_1$ with $\hat{V}_1$ satisfying $\nabla^2\hat{V}_1 = 0$. The solution for $\hat{V}_1$ corresponding to a conducting wall boundary condition at $r = b$ [i.e., $(\partial\hat{V}_1/\partial r)_b = 0$] is given by

$$\hat{V}_1 = A\left[K_r - \left(\frac{K_b'}{I_b'}\right)I_r\right]\exp[i(m\theta + kz)] \tag{9.37}$$

where $K_z = K_m(kz)$, $I_z = I_m(kz)$ and K_m, I_m are modified Besssel functions. The amplitude A is a free constant related to ξ_a through the boundary condition $\hat{B}_{1r}|_a = [\hat{\mathbf{B}}\cdot\nabla\xi - \xi\mathbf{n}\cdot(\mathbf{n}\cdot\nabla)\hat{\mathbf{B}}]_a$. See Eq. (8.99). This condition reduces to

$$\left.\frac{\partial\hat{V}_1}{\partial r}\right|_a = iF(a)\xi_a \tag{9.38}$$

and gives

$$A = \frac{iF(a)\xi_a}{K_a}\left[1 - \left(\frac{K_b'}{I_b'}\right)\left(\frac{I_a}{K_a}\right)\right]^{-1} \tag{9.39}$$

The last step in the calculation is to make use of the fact that

$\nabla^2 V_1 = 0$ and transform δW_V into a surface integral. The result is

$$\delta W_V = \frac{1}{2}\int_V \frac{|\hat{\mathbf{B}}_1|^2}{\mu_0}\,d\mathbf{r} = \frac{1}{2\mu_0}\int_V \nabla\cdot(\hat{V}_1^*\nabla\hat{V}_1)\,d\mathbf{r} = -\frac{1}{2\mu_0}\int_S dS\,\hat{V}_1^*\mathbf{n}\cdot\nabla\hat{V}_1 \tag{9.40}$$

Substituting Eqs. (9.37) and (9.39) into Eq. (9.40) yields the desired form of δW_V.

$$\frac{\delta W_V}{2\pi R_0} = \frac{\pi}{\mu_0}\left[\frac{r^2F^2\Lambda}{|m|}\right]_a \xi_a^2 \tag{9.41}$$

where

$$\begin{aligned}\Lambda &= -\frac{|m|\,K_a}{kaK_a'}\left[\frac{1-(K_b'I_a)/(I_b'K_a)}{1-(K_b'I_a')/(I_b'K_a')}\right]\\ &\approx \frac{1+(a/b)^{2|m|}}{1-(a/b)^{2|m|}}, \qquad kb \ll 1\end{aligned} \tag{9.42}$$

Summary. In summary, the Energy Principle for the general screw pinch reduces to a quadratic form in one dependent variable ξ. For internal modes this form can be written as

$$\frac{\delta W}{2\pi^2 R_0/\mu_0} = \int_0^a (f\xi'^2 + g\xi^2)\,dr \tag{9.43}$$

where ξ must satisfy $\xi(a) = 0$. For external modes, δW has the form

$$\frac{\delta W}{2\pi^2 R_0/\mu_0} = \int_0^a (f\xi'^2 + g\xi^2)\,dr + \left[\left(\frac{krB_z - mB_\theta}{k_0^2 r^2}\right)rF + \frac{r^2\Lambda F^2}{|m|}\right]_a \xi_a^2 \tag{9.44}$$

with $\xi(a)$ arbitrary. In these expressions Λ is given by Eq. (9.42) and f and g by Eq. (9.36).

The evaluation of δW is exact in that the minimizations with respect to $\xi_\parallel$ and η have been carried out without approximation. Note that $f(r)$ is positive while $g(r)$ can have either sign. Consequently, both terms are competitive and further simplifications, as occurred for the θ pinch and Z pinch, are not possible for the general screw pinch.

9.4.2. The Full Normal-Mode Formulation

Upon examining δW as given by Eq. (9.43) or (9.44), readers may be somewhat suspicious of earlier claims regarding the "simplicity" of the Energy Principle, particularly in view of the complex form of the function

$g(r)$. To put the problem in perspective it is useful to present the governing equation for the full normal-mode eigenvalue problem. This equation is obtained by the systematic reduction of Eq. (8.26). As might be expected, the analysis requires a substantial amount of algebra. The details are given in Appendix C. The result is a second-order, ordinary differential equation describing the eigenvalue problem, given by (Hain and Lüst, 1958)

$$\frac{d}{dr}\left[A\frac{d}{dr}(r\xi)\right] - C(r\xi) = 0 \tag{9.45}$$

where

$$\begin{aligned} A(r) &= \left[\frac{\rho(V_a^2 + V_s^2)}{r}\right]\frac{(\omega^2 - \omega_a^2)(\omega^2 - \omega_h^2)}{(\omega^2 - \omega_f^2)(\omega^2 - \omega_s^2)} \\ C(r) &= -\frac{\rho}{r}(\omega^2 - \omega_a^2) + \left(\frac{4k^2V_a^2B_\theta^2}{\mu_0 r^3}\right)\frac{(\omega^2 - \omega_g^2)}{(\omega^2 - \omega_f^2)(\omega^2 - \omega_s^2)} \\ &\quad + \frac{d}{dr}\left[\frac{B_\theta^2}{\mu_0 r^2} - \left(\frac{2kB_\theta G}{\mu_0 r^2}\right)\frac{(V_a^2 + V_s^2)(\omega^2 - \omega_h^2)}{(\omega^2 - \omega_f^2)(\omega^2 - \omega_s^2)}\right] \end{aligned} \tag{9.46}$$

and

$$\begin{aligned} \omega_a^2 &= F^2/\mu_0\rho \\ \omega_h^2 &= [V_s^2/(V_s^2 + V_a^2)]\omega_a^2 \\ \omega_g^2 &= [V_s^2/V_a^2]\omega_a^2 \\ \omega_{f,s}^2 &= \tfrac{1}{2}k_0^2(V_a^2 + V_s^2)[1 \pm (1 - \alpha^2)^{1/2}] \end{aligned} \tag{9.47}$$

Also, V_s^2, V_a^2, and α^2 are local quantities defined by

$$\begin{aligned} V_s^2 &= \gamma p/\rho \\ V_a^2 &= B^2/\mu_0\rho \\ \alpha^2 &= 4V_s^2\omega_a^2/[k_0^2(V_a^2 + V_s^2)^2], \qquad 0 \le \alpha^2 \le 1 \end{aligned} \tag{9.48}$$

In terms of complexity, Eq. (9.45) should be compared with the differential equation that determines the function $\xi(r)$ which minimizes δW. This equation is easily obtained by letting $\xi \to \xi + \delta\xi$ in Eq. (9.43) and setting the variation to zero:

$$\frac{d}{dr}\left(f\frac{d\xi}{dr}\right) - g\xi = 0 \tag{9.49}$$

with f and g given by Eq. (9.36). The complexity of the full normal-mode problem is apparent, particularly the complicated manner in which the eigenvalue ω^2 appears.

It is worth noting that after another lengthy calculation it can be shown that Eq. (9.45) reduces to Eq. (9.49) when ω^2 is set to zero in the coefficients A and C. This is to be expected, since the displacement that minimizes δW corresponds to the marginal stability eigenfunction.

Although simpler than the full normal-mode problem, the potential energy for the screw pinch is still difficult to analyze in a complete and general manner because of the complex form of the function $g(r)$. Nevertheless, there are some general properties that can be proven for arbitrary profiles. First, it is possible to derive a purely analytic criterion, depending only upon the equilibrium profiles, which provides a test of stability against localized interchange perturbations. This is known as Suydam's criterion (Suydam, 1958). Second, if the solution to Eq. (9.49) is known either analytically or numerically, then necessary and sufficient conditions for stability, based on the radial structure of $\xi(r)$, have been given by Newcomb (1960). Third, a general oscillation theorem proving the Sturmean and anti-Sturmean behavior of the eigenvalues has been given by Goedbloed and Sakanaka (1974). The last two properties are valuable, not only for their physics content, but for their practical usefulness in deriving numerical schemes for testing stability. These properties are discussed in detail in the sections that follow.

9.4.3. Suydam's Criterion

Suydam's criterion is derived under the assumption that $\xi(r)$ is a highly localized function of radius in the evaluation of δW. Since this corresponds to a special subclass of trial functions, Suydam's criterion is a necessary but not sufficient condition for stability. The motivation for a localized eigenfunction follows from the observation that if $F = 0$ at some radius $r = r_s$, then each term in f and g vanishes except the p' term, which is almost always negative (i.e., destabilizing). Thus, a perturbation localized about $r = r_s$ has a reasonable likelihood of causing a pressure-driven instability. Note that $F = \mathbf{k} \cdot \mathbf{B} = 0$ corresponds to $k_\parallel \approx 0$ so that such perturbations tend to minimize the bending of the magnetic lines; the perturbations represent internal localized interchange modes.

It is important to recognize that a localized perturbation does not automatically imply instability when $p' < 0$. If the equilibrium magnetic field has shear, then away from the resonant surface, F is no longer zero: $F \approx F'(r_s)(r - r_s)$. Even though this term is small, ξ' is large because of the localization. Hence, the $f\xi'^2$ term in δW produces a stabilizing contribution.

The first step in the derivation is to assume $\xi(r)$ is localized about the singular surface, $F(r_s) = 0$. All quantities in δW_F are then expanded about $x = r - r_s$. The leading order contributions to f and g are given by

$$f \approx \left[\frac{rF'^2}{k_0^2}\right]_{r_s} x^2$$
$$g \approx \left[\frac{2\mu_0 k^2 p'}{k_0^2}\right]_{r_s} \tag{9.50}$$

This leads to the following expression for δW_F:

$$\frac{\delta W_F}{2\pi^2 R_0/\mu_0} = \left[\frac{rF'^2}{k_0^2}\right]_{r_s} \delta W_n$$
$$\delta W_n = \int_{-\Delta}^{\Delta} dx \left[x^2\left(\frac{d\xi}{dx}\right)^2 - D_s\xi^2\right] \tag{9.51}$$
$$D_s = -\left[\frac{2\mu_0 k^2 p'}{rF'^2}\right]_{r_s}$$

Here, δW_n is a normalized form of δW_F and $\Delta \ll a$ is a measure of the localization. If one uses the fact that at the resonant surface

$$k = -\left[\frac{mB_\theta}{rB_z}\right]_{r_s} \tag{9.52}$$

then a simple calculation leads to an expression for D_s only in terms of the equilibrium fields

$$D_s = -\left[\frac{2\mu_0 p' q^2}{rB_z^2 q'^2}\right]_{r_s} \tag{9.53}$$

where $q = rB_z/R_0B_\theta$ and R_0 is the major radius of the equivalent torus.

The function ξ that minimizes δW_n satisfies the Euler–Lagrange equation

$$\frac{d}{dx}\left(x^2\frac{d\xi}{dx}\right) + D_s\xi = 0 \tag{9.54}$$

and has as its solution

$$\xi = c_1 x^{p_1} + c_2 x^{p_2}$$
$$p_{1,2} = -\tfrac{1}{2} \pm \tfrac{1}{2}(1 - 4D_s)^{1/2} \tag{9.55}$$

Since at least one solution for ξ is always singular, one cannot simply use the solution to Eq. (9.55) as a trial function. If one were solving the more complicated full eigenvalue problem for unstable modes, the singularity would disappear [i.e., the coefficient A in Eq. (9.46) remains finite when

$F = 0$, $\omega^2 < 0$]. The power of the Energy Principle is that a simpler question can be posed rather than analyzing the full eigenvalue problem. Specifically, one wants to know whether or not ξ can be modified very near $x = 0$ so that it becomes a well-behaved, allowable trial function capable of making $\delta W_n < 0$. As shown below, the answer depends upon the sign of $1 - 4D_s$.

For $1 - 4D_s < 0$ the roots are complex and

$$\xi = \frac{1}{|x|^{1/2}}[c_1 \sin(k_r \ln|x|) + c_2 \cos(k_r \ln|x|)] \qquad (9.56)$$

$$k_r = \tfrac{1}{2}(4D_s - 1)^{1/2}$$

The solution as $x \to 0$ oscillates infinitely rapidly with a diverging envelope proportional to $1/|x|^{1/2}$. Because of the oscillatory behavior it is possible to construct a well-behaved ξ such that δW_n is always negative (i.e., unstable). The modified ξ is illustrated in Fig. 9.9. In regions I and V, $\xi = 0$ so that $\delta W_n(I) = \delta W_n(\mathrm{V}) = 0$. In regions II and IV, ξ satisfies the Euler–Lagrange equation. The corresponding contributions to δW_n are obtained by multiplying Eq. (9.54) by ξ and integrating over each region. Since either ξ or ξ' is zero at the end points of both regions, one

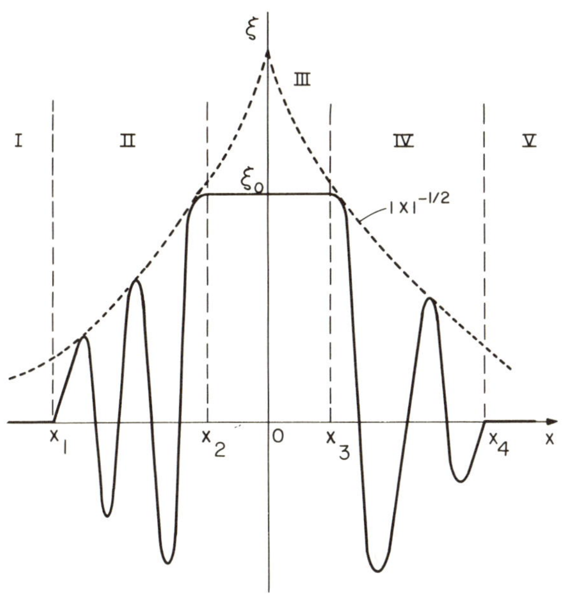

Figure 9.9. Trial function leading to Suydam instability.

finds

$$\delta W_n(\text{II}) = \int_{x_1}^{x_2} (x^2\xi'^2 - D_s\xi^2)\, dx = x^2\xi\xi'\Big|_{x_1}^{x_2} = 0$$
$$\delta W_n(\text{IV}) = \int_{x_3}^{x_4} (x^2\xi'^2 - D_s\xi^2)\, dx = x^2\xi\xi'\Big|_{x_3}^{x_4} = 0 \tag{9.57}$$

The total contribution to δW_n arises from region III where $\xi = \xi_0 =$ const and is given by

$$\delta W_n = \int_{x_2}^{x_3} (x^2\xi'^2 - D_s\xi^2)\, dx = -D_s\xi_0^2(x_3 - x_2) \tag{9.58}$$

By assumption $D_s > 1/4$. Therefore, $\delta W < 0$ and the system is unstable.

When $1 - 4D_s > 0$ the indicial roots are real and at least one solution is singular. In this case oscillatory solutions do not exist and a localized, well-behaved trial function cannot be constructed; that is, if $1 - 4D_s > 0$ the system is stable to localized interchange perturbations.

The stability boundary $1 - 4D_s > 0$ is known as Suydam's criterion. If it is violated for any r_s in the interval $(0, a)$ the plasma is unstable. The usual form for Suydam's criterion is given by (Suydam, 1958)

$$\frac{rB_z^2}{\mu_0}\left(\frac{q'}{q}\right)^2 + 8p' > 0 \tag{9.59}$$

Equation (9.59) is a necessary condition for stability. Suydam's criterion indicates that the stability boundary involves two competing effects: the destabilizing term, corresponding to the interchange drive, results from the combination of a negative pressure gradient and the unfavorable curvature of the B_θ field. The stabilizing term, proportional to q'^2, represents the work done in bending the field lines when interchanging two flux tubes in a system with shear.

One final question to address is whether a mode with such highly localized structure would be important in an actual experiment and not be dominated by other physics. The answer is connected with the oscillation theorem discussed later. Briefly, the theorem shows that violation of Suydam's criterion corresponds to the existence of an accumulation point of discrete eigenvalues at $\omega^2 = 0$ from the unstable side of the spectrum. In addition, the theorem proves that the unstable eigenvalues and eigenfunctions have Sturmean behavior; that is, if a highly oscillatory localized mode with a certain growth rate exists (as implied by violation of Suydam's criterion), a gross mode with zero radial nodes must also exist and will have the maximum growth rate. Thus, it is the guarantee that such a large-scale mode must exist that makes the violation of Suydam's criterion important.

9.4.4. Newcomb's Analysis

A general analysis of the screw-pinch Energy Principle has been given by Newcomb (1960). It is shown how a knowledge of the radial structure of $\xi(r)$, as determined from the minimizing Euler–Lagrange equation

$$\frac{d}{dr}\left(f\frac{d\xi}{dr}\right) - g\xi = 0 \tag{9.60}$$

can be used to derive a set of necessary and sufficient conditions for the stability of arbitrary internal modes. Although the end results do not yield analytical criteria, such as Suydam's criterion, they provide a simple general prescription, which can easily be implemented numerically, for testing stability of general screw-pinch profiles. A key feature of the analysis is the use of ideas associated with Sturm's separation theorem. A discussion of Newcomb's theory is given below.

To begin, consider Eq. (9.60) and assume that m and k are such that $F(r) \neq 0$ in the interval $(0, a)$. In this case, the Euler–Lagrange equation is nonsingular and has two independent solutions:

$$\xi = c_1\xi_1(r) + c_2\xi_2(r) \tag{9.61}$$

Here, $\xi_1(r)$ can be chosen regular at $r = 0$ but $\xi_1(a)$ is not in general zero. Similarly $\xi_2(r)$ can be chosen so that $\xi_2(a) = 0$ (corresponding to internal modes) but $\xi_2(0)$ is not in general regular at $r = 0$.

In the first part of the analysis it is shown that when $\xi_1(r)$ has a zero anywhere in the interval $(0, a)$, a trial function can be constructed which makes $\delta W_F < 0$. If no zero exists, $\delta W_F > 0$ and the system is stable to the given m and k. The analysis is based on the separation theorem which proves that the zeros of any two independent solutions of Eq. (9.60) [$\xi_1(r)$ and $\xi_2(r)$, for instance] must alternate.

This is easily demonstrated by assuming the opposite; that is, $\xi_1(r)$ has two consecutive zeros, r_1 and r_2 and $\xi_2(r) \neq 0$ in the interval (r_1, r_2). See Fig. 9.10. Clearly, $\xi_1' = 0$ for some intermediate r_0, $r_1 < r_0 < r_2$. Therefore, the well-behaved function $Y = \xi_1(r)/\xi_2(r)$ also has zeros at r_1 and r_2 and an intermediate zero derivative. If one now computes

$$Y' = \frac{\xi_2\xi_1' - \xi_1\xi_2'}{\xi_2^2} \tag{9.62}$$

it follows that Y' can equal zero only if the Wronskian $\xi_2\xi_1' - \xi_1\xi_2' = 0$ somewhere in the interval. However, should this occur, the assumption that ξ_1 and ξ_2 are independent is contradicted. Specifically, if one wants to integrate Eq. (9.60) starting with two arbitrary initial conditions (ξ_i, ξ_i'), then it must be possible to find a set of coefficients (c_1, c_2) such

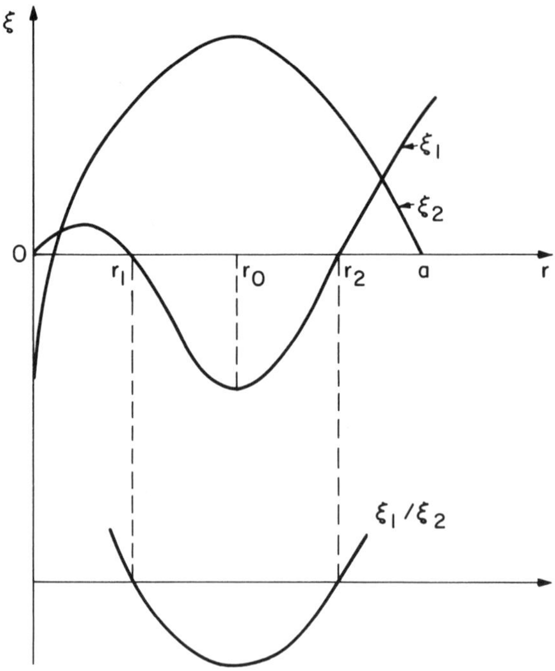

Figure 9.10. Illustration of the separation theorem.

that

$$\begin{aligned} \xi_i &= c_1\xi_1 + c_2\xi_2 \\ \xi_i' &= c_1\xi_1' + c_2\xi_2' \end{aligned} \tag{9.63}$$

if ξ_1 and ξ_2 are independent functions. The condition for independence requires that the determinant of Eq. (9.63), given by $\xi_1\xi_2' - \xi_1'\xi_2$, not vanish. Consequently, the zeroes of ξ_1 and ξ_2 must alternate.

Using this result, a trial function can be constructed using $\xi_1(r)$ and $\xi_2(r)$ in different parts of the interval $(0, a)$ as shown in Fig. 9.11a. Here it is assumed that $\xi_1(r)$ has one zero at r_1. (The results can easily be generalized to include more than one zero.) The separation theorem guarantees that an r_2 must exist such that $\xi_2(r_2) = 0$ with $0 < r_2 < r_1$. Likewise, in the interval $r_2 < r < a$, $\xi_2(r)$ cannot have a second zero: $\xi_2(r) \neq 0$, $r_2 < r < a$. The trial function is constructed using $\xi_1(r)$ to the left of the intersection point r_0 and $\xi_2(r)$ to the right. The value of δW_F is obtained by multiplying Eq. (9.60) by ξ and integrating over the separate

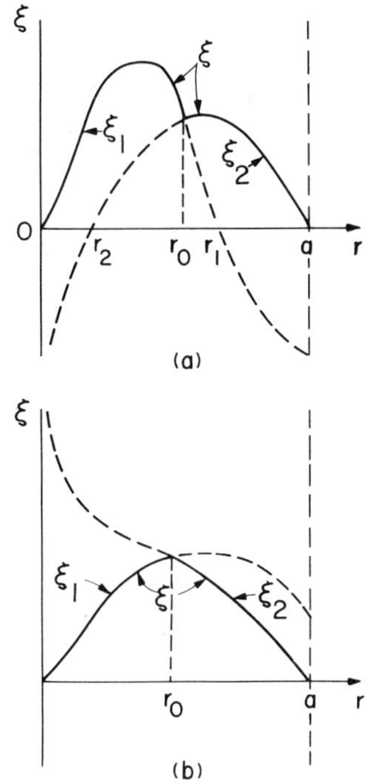

Figure 9.11. Trial functions showing (a) instability, and (b) stability by means of Newcomb's analysis for the case where $F(r) \neq 0,\ 0 \leq r \leq a$.

regions. The result is

$$\begin{aligned}\frac{\delta W_F}{2\pi^2 R_0/\mu_0} &= f\xi_1\xi_1'\Big|_0^{r_0} + f\xi_2\xi_2'\Big|_{r_0}^{a} \\ &= f(r_0)\xi(r_0)[\xi_1'(r_0) - \xi_2'(r_0)]\end{aligned} \tag{9.64}$$

Because of the separation theorem $\xi_1'(r_0) - \xi_2'(r_0) < 0$, implying $\delta W_F < 0$; the system is unstable. If, however, $\xi_1(r_0)$ has no zeros in the interval $(0, a)$ the only type of trial function that can be constructed consistent with the separation theorem and the boundary conditions is shown in Fig. 9.11b. In this case Eq. (9.64) still applies but $\xi_1'(r_0) - \xi_2'(r_0) > 0$ and the system is stable. Note that it is not possible to construct a situation such as shown in Fig. 9.12a, where neither ξ_1 nor ξ_2 has a zero, but one function, ξ_2 for instance, oscillates such that $\xi_1'(r_0) - \xi_2'(r_0) < 0$, implying instability. If neither function has a zero, there must be additional intersections of ξ_1 and ξ_2 as shown at r_1 and r_2. Consequently, if one replaces ξ_1 and ξ_2 with a new set of independent functions ξ_1 and $\xi_3 = \xi_2 - \xi_1$ then in the region $r_1 < r < r_2$, ξ_1 and ξ_3 have the form illustrated in Fig. 9.12b. Clearly the separation theorem is violated.

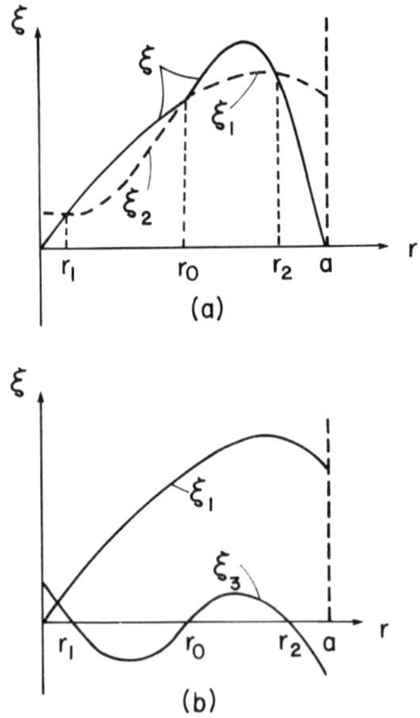

Figure 9.12. Graphical demonstration that an apparently unstable eigenfunction as shown in (a) actually violates the separation theorem as shown in (b).

The first part of Newcomb's analysis can thus be summarized as follows. For values of k and m such that $F(r) \neq 0$ in the interval $(0, a)$, the screw pinch is stable if and only if the nontrivial solution of the Euler–Lagrange equation [i.e., Eq. (9.60)] which is regular at $r = 0$ does not have a zero for $0 < r < a$.

The situation is slightly more complicated if $F(r_s) = 0$, $0 < r_s < a$. In this case, the Euler–Lagrange equation is singular and the proper prescription for calculating ξ in the vicinity of the singular surface must be determined. Recall that in the vicinity of r_s the behavior of ξ is given by $\xi = c_1 x^{p_1} + c_2 x^{p_2}$, where $x = r - r_s$ and $p_{1,2} = -(1/2) \pm (1/2)(1 - 4D_s)^{1/2}$ [see Eq. (9.55)]. To determine the conditions for stability one must assume $(1 - 4D_s)_{r_s} > 0$ or else Suydam's criterion would be violated, immediately implying instability. The proper choice of ξ is then determined by the physical condition that the potential energy remain bounded. Hence, the root $p_1 > -1/2$ is allowable while $p_2 < -1/2$ leads to a divergent energy; that is,

$$\begin{aligned} \int (x^2 \xi'^2 - D_s \xi^2)\, dx &= p x^{2p+1}\big|_{x \to 0} \\ &= \begin{cases} 0 & \text{for } p_1 > -\frac{1}{2} \\ \infty & \text{for } p_2 < -\frac{1}{2} \end{cases} \end{aligned} \tag{9.65}$$

The conclusion is that through a singular point one must choose trial functions containing only the small solution (i.e., the root p_1). In effect the singularity acts like a regularity condition at the origin in that it singles out one of the two independent solutions as admissible. Because of this, the interval $(0, a)$ must be broken into subintervals whose boundaries correspond to the successive zeros of $F(r) = 0$. The first and last intervals are bounded by $r = 0$ and $r = a$, respectively.

Each subinterval must be tested for stability separately. The analysis given for $F(r) \neq 0$, $0 < r < a$ can be directly applied to each subinterval. Typical trial functions for the case of one singular surface are illustrated in Fig. 9.13. The corresponding value of δW_F is given by

$$\begin{aligned}\frac{\delta W_F}{2\pi^2 R_0/\mu_0} &= f\xi_1\xi_1'\big|_{r_s}^{r_0} + f\xi_2\xi_2'\big|_{r_0}^{a} \\ &= f(r_0)\xi(r_0)[\xi_1'(r_0) - \xi_2'(r_0)] \qquad (9.66)\end{aligned}$$

Note that the requirement that $\xi_1(r)$ contain only the small solution

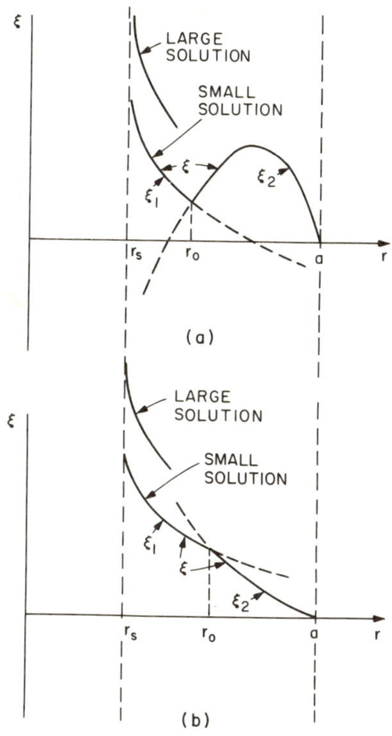

Figure 9.13. Trial function showing (a) instability and (b) stability by means of Newcomb's analysis for the case where $F(r) = 0$ at $r = r_s$.

guarantees that the boundary contribution at $r = r_s$ vanishes. As before, if $\xi_1(r)$ has a zero in the subinterval, $\delta W_F < 0$. If not, $\delta W_F > 0$.

The second part of Newcomb's analysis can be summarized as follows. For values of k and m such that $F(r_s) = 0$, $0 < r_s < a$, the screw pinch is stable if and only if (1) Suydam's criterion is satisfied at $r = r_s$ and (2) the nontrivial solution of the Euler–Lagrange equation containing only the small solution at $r = r_s$ does not have a zero either in $r_s > r > 0$ or $r_s < r < a$.

Newcomb's analysis clearly shows how the stability of a given screw-pinch profile is related to the radial structure of ξ as determined by the Euler–Lagrange equation. For any specific application one can easily solve Eq. (9.60) numerically and then ascertain stability by Newcomb's procedure.

A final practical point worth noting is that in large aspect ratio systems where k can essentially be considered a continuous rather than quantized mode number, the range of (m, k) parameter space that needs to be tested for stability is significantly reduced from the full range $-\infty < m < \infty$, $-\infty < k < \infty$. This follows from the form of f and g given in Eq. (9.36). For $m \neq 0$ consider $k' = k/m$ and m as the two independent mode numbers. Under this transformation f is independent of m and g contains only one term with explicit m dependence given by

$$g(m, k', r) = \hat{g}(k', r) + m^2 r(k' B_z + B_\theta/r)^2 \tag{9.67}$$

Since the m-dependent term is always stabilizing, the most unstable case corresponds to $m = 1$. Similarly, for $m = 0$, f is independent of k and g contains only one term with explicit k dependence:

$$g(m = 0, k, r) = \hat{g}(r) + rk^2 B_z^2 \tag{9.68}$$

In this case, the most unstable mode corresponds to $k^2 \to 0$.

Consequently, when testing for stability it is necessary and sufficient to examine only $m = 1$, $-\infty < k < \infty$ and $m = 0$, $k^2 \to 0$, a significant reduction of the overall parameter space.

9.4.5. The Oscillation Theorem

An important result concerning the stability of arbitrary screw pinches has been derived by Goedbloed and Sakanaka (1974) which demonstrates the existence of an oscillation theorem for the full eigenvalue problem. The theorem implies that for fixed m and k, as the number of radial nodes in a sequence of unstable eigenfunctions monotonically decreases, the corresponding eigenvalues ω^2 also decrease; that is, the most unstable mode is the one with zero nodes. The problem is conceptually similar to that of the classic Sturm–Liouville system but is significantly more difficult in practice because of the complex manner in which the eigenvalue enters the equations. A discussion of the derivation and conclusions of Goedbloed and Sakanaka is given below.

Since the calculation involves the behavior of ω^2, the full eigenvalue equation is required, and not simply the Euler–Lagrange equations corresponding to δW. This equation, derived in Section 9.4.2 and Appendix C, is given by

$$\frac{d}{dr}\left[A\frac{d}{dr}(r\xi)\right] - C(r\xi) = 0 \tag{9.69}$$

where the complicated coefficients $A(r, \omega^2)$ and $C(r, \omega^2)$ are defined by Eq. (9.46).

The oscillation theorem is derived by examining the behavior of $\xi(a)$ as a function of ω^2 for the solution to Eq. (9.69) which is regular at the origin. Assume that two arbitrary values of ω^2, denoted by ω_a^2 and ω_b^2, are specified. Equation (9.69) is then integrated twice yielding the two corresponding solutions $\chi = r\xi_a$ and $\zeta = r\xi_b$ satisfying $\chi(0) = \zeta(0) = 0$. Since ω_a^2 and ω_b^2 are arbitrary, neither χ nor ζ will in general satisfy the internal mode boundary condition $\chi(a) = 0$ or $\zeta(a) = 0$.

Assume now that ω_a^2 and ω_b^2 have neighboring values: $\omega_b^2 = \omega_a^2 + \delta\omega^2$, $\zeta = \chi + \delta\chi$. The equations of motion determining χ and $\delta\chi$ are given by

$$\frac{d}{dr}\left(A\frac{d\chi}{dr}\right) - C\chi = 0 \tag{9.70}$$

$$\frac{d}{dr}\left(A\frac{d}{dr}\delta\chi\right) - C\delta\chi = -\delta\omega^2\left[\frac{d}{dr}\left(\frac{\partial A}{\partial\omega_a^2}\frac{d\chi}{dr}\right) - \frac{\partial C}{\partial\omega_a^2}\chi\right] \tag{9.71}$$

where $A = A(r, \omega_a^2)$, $C = C(r, \omega_a^2)$.

Next, assume that $\chi(r_1) = 0$, $0 < r_1 \leq a$ (see Fig. 9.14). The relationship between $\delta\chi(r_1)$ and $\delta\omega^2$ is obtained by multiplying Eq. (9.70) by $\delta\chi$, Eq. (9.71) by χ, integrating over the range $0 \leq r \leq r_1$ and then subtracting. The result is

$$[A\chi'\delta\chi]_{r_1} = -\delta\omega^2\int_0^{r_1}\left(\frac{\partial A}{\partial\omega_a^2}\chi'^2 + \frac{\partial C}{\partial\omega_a^2}\chi^2\right)dr \tag{9.72}$$

Although the integral in Eq. (9.72) is straightforward to evaluate in principle, it is difficult in practice because of the complexity of A and C. Goedbloed and Sakanaka pointed out a simple method of evaluation by transforming to the original set of dependent variables ξ, η, $\xi_\parallel$ as

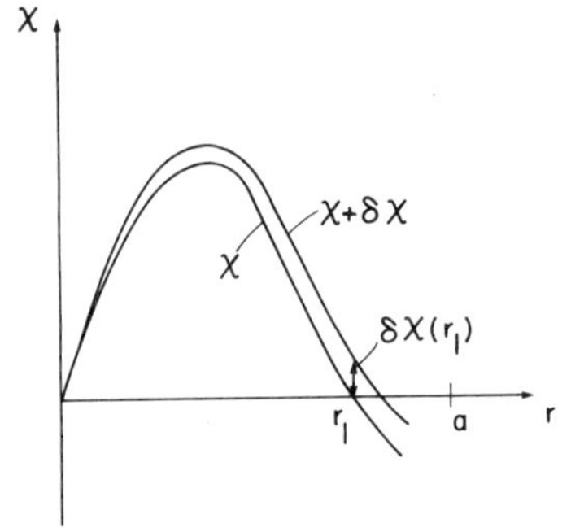

Figure 9.14. χ solution used in the derivation of the oscillation theorem.

follows:

$$\begin{aligned} I &\equiv \int_0^{r_1}\left(\frac{\partial A}{\partial \omega_a^2}\chi'^2 + \frac{\partial C}{\partial \omega_a^2}\chi^2\right) dr \\ &= \left[\frac{\partial}{\partial \omega^2}\int_0^{r_1}(A\chi'^2 + C\chi^2)\, dr\right]_{\omega^2=\omega_a^2} \\ &= \left[\frac{1}{2\pi^2 R_0/\mu_0}\frac{\partial}{\partial \omega^2}(\delta W - \omega^2 K)\right]_{\omega^2=\omega_a^2} \\ &= -\frac{K}{2\pi^2 R_0/\mu_0} = -\int_0^{r_1} r\, dr \mu_0 \rho(|\xi|^2 + |\eta|^2 + |\xi_\parallel|^2) < 0 \end{aligned} \tag{9.73}$$

Equation (9.72) now has the form

$$[A\chi'\delta\chi]_{r_1} = \left(\frac{K}{2\pi^2 R_0/\mu_0}\right)\delta\omega^2 \tag{9.74}$$

which implies that a given change in $\delta\omega^2$ causes a corresponding change in $\delta\chi(r_1)$, the sign of which is determined by the sign of $A(r_1)\chi'(r_1)$. In the unstable portion of the spectrum, $\omega^2 < 0$ so that $A(r) > 0$, $0 < r < a$. Consequently, if $\delta\omega^2 < 0$ the radial node of χ at r_1 always moves outward.

Continuous application of this result for a sequence of negative $\delta\omega^2$'s leads to the oscillation theorem. If an unstable eigenfunction exists with n radial nodes and corresponding eigenvalue $\omega^2 = \omega_n^2 < 0$, then a continual decrease in ω^2 (i.e., increase in instability) eventually gives rise to unstable eigenfunctions with $n - 1$, $n - 2$, $n - 3, \ldots, 0$ radial nodes. In all cases, the zero node mode has the fastest growth rate.

This has important implications with regard to Suydam's criterion. If Suydam's criterion is violated the oscillation theorem guarantees the existence of a macroscopic zero node mode with maximum growth rate

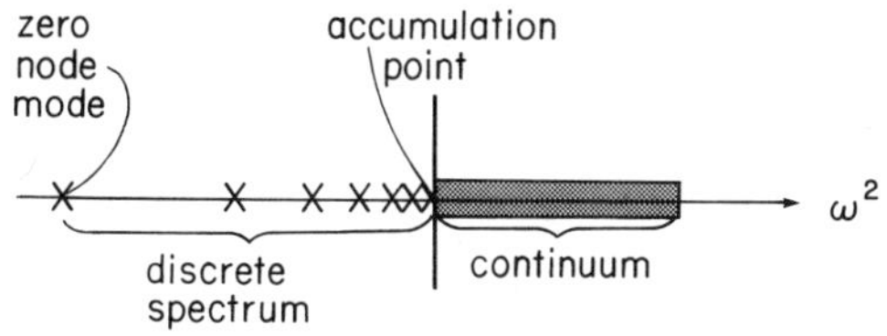

Figure 9.15. ω^2 spectrum for a plasma violating the Suydam criterion.

(for the given m and k). The oscillation theorem also shows that $\omega^2 = 0$ is a point of accumulation from the unstable side of the spectrum when Suydam's criterion is violated. See Fig. 9.15. A complete derivation of the oscillation theorem, including its implications on the stable side of the spectrum, can be found in the lecture notes of Goedbloed (1979).

Finally, it is useful to note that as a consequence of the oscillation theorem numerical parameter studies of maximum growth rates are significantly simplified by restricting attention only to the zero-node mode.

9.4.6. The Effect of a Resistive Wall

A final topic of interest related to general screw-pinch stability theory concerns the situation where the vacuum region surrounding the plasma is bounded by a resistive rather than a perfectly conducting wall. This has important practical experimental applications, particularly to the reversed-field pinch.

The situation arises as follows. In many experiments the plasma edge is defined by a metallic limiter. Just outside the limiter is a thin metallic vacuum chamber. To a certain approximation the vacuum chamber behaves as a perfect conductor. Because of its close proximity to the plasma it can thus have a dramatic stabilizing influence on external kink modes. In reality, vacuum chambers are deliberately designed to have a significant resistivity; this allows the externally applied vertical fields to penetrate the chamber sufficiently rapidly to provide effective feedback control of the equilibrium position of the plasma. Also, with a higher resistivity smaller currents are induced in the vacuum chamber during transients, alleviating power supply requirements.

Consequently, one might wonder whether the resistivity of the wall could have a detrimental effect on wall stabilization. This is indeed the case. In the analysis presented below it is shown that a system that is unstable to external kink modes without a wall, but that is stabilized with a perfectly conducting wall near the plasma, again becomes unstable if the wall is resistive. The growth rates are slower, comparable to the resistive diffusion time of the wall.

The calculation is carried out in four parts: (1) a reference is established by examining the potential energies for the case of the perfectly conducting wall; (2) the magnetic fields for the various regions outside the plasma are derived for the case of the resistive wall; (3) the boundary conditions connecting the different regions are evaluated; and (4) the resulting system of coupled boundary conditions is solved, yielding the dispersion relation.

The Perfectly Conducting Wall Case. The potential energy for the general screw pinch has been derived in Section 9.4.1. If there is a vacuum region surrounded by a perfectly conducting wall then δW is given by Eq. (9.44), repeated here for convenience:

$$\frac{\delta W}{2\pi^2 R_0/\mu_0} = \int_0^a (f\xi'^2 + g\xi^2)\,dr + \left[\frac{F\hat{F}}{k_0^2} + \frac{r^2\Lambda F^2}{|m|}\right]_a \xi_a^2 \tag{9.75}$$

where $\hat{F} = kB_z - mB_\theta/r$ and all other quantities have been previously defined. The effect of the wall is contained in the quantity Λ, which has the form

$$\Lambda = -\frac{|m|\,K_a}{kaK_a'}\left[\frac{1 - (K_b'I_a)/(I_b'K_a)}{1 - (K_b'I_a')/(I_b'K_a')}\right] \tag{9.76}$$

where $K_r \equiv K_m(kr)$, $I_r \equiv I_m(kr)$ are modified Bessel functions.

Assume that in the plasma region the minimizing ξ, which satisfies

$$(f\xi')' - g\xi = 0 \tag{9.77}$$

is known either analytically or numerically. By multiplying Eq. (9.77) by ξ and integrating over the plasma, one can then express δW corresponding to the wall at ∞ as follows:

$$\begin{aligned} \frac{\delta W_\infty}{2\pi^2 R_0/\mu_0} &= \left[\frac{F\hat{F}}{k_0^2} + \frac{r^2\Lambda_\infty F^2}{|m|} + \frac{F^2}{k_0^2}\left(\frac{r\xi'}{\xi}\right)\right]_a \xi_a^2 \\ \Lambda_\infty &= -\frac{|m|\,K_a}{kaK_a'} \end{aligned} \tag{9.78}$$

Similarly, if there is a perfectly conducting wall located at $r = b$, one obtains

$$\begin{aligned} \frac{\delta W_b}{2\pi^2 R_0/\mu_0} &= \frac{\delta W_\infty}{2\pi^2 R_0/\mu_0} + \left[\frac{r^2F^2}{|m|}\right]_a (\Lambda_b - \Lambda_\infty)\xi_a^2 \\ \Lambda_b - \Lambda_\infty &= \Lambda_\infty \frac{(I_a'/K_a') - (I_a/K_a)}{(I_b'/K_b') - (I_a'/K_a')} > 0 \end{aligned} \tag{9.79}$$

The interesting situation occurs when

$$\delta W_\infty < 0 < \delta W_b \tag{9.80}$$

That is, without a wall the system is unstable to external kinks while with a wall the modes become stabilized.

Resistive Wall Solution. Consider now the case of a resistive wall as illustrated in Fig. 9.16. It is important to realize that the introduction of dissipation, even through a resistive wall in the vacuum region, destroys the self-adjoint property of the system; ω enters the boundary conditions in a non-energy-conserving way. Thus, one must in principle solve the full eigenvalue problem. However, by making a number of assumptions that are well satisfied in practical experimental situations, much of the previous analysis can be retained.

The main assumption, which is verified *a posteriori,* is to assume that the modes of interest have a characteristic growth time comparable to the resistive diffusion time of the wall and that this time is much longer than the characteristic MHD time. For example, consider a deuterium plasma with minor radius $a = 0.3\,\text{m}$, major radius $R_0 = 1\,\text{m}$, and temperature $T_e = T_i = 2\,\text{keV}$. Assume the vacuum chamber has a minor radius $b \approx a$. Two cases of interest correspond to a thin stainless steel (SS) chamber of thickness $d = 1\,\text{mm}$ and resistivity $\eta = 11 \times 10^{-8}\,\Omega\,\text{m}$, and a thick copper (CU) chamber with $d = 1\,\text{cm}$ and $\eta = 1.7 \times 10^{-8}\,\Omega\,\text{m}$. The characteristic MHD and resistive diffusion times are given by $\tau_M = R_0/V_{T_i}$ and $\tau_D = \mu_0 bd/\eta$, respectively, and have the following numerical values:

$$\begin{aligned} \tau_M &= 2.3 \times 10^{-6}\,\text{s} \\ \tau_D &= 3.4 \times 10^{-3}\,\text{s (SS)} \\ \tau_D &= 0.22\,\text{s (CU)} \end{aligned} \tag{9.81}$$

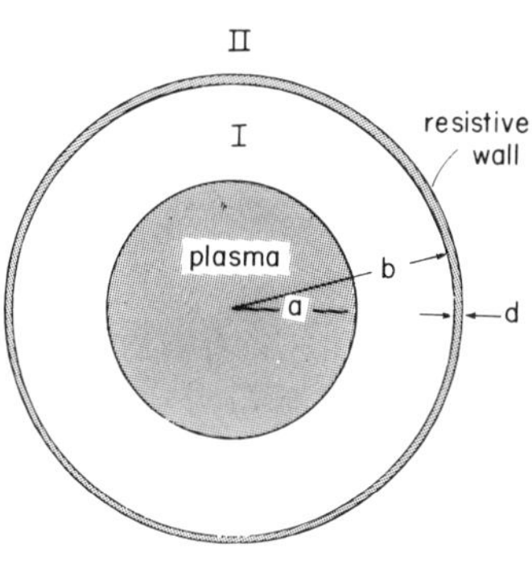

Figure 9.16. Geometry for the resistive wall stability analysis.

Observe the wide separation in time scales. For devices of the size described above, τ_D (SS) is typically comparable to or shorter than experimental pulse lengths, while τ_D(CU) is longer. In both cases $\tau_D \gg \tau_M$.

The disparity in time scales between τ_D and τ_M helps as follows. For the exact problem one must solve the full eigenvalue equation in the plasma region given by Eq. (9.45), containing the two complicated, frequency-dependent coefficients A and C. However, for slowly growing modes the frequency dependence in A and C can be neglected since $1/\tau_D \ll \omega_a^2, \omega_f^2, \omega_s^2, \omega_h^2, \omega_g^2$. Consequently, by setting $\omega^2 = 0$, one finds that the plasma displacement $\xi(r)$ satisfies the familiar minimizing equation associated with δW given by Eq. (9.77).

With this as background one can write the general solutions for the magnetic fields as follows. Regions I and II correspond to vacuum regions: $\nabla \times \hat{\mathbf{B}}_1 = \nabla \cdot \hat{\mathbf{B}}_1 = 0$. The fields can thus be written as

$$\text{Region I:} \qquad \begin{aligned} \hat{\mathbf{B}}_\text{I} &= \nabla \hat{\phi}_\text{I} \\ \hat{\phi}_\text{I} &= (c_1 I_r + c_2 K_r) \exp[-i(\omega t - m\theta - kz)] \end{aligned} \tag{9.82}$$

$$\text{Region II:} \qquad \begin{aligned} \hat{\mathbf{B}}_\text{II} &= \nabla \hat{\phi}_\text{II} \\ \hat{\phi}_\text{II} &= c_5 K_r \exp[-i(\omega t - m\theta - kz] \end{aligned} \tag{9.83}$$

Note that region II has only the K_r solution to guarantee boundedness of ϕ as $r \to \infty$.

In the wall region $\mathbf{B}$ satisfies a magnetic diffusion equation. Specifically, if one writes $\mathbf{J} = (1/\eta)\mathbf{E}$ with η the resistivity of the wall, then Ampere's law and Faraday's law reduce to

$$\begin{aligned} \nabla \times \mathbf{E}_w &= -\frac{\partial \mathbf{B}_w}{\partial t} \\ \nabla \times \mathbf{B}_w &= \frac{\mu_0}{\eta} \mathbf{E}_w \end{aligned} \tag{9.84}$$

or

$$\frac{\partial \mathbf{B}_w}{\partial t} = \frac{\eta}{\mu_0} \nabla^2 \mathbf{B}_w \tag{9.85}$$

The solution to Eq. (9.85) is found easily by assuming a thin wall: $d \ll b$. In this limit, which is well satisfied experimentally, Eq. (9.85) can be solved accurately in a local slab coordinate system $x = r - b$, $y = b\theta$, $z = z$. The x component of Eq. (9.85) becomes

$$\frac{d^2 B_{wx}}{dx^2} - \lambda^2 B_{wx} = 0 \tag{9.86}$$

where $\lambda^2 = \mu_0\omega_i/\eta + k^2 + m^2/b^2$ and $\omega \equiv i\omega_i$ (Re ω_i is the growth rate). If one makes use of the assumption that $\omega_i \sim 1/\tau_D = \eta/\mu_0 bd$, then $\mu_0\omega_i/\eta \sim 1/bd \gg k^2 + m^2/b^2$. Hence,

$$\lambda^2 = \frac{\mu_0\omega_i}{\eta} \tag{9.87}$$

The solution to Eq. (9.86) can be written as

$$B_{wx} = c_3 e^{\lambda x} + c_4 e^{-\lambda x} \tag{9.88}$$

The remaining components of $\mathbf{B}_w$ are given the two relations

$$\frac{m}{b} B_{wz} - kB_{wy} = 0 \tag{9.89}$$

$$i\left(\frac{m}{b} B_{wy} + kB_{wz}\right) = -\lambda(c_3 e^{\lambda x} - c_4 e^{-\lambda x}) \tag{9.90}$$

Equation (9.89) follows from tangential continuity arguments and the fact that $(m/r)\hat{B}_{1z} = k\hat{B}_{1y}$ in both regions I and II. Equation (9.90) is a consequence of $\nabla \cdot \mathbf{B}_w = 0$.

Equations (9.82), (9.83), (9.88)–(9.90) specify the general form of the magnetic fields exterior to the plasma. The fields are expressed in terms of the five unknown coefficients c_1–c_5 and the eigenvalue ω_i (appearing in λ).

Boundary Conditions. The next step in the analysis is to apply the boundary conditions, of which there are six: two at the region II–wall interface, two at the region I–wall interface, and two at the region I–plasma interface. These conditions form a set of six inhomogeneous equations for the quantities c_1–c_5 and ω_i. Ultimately c_1–c_5 are eliminated, giving rise to a dispersion relation for ω_i. The boundary conditions at each interface are described below.

At the region II–wall interface $x = d$, Maxwell's equations require continuity of the normal and tangential components of $\hat{\mathbf{B}}_1$. In view of the fact that $m\hat{B}_{1z}/r = k\hat{B}_{1\theta}$ in each region, these conditions can be expressed as $[\![\hat{B}_{1r}]\!] = 0$, $[\![i\mathbf{k} \cdot \hat{\mathbf{B}}_1]\!] = 0$, where $\mathbf{k} = (m/r)\mathbf{e}_\theta + k\mathbf{e}_z$. Upon substituting the fields one finds

$$c_5 k K_b' = c_3 e^{\lambda d} + c_4 e^{-\lambda d} \tag{9.91}$$

$$c_5 k_b^2 K_b = \lambda(c_3 e^{\lambda d} - c_4 e^{-\lambda d}) \tag{9.92}$$

where $k_b^2 = k^2 + m^2/b^2$.

At the region I–wall interface $x = 0$, the same conditions apply:

$[\![\hat{B}_{1r}]\!] = 0$, $[\![i\mathbf{k}\cdot\hat{\mathbf{B}}_1]\!] = 0$. The corresponding conditions are given by

$$k(c_1 I_b' + c_2 K_b') = c_3 + c_4 \tag{9.93}$$

$$k_b^2(c_1 I_b + c_2 K_b) = \lambda(c_3 - c_4) \tag{9.94}$$

Across the region I–plasma interface, the boundary conditions require continuity in (1) the normal component of magnetic field $[\![B_{1r}]\!] = 0$ and (2) perpendicular pressure balance $[\![p_1 + \mathbf{B}\cdot\mathbf{B}_1/\mu_0 + \boldsymbol{\xi}\cdot\nabla(p + B^2/2\mu_0)]\!] = 0$. To evaluate these conditions it is necessary to express each of the plasma quantities in terms of the value of ξ on the boundary. For the first condition, B_{1r} evaluated on the boundary $r = a$ is given by $B_{1r}|_a = \mathbf{e}_r\cdot(\mathbf{B}\cdot\nabla\boldsymbol{\xi}_\perp - \boldsymbol{\xi}_\perp\cdot\nabla\mathbf{B} - \mathbf{B}\nabla\cdot\boldsymbol{\xi}_\perp)|_a = i[F\xi]_a$.

The pressure balance condition is more complicated. If no surface currents are present and the plasma pressure decays smoothly to zero at the plasma edge, then the pressure balance condition has the form $[\![\mathbf{B}\cdot\mathbf{B}_1]\!] = 0$. In the plasma region a simple calculation shows that

$$\begin{aligned}\mathbf{B}\cdot\mathbf{B}_1|_a &= \mathbf{B}\cdot(\mathbf{B}\cdot\nabla\boldsymbol{\xi}_\perp - \boldsymbol{\xi}_\perp\cdot\nabla\mathbf{B} - \mathbf{B}\nabla\cdot\boldsymbol{\xi}_\perp)|_a \\ &= [-B^2(r\xi)'/r + 2(B_\theta^2/r)\xi - iGB\eta]_a.\end{aligned}$$

The quantity η is eliminated by the minimizing condition given by Eq. (9.34). After a short calculation, the boundary conditions across the region I–plasma interface can be written as

$$k(c_1 I_a' + c_2 K_a') = iF_a\xi_a \tag{9.95}$$

$$iF_a(c_1 I_a + c_2 K_a) = -\frac{F_a}{ak_0^2}\left[\hat{F}_a + F_a\left(\frac{r\xi'}{\xi}\right)_a\right]\xi_a \tag{9.96}$$

where $F_a = F(r)|_a$ and $\hat{F}_a = \hat{F}(r)|_a$. Equations (9.91)–(9.96) constitute a set of six inhomogeneous equations for the unknowns c_1–c_5 and ω_i.

Dispersion Relation. The objective now is to eliminate the constants c_1–c_5 from the boundary condition relations, thereby leading to a dispersion relation for ω_i. This task can be conveniently carried out as follows. First, Eq. (9.91) is divided by Eq. (9.92) yielding

$$\frac{\lambda k K_b'}{k_b^2 K_b} = \frac{c_3/c_4 + e^{-2\lambda d}}{c_3/c_4 - e^{-2\lambda d}} \tag{9.97}$$

Using the assumptions $\omega_i \sim 1/\tau_D$, $d/b \ll 1$ the left-hand side of Eq. (9.97) is of order $\lambda b \sim (b/d)^{1/2} \gg 1$. Similarly, $\lambda d \sim (d/b)^{1/2} \ll 1$. This implies that (1) the exponentials on the right-hand side of Eq. (9.97) can be Taylor expanded, $e^{-2\lambda d} \approx 1 - 2\lambda d$ and (2) the quantity $c_3/c_4 \approx 1 +$

$\delta(c_3/c_4)$, $\delta(c_3/c_4) \ll 1$. Under these conditions one can solve for c_3/c_4, obtaining

$$\frac{c_3}{c_4} \approx 1 + \frac{2k_b^2 K_b}{\lambda k K_b'}(1 + \Omega_i)$$
$$\Omega_i \equiv -\frac{kdK_b'\lambda^2}{k_0^2 K_b} \sim O(1) \tag{9.98}$$

Here, $\Omega_i \propto \omega_i$ is a normalized form of the eigenvalue.

The next step is to divide Eq. (9.93) by (9.94). This gives a relation for c_1/c_2 in terms of c_3/c_4

$$\frac{k\lambda K_b'}{k_b^2 K_b}\left[\frac{(c_1/c_2)(I_b'/K_b') + 1}{(c_1/c_2)(I_b/K_b) + 1}\right] = \frac{c_3/c_4 + 1}{c_3/c_4 - 1} \tag{9.99}$$

Substituting for c_3/c_4 from Eq. (9.98) yields

$$\frac{c_1}{c_2} \approx \frac{\Omega_i}{(I_b/K_b) - (1 + \Omega_i)(I_b'/K_b')} \tag{9.100}$$

The quantity c_1/c_2 is eliminated by dividing Eq. (9.96) by (9.95)

$$\frac{(c_1/c_2)(I_a/K_a) + 1}{(c_1/c_2)(I_a'/K_a') + 1} = T \tag{9.101}$$

where

$$T = \frac{kK_a'}{ak_0^2 K_a F_a}\left[\hat{F}_a + F_a\left(\frac{r\xi'}{\xi}\right)_a\right] \tag{9.102}$$

Solving for c_1/c_2 gives

$$\frac{c_1}{c_2} = \frac{T - 1}{(I_a/K_a) - T(I_a'/K_a')} \tag{9.103}$$

The dispersion relation is obtained by equating Eqs. (9.100) and (9.103) and then solving for Ω_i. The result is

$$\Omega_i = \frac{[(I_b/K_b) - (I_b'/K_b')](T - 1)}{(I_a/K_a) - (I_a'/K_a') + (T - 1)[(I_b'/K_b') - (I_a'/K_a')]} \tag{9.104}$$

The dispersion relation can be cast in a far more convenient form by expressing the numerator and denominator in terms of the perfectly conducting potential energies δW_∞ and δW_b given by Eqs. (9.78) and (9.79). A short calculation leads to the desired form of the dispersion relation

$$\omega_i \tau_D = \frac{k^2 b^2 + m^2}{k^2 b^2 K_b' I_b'[1 - (I_a' K_b'/I_b' K_a')]} \frac{\delta W_\infty}{\delta W_b} \tag{9.105}$$

In the interesting limit where $kb \ll 1$, Eq. (9.105) reduces to

$$\omega_i \tau_D \approx -\left[\frac{2|m|}{1-(a/b)^{2|m|}}\right]\frac{\delta W_\infty}{\delta W_b} \tag{9.106}$$

Discussion. There are a number of points to observe. First, the eigenfrequency ω_i is real, indicating pure growth or decay depending upon the sign of $\delta W_\infty/\delta W_b$. Second, as assumed ω_i has a characteristic frequency of order $1/\tau_D$. Third, in the interesting case which is unstable with the wall at infinity ($\delta W_\infty < 0$), but is stabilized by a perfectly conducting wall at $r = b$, ($\delta W_b > 0$) then $\omega_i > 0$; the presence of a resistive wall introduces a new slowly growing mode into the system so that the stability boundary reverts back to that corresponding to the wall at infinity.

The spectral picture is illustrated in Fig. 9.17. In Fig. 9.17a the trajectory of an initially unstable kink mode is traced in the complex ω plane as a perfectly conducting wall is moved in from infinity. Since ω^2 is real for this situation the roots must lie on either the real or imaginary axis. In Fig. 9.17b the roots are traced if the wall at $r = b$ becomes resistive. Although it has not been proven here, it can be shown that the ideal wall stabilized roots become slightly damped. The instability described above corresponds to the development of a new mode, growing out of the origin $\omega^2 = 0$.

The physics of the resistive wall "destabilization" can be understood as follows. As the plasma moves under the action of a potentially unstable perturbation, currents are induced in the conducting wall. In

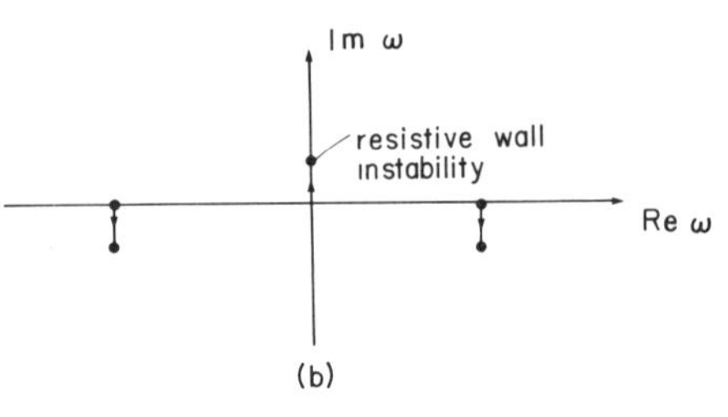

Figure 9.17. Spectral behavior of resistive wall instabilities: (a) ideal wall stabilization of an unstable mode with the wall at ∞; (b) growth of a resistive wall instability from the origin $\omega^2 = 0$.

general these currents flow in a direction to oppose the plasma motion and thus provide stabilization. For a perfectly conducting wall these currents can exist *ad infinitum*. For a resistive wall the currents will decay on a diffusion time scale τ_D. Since the characteristic time of a stable ideal MHD oscillation is much shorter than τ_D the rapidly oscillating wall stabilized mode is only slightly affected by the presence of a resistive wall (i.e., it is slowly damped). If, however, the plasma develops an unstable perturbation on the slower τ_D time scale, then stabilizing wall currents are not able to develop and the disturbance continues to grow. This corresponds to the mode calculated above.

The issue of the resistive wall is very important in certain configurations, particularly the reversed field pinch. As is shown in the next section, a perfectly conducting wall is required to stabilize a rather large number of external kink modes. If the wall is resistive a sophisticated multimode feedback system may be required to provide stabilization. This problem is one that must be addressed in future, longer pulse RFP experiments.

9.5. THE REVERSED FIELD PINCH

9.5.1. Introduction

In this section the general screw-pinch analysis is applied to the reversed field pinch configuration. Recall that the appropriate inverse aspect ratio expansion for the RFP requires $B_z/B_\theta \sim 2\mu_0 p/B_\theta^2 \sim 1$. Consequently, the toroidal corrections are small with respect to all important stability questions and an essentially complete MHD stability description is obtained from an analysis of the one-dimensional cylindrical model.

By making use of Suydam's criterion and various trial functions it is possible to develop an intuition as to why RFP profiles have favorable stability properties with respect to internal current-driven and pressure-driven modes. Equally important, the analysis shows why the presence of a perfectly conducting wall is crucial for RFP stability.

Cylindrical MHD stability studies have been carried out by Robinson (1971) and by Sakanaka and Goedbloed (1974). They have shown that of all plausible B_θ and B_z profiles containing an isolating, current-free region near the conducting wall, the RFP profile is the one most capable of sustaining complete ideal MHD stability at high values of β. In each case, intuitive arguments are supported by numerical calculations of the exact equations. As an example, Robinson showed that with a reasonably located conducting wall, $b/a \sim 2$, completely stable RFP profiles exist for $\beta \approx 30\%$. This is indeed an optimistic theoretical result.

Next, a short discussion is presented of the so-called "force-free paramagnetic model." This calculation shows that the profiles corresponding to the natural ohmic states of an RFP are reasonably close to those desired for good MHD stability behavior. A detailed examination shows, however, that instabilities always persist in this model.

The relaxation of such unstable states to stable profiles has been investigated by Taylor (1975) and a brief outline of his theory is presented at the end of Section 9.5. His theory suggests that under a wide range of conditions, many configurations will naturally evolve to low β "RFP-like" profiles, including self-reversal of the toroidal field.

9.5.2. Internal Pressure-Driven Modes

To determine the influence of internal pressure-driven modes on the RFP configuration two stability tests are performed as originally suggested by Robinson (1971). First, an examination of Suydam's criterion shows that the profiles must possess certain local features for stability. Second, the use of appropriate trial functions gives reasonable bounds for the overall limit on β.

Consider first Suydam's criterion [Eq. (9.59)]. If the toroidal current density on axis $J_z(0)$ is nonzero, then near $r = 0$ Suydam's criterion has the form

$$\frac{B_z^2(0)}{4\mu_0}\left[\frac{q''(0)}{q(0)}\right]^2 r^3 + 8p''(0)r > 0 \tag{9.107}$$

Near the axis there is insufficient shear to stabilize the interchange perturbation. To avoid such instabilities it is necessary that $p''(0) \geq 0$, implying that for small r the pressure profile must be hollow or extremely flat.

Suydam's criterion also points out the importance of the reversal in B_z. If there is a current-free region near the wall then $B_z \sim$ const, $B_\theta \sim 1/r$ and $q \sim r^2$. In addition, near the axis B_z decreases more rapidly than J_z (to keep the pressure profile hollow) so that q is a decreasing function of r for small r. Consequently, if B_z does not reverse, there must be a minimum in q near the outside of the plasma as shown in Fig. 9.18a. Suydam's criterion indicates that an instability will occur when $q' = 0$ and $p' < 0$. This situation is avoided if B_z reverses, in which case q passes through zero and $q' \neq 0$. See Fig. 9.18b. Thus, in the RFP configuration, stability to interchange perturbations requires a hollow or very flat

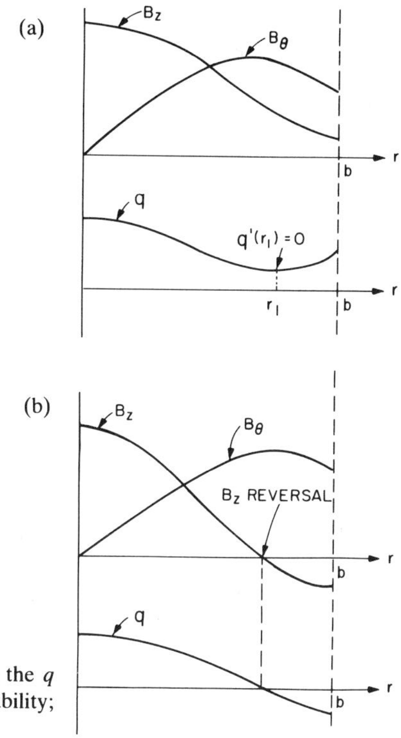

Figure 9.18. Avoidance of a pitch minimum in the q profile by reversing B_z: (a) no reversal → instability; (b) reversal → stability.

pressure profile to avoid modes near $r = 0$ and high shear caused by the B_z reversal to avoid modes on the outside of the profile.

Although Suydam's criterion provides valuable insight into the local properties of the RFP it does not provide an overall β limit. Such limits can be estimated by substituting appropriate trial functions into δW for the $m = 1$ and $m = 0$ modes. For the $m = 1$ mode a trial function likely to cause instability is illustrated in Fig. 9.19a. By letting ξ drop to zero [as it must since $\xi(a) = 0$] in a thin layer about r_s where $F(r_s) = 0$, the stabilizing contribution, $\int f\xi'^2\,dr \to 0$ as $\epsilon \to 0$. See Problem 9.2. Furthermore, one can show that the destabilizing region is maximized when $r_s \to a$. Using this trial function the energy principle implies that

$$\frac{\delta W_F}{2\pi^2 R_0/\mu_0} < \xi_0^2 \int_0^a g\,dr \tag{9.108}$$

After some simple rearranging, $g(r)$ can be rewritten as

$$g(r) = \frac{d}{dr}\left[k^2 r^2\left(\frac{k^2}{k_0^2}B^2 + 2\mu_0 p\right)\right] - 4k^2 r\mu_0 p - \frac{rk^2}{k_0^2}\left(kB_z - \frac{B_\theta}{r}\right)^2 \tag{9.109}$$

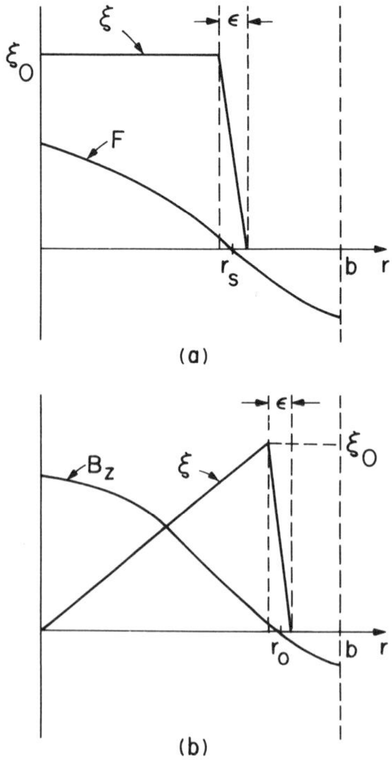

Figure 9.19. Trial functions for (a) the $m = 1$ and (b) the $m = 0$ macroscopic pressure-driven modes.

Since the last term is always negative a necessary condition for stability is that the remaining contribution be positive. Setting $k = -B_\theta(a)/aB_z(a)$ at the resonant surface $r_s \to a$ and using the RFP definition of β_p given by Eq. (6.53) it then follows that

$$\beta_p < 1 \tag{9.110}$$

for stability. Clearly the $m = 1$ limit is a very weak restriction on internal pressure-driven modes.

A more stringent condition arises from the $m = 0$ mode. In this case the appropriate trial function is illustrated in Fig. 9.19b. Note that the jump in ξ occurs at the reversal point $r = r_0$ where $f(r_0) \sim B_z^2(r_0) = 0$. Hence, the potentially large stabilizing contribution from the layer where ξ'^2 is large vanishes as $\epsilon \to 0$. For this trial function,

$$\frac{\delta W_F}{2\pi^2 R_0/\mu_0} = \frac{\xi_0^2}{r_0^2}\int_0^{r_0} (f + r^2 g)\, dr \tag{9.111}$$

Upon setting $m = 0$, $k^2 \to 0$ (the most unstable k^2) one can write

$$f + r^2 g = 2\mu_0 r^2 p' + 2rB_z^2 \tag{9.112}$$

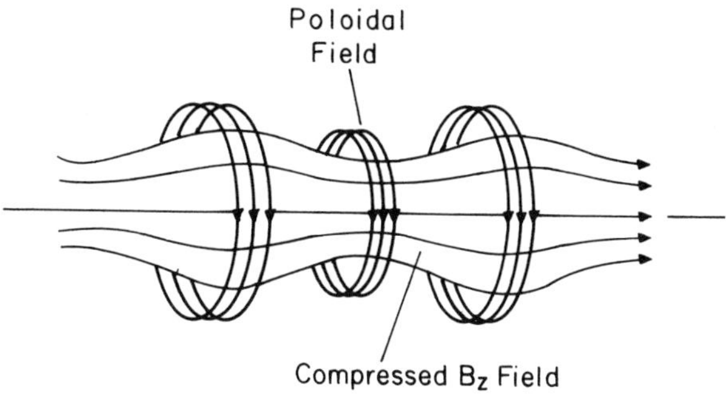

Figure 9.20. "Stiffening" of the plasma against $m = 0$ perturbations by an internal B_z field.

Usually the reversal point is relatively close to the current-free region. Thus, to within a small error substitution of Eq. (9.112) into Eq. (9.111) with $r_0 \approx b$ yields the following condition on β_p:

$$\beta_p < \tfrac{1}{2} \tag{9.113}$$

for stability. The $m = 0$ condition is more stringent than $m = 1$, but still represents only a weak overall limitation on β.

The reduction of β_p away from the value of unity reflects the physical fact that near the center of the profile there must be a sufficient B_z field to "stiffen" the plasma against $m = 0$ sausage perturbations; in an RFP an $m = 0$, $k \to 0$ perturbation gives rise to a substantial compression of the internal B_z field which is stabilizing. See Fig. 9.20.

In summary, a combination of profile shaping and a weak β_p limit is required to stabilize the RFP configuration against pressure-driven modes.

9.5.3. Internal Current-Driven Modes

The influence of internal current-driven modes on the RFP configuration can be determined using the procedure suggested by Robinson (1971). In this procedure a necessary condition for stability is derived by examining the sign of $g(r)$ vs. r and determining under which circumstances it is negative for $0 < r < a$, consistent with a trial function that makes the $f\xi'^2$ contribution vanish.

To begin consider $g(r)$ as given by Eq. (9.36). For low-β systems whose stability is dominated by current-driven modes, p' can be set to zero and g can be written as

$$g = \frac{rk^2B_\theta^2}{k_0^2}(kQ + 1)[(3 + k^2r^2)kQ - 1 + k^2r^2] \tag{9.114}$$

where $Q \equiv R_0 q = rB_z/B_\theta$. Equation (9.114) corresponds to the most unstable situation, $m = 1$.

A useful way to ascertain the sign of $g(r)$ for a given k is to plot a diagram of k vs. r where $g(r)$ vanishes:

$$\begin{aligned} kQ &= -1 \\ kQ &= (1 - k^2r^2)/(3 + k^2r^2) \end{aligned} \tag{9.115}$$

This is illustrated in Fig. 9.21a. In the unshaded region $g < 0$ (i.e., destabilizing). For the trial function shown in Fig. 9.19a, δW satisfies

$$\frac{\delta W_F}{2\pi^2 R_0/\mu_0} < \xi_0^2 \int_0^{r_s} g(r)\,dr \tag{9.116}$$

Note that the trial function for the $m = 1$ pressure-driven and current-driven modes are identical, having been chosen to minimize the stabilizing line-bending magnetic energy. At the singular surface $F(r_s) = 0$ corresponding to a wave number $k = -1/Q(r_s)$. If the conducting wall is sufficiently far out ($a = a_1$) a value of r_s (i.e., k) can be chosen so that $k(0) > k > k(a)$, where $k(0) = 1/3Q(0)$ and $k(a) = -1/Q(a)$. See the curve $k = k_1$ in Fig. 9.21b. In this case $g(r)$ is negative over the whole range $0 < r < r_s$ and instability occurs. As the wall is moved closer in ($a = a_2$), the range of k's for which $g(r)$ is always negative diminishes. Eventually, when $k(a) > k(0)$, the most unstable k becomes $k = k(a)$, since r_s cannot exceed a for the trial function to satisfy the internal mode boundary condition $\xi(a) = 0$. In this situation a stabilizing contribution develops near the axis which increases as k increases (i.e., as the wall moves further in). See the curve $k = k_2$ in Fig. 9.21c. It thus follows that a necessary condition for stability is $k(0) > k(a)$ or

$$Q(0) > -\tfrac{1}{3}Q(a) \tag{9.117}$$

There are two alternate forms in which Eq. (9.117) can be written

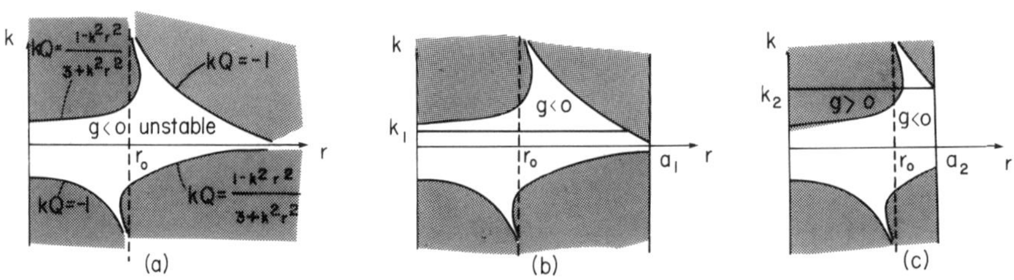

Figure 9.21. Stability diagram for $m = 1$ internal modes: (a) region where $g < 0$; (b) for the wall at $r = a_1$ and $k = k_1$, g is negative everywhere; (c) for the wall at a_2 and $k = k_2$ there are regions where g is both positive and negative.

which point out the influence of current-driven modes on RFP profiles. First, the average radius of the current channel and hence the plasma can be defined as follows: $\langle a\rangle^2 \equiv I/\pi J_z(0)$. In this case Eq. (9.117) reduces to

$$\frac{a^2}{\langle a\rangle^2} < -3\frac{B_z(0)}{B_z(a)} \tag{9.118}$$

For a given field reversal the conducting wall must be sufficiently close to the plasma core to prevent current-driven instabilities. The conducting wall prohibits potentially unstable singular surfaces from existing in the plasma.

The second form of Eq. (9.117) is an approximate relation obtained by assuming $B_z \approx B_z(0)$ for $0 < r < \langle a\rangle$ and $B_z \approx B_z(a)$ for $\langle a\rangle < r < a$. Under this assumption the total toroidal flux is given by $\psi_t = \pi[B_z(0)\langle a\rangle^2 + B_z(a)(a^2 - \langle a\rangle^2)]$. Equation (9.118) can then be written as $\psi_t > \pi a^2 B_z(0)[3\langle a\rangle^2/a^2 - 2]$.

This condition implies that the total toroidal flux cannot become too negative despite the need for a reversed B_z to stabilize interchanges. As a practical matter numerical calculations have shown that a general guideline is that the total toroidal flux must remain positive:

$$\psi_t > 0 \tag{9.119}$$

In other words, the region of reversed field cannot be too large.

A final point worth noting is as follows. Consider the situation where B_z does not reverse and there is a minimum in Q on the outside of the profile as shown in Fig. 9.22. One now chooses a value of k so that $kQ = -1$ at two neighboring points about the minimum and evaluates δW_F for the trial function illustrated in Fig. 9.22. Since $f(r) = 0$ when ξ' is nonzero, the line-bending contribution of δW_F again vanishes at both endpoints of the trial function. Hence, the bound on δW_F given by Eq. (9.116) is valid for this case as well. The function $g(r)$, given by Eq.

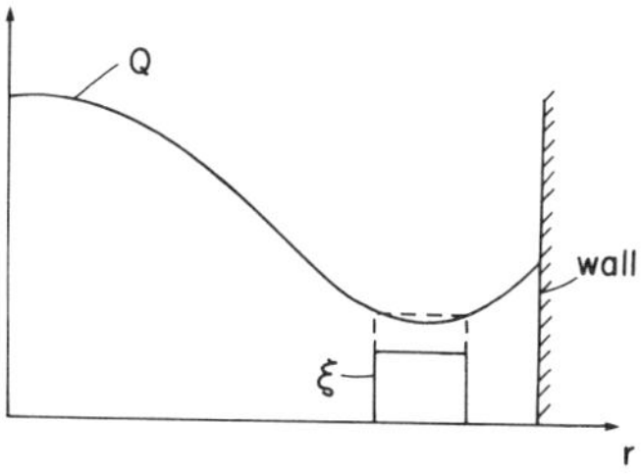

Figure 9.22. Trial function that makes $\delta W_F < 0$ for an $m = 1$ current driven internal kink mode when Q has a minimum.

(9.114), can be written as

$$g = \frac{rk^2B_\theta^2}{k_0^2}(1 + kQ)[3kQ - 1 + (1 + kQ)k^2r^2] \tag{9.120}$$

To evaluate Eq. (9.116) observe that over the range of integration of the trial function the factor $1 + kQ$ is small and positive. Similarly the term $3kQ - 1 \approx -4$ is finite and negative while the term $(1 + kQ)k^2r^2$ can be neglected compared to $(3kQ - 1)$ because of the smallness of $(1 + kQ)$. Consequently, $g(r)$ is negative over the entire range of integration, implying that $\delta W_F < 0$; that is, a minimum in the q profile can lead to both pressure-driven and current-driven instabilities.

9.5.4. External Modes

The analysis presented in Sections 9.5.2 and 9.5.3 provides a strong indication that the RFP configuration should be capable of achieving stability to all internal current-driven and pressure-driven modes at quite high values of $\beta_p \sim 0.5$. When the plasma is bounded by a perfectly conducting wall, the internal mode analysis constitutes a complete description of ideal MHD stability. In this section it is shown that the presence of a conducting wall is crucial; without a wall an RFP is very unstable to current-driven and pressure-driven external modes.

The analysis begins with a discussion of the $m = 1$ external mode as described by the Extended Energy Principle [Eq. (9.44)]. Since ξ_a is nonzero for an external kink, a simple choice of trial function is $\xi(r) = \xi_a = \text{const}$, $0 \le r \le a$. This choice automatically guarantees that the stabilizing $f\xi'^2$ term in the plasma energy vanishes. The contribution to δW from the plasma integration is thus given by

$$\int_0^a (f\xi'^2 + g\xi^2)\,dr = \xi_a^2\int_0^a g\,dr$$

$$g(r) = 2\frac{k^2}{k_0^2}(\mu_0 p)' + \frac{rk^2B_\theta^2}{k_0^2}(kQ + 1)[(3 + k^2r^2)kQ - 1 + k^2r^2] \tag{9.121}$$

From the analysis in Section 9.5.3 it follows that $g(r) < 0$ over the entire region $0 < r < a$ if $p' = 0$ and k is restricted to lie in the range

$$k < \text{Min}\begin{cases} -1/Q(a) \\ 1/3Q(0) \end{cases} \tag{9.122}$$

Consequently, from Eq. (9.44) one obtains a simple upper bound on δW

by ignoring the plasma integral contribution:

$$\frac{\delta W}{2\pi^2 R_0/\mu_0} < \xi_a^2 \left\{ \frac{r^2}{k_0^2} \left(kB_z + \frac{B_\theta}{r} \right) [(2 + k^2 r^2) kB_z + k^2 r B_\theta] \right\}_{r=a} \tag{9.123}$$

Equation (9.123) has been evaluated assuming that the conducting wall is at infinity: $\Lambda \approx 1$. Clearly, the system is unstable for

$$-B_{\theta a} < kaB_{za} < -\left(\frac{k^2 a^2}{2 + k^2 a^2} \right) B_{\theta a} \tag{9.124}$$

If one now identifies an equivalent torus of length $2\pi R_0$, then the wave number k becomes quantized: $k = n/R_0$ where n is an integer. Combining the requirements of Eqs. (9.122) and (9.124), one finds that for $-q_a > \epsilon/\sqrt{a}$ an RFP is unstable to external current-driven modes for

$$1 \le n < \mathrm{Min} \begin{cases} -1/q_a \\ 1/3q_0 \end{cases} \tag{9.125}$$

Since q_a and q_0 are of order $\epsilon \equiv a/R_0$ in an RFP, there are at least several modes that are always unstable.

The inclusion of plasma pressure further destabilizes the situation. The pressure contribution to Eq. (9.121) can be easily bounded as follows:

$$\int_0^a dr \frac{2k^2}{k_0^2} (\mu_0 p)' < \left[\frac{2k^2}{k_0^2} \right]_a \int_0^a dr (\mu_0 p)' = -\left(\frac{2k^2}{1 + k^2 a^2} \right) \mu_0 p(0) < 0 \tag{9.126}$$

Thus, the plasma pressure further destabilizes external modes.

A physical interpretation of the external kink mode is illustrated in Fig. 9.23. As in the case of a pure Z pinch, the tension in the poloidal field increases in the tighter portion of the column when the plasma develops an $m = 1$ kink perturbation. The resulting force is in the direction to enhance the perturbation, thus causing an instability. The addition of a toroidal field has two important effects. First, when the toroidal field is substantial, the J_z current driving the instability is largely flowing parallel to the field (as opposed to purely perpendicular in a Z

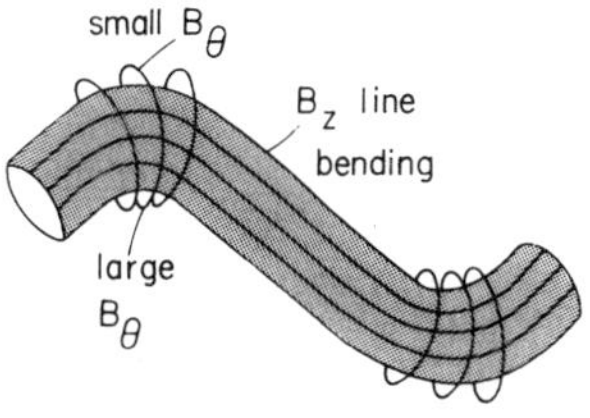

Figure 9.23. Physical picture of the external $m = 1$ kink mode.

pinch). Hence, the mode is more accurately described as a current-driven, rather than pressure-driven, instability. Second, in addition to producing the poloidal destabilizing force, the $m = 1$ kink perturbation bends the toroidal magnetic lines, a strong stabilizing effect. In order for line bending not to dominate, the modes can only become unstable for sufficiently long wavelengths. This physical description is in good agreement with the analytic results.

The $m = 1$, low n modes are dangerous instabilities. It is to avoid such modes that an RFP requires a perfectly conducting wall on the plasma boundary. Recall, however, from the discussion in the previous section that if the wall is resistive, the plasma again becomes unstable, although with a much slower growth rate. The consequences of these slowly growing modes on the confinement properties of long-pulse RFP experiments remains a topic of future theoretical and experimental research.

9.5.5. Discussion of Ideal Stability Results

In the previous sections it has been shown on the basis of various necessary conditions that stable high-β RFP profiles exist possessing the following properties:

1. A hollow or very flat pressure profile to suppress interchange modes near the magnetic axis.
2. A B_z field which reverses on the outside of the plasma. This provides shear to stabilize interchange modes and prevents the formation of $m = 1$ current-driven modes, both of which would occur in the presence of a minimum in the q profile.
3. A value of $\beta_p < 1/2$ to suppress $m = 0$, pressure-driven "sausage" modes.
4. A conducting wall sufficiently close to the plasma core (corresponding to a net positive toroidal flux) to suppress $m = 1$ current-driven internal kinks and to completely avoid all external modes.

Since the above properties were obtained by the use of trial functions one cannot guarantee that they are complete or sufficient for stability. However, extensive numerical computations using the exact δW and the full eigenvalue equation indicate that these properties provide a very reliable set of guidelines for obtaining stable configurations (Robinson, 1971; and Sakanaka and Goedbloed, 1974). One early example of physically interesting RFP profiles, completely stable to all internal ideal

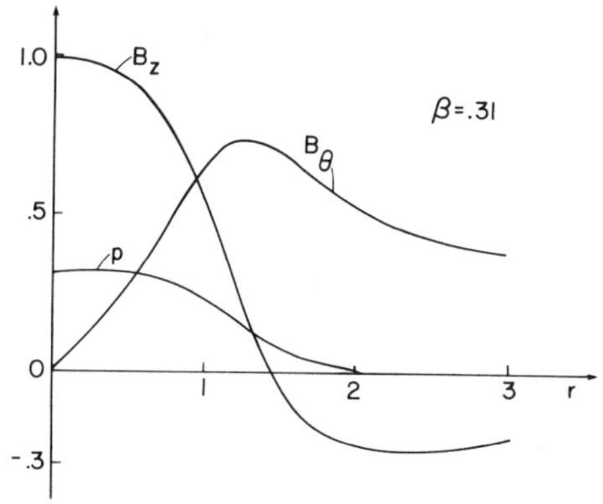

Figure 9.24. A set of RFP profiles completely stable to all internal ideal MHD modes for $\bar{\beta} \approx 0.31$ [from Robinson (1971)].

MHD modes, has been given by Robinson and is illustrated in Fig. 9.24. All of the features of the guidelines are present. Typically, the wall-to-plasma ratio must lie in the range $2 \lesssim a/\langle a \rangle \lesssim 3$ to provide both stability and isolation. In addition, the pressure profile corresponds to $\beta_p = 0.31$, a very optimistic result.

9.5.6. The Force-Free Paramagnetic Model

Having established the desirable profiles for a stable RFP it is of interest to investigate the plausibility of achieving such profiles in an actual experiment. Some insight into this question follows from an examination of the natural profiles that arise in a low-β ohmically heated RFP. These natural profiles are consistent with both an Ohm's law that includes resistivity and the ideal MHD radial pressure-balance relation. In the absence of any instabilities one would expect the plasma to evolve to such a state on a resistive diffusion time scale. This section demonstrates that ohmic profiles closely approximate in many, but not all, ways the desirable profiles of stable RFPs.

To carry out the analysis it is assumed for simplicity that the plasma β is small. In this limit the ∇p term can be neglected in the MHD radial pressure-balance relation, implying that the currents must flow parallel to the magnetic field line; that is, the system is force free:

$$\mu_0 \mathbf{J} = \lambda(r)\mathbf{B} \tag{9.127}$$

Within the context of ideal MHD, $\lambda(r)$ is a free function.

However, if one now requires consistency with an Ohm's law that includes resistivity, then this constraint serves to determine $\lambda(r)$. The simplest form of Ohm's law is given by

$$\mathbf{E} + \mathbf{v} \times \mathbf{B} = \eta \mathbf{J} \tag{9.128}$$

where $\eta(r)$ is the resistivity, assumed arbitrary but positive: $\eta(r) > 0$. In a steady-state equilibrium situation $\nabla \times \mathbf{E} = -(\partial \mathbf{B}/\partial t) = 0$. For cylindrical geometry this implies that $\mathbf{E}$ can be written as

$$\mathbf{E} = E_0 \mathbf{e}_z - \phi'(r)\mathbf{e}_r \tag{9.129}$$

In general, there is an electrostatic radial electric field and a uniform dc electric field $E_0 = \text{const}$ in the z direction. There is no E_θ field in steady state.

Equation (9.128) is now decomposed into parallel and perpendicular components:

$$\mathbf{v}_\perp = \frac{\mathbf{E} \times \mathbf{B}}{B^2} - \eta \frac{\mathbf{J} \times \mathbf{B}}{B^2} \tag{9.130}$$

$$E_\parallel = \eta J_\parallel \tag{9.131}$$

The $\mathbf{v}_\perp$ relation describes $\mathbf{E} \times \mathbf{B}$ motion and radial diffusion. The $E_\parallel$ equation is the important one for the present analysis and relates the $J_\parallel$ profile to the resistivity and the applied electric field. Upon substituting Eq. (9.127) and (9.129) into Eq. (9.131) one obtains the following expression for λ in terms of $\mathbf{B}$ and η:

$$\lambda = \left(\frac{\mu_0 E_0}{\eta}\right) \frac{B_z}{B^2} \tag{9.132}$$

The final equations determining $\mathbf{B}$ are obtained by substituting Eq. (9.132) into Ampere's law: $\nabla \times \mathbf{B} = \mu_0 \mathbf{J} = \lambda \mathbf{B}$. The result is

$$\frac{dB_z}{dr} = -\left(\frac{\mu_0 E_0}{\eta}\right) \frac{B_z B_\theta}{B^2} \tag{9.133}$$

$$\frac{1}{r}\frac{d}{dr}(rB_\theta) = \left(\frac{\mu_0 E_0}{\eta}\right) \frac{B_z^2}{B^2} \tag{9.134}$$

For a given $\eta(r)$ these coupled nonlinear equations must be solved numerically. However, some analytic insight can be gained by choosing the plausibly shaped but artificial form for η given by

$$\frac{\eta}{\mu_0 E_0} = \frac{r_0}{\alpha}\left(\frac{B_\theta}{B^2}\right)\left[\frac{1 + \alpha e^{-x^2}}{x e^{-x^2}}\right] \tag{9.135}$$

where $\alpha > -1$ is an arbitrary constant and $x = r/r_0$. For standard B_θ profiles, η is constant at the origin and increases monotonically with r, the typical experimental situation. The exact dependence of η is obtained *a posteriori*, once the solutions for B_z and B_θ are derived.

The next step is to introduce normalized variables U, W, x defined by $B_z = B_0 U$, $B_\theta = B_0 W$, $r = r_0 x$, where $B_0 = B_z(0)$. Equations (9.133) and (9.134) reduce to

$$\frac{dU}{dx} = -\frac{\alpha x e^{-x^2} U}{1 + \alpha e^{-x^2}} \tag{9.136}$$

$$\frac{d}{dx}(xW)^2 = \frac{2\alpha x^3 e^{-x^2} U^2}{1 + \alpha e^{-x^2}} \tag{9.137}$$

The first equation can be easily integrated, yielding

$$\frac{B_z}{B_0} = \left(\frac{1 + \alpha e^{-x^2}}{1 + \alpha}\right)^{1/2} \tag{9.138}$$

Substituting into Eq. (9.137) one can then evaluate W

$$\frac{B_\theta}{B_0} = \frac{1}{x}\left\{\frac{\alpha}{1 + \alpha}[1 - (1 + x^2)e^{-x^2}]\right\}^{1/2} \tag{9.139}$$

Knowing B_z and B_θ it is straightforward to evaluate η from Eq. (9.135) and the q profile from the definition $q = rB_z/R_0 B_\theta$. These quantities are illustrated in Fig. 9.25. Observe that the profile corresponding to the artificial form of η is quite plausible from an experimental point of view. The B_θ profile is also typical of experimental observations; B_θ increases linearly with r near the origin and decays inversely with r at large radii. The B_z profile is large near the origin and decreases towards zero at large radii; that is, the ohmic profiles are characterized by a substantial natural paramagnetism, an important and desirable feature of the RFP configuration.

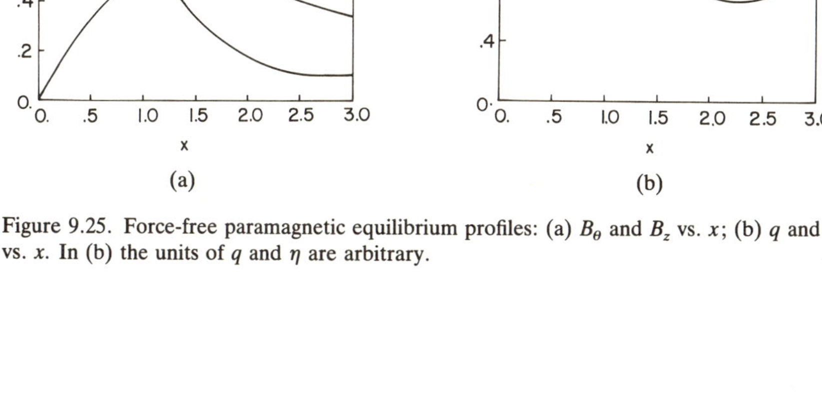

Figure 9.25. Force-free paramagnetic equilibrium profiles: (a) B_θ and B_z vs. x; (b) q and η vs. x. In (b) the units of q and η are arbitrary.

Note, however, that while the B_z profile decreases with r, it does not reverse, at least for the special η profile under consideration. This gives rise to a minimum in the q profile which leads to current-driven and pressure-driven internal MHD instabilities. The nonreversal in B_z is in fact a general property of the force-free paramagnetic model for any arbitrary resistivity profile satisfying the physical requirement $\eta(r) > 0$. This can be seen in two steps. First, one integrates Eq. (9.134) obtaining

$$rB_\theta = \mu_0 E_0 \int_0^r \frac{rB_z^2}{\eta B^2}\,dr \tag{9.140}$$

Since the integrand is positive, B_θ is always of one sign; B_θ cannot reverse at any point in the plasma. Next, Eq. (9.133) is integrated, yielding

$$\begin{aligned} B_z &= B_0 \exp\left[-\int_0^r \frac{\mu_0 E_0 B_\theta}{\eta B^2}\,dr'\right] \\ &= B_0 \exp\left[-\mu_0^2 E_0^2 \int_0^r \frac{dr'}{r'\eta B^2}\int_0^{r'} \frac{r'' B_z^2}{\eta B^2}\,dr''\right] \end{aligned} \tag{9.141}$$

Equation (9.141) implies that B_z is in general a decreasing function of radius which does not change sign. This property is true even for more general ohmic equilibria where $p' \neq 0$. (See Problem 6.6.)

To summarize, the force-free paramagnetic model gives rise to profiles that are naturally close to those desirable for stable RFPs. Although close, the ohmic states cannot exactly describe a stable RFP since B_z is not allowed to reverse. In practice, ohmic states excite a variety of instabilities which, under certain conditions, cause the plasma to relax to a stable field reversed configuration at the expense of losing some plasma energy. This relaxation is a crucial aspect of RFP operation, observed in many experiments. A theoretical explanation has been given by Taylor (1976), which is discussed in the next section.

9.5.7. Taylor's Theory

Stable high-β RFP profiles have been shown to exist theoretically. These profiles are reasonably close to the natural states that one would expect from ohmically heated discharges. However, as just discussed, ohmic profiles do not allow a reversal in B_z. Hence, such discharges should be MHD unstable.

A possible resolution of this problem has been suggested by some of the data from the Zeta experiment. [See, for instance, Bodin and Newton (1980).] Here, after a rather turbulent initial phase, the plasma under

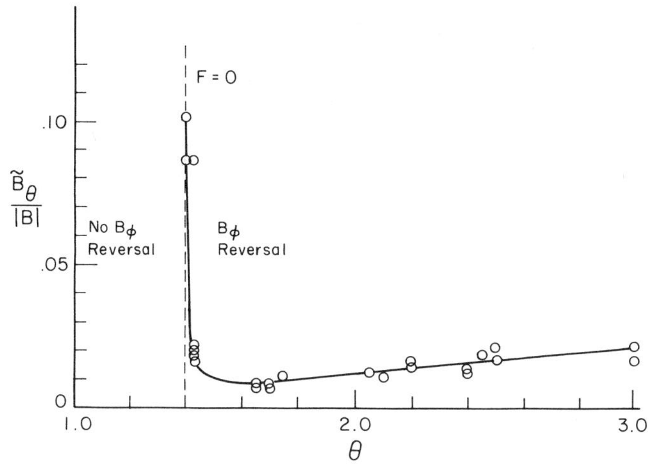

Figure 9.26. Experimental observations of B_θ fluctuations vs. $\theta \propto I_0$ in the HBTX-1A reversed field pinch at Culham Laboratory, Great Britain. Experimentally, field reversal occurs at $\theta = 1.4$. Observe the low fluctuation, quiescent operation for $\theta > 1.4$. Courtesy I. Hutchinson, MIT.

certain conditions remarkably evolves naturally to a stable quiescent RFP state including a spontaneous self-reversal of the toroidal field. See Fig. 9.26. A theory explaining the evolution to "RFP-like" profiles and the conditions under which self-reversal occurs has been given by Taylor (1974, 1975, and 1976), and is outlined below.

The basic idea is as follows. Consider a slightly dissipative plasma surrounded by a perfectly conducting shell which initially is not in a state of stable equilibrium. As the plasma turbulently evolves from its initial state, perhaps violently, it dissipates energy (e.g., through thermal contact with the wall, thermal and particle diffusion, etc.). It will continue to do so until it reaches a state of minimum energy after which it is incapable of further motion. The properties of the final energy state are determined by the constraints governing the evolution. Clearly, the crucial physics issue, resolved by Taylor, is the determination of the appropriate constraints. Once the constraints are known, it is relatively straightforward to obtain the final profiles. The presence of dissipation is important as it provides accessibility to a much wider range of states than those allowed by ideal MHD.

The calculation begins with the following exact nonlinear form of the potential energy given by [see Eq. (3.31)]

$$W = \frac{1}{2}\int d\mathbf{r}\left[\frac{B^2}{2\mu_0} + \frac{p}{\alpha - 1}\right] \tag{9.142}$$

For simplicity, it is assumed initially that the plasma energy is much smaller than the magnetic energy (i.e., $\beta \ll 1$). The effect of finite β is discussed later. In this limit Eq. (9.142) reduces to

$$W = \frac{1}{2\mu_0} \int B^2 \, d\mathbf{r} \tag{9.143}$$

The goal of the theory is to determine those profiles which minimize W (i.e., correspond to a minimum energy state) while satisfying all appropriate constraints.

Ideal MHD Constraint. Consider now the constraints. First, any physically acceptable magnetic field must satisfy the condition $\nabla \cdot \mathbf{B} = 0$. Second, if the magnetic field is embedded in a perfectly conducting ideal MHD plasma, then any variation of the field away from its initial value must be consistent with the constraint

$$\frac{\partial \mathbf{B}}{\partial t} = \nabla \times (\mathbf{v} \times \mathbf{B}) \tag{9.144}$$

That is, magnetic field variations must result from physically allowable plasma displacements that preserve the topology.

Taylor has shown that Eq. (9.144) can be replaced by an infinite set of integral constraints involving the helicity K defined by

$$K = \int_\psi \mathbf{A} \cdot \mathbf{B} \, d\mathbf{r} \tag{9.145}$$

where $\mathbf{A}$ is the vector potential and the integration is carried out over the volume enclosed by any arbitrary flux surface. In particular, it has been shown that K is a constant of the motion on every flux surface in an ideal MHD plasma: $\dot{K} = 0 \rightarrow K(\psi, t) = K(\psi)$. Furthermore, the minimum energy state obtained by varying $\mathbf{B}$ is the same whether one uses Eq. (9.144) as a local constraint, or instead uses $\dot{K} = 0$ as an infinite set of integral constraints (i.e., one on each flux surface). The corresponding minimizing magnetic field satisfies

$$\nabla \times \mathbf{B} = \lambda(\psi)\mathbf{B} \tag{9.146}$$

where $\lambda(\psi)$ is a free function related to $K(\psi)$. See Appendix D.1 for details.

The function $\lambda(\psi)$ is determined in principle by solving $\nabla \times \mathbf{B} = \lambda\mathbf{B}$ for $\mathbf{B}$ and $\mathbf{A}$ assuming arbitrary λ. The resulting $\mathbf{B}(\mathbf{r}, \lambda)$, $\mathbf{A}(\mathbf{r}, \lambda)$ are then used to compute $K(\psi, \lambda)$. Since K is a constant of the motion $K(\psi, \lambda) = K_i(\psi)$ where $K_i(\psi)$ is the prescribed helicity of the initial state. The equation $K(\psi, \lambda) = K_i(\psi)$ can then be inverted to give $\lambda = \lambda(\psi, K_i)$. Consequently the final minimum energy state as described by λ is directly dependent upon the details of the initial conditions.

The strong correlation between final and initial states is highly artificial and is unlikely to occur in a realistic plasma with dissipation. It is in fact a direct consequence of the perfect conductivity assumption which prohibits changes in topology. Therefore, many nearby lower energy states are inaccessible and there is a one-to-one correspondence between each flux surface in the initial and final states.

Taylor's Constraint. With the admittedly unrealistic ideal case serving as a reference, consider now the appropriate constraints in a plasma with small but nonzero dissipation. In such systems field line topology need no longer be preserved as the plasma moves so that a much wider class of lower energy states is now accessible. Furthermore, small changes in the magnetic field can often lead to finite changes in the magnetic topology. Consequently, even when the dissipation is small it is no longer correct to assume that K is conserved on every flux surface during the evolution since the continual tearing and coalescing of field lines destroys the identity of most surfaces.

Taylor pointed out that in systems with small dissipation, the only flux surface that would preserve its identity during the entire evolution is the plasma boundary since it is assumed rigid and perfectly conducting. He thus asserted that it was inappropriate to minimize the energy requiring K on every surface to remain invariant. Instead, the proper minimization requires only the single constraint that K_a remain invariant, where K_a is the helicity associated with the total volume.

Using the method of Lagrange multipliers, one applies Taylor's constraint by minimizing

$$W = \frac{1}{2\mu_0}\left\{\int (B^2 - \mu \mathbf{A}\cdot\mathbf{B})\, d\mathbf{r} + \mu K_a\right\} \tag{9.147}$$

where the integration is over the entire plasma volume. Since there is only one constraint to satisfy, the Lagrange multiplier μ is a *constant* having the same value on every surface. The minimization of W is described in Appendix D.2. One finds that the lowest energy state is described by

$$\nabla \times \mathbf{B} = \mu \mathbf{B} \tag{9.148}$$

Since the final equilibrium is characterized by only a single constant μ, this indicates that the evolved state is almost independent of initial conditions. The energy of this state is lower than that of the corresponding ideal case with similar initial conditions since all but one of the infinity of K constraints has been relaxed. The equilibria satisfying Eq. (9.148) are states of minimum potential energy thus guaranteeing ideal MHD stability (and resistive MHD stability as well).

Solution for **B**. The next step in the analysis is to solve for **B** and then **A** assuming μ is an arbitrary constant. In general the solution is not unique. For each solution, the corresponding **B** and **A** can be used to evaluate K_a, which then determines $\mu = \mu(K_a)$ (for that solution). Substituting this value of μ, one can evaluate $W(K_a)$ for each solution. Taylor conjectures that the plasma will evolve to that particular state which has the lowest absolute value of W.

Taylor (1975) and Reiman (1980) have analyzed the solutions to Eq. (9.148). Their studies show that in the limit of large aspect ratio (i.e., straight cylinder) there are only two configurations that can have absolute minimum energy: an $m = 0$ cylindrically symmetric solution and an $m = 0, 1$ mixed helicity state. These are given by (see Appendix D.3)

$m = 0$ *cylindrical state*:

$$\begin{aligned} B_z/B_0 &= J_0(\mu r) \\ B_\theta/B_0 &= J_1(\mu r) \end{aligned} \tag{9.149}$$

$m = 0$, $m = 1$ *helically symmetric state*:

$$\begin{aligned} B_z/B_0 &= J_0(\mu r) + cJ_1(\alpha r)\cos\chi \\ B_\theta/B_0 &= J_1(\mu r) - \frac{c}{\alpha}\left[\mu J_1'(\alpha r) + \frac{k}{\alpha r}J_1(\alpha r)\right]\cos\chi \\ B_r/B_0 &= -\frac{c}{\alpha}\left[kJ_1'(\alpha r) + \frac{\mu}{\alpha r}J_1(\alpha r)\right]\sin\chi \end{aligned} \tag{9.150}$$

Here, $\chi = \theta + kz$, $\alpha = (\mu^2 - k^2)^{1/2}$, c is an arbitrary helical amplitude, k is the helical wave number, and J_0 and J_1 are Bessel functions. The quantity B_0 is equal to the toroidal field on axis. It is related to the toroidal flux in the plasma which is conserved during the entire evolution because of the presence of the conducting wall. A simple calculation shows that B_0 is the same for both equilibria and is given by

$$B_0 = \left[\frac{\mu a}{2J_1(\mu a)}\right]\left(\frac{\psi}{\pi a^2}\right) \tag{9.151}$$

where ψ is the toroidal flux.

Note that for the cylindrical state B_z reverses on the outside if $\mu a > 2.40$. Also, for the helical state, the wavelength k is related to μ

through the boundary condition $B_r(a, \theta, z) = 0$ on the conducting wall. This condition is trivially satisfied for the cylinder since $B_r \equiv 0$.

The corresponding vector potentials for either equilibrium can be written as

$$\mathbf{A} = \frac{1}{\mu}[\mathbf{B} - B_0 J_0(\mu a)\mathbf{e}_z] \tag{9.152}$$

The constant term in Eq. (9.152) corresponds to the gauge condition $\int A_z(a, \theta)\, d\theta = 0$. Reiman (1980) has pointed out the importance of choosing a proper gauge condition (i.e., one which is independent of μ) since the evaluation of K_a depends upon the value of the constant in A_z. An improper gauge condition yields misscaled values of $\mu(K_a)$ and hence W when comparing different equilibria.

Minimum Energy States. Using these solutions one can evaluate $W(K_a)$ for each equilibrium. (See Appendix D.4 for details.) For the $m = 0$ state W is parametrically related to K_a through $x \equiv \mu a$. The relevant equations are given by

$$\begin{aligned} \hat{W} &= \frac{x^2}{J_1^2}\left(J_0^2 + J_1^2 - \frac{J_0 J_1}{x}\right) \\ \hat{K} &= \frac{x}{J_1^2}\left(J_0^2 + J_1^2 - \frac{2J_0 J_1}{x}\right) \end{aligned} \tag{9.153}$$

where $J_0 = J_0(x)$, $J_1 = J_1(x)$, and $\hat{W} = 2a^2\mu_0 W/R_0\psi^2$, $\hat{K} = aK_a/R_0\psi^2$ are normalized forms of the quantities W and K_a. The function $\hat{W}(\hat{K})$ is illustrated in Fig. 9.27. Observe that $\hat{W}$ is a rapidly increasing function of $\hat{K}$ for the $m = 0$ solution.

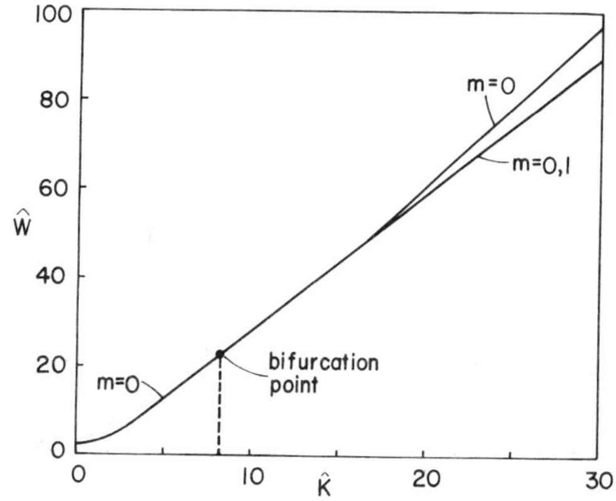

Figure 9.27. Energy versus helicity for the $m = 0$ cylindrical state and the $m = 0, 1$ mixed helicity state.

A similar calculation can be carried out for the $m = 0, 1$ mixed helicity state. In this case $\hat{W}$ is related to $\hat{K}$ through a set of coupled equations given by

$$\hat{W} = x\left[\hat{K} + \frac{J_0(x)}{J_1(x)}\right] \tag{9.154}$$

$$\hat{k}yJ_1'(y) + xJ_1(y) = 0 \tag{9.155}$$

where $\hat{k} = ka$ and $y = (x^2 - \hat{k}^2)^{1/2}$. Equation (9.155) is a transcendental relation that determines $x = x(\hat{k})$. Since the energy is an increasing function of x in the regime of interest, the lowest energy state corresponds to that value of $\hat{k}$ that minimizes x. A numerical calculation shows that $\hat{k} = 1.2$ and gives $x = 3.11$. Substituting $x = x_0 = 3.11$ into Eq. (9.154) gives the desired $m = 0, 1$ relation for $\hat{W}(\hat{k})$:

$$\hat{W} = x_0\left[\hat{K} + \frac{J_0(x_0)}{J_1(x_0)}\right] \tag{9.156}$$

Equation (9.156) is also illustrated in Fig. 9.27.

Observe the following points. For low values of K_a only the $m = 0$ solution can exist. Hence, in this regime the plasma is cylindrical. At higher values of K_a both the $m = 0$ and the $m = 0,1$ solutions can exist. The $m = 0,1$ solution has the lowest energy and in this regime the plasma develops a helical deformation. The helical solution cannot exist for low K_a because the radial boundary condition given by Eq. (9.155) can only be satisfied for $x > x_0 = 3.11$ corresponding to a minimum required value of K_a. Physically, for small K_a there is insufficient parallel current to sustain a helical equilibrium. The transition from the $m = 0$ to $m = 0,1$ state occurs at the bifurcation point $\hat{K} = aK_a/R_0\psi^2 = 8.21$. Note also that field reversal which occurs at $x = 2.40$ in the cylindrical state, corresponds to $\hat{K} = 2.40$. Thus, in principle a plasma discharge that is initially characterized by a value of helicity in the range $2.40 < \hat{K} < 8.21$ should relax to a cylindrical RFP configuration including a reversal in the B_z field.

Discussion. There are a number of important applications of Taylor's theory of plasma relaxation to actual experimental situations. First, consider the result that field reversal occurs for $\hat{K} > 2.40$. Strictly speaking, field reversal should not play a crucial role in Taylor's theory; whether $\hat{K}$ is greater or less than 2.40, the Taylor equilibria are minimum energy states and are thus MHD stable. Note, however, that these equilibria are in general characterized by $J_z(a) \neq 0$. This is in contrast to actual experiments where the plasma temperature near the wall is very low, implying that $J_z(0) \approx 0$. Such behavior is not included in Taylor's theory. The consequence is that the minimum energy profiles are somewhat modified near the outside of the plasma. This has a major impact on non-field-reversed cases. When $J_z(a) \approx 0$, it is straightforward to show that the q profile develops a minimum when $B_z(a)$ does not reverse. As discussed, such profiles are unstable to current-driven and pressure-driven modes. For the field-reversed case, $J_z(a) \approx 0$ does not lead to a minimum in q. The experimental data are in general consistent with these predictions. Plasma confinement, temperature, and β are seriously degraded in non-field-reversed discharges. Consequently, the onset of field reversal as predicted by Taylor's theory plays a far more important role in RFP operation than one might have originally anticipated.

A second application of Taylor's theory concerns the evolution of RFP discharges. In general, one does not start with a reasonably well established plasma with a given $\hat{K}$ and then watch it evolve to a Taylor state. Instead, RFPs are initiated with a small uniform vacuum B_z field. The toroidal current starts from zero and is increased until it reaches its final value. If the current rise time is slow (as it is in most discharges), then Taylor's theory predicts that the RFP should adiabatically evolve through a sequence of minimum energy states; the relaxation occurs rapidly with respect to the current rise time. This is in quite good agreement with the experimental data. The usual way to compare the theory and experiment is by means of an F vs. θ diagram, where

$$F \equiv \frac{B_z(a)}{\langle B_z \rangle} = \frac{\pi a^2 B_z(a)}{\psi} \tag{9.157}$$

is a measure of B_z at the plasma edge and

$$\theta \equiv \frac{B_\theta(a)}{\langle B_z \rangle} = \frac{\mu_0 a I_0}{2\psi} \tag{9.158}$$

known as the pinch parameter, is a measure of the toroidal current. For the Taylor minimum energy cylindrical state one finds F and θ related parametrically through x as follows:

$$\begin{aligned} F &= \frac{xJ_0(x)}{2J_1(x)} \\ \theta &= \frac{x}{2} \end{aligned} \tag{9.159}$$

A plot of F vs. θ is illustrated in Fig. 9.28. Here one can view increasing θ as equivalent to increasing time during the discharge

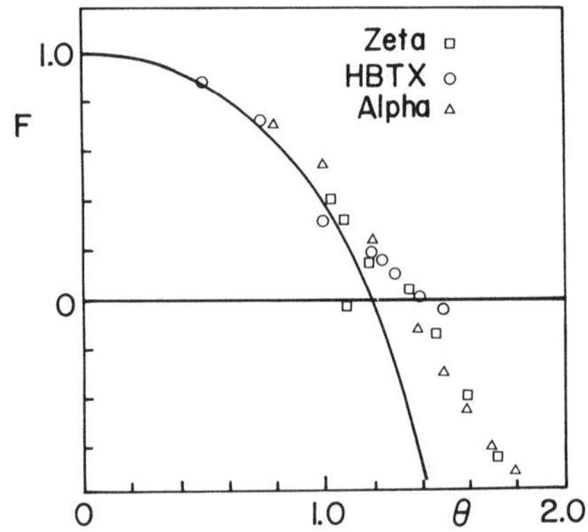

Figure 9.28. F vs. θ diagram comparing theoretical and experimental data [from Taylor (1985)].

evolution. Note that field reversal occurs at $\theta = 1.2$ and that the $m = 0$, $m = 0$, 1 bifurcation point corresponds to $\theta = 1.56$. Also shown is a sequence of experimental data points. There is excellent agreement between theory and experiment.

A third point worth noting concerns the fact that Taylor's minimum energy state with a B_z field reversal cannot arise from purely ohmic discharges. The implication is that there must be a certain background level of turbulence present in the plasma at all times which maintains the reversal by a combination of anomalous transport and plasma convection. How seriously this turbulence affects the overall plasma confinement is a major concern of current RFP research.

In summary, Taylor's hypothesis concerning the constancy of K_a during the plasma evolution has been shown to provide a reliable qualitative and semiquantitative guide to the behavior of reversed field pinches. Within the enormous space of possible equilibrium profiles, most discharges should evolve adiabatically through a sequence of minimum energy states uniquely determined only by the helicity, toroidal flux, and geometric dimensions. Furthermore, for the final configuration to remain cylindrically symmetric and possess a field reversal (which is important when realistic conditions at the wall are included) the pinch parameter must lie in the range $1.20 < \theta < 1.56$ or equivalently $2.40 < \hat{K} < 8.21$.

The result that the final state is force-free should not be surprising in view of the fact that the plasma energy was neglected in the minimization of W. However, to the extent that the tearing and reconnecting of field lines allows the pressure to redistribute and equalize itself, the force-free result would apply even for finite β initial states (i.e., the variation of δp would be independent of δB and upon separate minimization would lead to $\nabla p = 0$). This would be a very pessimistic result. In a real plasma, however, the situation is probably somewhere intermediate between the perfect coupling of δp and δB as in ideal MHD to the complete decoupling just discussed. The actual pressure profiles will depend upon the details of the plasma turbulence and its effect on the resistivity and thermal conductivity transport coefficients as well as fluid convection which in combination determine the energy balance. It is likely that the final β's will be considerably less than the maximum β's that can be theoretically constructed. The determination of the evolved pressure profiles and final β_p remains one of the important areas of RFP research.

9.5.8. Overview of the RFP

By combining the equilibrium and stability results one can obtain an overview of the RFP, at least with respect to ideal MHD.

The RFP is an axisymmetric toroidal configuration with $B_\theta^2 \sim B_z^2 \sim \mu_0 p$ which achieves MHD equilibrium and stability with optimally efficient profiles requiring a minimum of toroidal field. Radial pressure-balance and toroidal force-balance are similar to those of a Z pinch. Because of the relatively large B_θ, toroidicity is required only to close the field lines, but does not play a major role in MHD stability. Hence, the device can have the technological advantage of a large aspect ratio. Another attractive feature is that the toroidal shift of the plasma is very small. Perhaps the main advantage of large B_θ is that it is possible in principle to joule heat an RFP to ignition, although such strong heating has yet to be observed experimentally.

Since $\beta \sim 1$ and $q \ll 1$ the stability of the RFP is in general quite sensitive to profiles. For stability at high β the pressure profile must be hollow or very flat on axis and the B_z field reversed on the outside. Equally important, a perfectly conducting shell relatively near the main plasma core is required. This is a disadvantage technologically.

The rather complicated profiles required for stability are difficult to program externally. However, experimental data and a recent theory suggest that over a certain range of the pinch parameter θ, the system will evolve naturally to an "RFP-like" profile, although probably with β_p lower than the theoretically predicted maximum.

These minimum energy states require a background level of turbulence to maintain the field reversal, but also lead to a degradation of the confinement properties of the plasma, a phenomenon not well understood at the present time. Even so, the possibility of producing an MHD stable, high-β plasma, possibly capable of being joule heated to ignition, makes the RFP an interesting and important concept to investigate in the magnetic fusion program.

9.6. THE "STRAIGHT" TOKAMAK

9.6.1. Introduction

The final one-dimensional configuration examined by means of the general screw-pinch analysis is the "straight" tokamak, a circular column of plasma of length $2\pi R_0$ in which the fields satisfy the tokamak inverse aspect ratio expansion $B_\theta/B_z \sim \epsilon$, $q \sim 1$, $2\mu_0 p/B_z^2 \sim \epsilon^2$ (ohmically heated tokamak) or $2\mu_0 p/B_z^2 \sim \epsilon$ (high-β tokamak). Although a reasonably complete stability picture is obtained from this model, the results can be unreliable because the geometric effects associated with toroidicity are of the same order as the field ratio. Thus, toroidicity sometimes gives rise to contributions in δW that are of the *same* order as those of the straight tokamak. Despite these uncertainties the model provides a very useful frame of reference for the understanding of tokamak stability.

It is shown that the most dangerous instabilities in a straight tokamak are internal and external current-driven modes. While a conducting wall helps to stabilize external modes, the tokamak, unlike the RFP, does not depend on the presence of such a wall. Consequently, discussions of overall tokamak stability usually take into account external modes with the wall at infinity. Pressure-driven modes do not play a major role in the straight tokamak because of the combination of finite shear and low β.

The straight tokamak is predicted to have good MHD stability properties if the total toroidal current is limited and the current profile is peaked. This result is qualitatively and often quantitatively valid for the fully toroidal case. The stability of the straight tokamak imposes no important restrictions on β or the pressure profile. This, however, is qualitatively different from the toroidal case where ballooning and ballooning-kink modes set important limits on the maximum stable β. Toroidal modifications to tokamak stability are discussed in Chapter 10.

As in the case of the RFP, much of the intuition concerning stability is derived from the use of trial functions and by exploitation of the tokamak expansion in δW. This intuition is supported by a large number of numerical calculations carried out in recent years.

9.6.2. Internal Pressure-Driven Modes

The internal pressure-driven modes in a straight tokamak correspond to localized interchange instabilities. These can be investigated by substituting the tokamak expansion into Suydam's criterion which is valid for arbitrary profiles. Using the definition $\beta(r) \equiv 2\mu_0 p(r)/B_0^2$, Suydam's criterion has the form

$$\left(\frac{rq'}{q}\right)^2 + 4r\beta' > 0 \tag{9.160}$$

for stability.

Since β is either of order ϵ^2 or ϵ and q is a monotonically increasing function of radius with $rq'/q \sim 1$ (i.e., finite nonzero shear), Suydam's criterion is easily satisfied over almost the entire profile. The only exception is very near the axis where the shear is small, a situation similar to that in the RFP [see Eq. (9.107)]. Within the framework of the straight tokamak these modes can be stabilized by keeping the pressure flat over a narrow region near $r = 0$. See Fig. 9.29. The width of this region can be estimated by expanding Suydam's criterion in the vicinity of the axis:

$$\left(\frac{q_0''}{q_0}\right)^2 r^2 + 4\beta_0'' > 0 \tag{9.161}$$

Thus, if the pressure profile is flattened over a distance Δr given by

$$\frac{\Delta r}{a} = \frac{2q_0}{aq_0''}(-\beta_0'')^{1/2} \sim \beta^{1/2} \ll 1 \tag{9.162}$$

then the modes are stabilized. This represents only a minor restriction on the profile.

Furthermore, as shown in Chapter 10, there are important toroidal modifications to Suydam's criterion when $q \sim 1$. These are unimportant for the RFP since $q \sim \epsilon$. However, for the tokamak they imply that under reasonable conditions, even the small flattening of the pressure profile is no longer required for stability.

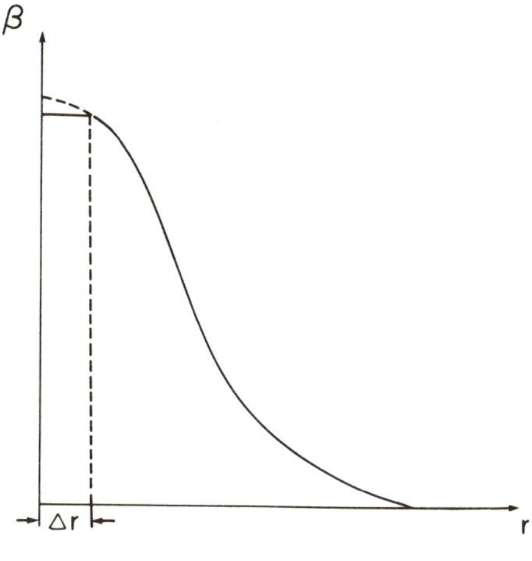

Figure 9.29. $\beta(r)$ vs. r for a straight tokamak. The flattened profile near $r=0$ is required to satisfy Suydam's criterion.

In summary, the combination of low β and finite shear implies that localized interchange perturbations are not very dangerous in a straight tokamak.

9.6.3. Internal Current-Driven Modes

The stability of general nonlocalized internal current-driven modes can be investigated by substituting the tokamak expansion into δW_F as given by Eq. (9.43). This gives rise to an asymptotic expansion of the form $\delta W_F = \delta W_0 + \delta W_2 + \delta W_4 + \cdots$, where $\delta W_n \sim \epsilon^n$. The calculation is straightforward, although one must include the toroidal periodicity constraint by setting

$$k = -n/R \tag{9.163}$$

with n an integer representing the toroidal mode number. Toroidal periodicity is important in tokamaks because the unstable modes correspond to long wavelengths, typically $n \sim 1$. (This is not important in an RFP where the instabilities occur for $ka \sim 1$ or $n \sim 1/\epsilon$; hence, n can effectively be considered continuous.)

The first contribution to δW_F that does not trivially vanish is of second order and is given by (Shafranov, 1970)

$$\frac{\delta W_2}{W_0} = \frac{1}{a^2}\int_0^a \left(\frac{n}{m} - \frac{1}{q}\right)^2 [r^2\xi'^2 + (m^2 - 1)\xi^2] r\,dr \tag{9.164}$$

where $W_0 = 2\pi^2 a^2 B_0^2/\mu_0 R_0$. For $m \geq 2$ and n arbitrary both terms in the integrand are positive and nonzero. Hence $\delta W_2 > 0$ and these modes are stable. For $m = 1$ the second term vanishes. This mode is stable in systems with an increasing q profile if $nq_0 > q_0 > 1$ since the quantity $n/m - 1/q$ never vanishes.

However, if a $q = 1$ surface exists in the plasma [i.e., $q_0 < 1 < q_a$] a trial function can be constructed as shown in Fig. 9.30, which in the limit $\delta \to 0$ causes $\delta W_2 \to 0$. For this case stability is determined by the higher-order corrections to δW_F. The next contribution is of order ϵ^4 and for $n = 1$ can be written as (Rosenbluth *et al.*, 1973)

$$\frac{\delta W_4}{W_0} = \frac{\xi_0^2}{a^2}\int_0^{r_s} \left[r\beta' + \frac{r^2}{R_0^2}\left(1 - \frac{1}{q}\right)\left(3 + \frac{1}{q}\right)\right] r\,dr \tag{9.165}$$

where r_s is the location of the $q(r_s) = 1$ surface. In Eq. (9.165) the contributions from both the pressure and parallel current are destabilizing and thus the system is unstable.

The instability is often referred to as the $m = 1$ internal kink mode,

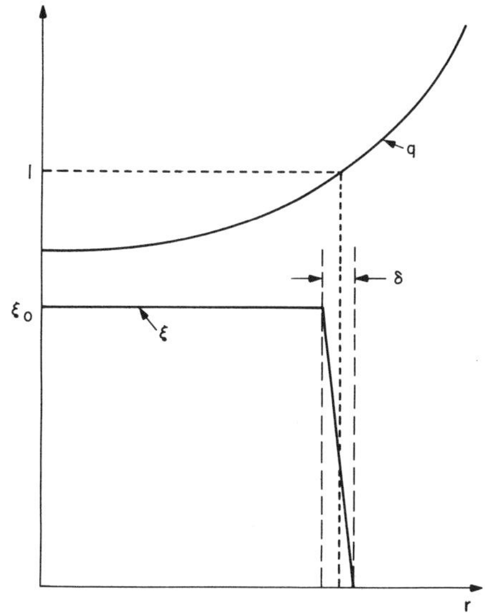

Figure 9.30. Trial function for an $m = 1$ internal kink mode in a straight tokamak.

a slightly misleading nomenclature since the mode can also be driven by the pressure gradient, particularly in the high-β tokamak regime. Many of the qualitative features of the mode persist in the toroidal case, although there are important corrections to the $n = 1$ driving term. These are discussed in Chapter 10. It is also worth noting that the nonlinear evolution of the $m = 1$ internal kink mode, including the effects of dissipation, is believed to be an important component in "sawtooth" oscillations observed in many tokamaks. Also, if the effects of a species of energetic particles is included, such as generated by neutral beam injection, one finds a modified $m = 1$ instability thought to be responsible for the so-called "fish bone" oscillations observed in certain experiments.

Summarizing, in a straight tokamak $m \geq 2$ internal modes are stable for both the ohmically heated and high-β orderings. The $m = 1$ internal mode exhibits a weak instability (i.e., $\delta W \sim \epsilon^4$) which can be prevented if

$$q_0 > 1 \tag{9.166}$$

In practice, the $m = 1$ mode and its nonideal modifications play an important role in tokamak operation through the sawtooth oscillations. These oscillations are generated by a two-stage process. First, the temperature dependence of electrical conductivity $\sigma \sim T^{3/2}$ tends to channel the current into the center of the discharge. This in turn preferentially heats the center of the plasma leading to a further channeling of the current. The resulting "thermal" instability causes q_0 to decrease with time as J_0 increases. Ultimately, $q_0 = 1$, at which point the $m = 1$ internal mode is triggered. This is the second stage. The sawtooth oscillation is not catastrophic in that its main consequence consists of a redistribution of plasma energy and current within a relatively small region from the origin to the $q = 1$ surface after which the cycle repeats. In effect, sawtooth oscillations anchor the average value of the safety factor on axis to its threshold value $q_0 = 1$, thereby limiting the maximum achievable value of toroidal current density on axis.

9.6.4. External Modes (The $m = 1$ Kruskal–Shafranov Limit)

The most dangerous ideal MHD instabilities in a straight tokamak are the external kink modes. These are current-driven modes in which the $\mathbf{k} \cdot \mathbf{B} = 0$ singular surface lies in the vacuum region. External kinks are more dangerous energetically than internal modes in that their stability is determined solely by the second-order contribution to δW. One consequence of this fact is that the stability boundaries are unaffected by the pressure gradient (in the straight model).

Much of the basic insight concerning external kinks has been given by Shafranov (1970) and has been summarized by Wesson (1978). The essence of these results can be obtained by examining the first nonvanishing contribution to δW which is of order ϵ^2. From Eq. (9.44) a short calculation shows that in the limit where the conducting wall moves to infinity (i.e., $b/a \to \infty$), δW for either the ohmically heated or high-β expansion can be written as

$$\begin{aligned}\frac{\delta W_2}{W_0} = \frac{1}{a^2}\int_0^a \left(\frac{n}{m} - \frac{1}{q}\right)^2 [r^2\xi'^2 + (m^2 - 1)\xi^2] r\,dr \\ + \xi_a^2\left(\frac{n}{m} - \frac{1}{q_a}\right)\left[|m|\left(\frac{n}{m} - \frac{1}{q_a}\right) + \left(\frac{n}{m} + \frac{1}{q_a}\right)\right]\end{aligned} \tag{9.167}$$

One can gain some insight into the possible range of unstable wave numbers by noting that the integral contribution is always positive or zero and by assuming, without loss in generality, that $q_a > 0$, $m > 0$, and $-\infty < n < \infty$. By examining the boundary term it then follows that sufficient conditions for stability to a given n, m mode are given by

$$\frac{n}{m} > \frac{1}{q_a} \tag{9.168a}$$

$$\frac{n}{m} < 0 \tag{9.168b}$$

Similarly, a necessary condition for instability is that

$$0 < \frac{n}{m} < \frac{1}{q_a} \tag{9.169}$$

If the q profile is an increasing function of radius (as it is in a tokamak) then Eq. (9.169) implies that the resonant surface for an external mode $q = m/n$ lies in the vacuum region.

Consider now the $m = 1$ mode. For this case the minimizing eigenfunction in the plasma is given by $\xi(r) = \xi_a = \text{const}$ and causes the integral contribution to vanish. Note that such an eigenfunction can only be chosen for external modes, since the internal-mode boundary condition, $\xi(a) = 0$, is no longer required. Furthermore, the minimizing function, $\xi = \text{const}$ is independent of the q profile. This leads to the important conclusion that the $m = 1$ stability boundary is a function only of q_a (i.e., it depends only on the total current and not the current profile).

$$\frac{\delta W_2}{W_0} = 2\xi_a^2\left[n\left(n - \frac{1}{q_a}\right)\right] \tag{9.170}$$

From Eq. (9.170) it follows that the condition to stabilize the $m = 1$ external kink mode for the worst case, $n = 1$, is given by

$$q_a > 1 \tag{9.171}$$

This criterion is known as the Kruskal–Shafranov condition (Kruskal and Schwarschild, 1954; and Shafranov, 1956). It imposes an important constraint on tokamak operation in that it limits the total toroidal current that can flow to $I < I_{KS}$ where

$$I_{KS} \equiv 2\pi a^2 B_0/\mu_0 R_0 = 5a^2 B_0/R_0(\text{MA}) \tag{9.172}$$

For $a = 0.2\,\text{m}$, $R = 1\,\text{m}$ and $B_0 = 5\,\text{T}$, $I_{KS} = 1\,\text{MA}$.

The driving mechanism for the external kink instability has been discussed in Section 9.5.4 and Fig. 9.23. Basically, the destabilizing force is similar to that of the $m = 1$ mode in a Z pinch. However, the modes are restricted to long wavelengths in order to prevent a strong stabilizing contribution from B_z line bending. In fact, the criterion $q_a > 1$ can be viewed as a condition on the geometry (i.e., the inverse aspect ratio) which prohibits the formation of potentially unstable long-wavelength modes because of toroidal periodicity requirements.

To summarize, the $m = 1$ Kruskal–Shafranov instability is an external current-driven kink mode. The stability boundary $q_a = 1$ does not depend on the existence of a pressure gradient, the shape of the J_z

profile, or the position of the wall (see Problem 9.10). As such it is a potentially very dangerous instability. In practical situations the $m = 1$ external kink is difficult to observe experimentally because higher m modes, ideal and resistive, pose even stronger requirements on q_a. These modes are more sensitive to profile and have somewhat slower growth rates. Even so, the resistive tearing mode is thought to be responsible for the rapid termination of tokamak discharges by means of a "major disruption" when $q_a \sim 2$–3, well below the current required to excite the Kruskal–Shafranov mode.

9.6.5. External Modes ($m \geq 2$ External Kinks)

In general, the $m \geq 2$ external kinks impose more stringent conditions on q_a and the current profile than those corresponding to the $m = 1$ mode.

The $m \geq 2$ stability boundaries also follow from an analysis of Eq. (9.167). Clearly, a sufficient condition for stability to a given (m, n) mode is $q_a > m/n$. This is not a very useful criterion because it imposes unnecessarily low limits on the toroidal current for high-m, low-n modes.

A slightly improved stability estimate can be obtained by noting that for monotonically increasing q profiles a lower bound on δW is obtained by setting $q(r) = q_a$ in the integral contribution; that is, for external modes the factor $n/m - 1/q$ is nonvanishing in the plasma region and assumes its smallest value at the boundary $r = a$. The factor $(n/m - 1/q_a)^2$ can then be removed from the integrand. The remaining integral is minimized by choosing $\xi(r) = \xi_a(r/a)^{m-1}$. The resulting form of δW is given by

$$\frac{\delta W}{W_0} \geq \frac{2\xi_a^2}{q_a^2}(nq_a - m)(nq_a - m + 1) \tag{9.173}$$

Consequently, for a given m and n the regions of stability are bounded by (Shafranov, 1970)

$$\frac{m-1}{n} \leq q_a \leq \frac{m}{n} \tag{9.174}$$

Equation (9.174) is not a very useful estimate because upon setting $n = 1$ and continuously incrementing m by unity, it follows that the lower bound on δW predicts instability over the entire range of q_a, $0 < q_a < \infty$. It is worth noting that the inequality given by Eq. (9.173) becomes equality for the special profile characterized by $B_z = \text{const}$, $B_\theta \propto r$. This case corresponds to a J_z profile which is constant for $0 < r < a$ and jumps to zero just outside the plasma surface. The resulting q profile is

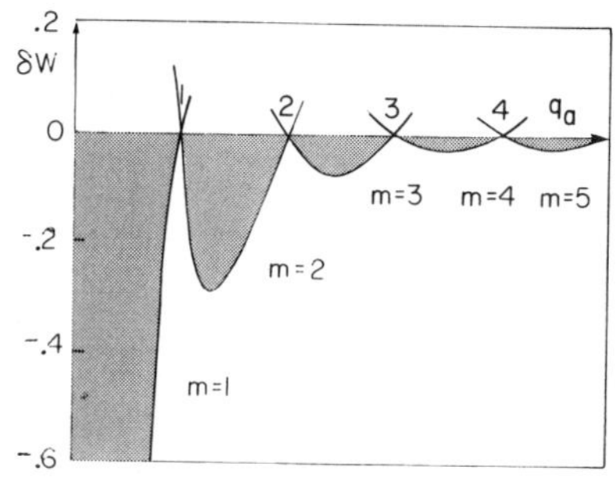

Figure 9.31. Stability diagram for the uniform current model corresponding to $n = 1$. The shaded areas represent regions of unstable q_a values.

shearless: $q(r) = q_a = \text{const}$. Consequently, the constant J_z profile is indeed unstable over the entire range of q_a as described by Eq. (9.174). See Fig. 9.31.

Shafranov (1970) first pointed out that this pessimistic result is strongly dependent upon the shape of the current profile. He showed both analytically and numerically that for more realistic profiles where the current goes smoothly to zero at the plasma surface, the stability properties are more optimistic. In essence, as m increases, the mode structure for external kinks becomes progressively more localized in the vicinity of the plasma surface. As a consequence, the plasma contribution to δW depends sensitively on the local behavior of the current density near $r = a$ and when $J(a)$, $J'(a)$, etc. are small, the instabilities diminish. In contrast, recall that the $m = 1$ instability is determined solely by the total current.

Asymptotic Analysis. A convenient way to obtain a quantitative analytic theory which describes these results has been given by Laval and Pellat (1973) and Laval, Pellat, and Soulé (1974). They considered the external kink mode in the limit of large m, including the effects of noncircularity as well. For the present purposes, it suffices to outline their theory for the circular case.

The analysis begins with the expression for δW describing external kinks within the tokamak expansion as given by Eq. (9.167). The main objective of the calculation is to obtain an accurate evaluation of the plasma contribution from which it is then straightforward to determine the stability boundaries.

The trial function which minimizes the plasma term satisfies

$$\frac{d}{dr}\left[r^3\left(\frac{n}{m}-\frac{1}{q}\right)^2\frac{d\xi}{dr}\right] - r(m^2-1)\left(\frac{n}{m}-\frac{1}{q}\right)^2\xi = 0 \qquad (9.175)$$

Assuming that Eq. (9.175) can be solved for ξ, one can rewrite δW as follows:

$$\frac{\delta W_2}{W_0} = \frac{\xi_a^2}{q_a^2}\left(1 - \frac{nq_a}{m}\right)\left[\left(m + \frac{a\xi_a'}{\xi_a}\right)\left(1 - \frac{nq_a}{m}\right) - \left(1 + \frac{nq_a}{m}\right)\right] \quad (9.176)$$

where without loss of generality $q_a > 0$, $m > 0$ and $-\infty < n < \infty$. To evaluate δW_2 one must solve Eq. (9.175) and then calculate $a\xi_a'/\xi_a$.

The analysis is greatly simplified by considering the limit of large m. For these modes it is possible to exploit the fact that instability occurs only when the resonant surface lies just outside the plasma. The corresponding plasma displacement is highly localized just inside the plasma, allowing one to perform a high m asymptotic analysis of the Euler–Lagrange equation. This leads to an accurate evaluation of the plasma contribution to δW. See Fig. 9.32.

The first step in the solution to Eq. (9.175) is to introduce a new dependent variable defined by

$$b(r) = K\xi$$
$$K(r) = r^{3/2}\left(\frac{n}{m} - \frac{1}{q}\right) \quad (9.177)$$

Substituting into Eq. (9.175) yields

$$\frac{d^2b}{dr^2} - \left[\frac{m^2 - 1}{r^2} + \frac{1}{K}\frac{d^2K}{dr^2}\right]b = 0 \quad (9.178)$$

The next step is to expand Eq. (9.178) in the vicinity of the plasma

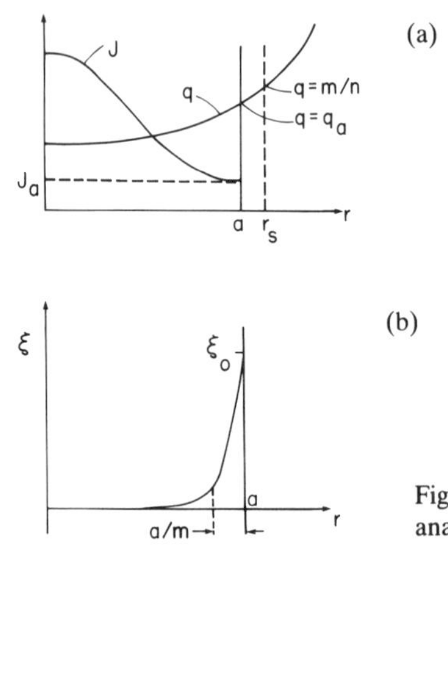

Figure 9.32. Profiles for the high m external kink analysis. (a) Equilibrium J and q; (b) eigenfunction ξ.

surface assuming $m \gg 1$. Specifically, one writes

$$\frac{r}{a} = 1 - \frac{z}{2m} \tag{9.179}$$

where z is a new independent variable. Localization of the mode requires that $0 \lesssim z \lesssim 1$. Similarly, if the resonant surface is near the plasma then $1 - nq_a/m$ is small. Is is assumed initially that $1 - nq_a/m \sim 1/m$. Under these conditions the function K can be expanded as follows:

$$K = K_1 + K_2 + \cdots \tag{9.180}$$

where $K_1 \sim 1/m$ and $K_2 \sim 1/m^2$ are given by

$$\begin{aligned} K_1 &= a^{3/2}\left[\left(\frac{n}{m} - \hat{\iota}_a\right) + \left(\frac{a\hat{\iota}'_a}{2m}\right)z\right] \\ K_2 &= -a^{3/2}\left[\left(\frac{a^2\hat{\iota}''_a}{8m^2}\right)z^2 + \frac{3z}{4m}\left(\frac{n}{m} - \hat{\iota}_a + \frac{a\hat{\iota}'_a}{2m}z\right)\right] \end{aligned} \tag{9.181}$$

Here, $\hat{\iota}_a = 1/q_a$ and prime denotes differentiation with respect to r. Upon substituting Eqs. (9.179) and (9.181) into Eq. (9.178) one finds, after a short calculation, that

$$\frac{d^2b}{dz^2} - \left[\frac{1}{4} - \frac{\lambda}{m(z + z_0)}\right]b = 0 \tag{9.182}$$

The coefficient multiplying b has been expanded as a series in $1/m$ and is correct to order $(1/m^2)$. The parameters λ and z_0 are each of order unity and are given by

$$\begin{aligned} \lambda &= \frac{1}{2}\left(\frac{a^2\hat{\iota}''_a + 3a\hat{\iota}'_a}{a\hat{\iota}'_a}\right) \\ z_0 &= \frac{2m}{a\hat{\iota}'_a}\left(\frac{n}{m} - \hat{\iota}_a\right) \end{aligned} \tag{9.183}$$

λ and z_0 can be more conveniently expressed in terms of the properties of the current profile at the plasma edge. In particular, if one notes that $\hat{\iota} = R_0B_\theta/rB_0$ and $(rB_\theta)'/r = \mu_0 J$ then a simple calculation shows that

$$\begin{aligned} a\hat{\iota}'_a &= -\frac{2}{q_a}\left(1 - \frac{J_a}{\langle J\rangle}\right) \\ a^2\hat{\iota}''_a + 3a\hat{\iota}'_a &= \frac{2aJ'_a}{q_a\langle J\rangle} \end{aligned} \tag{9.184}$$

where $\langle J \rangle \equiv I_0/\pi a^2 = 2B_0\hat{\iota}_a/\mu_0 R_0$ is the average current density in the plasma and $J_a = J(a)$ and $J'_a = J'(a)$ are the current density and current density gradient at the plasma edge, respectively. This gives

$$\begin{aligned} \lambda &= -\frac{1}{2}\left(\frac{aJ'_a}{\langle J \rangle - J_a}\right) \\ z_0 &= m\left(1 - \frac{nq_a}{m}\right)\left(\frac{\langle J \rangle}{\langle J \rangle - J_a}\right) \end{aligned} \tag{9.185}$$

The final step in the formulation of the stability problem is to substitute b for ξ in the expression for δW [Eq. (9.176)]. The result is

$$\frac{\delta W_2}{W_0} = \frac{\xi_a^2}{q_a^2}\left(1 - \frac{nq_a}{m}\right)\left[m\left(1 - \frac{nq_a}{m}\right)\left(1 - \frac{2b'_0}{b_0} - \frac{1}{m}\right) - \frac{2J_a}{\langle J \rangle}\right] \tag{9.186}$$

where $b'_0/b_0 = [(db/dz)/b]_{z=0}$. Equations (9.182) and (9.186) represent the desired form of the high-m external kink stability problem.

The Case $J_a \neq 0$. The analysis of Eq. (9.186) separates into two cases depending upon whether or not J_a vanishes. Consider first the situation where the current profile is a decreasing function of r which remains finite at $r = a$: $J_a \neq 0$. For such profiles one need only calculate the leading-order contribution to b from Eq. (9.182); that is, the λ term can be neglected. The solution to Eq. (9.182) that corresponds to a localized eigenfunction (i.e., the one that decays away from the plasma surface as $z \to \infty$) is given by

$$b(z) = e^{-z/2} \tag{9.187}$$

Thus $b'_0/b_0 = -1/2$. This yields the following leading order (in $1/m$) expression for δW:

$$\frac{\delta W_2}{W_0} = \frac{2m\xi_a^2}{q_a^2}\left(1 - \frac{nq_a}{m}\right)\left(1 - \frac{nq_a}{m} - \frac{J_a}{m\langle J \rangle}\right) \tag{9.188}$$

Equation (9.188) predicts that instabilities exist for

$$\frac{1}{n}\left(m - \frac{J_a}{\langle J \rangle}\right) \leq q_a \leq \frac{m}{n} \tag{9.189}$$

Consequently, any profile with a current jump at the surface is unstable to high m external kinks over bands of q_a values as illustrated in Fig. 9.33. The most unstable case occurs for uniform current, $J_a/\langle J \rangle = 1$, in which neighboring bands touch and all q_a values are unstable. As the current at the boundary decreases, the unstable bands become narrower, forming windows of stability.

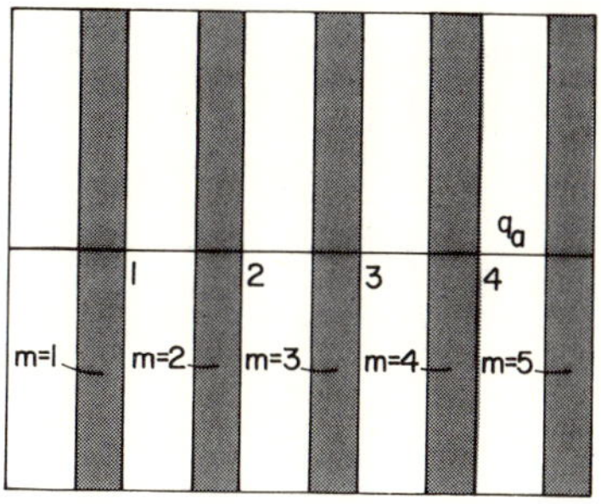

Figure 9.33. Stability diagram showing bands of instability for the case where $J_a \neq 0$. The shaded areas represent regions of unstable q_a. The curves correspond to $n = 1$.

The Case $J_a = 0$. Equation (9.189) suggests that when the current density on the boundary vanishes ($J_a = 0$) the bands of instability shrink to zero and the system becomes stable. This is not quite the situation, because the $1/m$ corrections to $b(r)$ must now be taken into account. In fact, an examination of Eq. (9.186) for $J_a = 0$ indicates that these "corrections" must be finite for instabilities to exist. Specifically,

$$\frac{\delta W_2}{W_0} = \frac{m\xi_a^2}{q_a^2}\left(1 - \frac{nq_a}{m}\right)^2\left(1 - \frac{2b_0'}{b_0} - \frac{1}{m}\right) \tag{9.190}$$

As $m \to \infty$ the condition for instability is $2b_0'/b_0 > 1$. The "small corrections" to b must somehow change $2b_0'/b_0$ a finite amount from -1 to $+1$. The structure of Eq. (9.182) is such that the correction term can only make a finite contribution to b when z_0, rather than being of order unity is assumed much smaller: $z_0 \ll 1$. In this case the solution for b is strongly modified near $z = 0$. The condition $z_0 \ll 1$ corresponds to the requirement that the resonant surface be extremely close to the plasma surface: $m(1 - nq_a/m) \ll 1$.

With this mathematical insight, one can solve for b as follows. New variables y and $U(y)$ defined by

$$\begin{aligned} z + z_0 &= y \\ b &= ye^{-y/2}U \end{aligned} \tag{9.191}$$

are introduced into Eq. (9.182), yielding

$$y\frac{d^2U}{dy^2} + (2 - y)\frac{dU}{dy} - \left(1 - \frac{\lambda}{m}\right)U = 0 \tag{9.192}$$

The solution for U can be expressed in terms of confluent hypergeometric functions. If one requires that $b \to 0$ as $z \to \infty$ then b can be written as

$$b = ye^{-y/2}\Psi(1 - \lambda/m, 2; y) \tag{9.193}$$

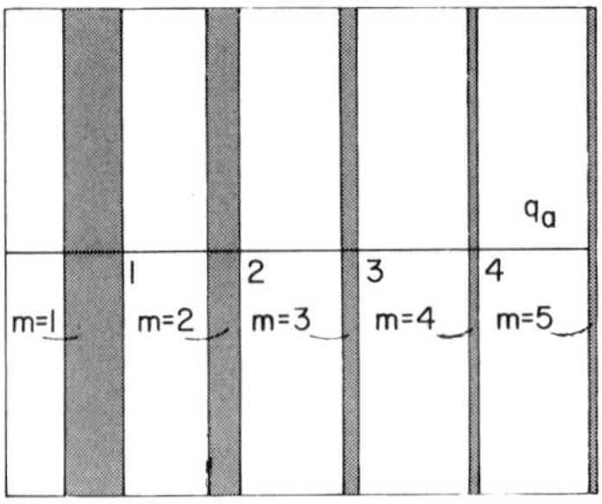

Figure 9.34. Stability diagram showing exponentially narrowing bands of instability for the case where $J_a = 0$. The shaded regions correspond to instability.

where Ψ is the branch satisfying $\Psi \to y^{-1+\lambda/m}$ for large y and $\Psi \to \ln y$ for small y (Erdélyi, 1953).

To determine the stability boundaries one needs to evaluate b'/b at $y = z_0$. Since the regime of interest corresponds to $z_0 \ll 1$ one can make use of the small argument expansion for Ψ which gives $(\Psi'/\Psi)_{z_0} \approx -1/z_0 - (\lambda/m)\ln z_0 + \cdots$. The condition for instability thus reduces to

$$1 + \lambda \ln z_0 > 0 \tag{9.194}$$

or, in terms of the standard notation,

$$\frac{1}{n}\left[m - \exp\left(\frac{2m\langle J\rangle}{aJ_a'}\right)\right] < q_a < \frac{m}{n} \tag{9.195}$$

After this rather complicated analysis one sees that instability can only exist over exponentially narrow bands of q_a as illustrated in Fig. 9.34. The conclusion is that when the current vanishes at $r = a$ the range of unstable q_a values is greatly reduced.

Numerical Results. A further implication of Eq. (9.195) is that if J_a' as well as J_a vanishes, the high-m external kinks can be completely stabilized. Exactly how fast the current gradient must vanish is difficult to calculate, particularly for the low-m modes where the eigenfunctions are not highly localized.

This question has been addressed numerically by Wesson (1978) who studied the family of profiles $J = J_0(1 - r^2/a^2)^\nu$, corresponding to $q = q_a u[1 - (1-u)^{\nu+1}]^{-1}$ with $\nu + 1 = q_a/q_0$ and $u = r^2/a^2$. Note that the parameter q_a/q_0 is a measure of the steepness of the current profile. The results of Wesson's calculations are illustrated in Fig. 9.35 as marginal stability curves of q_a/q_0 vs. q_a. Also superimposed is the stability boundary for ideal internal kink modes, $q_0 \geq 1$.

Observe that the unshaded region is stable to both internal and external ideal MHD modes. Most existing tokamaks operate in this

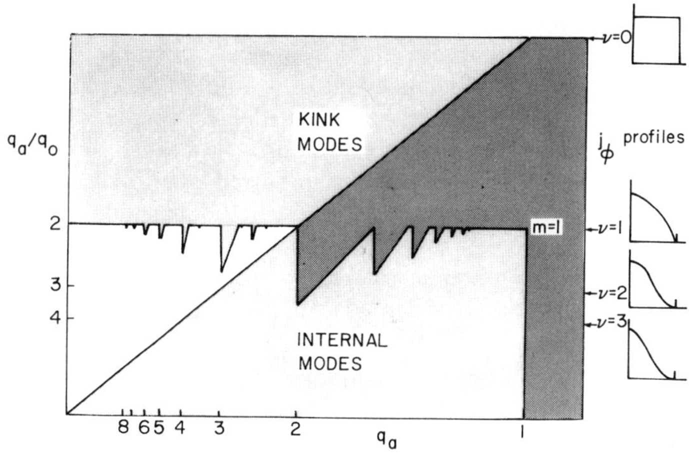

Figure 9.35. Marginal stability diagram for external kinks in a straight tokamak for a current profile $J(r) = J_0(1 - r^2/a^2)^\nu$. The stable region is unshaded. [From Wesson (1978).]

regime. With regard to external kinks, Wesson finds that for $\nu > 2.5$ or $q_a/q_0 \gtrsim 3.5$ all $m \geq 2$ modes are stable for any q_a. Windows of stability in q_a also exist for $1 < \nu < 2.5$. Complete external-mode stability is achieved by the additional requirement $q_a > 1$ to stabilize the $m = 1$ mode. In fact if it were not for the internal-mode criterion, tokamaks could operate at significantly higher currents (i.e., lower q_a) than they normally do; for instance, $q_a = 1$, $q_0 = 1/4$ satisfies all the external-kink conditions.

To summarize, it has been shown that $m \geq 2$ external-kink modes are sensitive to the features of the current density profile near the edge of the plasma. A combination of rapidly vanishing current gradient and a limit on the total current is sufficient to completely stabilize all high- and low-m number external modes, respectively. The dual requirements of $q_0 > 1$ for internal-mode stability and peaked current profiles for high-m external-kink stability can only be simultaneously satisfied if $q_a \gtrsim 3$. It is for this reason that $m \geq 2$ kinks impose stronger constraints on q_a than the $m = 1$ Kruskal–Shafranov mode.

9.6.6. SUMMARY

The one-dimensional "straight" tokamak provides a useful if sometimes unreliable guide to the stability of the tokamak configuration. The model predicts that the most dangerous ideal MHD instabilities are current-driven, external-kink modes. The $m = 1$ mode becomes unstable if the total toroidal current exceeds the Kruskal–Shafranov limit; that is, $q_a < 1$ leads to instability independent of profile. High-m external kinks are driven unstable if a significant current density gradient exists at the edge of the plasma. These modes require a peaked current profile, $q_a/q_0 \sim 3$ [i.e., $J'_a \to 0$] in order to be stabilized.

Of the internal modes only $m = 1$ can be unstable in a straight tokamak. This is a weaker instability than the external kink but can be driven by either the parallel current or the pressure gradient. The condition to stabilize the $m = 1$ mode requires $q_0 > 1$. In practice, most experimental tokamaks exhibit a form of relaxation oscillation near the magnetic axis as a result of the alternating competition between the $m = 1$ internal mode (and its nonlinear resistive evolution) and a thermal instability resulting from the temperature dependence of the resistivity which causes the current to peak. The net effect of these "sawtooth" oscillations is to maintain q_0 at unity during most of the duration of the experiment.

In general the straight tokamak provides an unreliable description of the influence of β on ideal MHD stability. Specifically, most of the stability results are only weakly dependent on β. Consequently, even in the regime of the high-β tokamak, no β limits arise within the context of the model. This is perhaps the main deficiency of the straight tokamak since there are important β limits when one considers the full toroidal geometry. These are discussed in the next chapter.

In summary,the combination of peaked profile $q_a/q_0 \gtrsim 3$ and limited current density on axis $q_0 \geq 1$, is sufficient to stabilize the straight tokamak to ideal MHD instabilities. Even so, it should be noted that there are important resistive MHD modes (i.e., tearing modes) which can also occur. These modes further decrease the stable region of parameter space illustrated in Fig. 9.35. Most tokamaks in fact actually operate in the regime that is resistively unstable. Such operation is possible because resistive modes are inherently much weaker instabilities than those of ideal MHD. Their nonlinear evolution often produces only relatively minor turbulence while permitting operation at much higher currents than allowed by "completely stable" operation. This is an important topic but lies beyond the scope of ideal MHD.

9.7. SUMMARY

In Chapter 9 the Energy Principle has been used to investigate the stability of one-dimensional, cylindrically symmetric equilibria. A summary of these results is given below.

The θ Pinch. The straight θ pinch is shown to be stable to all (m, k) MHD modes for arbitrary values of β. δW is either positive or approaches zero when $k \to 0$. This optimistic result is a consequence of the fact that a θ pinch has neither parallel currents nor field line curvature to drive instabilities. Unfortunately, the θ pinch cannot be bent into a torus unless additional fields are added to provide toroidal force balance.

The Z Pinch. The straight Z pinch is unstable to two dangerous MHD instabilities, each driven by the pressure gradient and unfavorable field line curvature. The first instability is the short wavelength $m = 1$ mode while the second corresponds to the $m = 0$ sausage mode. Either of these instabilities can lead to a catastrophic termination of the discharge. This poor stability behavior more than counterbalances the ease with which a Z pinch can be bent into a toroidal equilibrium.

The General Screw Pinch. The Energy Principle for the general screw pinch reduces to a one-dimensional quadratic integral in one unknown. An analysis of this energy integral leads to several general results. (1) Suydam's criterion provides a simple necessary condition for stability against localized interchange modes. The criterion depends only upon the equilibrium fields and quantifies the balance between shear and unfavorable curvature. (2) Newcomb's analysis demonstrates how the marginal stability differential equation associated with the variation of δW can be used to determine the necessary and sufficient conditions for stability of arbitrary screw pinch configurations. (3) An oscillation theorem has been derived for the general screw pinch which shows that the most unstable (i.e., fastest growing) MHD perturbation always has the structure of a macroscopic zero node mode. The oscillation theorem stresses the importance of satisfying Suydam's criterion. (4) A perfectly conducting wall, close to the plasma, can stabilize external kink modes. If, however, the wall is resistive, the stability boundary reverts back to the situation where the wall is at infinity. The growth rates are much slower, comparable to the resistive diffusion time of the wall.

The Reversed-Field Pinch. The ideal MHD stability properties of the reversed-field pinch are accurately described by the one-dimensional cylindrical model. A combination of flat or hollow pressure profile and a reversal in B_z producing high shear are necessary to suppress internal pressure-driven instabilities. Field reversal and a perfectly conducting wall near the plasma core are required to stabilize internal current-driven modes. Numerical studies incorporating these features lead to the optimistic result that internally stable RFP profiles can be found with β values as high as 30%–40%. The perfectly conducting wall on the plasma surface is crucial to suppress external kinks. The somewhat unusual profiles characterizing the RFP arise naturally from force free ohmic discharges with large currents, although such equilibria can never sustain a field reversal in steady state. A theory predicting the relaxation of ohmic states to field reversed states has been given by Taylor. This theory, based on the conservation of total helicity, implies that the relaxed states require a background level of turbulence to maintain the field reversal. The turbulence also leads to a degradation in plasma confinement. The seriousness of the anomalous transport with respect to the long-term objectives of the RFP is not well understood at the present time.

The Straight Tokamak. The straight tokamak model provides reliable information about current-driven kink modes. However, its predictions concerning pressure-driven modes are overly optimistic. Regarding internal current-driven modes, only the $m = 1$ internal kink can be driven unstable. This mode and its nonlinear resistive counterpart are thought to be responsible for sawtooth oscillations in tokamaks. These oscillations effectively clamp the value of the safety factor on axis to $q_0 = 1$. The $m = 1$ external kink sets a limit on the total current that can flow in the plasma. The stability condition is $q_a > 1$ and is known as the Kruskal–Shafranov limit. A more severe limit on q_a arises from $m \geq 2$ external kinks. These modes can be stabilized by sufficiently peaked current density profiles which cause J and J' to vanish at the edge of the plasma. Typically, $q_a/q_0 \gtrsim 3$ to stabilize $m \geq 2$ external kinks. Thus, the combination $q_0 = 1$ and peaked current profiles corresponding to $q_a \approx 3$ are sufficient to stabilize the straight tokamak. There are no significant pressure-driven modes in either the ohmically heated or high-β tokamak orderings in the straight tokamak.

REFERENCES

Bodin, H. A. B., and A. A. Newton (1980). *Nucl. Fusion* **20,** 1255.

Erdelyi, A. (1953). *Higher Transcendental Functions,* Bateman Manuscript Project, McGraw-Hill, N.Y., Vol. II.

Goedbloed, J. P. (1979). *Lecture Notes on Ideal Magnetohydrodynamics,* Fom-Instituut voor Plasmafysica, Nieuwegein, Netherlands.

Goedbloed, J. P., and P. H. Sakanaka (1974). *Phys. Fluids* **17,** 908.

Hain, K., and R. Lust (1958). *Z. Naturforsch. Teil A* **13,** 936.

Kadomtsev, B. B. (1966). In *Reviews of Plasma Physics,* edited by M. A. Leontovich, Consultants Bureau, New York, Vol. II.

Kruskal, M. D., and M. Schwarzschild (1954). *Proc. R. Soc. London, Ser. A* **223,** 348.
Laval, G., and R. Pellat (1973). In *Controlled Fusion and Plasma Physics,* Proceedings of the Sixth European Conference, Moscow, Vol. II, p. 640.
Laval, G., R. Pellat, and J. L. Soule (1974). *Phys. Fluids* **17,** 835.
Newcomb, W. A. (1960). *Ann. Phys. (NY)* **10,** 232.
Reiman, A. (1980). *Phys. Fluids* **23,** 230.
Robinson, D. C. (1971). *Plasma Phys.* **13,** 439.
Rosenbluth, M. N., R. Y. Dagazian, and P. H. Rutherford (1973). *Phys. Fluids* **11,** 1984.
Sakanaka, P. H., and J. P. Goedbloed (1974). *Phys. Fluids* **17,** 918.
Shafranov, V. D. (1956). *At. Energy* **5,** 38.
Shafranov, V. D. (1970). *Sov. Phys.-Tech.* **15,** 175.
Suydam, B. R. (1958). In *Proceedings of the Second United Nations International Conference on the Peaceful Uses of Atomic Energy,* United Nations, Geneva, Vol. 31, p. 157.
Taylor, J. B. (1974). *Phys. Rev. Lett.* **33,** 139.
Taylor, J. B.. (1975). In *Plasma Physics and Controlled Nuclear Fusion Research,* IAEA, Vienna, Vol. I, p. 161.
Taylor, J. B. (1976). In *Pulsed High Beta Plasmas,* edited by D. E. Evans, Pergamon, Oxford, p. 59.
Taylor, J. B. (1985). *Relaxation and Magnetic Reconnection in Plasmas,* Culham Report CLM-P765.
Wesson, J. A. (1978). *Nucl. Fusion* **18,** 87.

GENERAL REFERENCES

General Screw Pinch Stability Theory

Goedbloed, J. P. (1979). *Lecture Notes on Ideal Magnetohydrodynamics,* Fom-Instituut voor Plasmafysica, Nieuwegein, Netherlands.

The Reversed Field Pinch

Bodin, H. A. B., and A. A. Newton (1980). *Nucl. Fusion* **20,** 1255.
Robinson, D. C. (1971). *Plasma Phys.* **13,** 439.
Taylor, J. B. (1985). *Relaxation and Magnetic Reconnection in Plasmas,* Culham Report CLM-P765.

Tokamaks

Bateman, G. (1978). *MHD Instabilities* MIT, Cambridge, Massachusetts.
Furth, H. P. (1975). *Nucl. Fusion* **15,** 133.
Furth, H. P. (1985). *Phys. Fluids* **28,** 1595.
Mukhovatov, V. S., and V. D. Shafranov (1971). *Nucl. Fusion* **11,** 605.
Wesson, J. A. (1978). *Nucl. Fusion* **18,** 87.
White, R. B. (1986). *Rev. Mod. Phys.* **58,** 183.

PROBLEMS

9.1. Consider the marginal stability equation for the general screw pinch, given by

$$\mathbf{F}(\boldsymbol{\xi}) = 0$$

By direct reduction of this equation (i.e., do not use a variational minimization procedure) derive the one-dimensional radial equation for the eigenfunction $\xi = \xi_r(r)$. Show that this equation can be written as

$$(f\xi')' - g\xi = 0$$

where $f(r)$ and $g(r)$ are given by Eq. (9.36).

9.2. One term in δW that is always stabilizing for an arbitrary cylindrical screw pinch is given by

$$\delta W_a = \int f\xi'^2\, dr$$

where $f = r^3(kB_z + mB_\theta/r)^2/(k^2r^2 + m^2)$. Show that for an $m = 1$ mode, $\delta W_a \to 0$ as $\Delta \to 0$ for the trial function illustrated below. Assume the pinch represents an equivalent torus of length $2\pi R_0$, implying that $k = -n/R_0$. Note, that for $\delta W_a \to 0$ there must be a surface in the plasma where $nq(r_s) = 1$.

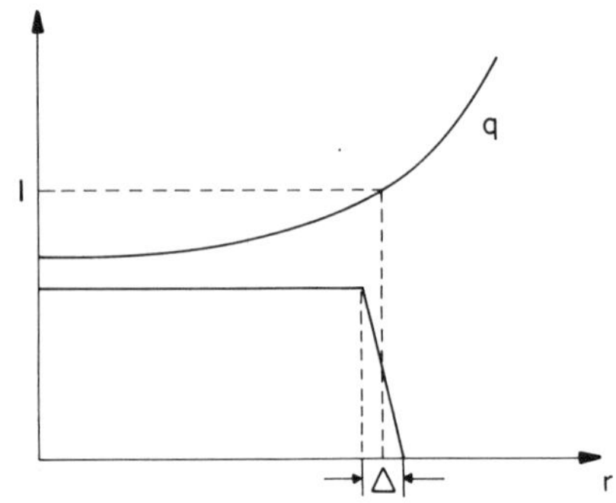

9.3. Consider a slab model of a plasma supported in equilibrium against gravity by a uniform magnetic field $\mathbf{B} = B\mathbf{e}_z$ as shown in the accompanying figure.

a. Calculate the one-dimensional radial pressure-balance relation describing the equilibrium.

b. Derive the one-dimensional eigenvalue equation describing linear stability. Assume the perturbations vary as $\xi = \xi(x)\exp(-i\omega t + ik_y y + ik_z z)$. For simplicity assume incompressible displacements.

c. Form an Energy Principle from the eigenvalue equation and show that $k_z^2 \to 0$ is the most unstable mode.

d. Assume that $\rho(x) = \rho_0 \exp(-x/a)$ and show that

$$\omega^2 = -g/a$$

This is the Raleigh–Taylor instability.

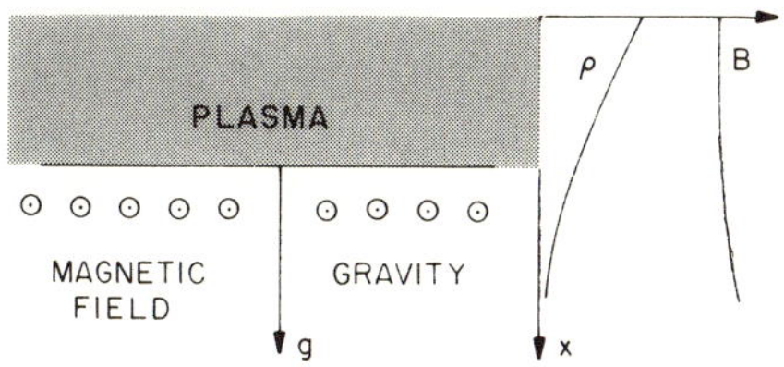

9.4. Consider a slab model of a plasma with a constant external gravitational force in the context of the collisionless MHD model.

$$\rho \frac{d\mathbf{v}_\perp}{dt} = \mathbf{J} \times \mathbf{B} - \nabla_\perp p + \rho g \mathbf{e}_x$$

Assume a one-dimensional equilibrium with $p(x)$, $B_y(x)$, $B_z(x)$, g = const, each nonzero.

a. Show that the Energy Principle for this system reduces to

$$\frac{\delta W}{L_y L_z} = \frac{1}{2}\int dx \left[\frac{F^2}{k^2}|\xi'|^2 + (F^2 + g\rho')|\xi|^2\right]$$

where $\mathbf{k} = k_y \mathbf{e}_y + k_z \mathbf{e}_z$, $F(x) = k_y B_y + k_z B_z$ and $\xi(x) = \xi_x$.

b. Derive the equivalent Suydam's criterion for this system. (The final criterion should be a function only of equilibrium quantities.)

c. By comparing the slab and cylindrical versions of Suydam's criterion, show how gravity can be used to "mock up" field line curvature.

9.5. In Section 9.5 it was shown that an RFP is always unstable to external modes if the conducting wall is moved to infinity. The purpose of this problem is to estimate how close the conducting wall must be moved in order to obtain stability against all external kinks. Consider the surface current model of an RFP illustrated below. Note that all currents flow only on the plasma surface.

a. Write the equilibrium pressure balance relation across the plasma surface: $[\![p + B^2/2\mu_0]\!] = 0$.

b. Starting with Eq. (8.83) evaluate δW_F, δW_S, and δW_V. Show that

$$\frac{\delta W}{2\pi R_0} = \frac{\pi B_\theta^2(a)\xi_a^2}{\mu_0}\left[(1-\beta)\frac{kaI_a}{I_a'} - 1 - \frac{m^2 K_a}{kaK_a'}\Lambda\right]$$

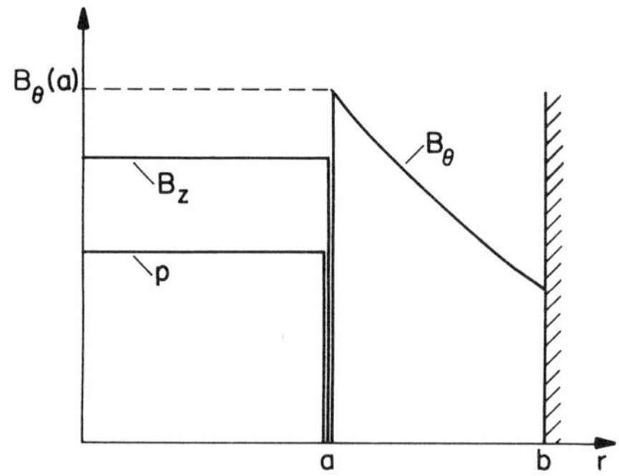

where $\beta = 2\mu_0 p / B_\theta^2(a)$ and

$$\Lambda = \frac{1 - I_a K_b' / K_a I_b'}{1 - I_a' K_b' / I_b' K_a'}$$

Also, $I_x \equiv I_m(kx)$, $K_x \equiv K_m(kx)$ are modified Bessel functions. The notation K_x', I_x' denotes differentiation with respect to the argument [e.g., $I_x' = dI_m(kx)/d(kx)$]. Hint: When evaluating δW_F treat $\mathbf{B}_1$ rather than ξ as the quantity to vary. Check that the ξ that results from minimizing $\mathbf{B}_1$ is well behaved.

c. Show that the most unstable mode for $m = 0$ corresponds to $k \to 0$ and requires $\beta < 1/2$ for stability.
d. Simplify Λ by assuming $kb \ll 1$. (This is only a marginally acceptable approximation, made for mathematical simplicity.)
e. For $m = 1$ approximate

$$\frac{zI_1(z)}{I_1'(z)} \approx \frac{z^2}{c_1 + c_2 z}$$

$$-\frac{K_1(z)}{zK_1'(z)} \approx \frac{1}{b_1 + b_2 z}$$

Determine c_1, c_2, b_1, b_2 so that each function assumes the correct limiting values as $z \to 0$ and $z \to \infty$.

f. Calculate the most unstable wave number and the corresponding farthest wall position that leads to stability against $m = 1$ external kinks for (1) $\beta = 0$ and (2) $\beta = 1/2$.

9.6. Show that δW for a general sharp-boundary screw pinch is given by

$$\frac{\delta W}{2\pi R_0} = \frac{\pi \xi_a^2}{\mu_0}\left[(B_0^2 + B_\theta^2 - 2\mu_0 p)\frac{kaI_a}{I_a'} - B_\theta^2 - \Lambda(kaB_0 + mB_\theta)^2 \frac{K_a}{kaK_a'}\right]$$

where B_0 is the external B_z field, $B_\theta = B_\theta(a)$, and Λ, K_a, I_a are defined in Problem 9.5.

9.7. Consider an arbitrary diffuse screw pinch. Prove that when $J_z(r)$ reverses sign at some value $r = r_0$, then Suydam's criterion is automatically satisfied for $r > r_0$ as long as B_θ itself does not reverse. Hint: Assume $rJ_z/B_\theta < 0$ for $r_0 < r < a$ and eliminate p' from Suydam's criterion using the equilibrium pressure balance relation.

9.8. Starting from the ideal MHD equations prove that

$$\frac{dK}{dt} = -2\int_V \mathbf{E}\cdot\mathbf{B}\,d\mathbf{r} + \int_S \mathbf{n}\cdot\mathbf{A}\times(\mathbf{E}+\mathbf{v}\times\mathbf{B})\,dS = 0$$

where the helicity K is given by

$$K = \int_V \mathbf{A}\cdot\mathbf{B}\,d\mathbf{r}$$

and the integration volume corresponds to the volume enclosed by a flux surface.

9.9. Estimate the growth rate for the $m = 1$, $n = 1$ internal kink mode in a low-β, large aspect ratio tokamak as follows. Start with the variational formulation of MHD given by

$$\omega^2 = \frac{\delta W}{K}$$

Use the ohmically heated tokamak expansion and write $\delta W = \delta W_2 + \delta W_4 + \cdots$. Substitute the trial function illustrated below and treat Δ as a variational parameter. Note that for the $m = 1$, $n = 1$ mode, $\delta W_2 \to 0$ as $\Delta \to 0$. Assume Δ is ordered so that $\delta W_2 \sim \delta W_4$ as $\Delta \to 0$ and for simplicity treat $\rho(r) \approx \rho_0 = \text{const}$ out to the $q = 1$ surface. Show that for the minimizing Δ

$$\omega^2 \approx -\frac{3V_a^2}{R_0^2}\left[\frac{W_4^2}{r_s^2 q'^2(r_s)}\right]$$

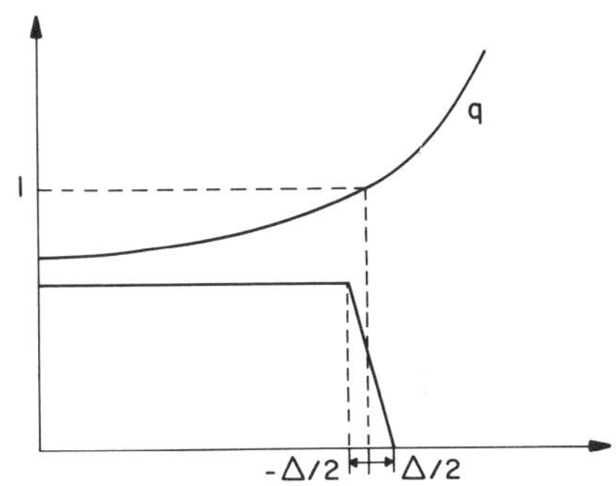

where $V_a^2 = B_0^2/\mu_0\rho_0$, r_s is the radius of the $q = 1$ surface and

$$W_4 = -\frac{1}{r_s^2}\int_0^{r_s} r\,dr\left[r\beta' + \frac{r^2}{R_0^2}\left(1 - \frac{1}{q}\right)\left(3 + \frac{1}{q}\right)\right]$$

9.10. The Kruskal–Shafranov condition for an $m = 1$, $n = 1$ external kink instability in a low-β large aspect ratio tokamak is given by

$$q_a \leq 1$$

In Section 9.6 it was shown that this condition is independent of the current profile. Show here that the condition is also independent of the conducting wall position as long as the wall is not on the plasma.

9.11. (Due to R. R. Parker, 1985). This problem examines the stability of a large aspect ratio, low-β tokamak against external kink modes. Recall that in the uniform current density model, instability exists for values of q_a in the range $(m - 1)/n < q_a < m/n$ for a given (m, n) mode with the wall at infinity. To simplify the analysis, attention is focused on the $m = 4$, $n = 1$ mode. The q profile is assumed to be

$$q(r) = \frac{4}{1 + 6J_1(\alpha r)/\alpha r}$$

where J_1 is a Bessel function and α is a parameter.

a. Sketch the $q(r)$ and $J(r)$ profiles for several values of αa in the range $0 \leq \alpha a \leq \alpha_0 a$. Here, $\alpha_0 a$ is the value of α for which $J(a) = 0$. Calculate the value of $\alpha_0 a$ and the corresponding value of q_a.
b. Find the perturbation which minimizes the plasma contribution to δW [given by Eq. (9.175)]. Hint: Make the substitution $b = r^{3/2}\Delta\xi$ where $\Delta(r) = n/m - 1/q(r)$.
c. Using the solution for ξ found in (b), substitute into δW as given by Eq. (9.176). Note that the range of unstable q_a values can be written as $q_1 \leq q_a \leq 4$ where $q_1 = q_1(\alpha a)$. Sketch q_1 for $0 < \alpha a < \alpha_0 a$. Show that as $\alpha a \to \alpha_0 a$ (corresponding to zero current density at the plasma edge) the range of unstable wave numbers shrinks to zero for the $m = 4$, $n = 1$ mode.

9.12. This problem demonstrates the powerful stabilizing effect that arises for external kink modes when the vacuum region is replaced by an ideal force free plasma. Consider external kink modes in a low-β, large aspect ratio ohmically heated tokamak.

a. Show that for the external kink mode, the general screw pinch δW [Eq. (9.44)] reduces to

$$\frac{\delta W}{W_0} = \int_0^a r\,dr\left(\frac{n}{m} - \frac{1}{q}\right)^2 [r^2\xi'^2 + (m^2 - 1)\xi^2] + BT$$

where $W_0 = 2\pi^2 B_0^2/\mu_0 R_0$,

$$BT = a^2\xi_a^2\left(\frac{n}{m} - \frac{1}{q_a}\right)\left[\frac{n}{m} + \frac{1}{q_a} + m\left(\frac{n}{m} - \frac{1}{q_a}\right)\Lambda\right]$$

and

$$\Lambda \approx \frac{1 + (a/b)^{2m}}{1 - (a/b)^{2m}}$$

For simplicity assume $m > 0$ and $nb/R \ll 1$ in the expansion of Λ.

b. Assume now that the vacuum region is replaced by an ideal force free plasma. In order for ξ to remain bounded, B_{1r} must equal zero at the resonant surface in the exterior region, rather than at the location of the conducting wall; that is, since $\xi = -iB_{1r}/F$, then $B_{1r}(r = r_s) = 0$ at the radius where $F(r_s) = 0$. This implies that the value of a/b appearing in Λ should correspond to a/r_s. Show that

$$\frac{a^2}{b^2} \to \frac{a^2}{r_s^2} = \frac{nq_a}{m}$$

c. Using this relation show that the boundary term is always positive. All external kinks have been stabilized for arbitrary current profiles in the plasma.

10

Stability

Multidimensional Configurations

10.1. INTRODUCTION

Chapter 10 is concerned with the MHD stability of multidimensional configurations with particular application to tokamaks, stellarators, and the Elmo bumpy torus. As might be expected, the complexities associated with multidimensionality make analysis quite difficult. Hence, stability results are often obtained numerically or through the use of simple models.

There is, however, one important general aspect of multidimensionality that has received considerable analytic study, and that is the issue of ballooning mode stability. These modes arise because of the existence of alternating regions of favorable and unfavorable field line curvature. Ballooning mode perturbations are concentrated in the unfavorable curvature regions and play a crucial role as they set upper limits on the maximum achievable value of β.

The first part of Chapter 10 discusses the general reduction of the Energy Principle to a form particularly convenient for analyzing ballooning modes. The derivation is valid for arbitrary three-dimensional geometry but actually comprises only a partial formulation of the problem. The full formulation is completed after specific geometries are introduced, and as will be shown, this leads to important but subtle effects when the equilibria of interest contain shear.

The second major topic of interest is a discussion of the relationship between "average curvature" and "magnetic well," alluded to earlier in the text. To make this comparison an analysis of the general ballooning mode formulation is performed for the special class of magnetic geometries possessing closed-line symmetry.

The third major topic investigated in Chapter 10 is the stability of the straight Elmo bumpy torus configuration. The analysis begins with a short discussion of the "rigid ring" assumption and its influence on MHD stability. This is followed by a calculation of the interchange and ballooning mode stability criteria, which set the most severe β limits. These results are combined to demonstrate how one can create a desirable EBT configuration encompassing both equilibrium and stability requirements.

The next major topic discussed is the stability of tokamaks. The analysis shows that toroidicity produces qualitative changes in the stability of internal pressure-driven modes. Specifically, the ballooning mode equation leads to the derivation of the Mercier criterion for interchange stability and the evaluation of ballooning mode β limits. Also demonstrated is the enticing prospect of high β by operation in the so-called "second region of stability." A discussion is then presented of the effects of toroidicity on low-n internal and external kink modes. When toroidicity is included, external kinks develop a strong ballooning mode component to the perturbation, setting limits on the maximum achievable β. The analytic results are compared with rather extensive numerical results, which include noncircularity as well as toroidicity. The numerical results are conveniently summarized in terms of the Troyon limit and the Sykes limit for stability. Also, a short discussion is presented of the $n = 0$ axisymmetric modes, an important class of instabilities that can arise in the practical operation of tokamak experiments.

The final topic investigated is the stability of the stellarator configuration. In current-free systems the most dangerous instabilities are high-n internal pressure-driven modes. A discussion is presented which demonstrates the importance of a favorable magnetic well. The conditions for achieving such a well are also discussed. As a specific example, a short description is given of the optimized design criteria used in the Oak Ridge ATF Torsatron. It is shown how both MHD equilibrium and stability studies have a major impact on the design of current generation fusion experiments.

10.2. GENERAL REDUCTION OF δW FOR BALLOONING MODES

Significant simplifications occur in the stability analysis of multidimensional MHD equilibria if one focuses attention on internal pressure-driven modes. The most dangerous modes, in terms of setting β limits, are usually characterized by a highly localized $k_\perp \to \infty$ perturbation. The simplifications that arise are a consequence of exploiting the localized nature of the instabilities. In essence, δW is reduced from its original three-dimensional form involving the three components of $\boldsymbol{\xi}$ into a more tractable one-dimensional form involving only the normal component of $\boldsymbol{\xi}$. Stability can then be tested separately, on each closed line or flux surface, one at a time. Thus, the original multidimensional problem is reduced to a series of independent one-dimensional problems representing an enormous savings in effort.

The derivation presented here is valid for arbitrary three-dimensional geometries. However, the formulation is not complete until a specific geometry is introduced. As will be shown, this last step leads to important but subtle difficulties in systems with shear. These difficulties are ultimately resolved by means of additional analysis.

The derivation proceeds as follows. Consider first the "intuitive" form of δW_F [i.e., Eq. (8.90)], repeated here for convenience:

$$\delta W_F = \frac{1}{2}\int d\mathbf{r}\left[\frac{|\mathbf{Q}_\perp|^2}{\mu_0} + \frac{B^2}{\mu_0}|\nabla\cdot\boldsymbol{\xi}_\perp + 2\boldsymbol{\xi}_\perp\cdot\boldsymbol{\kappa}|^2 + \gamma p\,|\nabla\cdot\boldsymbol{\xi}|^2 \right. \qquad (10.1)$$
$$\left. - 2(\boldsymbol{\xi}_\perp\cdot\nabla p)(\boldsymbol{\xi}_\perp^*\cdot\boldsymbol{\kappa}) - J_\parallel(\boldsymbol{\xi}_\perp^*\times\mathbf{b})\cdot\mathbf{Q}_\perp\right]$$

In order to exploit the $\mathbf{k}_\perp \to \infty$ limit, an eikonal representation is used for $\boldsymbol{\xi}_\perp$:

$$\boldsymbol{\xi}_\perp = \boldsymbol{\eta}_\perp e^{iS} \qquad (10.2)$$

where $\mathbf{k}_\perp$ is defined as

$$\begin{aligned}\mathbf{k}_\perp &= \nabla S\\ \mathbf{B}\cdot\nabla S &= 0\end{aligned} \qquad (10.3)$$

The quantity $\boldsymbol{\eta}_\perp$ is assumed to vary "slowly" on the equilibrium length scale: $|a\nabla\boldsymbol{\eta}_\perp|/|\boldsymbol{\eta}_\perp| \sim 1$. In contrast, the assumption $\mathbf{k}_\perp \to \infty$ implies that the variation of S is rapid: $|a\nabla S| \gg 1$.

The derivation assumes that a physically acceptable function S can be found that satisfies $\mathbf{k}_\perp \to \infty$ and $\mathbf{B}\cdot\nabla S = 0$. The problem of finding such a function is temporarily deferred. It is resolved case by case as different configurations are investigated.

The mathematical motivation for introducing separate length scales can be understood by substituting Eq. (10.2) into Eq. (10.1), making no approximations. A key feature of the substitution is that $\mathbf{Q}_\perp$ does not contain any explicit gradients in S:

$$\begin{aligned}\mathbf{Q}_\perp &= e^{iS}[\nabla \times (\boldsymbol{\eta}_\perp \times \boldsymbol{B}) + i\nabla S \times (\boldsymbol{\eta}_\perp \times \mathbf{B})]_\perp \\ &= e^{iS}[\nabla \times (\boldsymbol{\eta}_\perp \times \mathbf{B}) + i(\mathbf{k}_\perp \cdot \mathbf{B})\boldsymbol{\eta}_\perp - i(\mathbf{k}_\perp \cdot \boldsymbol{\eta}_\perp)\mathbf{B}]_\perp \\ &= e^{iS}\nabla \times (\boldsymbol{\eta}_\perp \times \mathbf{B})_\perp \end{aligned} \tag{10.4}$$

The exact form of δW_F can be expressed as

$$\delta W_F = \frac{1}{2\mu_0}\int d\mathbf{r}[|\nabla \times (\boldsymbol{\eta}_\perp \times \mathbf{B})_\perp|^2 + B^2\,|i\mathbf{k}_\perp \cdot \boldsymbol{\eta}_\perp + \nabla \cdot \boldsymbol{\eta}_\perp + 2\boldsymbol{\kappa} \cdot \boldsymbol{\eta}_\perp|^2 \\ - 2\mu_0(\boldsymbol{\eta}_\perp \cdot \nabla p)(\boldsymbol{\eta}_\perp^* \cdot \boldsymbol{\kappa}) - \mu_0 J_\parallel(\boldsymbol{\eta}_\perp^* \times \mathbf{b}) \cdot \nabla \times (\boldsymbol{\eta}_\perp \times \mathbf{B})_\perp] \tag{10.5}$$

Note that the plasma compressibility term is not included in Eq. (10.5). The analysis presented in Section 8.9 indicates that in systems with shear, a $\xi_\parallel$ can be found that makes the stabilizing contribution of compressibility vanish. Strictly speaking, however, this contribution must be maintained for closed-line systems such as EBT. Nevertheless, the contribution is neglected for these configurations (1) for reasons of mathematical simplicity, (2) because of the unreliability of the energy equation, (3) because the specific application considered, EBT, is quantitatively but not qualitatively affected by compressibility, and (4) because by so doing one is led to a conservative estimate of δW_F.

An examination of Eq. (10.5) indicates that the only explicit appearance of S (i.e., $\mathbf{k}_\perp$) occurs in the magnetic compression term. One is now motivated to consider the limit $\mathbf{k}_\perp \to \infty$ since δW_F can be systematically minimized by expanding

$$\boldsymbol{\eta}_\perp = \boldsymbol{\eta}_{\perp 0} + \boldsymbol{\eta}_{\perp 1} + \cdots \tag{10.6}$$

with $|\boldsymbol{\eta}_{\perp 1}|/|\boldsymbol{\eta}_{\perp 0}| \sim 1/k_\perp a$.

The zeroth-order contribution to δW_F reduces to

$$\delta W_0 = \frac{1}{2\mu_0}\int d\mathbf{r}\, B^2\,|\mathbf{k}_\perp \cdot \boldsymbol{\eta}_{\perp 0}|^2 \tag{10.7}$$

Clearly the minimizing perturbation satisfies $\mathbf{k}_\perp \cdot \boldsymbol{\eta}_{\perp 0} = 0$, implying that $\boldsymbol{\eta}_{\perp 0}$ can be written as

$$\boldsymbol{\eta}_{\perp 0} = Y\mathbf{b} \times \mathbf{k}_\perp \tag{10.8}$$

Here, Y is a scalar quantity, varying on the "slow" equilibrium length scale.

The first nonvanishing contribution to δW_F occurs in second order. In this expression the only appearance of the quantity $\boldsymbol{\eta}_{\perp 1}$ is in the magnetic compression term,

$$\delta W_2(\text{comp}) = \frac{1}{2\mu_0} \int d\mathbf{r}\, B^2 \left| i\mathbf{k}_\perp \cdot \boldsymbol{\eta}_{\perp 1} + \nabla \cdot \boldsymbol{\eta}_{\perp 0} + 2\boldsymbol{\kappa} \cdot \boldsymbol{\eta}_{\perp 0} \right|^2 \quad (10.9)$$

Clearly δW_F is minimized by choosing $i\mathbf{k}_\perp \cdot \boldsymbol{\eta}_{\perp 1} = -\nabla \cdot \boldsymbol{\eta}_{\perp 0} - 2\boldsymbol{\kappa} \cdot \boldsymbol{\eta}_{\perp 0}$. Thus, the most unstable modes for $\mathbf{k}_\perp \to \infty$ do not involve any compression of the magnetic field.

In the evaluation of δW_2 it is useful to simplify the quantity $\nabla \times (\boldsymbol{\eta}_\perp \times \mathbf{B})_\perp$ as follows:

$$\begin{aligned} \nabla \times (\boldsymbol{\eta}_\perp \times \mathbf{B})_\perp &= \nabla \times [YB(\mathbf{b} \times \mathbf{k}_\perp) \times \mathbf{b}]_\perp \\ &= \nabla \times (YB\nabla S)_\perp \\ &= (\nabla X \times \mathbf{k}_\perp)_\perp \end{aligned} \quad (10.10)$$

where $X \equiv YB$. If one now writes $\nabla X = \nabla_\perp X + \mathbf{b}(\mathbf{b} \cdot \nabla X)$ then the perpendicular component of $\nabla \times (\boldsymbol{\eta}_\perp \times \mathbf{B})$ can be expressed as

$$\nabla \times (\boldsymbol{\eta}_\perp \times \mathbf{B})_\perp = (\mathbf{b} \cdot \nabla X)\mathbf{b} \times \mathbf{k}_\perp \quad (10.11)$$

Using this relation one can show that the kink contribution to δW_2 is given by

$$\begin{aligned} \delta W_2(\text{kink}) &= -\frac{1}{2} \int d\mathbf{r}\, J_\parallel (\boldsymbol{\eta}_\perp^* \times \mathbf{b}) \cdot \nabla \times (\boldsymbol{\eta}_\perp \times \mathbf{B})_\perp \\ &= -\frac{1}{2} \int d\mathbf{r} [(J_\parallel / B) X^* (\mathbf{b} \cdot \nabla X)][\mathbf{k}_\perp \cdot (\mathbf{b} \times \mathbf{k}_\perp)] \\ &= 0 \end{aligned} \quad (10.12)$$

Thus, in the limit of $\mathbf{k}_\perp \to \infty$, the kink term makes no contribution to stability.

The remaining contributions to δW_2 describe a competition between the stabilizing effects of line bending and the destabilizing effects of unfavorable curvature:

$$\delta W_2 = \frac{1}{2\mu_0} \int d\mathbf{r} \left[k_\perp^2 \left| \mathbf{b} \cdot \nabla X \right|^2 - \frac{2\mu_0}{B^2} (\mathbf{b} \times \mathbf{k}_\perp \cdot \nabla p)(\mathbf{b} \times \mathbf{k}_\perp \cdot \boldsymbol{\kappa}) \left| X \right|^2 \right] \quad (10.13)$$

This is the desired form of the potential energy. It is a function of one variable X, and is essentially one-dimensional as only a unidirectional derivative appears along $\mathbf{B}$. The one-dimensional nature of δW_2 becomes explicit when specific coordinate systems are introduced corresponding to the applications of interest.

10.3. THE RELATIONSHIP BETWEEN THE "MAGNETIC WELL" AND "AVERAGE CURVATURE"

10.3.1. Introduction

With the aid of the general ballooning mode equation given by Eq. (10.13) it now becomes possible to determine the precise relationship between the magnetic well and the average curvature of a field line. This is a useful relation to ascertain since both quantities are often used as figures of merit in the evaluation of different magnetic geometries. Although widely applied to systems with and without shear, the figures of merit actually derive their validity from an analysis of shearless, closed-line systems.

The analysis, which is valid for three-dimensional systems, requires several steps. First, the proper definition of the flux function in a closed-line system is derived and is shown to be closely related to the magnetic well. Second, a general set of closed-line flux coordinates is introduced which greatly simplifies the ballooning mode potential energy. Next, the simplified δW_2 is examined for stability against the special class of trial functions corresponding to interchange perturbations. The natural stability criterion that arises involves the average field-line curvature. Finally, by making use of the MHD equilibrium equations, this criterion is related to the magnetic well.

10.3.2. Flux Function in a Closed-Line System

The basic question addressed is as follows. In a system with ergodic field lines the constant pressure surfaces can be traced out by following magnetic field-line trajectories many transits around the torus. In an axisymmetric torus these surfaces are labeled by the poloidal flux ψ and $p = p(\psi)$. Consequently, contours of constant flux also correspond to surfaces of constant pressure. In a closed-line system, however, a magnetic line by definition does not trace out a surface. The question that is then posed is whether or not an equivalent "flux function" for determining the pressure contours exists for closed-line systems. The answer is that there is such a function and its specific form is obtained from the following derivation (Shafranov, 1966).

Consider two contours of constant pressure in a closed-line system as illustrated in Fig. 10.1. Upon integrating $\nabla \cdot \mathbf{J} = 0$ over the shaded volume one obtains

$$\int \nabla \cdot \mathbf{J} \, d\mathbf{r} = \int \mathbf{J} \cdot d\mathbf{S} = 0 \tag{10.14}$$

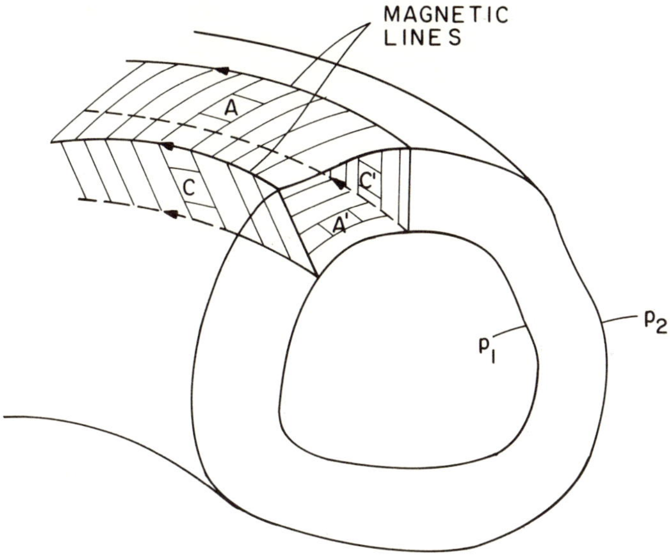

Figure 10.1. Constant pressure contours in a closed-line system.

On surfaces A and A', lying on the p = const contours the differential area element $d\mathbf{S}$ points in the ∇p direction. Since $\mathbf{J} \cdot \nabla p = 0$ in equilibrium the surface contributions on A and A' must vanish; that is, the current flows parallel to constant-pressure surfaces. Keeping in mind that the magnetic lines lie in constant p surfaces, the differential area element on C and C' can be written as

$$d\mathbf{S} = \frac{\mathbf{B} \times \nabla p}{B\,|\nabla p|^2}\,dp\,dl \tag{10.15}$$

Here $d\mathbf{l} = (\mathbf{B}/B)\,dl$ represents the arc length along the magnetic field and $d\mathbf{r} = (\nabla p/|\nabla p|^2)\,dp$ represents the "radial" line element expressed in terms of p. Using the fact that $\mathbf{J}_\perp = (\mathbf{B} \times \nabla p)/B^2$ it then follows that

$$\int_C \frac{dp\,dl}{B} = \int_{C'} \frac{dp\,dl}{B} \tag{10.16}$$

Equation (10.16) expresses the condition that the current I flowing into C' equals the current I flowing out of C. Consider the limit in which the two pressure surfaces are arbitrarily close to one another, $p_2 = p_1 + \delta p$, $\delta p/p_1 \to 0$. In this limit, the current I flowing out of C is given by

$$I = \int_C \frac{dp\,dl}{B} = \delta p \oint \frac{dl}{B} \tag{10.17}$$

where the line integral is taken along the magnetic line defined by the surface C as $\delta p \to 0$. Equation (10.16) implies that I will be the same for any other arbitrary surface C' bounded by p_1 and $p_1 + \delta p$. This can only be true if the quantity

$$U = \oint \frac{dl}{B} \tag{10.18}$$

is constant on every magnetic line lying on the constant pressure surface. $dU/d\psi$ will be shown to be proportional to the vacuum magnetic well.

Summarizing, in a closed-line system, U plays the role of the flux function. It is determined directly from the magnetic field, and surfaces of constant U correspond to surfaces of constant p so that $p = p(U)$.

10.3.3. Closed-Line Flux Coordinates

In order to further simplify the ballooning mode potential energy it is convenient to introduce a set of coordinates that exploits the closed-line symmetry. These are known as closed-line flux coordinates and are defined as follows.

The magnetic field can be written in the form

$$\mathbf{B} = \nabla\alpha \times \nabla\beta \tag{10.19}$$

where $\alpha(\mathbf{r})$, $\beta(\mathbf{r})$ are scalar functions of position. Equation (10.19) automatically guarantees that $\nabla \cdot \mathbf{B} = 0$. It is also straightforward to show

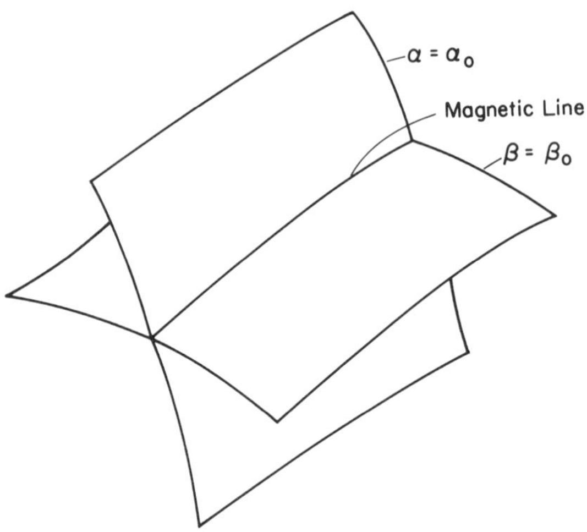

Figure 10.2. Magnetic line defined by the intersection of the two contours $\alpha = \alpha_0$, $\beta = \beta_0$.

that any divergence-free field can be expressed in the form of Eq. (10.19). See Problem 10.1. The magnetic lines lie in the contours of $\alpha = \text{const}$, $\beta = \text{const}$ since $\mathbf{B} \cdot \nabla\alpha = \mathbf{B} \cdot \nabla\beta = 0$. In fact the intersection of two contours $\alpha = \alpha_0$, $\beta = \beta_0$ defines the magnetic line. See Fig. 10.2. Thus, the functions α and β serve as the two transverse flux coordinates; if one specifies values $\alpha = \alpha_0$, $\beta = \beta_0$ the coordinate pair (α_0, β_0) define the transverse location of a given magnetic line at any point along its length.

In general α and β are not orthogonal, but this poses no great difficulty. Also, α and β are not unique. For instance, replacing $\beta \to \beta + g(\alpha)$ leaves $\mathbf{B}$ unchanged. In closed-line systems it is customary to specify α and β as follows. Without loss in generality one coordinate can be chosen as the magnetic flux ψ and represents a "radial-like" variable. The other, defined as χ, describes a poloidal "anglelike" variable. See Fig. 10.3. To show this consider a given $\alpha(\mathbf{r})$, $\beta(\mathbf{r})$ where the magnetic flux $\psi = \psi(\alpha, \beta)$. The transformation to (ψ, χ) coordinates is given by the simultaneous solution of

$$\psi = \psi(\alpha, \beta)$$

$$\chi = \frac{\partial}{\partial\psi} \int_0^\beta \alpha(\psi, \hat{\beta})\, d\hat{\beta} \tag{10.20}$$

Here $\alpha = \alpha(\psi, \beta)$ is the function obtained by inverting $\psi = \psi(\alpha, \beta)$. In terms of (ψ, χ) $\mathbf{B}$ can be written as

$$\mathbf{B} = \nabla\psi \times \nabla\chi \tag{10.21}$$

Hereafter, (ψ, χ) are the two transverse coordinates used in the reduction of the ballooning mode potential energy.

The third coordinate that needs to be defined is a "lengthlike" variable measuring distance along the magnetic line. This coordinate is denoted by ζ. For present purposes ζ is treated as a general variable; no further properties are required. The specialization of ζ to arc length l is carried out after the analysis is completed.

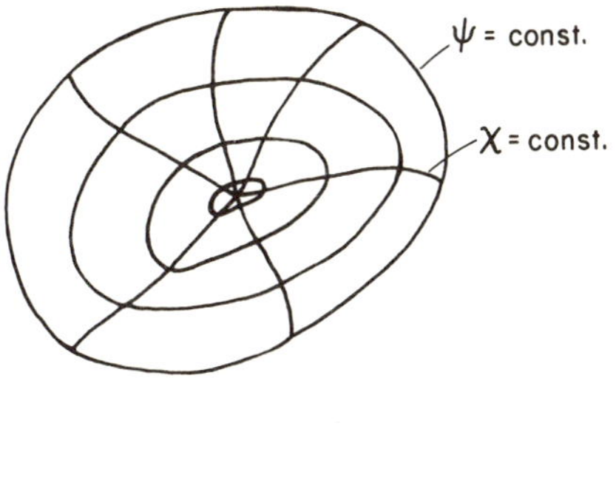

Figure 10.3. Transverse flux plot showing contours of constant ψ and χ.

The coordinates (ψ, χ, ζ) define the complete set of closed-line flux coordinates. Several useful relations required for the analysis can easily be derived and are given by

$$
\begin{aligned}
\mathbf{B}\cdot\nabla\psi &= \mathbf{B}\cdot\nabla\chi = 0 \\
\nabla &= \nabla\psi\frac{\partial}{\partial\psi} + \nabla\chi\frac{\partial}{\partial\chi} + \nabla\zeta\frac{\partial}{\partial\zeta} \\
\mathbf{B}\cdot\nabla &= \frac{1}{J}\frac{\partial}{\partial\zeta} \\
d\mathbf{r} &= J\,d\psi\,d\chi\,d\zeta
\end{aligned}
\tag{10.22}
$$

where J is the Jacobian of the transformation

$$J = \frac{1}{\nabla\psi \times \nabla\chi \cdot \nabla\zeta} = \frac{1}{\mathbf{B}\cdot\nabla\zeta} \tag{10.23}$$

With this information one can now proceed with the reduction of δW_2.

10.3.4. The Interchange Condition

The next step in the analysis is to substitute the flux coordinates into the expression for the ballooning mode potential energy given by Eq. (10.13). Consider first the evaluation of the eikonal function S. The requirement $\mathbf{B}\cdot\nabla S = 0$ implies that

$$S(\psi, \chi, \zeta) \rightarrow S(\psi, \chi) \tag{10.24}$$

In closed-line systems there is no difficulty in evaluating S. The wave vector $\mathbf{k}_\perp$ immediately follows:

$$\mathbf{k}_\perp = \nabla S = \frac{\partial S}{\partial\psi}\nabla\psi + \frac{\partial S}{\partial\chi}\nabla\chi \tag{10.25}$$

Similarly, since $\mathbf{b}\cdot\boldsymbol{\kappa} = 0$ it is convenient to expand the curvature vector as

$$\boldsymbol{\kappa} = \kappa_\psi\nabla\psi + \kappa_\chi\nabla\chi \tag{10.26}$$

where

$$
\begin{aligned}
\kappa_\psi &= -(\mathbf{b}\times\nabla\chi\cdot\boldsymbol{\kappa})/B \\
\kappa_\chi &= (\mathbf{b}\times\nabla\psi\cdot\boldsymbol{\kappa})/B
\end{aligned}
\tag{10.27}
$$

Substituting these results into the expression for δW_2 yields

$$dW_2 = \frac{1}{2\mu_0}\int d\psi\,d\chi\,W \tag{10.28}$$

$$W(\psi, \chi, \zeta) = \int J\,d\zeta\left[\frac{k_\perp^2}{J^2}\left(\frac{\partial X}{\partial\zeta}\right)^2 - 2\mu_0\frac{\partial S}{\partial\chi}\frac{dp}{d\psi}\left(\frac{\partial S}{\partial\chi}\kappa_\psi - \frac{\partial S}{\partial\psi}\kappa_\chi\right)X^2\right]$$

Observe that the derivatives appearing on X involve only ζ. Hence, ψ and χ enter W solely as parameters. Stability can thus be tested one magnetic line at a time (i.e., for fixed ψ and χ). The three-dimensional stability problem has been reduced to the solution of a sequence of one-dimensional problems, representing an enormous reduction in effort.

Attention is now focused on the stability of interchange perturbations. This type of disturbance corresponds to a trial function of the form $X(\psi, \chi, \zeta) = \hat{X}(\psi - \psi_0, \chi - \chi_0)$ where $\hat{X}$ is a localized function about $\psi = \psi_0$, $\chi = \chi_0$, the magnetic line under consideration. Since X is assumed independent of ζ the first term W vanishes; the interchange perturbation sets $|\mathbf{Q}_\perp|^2 = 0$, thereby eliminating the stabilizing effects of the line bending. For the interchange trial function W reduces to

$$W = -2\mu_0\hat{X}^2 \int J\, d\zeta \frac{\partial S}{\partial \chi}\frac{dp}{d\psi}\left(\frac{\partial S}{\partial \chi}\kappa_\psi - \frac{\partial S}{\partial \psi}\kappa_\chi\right) \tag{10.29}$$

This expression can be further simplified by noting that the parallel current relation given by Eq. (7.6) implies that

$$\frac{\partial}{\partial \zeta}\left(\frac{J_\parallel}{B}\right) = J\frac{dp}{d\psi}\kappa_\chi \tag{10.30}$$

Consequently, the second term in the integrand is an exact differential with respect to ζ. This term integrates to zero because of toroidal periodicity requirements. W now has the form

$$W = -2\mu_0\hat{X}^2\left(\frac{\partial S}{\partial \chi}\right)^2 \frac{dp}{d\psi}\int \kappa_\psi\, J\, d\zeta \tag{10.31}$$

Under the usual assumption $dp/d\psi < 0$, it follows that a necessary condition for stability in a closed-line system is

$$\int \kappa_\psi J\, d\zeta > 0 \tag{10.32}$$

That is, the average curvature as defined by Eq. (10.32) must be positive for interchange stability.

10.3.5. The Relation between Average Curvature and Magnetic Well

The relation between average curvature and magnetic well is now determined by introducing arc length as a coordinate in the interchange stability condition. The arc length l is defined by

$$\frac{d\mathbf{r}}{dl} = \mathbf{b} \tag{10.33}$$

This implies that

$$\mathbf{b} \cdot \nabla = \frac{d\mathbf{r}}{dl} \cdot \nabla = \frac{\partial}{\partial l} \tag{10.34}$$

where the $\partial/\partial l$ corresponds to fixed ψ and χ. Thus, setting $\zeta = l$ one finds that the Jacobian is given by

$$J = \frac{1}{\mathbf{B} \cdot \nabla l} = \frac{1}{B} \tag{10.35}$$

Combining these results leads to the usual form of the interchange stability criterion in terms of the average curvature

$$\langle \kappa_\psi \rangle > 0 \tag{10.36}$$

where for any quantity Q

$$\langle Q \rangle \equiv \oint \frac{Q\, dl}{B} \Big/ \oint \frac{dl}{B} \tag{10.37}$$

Equation (10.36) is valid for general three-dimensional closed-line systems with arbitrary β. Observe that only κ_ψ enters the stability criterion. The average curvature condition is quite useful in systems with one or two degrees of symmetry as κ_ψ is easily evaluated in these cases. For the general case $\kappa_\psi \equiv -(\mathbf{b} \times \nabla\chi \cdot \boldsymbol{\kappa})/B$ is somewhat cumbersome to evaluate because an explicit knowledge of $\nabla\chi$ is required.

Equation (10.36) can be cast in a form more convenient for evaluation in the limit $\beta \to 0$ by means of some algebraic manipulation. The simplification requires the introduction of a specially chosen potential function for the length coordinate ζ as an intermediate step before transforming to arc length. The resulting form of the interchange stability criterion is known as the vacuum magnetic well, and is derived below. A discussion of the effects of finite β follows.

Consider first the limit $\beta \to 0$. In this case $\mathbf{B}$ can be written as

$$\mathbf{B} = \nabla\psi \times \nabla\chi = \nabla\phi \tag{10.38}$$

For the vacuum limit $\mathbf{B}$ has an alternate representation in terms of a scalar potential ϕ. One now chooses $\zeta = \phi$ as the third coordinate. The Jacobian is given by

$$J = \frac{1}{\mathbf{B} \cdot \nabla\phi} = \frac{1}{B^2} \tag{10.39}$$

Next, in the low β limit $\boldsymbol{\kappa}$ can be written as

$$\boldsymbol{\kappa} = \frac{1}{B^2}\left[\nabla\frac{B^2}{2} - \mathbf{b}(\mathbf{B} \cdot \nabla B)\right] \tag{10.40}$$

Expanding

$$\nabla = \nabla\psi \frac{\partial}{\partial\psi} + \nabla\chi \frac{\partial}{\partial\chi} + \nabla\phi \frac{\partial}{\partial\phi} \tag{10.41}$$

it follows that

$$\kappa_\psi = \frac{1}{B}\frac{\partial B}{\partial\psi} \tag{10.42}$$

Substituting these results into Eq. (10.32) yields the stability condition

$$\int \frac{d\phi}{B^3}\frac{\partial B}{\partial\psi} = -\frac{1}{2}\frac{\partial}{\partial\psi}\int \frac{d\phi}{B^2} > 0 \tag{10.43}$$

One now introduces arclength in place of ϕ. The relation between Jacobians is given by

$$J\,d\zeta = \frac{d\zeta}{\mathbf{B}\cdot\nabla\zeta} = \frac{d\phi}{B^2} = \frac{dl}{B} \tag{10.44}$$

The stability condition reduces to

$$\frac{\partial}{\partial\psi}\oint \frac{dl}{B} < 0 \tag{10.45}$$

This is one form of the well-known condition for a vacuum magnetic well. Observe that the integral is in general a function of both ψ and χ. However, recall from the discussion in Section 10.3.2 that $U = \oint dl/B$ serves as the flux function in a closed-line system: $U = U(\psi)$. This implies that the vacuum magnetic well is not a function of χ.

There is one subtle point worth noting that leads to an alternate form of Eq. (10.45). The subtlety arises when comparing the integration limits in the ϕ and l coordinates. For any quantity Q it follows that

$$\int_0^{\phi_0} \frac{Q\,d\phi}{B^2} = \int_0^L \frac{Q\,dl}{B} \tag{10.46}$$

where ϕ_0 and L correspond to one complete circuit of the torus. The condition that $\mathbf{J}$ vanish as $\beta \to 0$ implies that ϕ_0 is independent of ψ; that is, along an arbitrary field line (ψ = const, χ = const)

$$\int_0^L \mathbf{B}\cdot d\mathbf{l} = \int_0^L \nabla\phi\cdot d\mathbf{l} = \int_0^{\phi_0} d\phi = \phi_0 \tag{10.47}$$

By choosing any closed path as shown in Fig. 10.4, the vanishing of $\mathbf{J}$

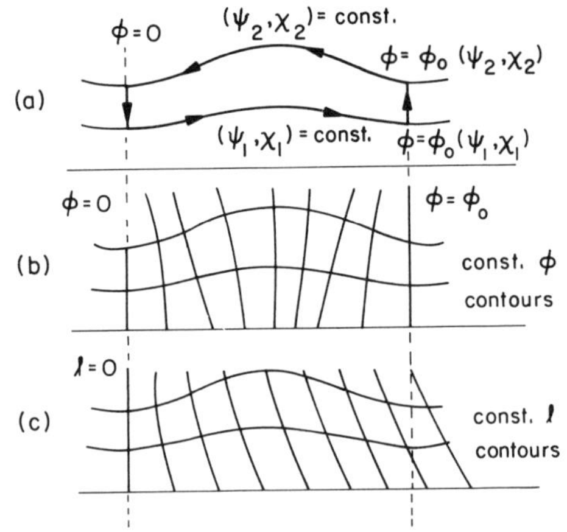

Figure 10.4. (a) Contours demonstrating that ϕ_0 is independent of (ψ, χ), (b) contours of constant ϕ and (c) contours of constant l.

requires that

$$\mu_0 I = \oint \mathbf{B} \cdot d\mathbf{l} = \phi_0(\psi_1, \chi_1) - \phi_0(\psi_2, \chi_2) = 0 \tag{10.48}$$

Since the field lines (ψ_1, χ_1), (ψ_2, χ_2) are arbitrary, Eq. (10.48) demonstrates that ϕ_0 is independent of ψ and χ.

The length L is related to ϕ_0 by

$$\phi_0 = \int_0^L B(\psi, \chi, l)\, dl \tag{10.49}$$

One sees that in general $L = L(\psi, \chi)$. Consequently, the integration limit L implicit in Eq. (10.45) must be differentiated when evaluating the stability criterion.

Equation (10.45) can be rewritten in a somewhat more convenient form where the explicit L dependence vanishes. Using the condition $\partial\phi_0/\partial\psi = 0$ one divides Eq. (10.45) by ϕ_0. The stability condition has the form

$$\frac{\partial}{\partial\psi} \langle B^2 \rangle > 0 \tag{10.50}$$

The concept of a "magnetic well" is now apparent. If $\langle B^2 \rangle$ increases with flux away from the origin, the system is interchange stable.

The conclusion from this analysis is that in the limit $\beta \to 0$ the vacuum magnetic well and the average curvature stability conditions are identical.

Consider now the effects of finite β. If $\mathbf{J} \neq 0$ a simple scalar potential can no longer be introduced for **B**. Analysis shows that the stability condition for general three-dimensional systems becomes quite complicated. Some insight can be gained by examining κ_ψ expressed in (ψ, χ, ζ) coordinates:

$$\kappa_\psi = \frac{1}{B^2}\frac{\partial}{\partial\psi}\left(\frac{B^2}{2} + \mu_0 p\right) + \frac{(\nabla\chi \cdot \mathbf{b} \times \nabla\zeta)}{B^2}\frac{\partial B}{\partial\zeta} \tag{10.51}$$

The $B^{2\prime}$ term represents the "vacuum" magnetic well contribution using the finite β **B** field. The p' term is a destabilizing contribution which cancels the stabilizing diamagnetic contribution from the "vacuum well." This cancellation is responsible for the often made statement that "a plasma cannot dig its own magnetic well by means of diamagnetic currents." The last term is a complicated contribution that is often destabilizing. For finite β, $\nabla\zeta$ cannot in general be chosen parallel to **B** so that this term does not vanish. Evaluation of the last term involves the same difficulties in calculating $\nabla\chi$ as in the original calculation of κ_ψ.

As a simple approximation to obtain a relatively easily evaluated figure of merit, the vacuum well given by Eq. (10.50) is modified as follows:

$$\frac{\partial}{\partial\psi}\langle B^2 + 2\mu_0 p\rangle > 0 \tag{10.52}$$

This cancels the plasma diamagnetism. It should be emphasized that Eq. (10.52) is neither necessary nor sufficient for interchange stability in closed-line systems. It is simply a useful figure of merit as introduced in Chapter 4.

10.3.6. Summary

In summary the average curvature and magnetic well are identical in low-β closed-line systems, giving a necessary condition for interchange stability. At finite β the two criteria differ. The average curvature is the more fundamental interchange condition and remains a necessary, although difficult to evaluate, stability criterion. The finite-β magnetic well as defined in Eq. (10.52) is relatively simple to evaluate but only represents a figure of merit: it is neither necessary nor sufficient for stability.

Finally, while the stability criteria have been derived on the basis of closed-line symmetry, they are also widely used for systems with shear. In these configurations the low-β criteria are no longer necessary for stability. Both the average curvature and magnetic well should be viewed only as figures of merit.

10.4. THE ELMO BUMPY TORUS (EBT)

10.4.1. Introduction

In Section 7.4 it was shown that EBT can be viewed as a closed-line toroidal configuration consisting of a series of linked mirror cells. In the bulge region of each mirror there is an ECRH (electron cyclotron resonance heated) hot electron ring. The combined effects of toroidicity and mirror bumpiness imply that EBT is a three-dimensional configuration. However, because of the large aspect ratio, toroidicity does not play an essential role in EBT stability and in this respect it is sufficient to treat the system as a straight, infinitely long series of two-dimensional linked mirrors. The MHD stability of such systems has been investigated extensively in recent years and this section presents a summary of these results.

The most dangerous MHD instabilities in EBT are $k_\perp \to \infty$, internal pressure-driven ballooning modes. This follows because $J_\| = 0$, implying that current-driven kinks are not excited. Furthermore, since the most severe stability limits arising from pure pressure-driven modes occur for $k_\perp \to \infty$, the corresponding eigenfunctions are highly localized in radius. Consequently, the vacuum region is decoupled and only internal modes need be considered.

The basic mechanism of the EBT instability is associated with the fact that the configuration has alternating regions of favorable and unfavorable curvature. A detailed analysis shows that without the hot electron rings, the unfavorable regions are more heavily weighted so that both interchange and ballooning modes are unstable even in the limit $\beta \to 0$.

It is the distinctly nonideal MHD-like behavior of the hot electron rings that provides the stability in EBT. The physical reason for this is related to the normal EBT operating condition $T_e/T_c \sim 10^3$, where T_e and T_c are the temperatures of the hot electron ring and the plasma core, respectively. Note the high value of T_e (typically hundreds of keV) due to the resonant heating. In this regime the characteristic diamagnetic drift velocity of the ring $V_D \sim T_e/eaB$ is much greater than the core ion thermal velocity $V_{T_i} \sim (T_i/m_i)^{1/2}$. Consequently, during a typical MHD growth time a/V_{T_i}, the hot electrons drift many times around the poloidal cross section and the flux contained within this poloidal trajectory is approximately an adiabatic invariant.

A complete analysis of EBT stability requires at least a partial kinetic treatment to properly account for the hot electron rings. The situation is further complicated by the fact that in addition to modifying the MHD stability of the core, the ring itself can become unstable. Such an analysis is beyond the scope of the present textbook. Instead, as a simplifying procedure attention is focused only on the MHD stability properties of the plasma core. The hot electron rings, on the basis of more general kinetic analysis, are approximated as fixed rigid external sources of current, as first suggested by Nelson and Hedrick (1979). With this model one can obtain a reasonably reliable picture of the MHD stability of the EBT plasma core.

The analysis proceeds in several steps. First, the Energy Principle must be modified to include fixed external currents. Next, since the most dangerous instabilities correspond to interchange and ballooning modes, the general ballooning mode equation derived in Section 10.2 must also be modified to include fixed external currents. EBT flux coordinates are introduced that reduce the ballooning mode equation into an explicit one-dimensional relation. The resulting equation is then analyzed with respect to interchange and ballooning mode instabilities. The analysis shows how the hot electron rings can provide MHD stability to the plasma core. The required parameters and profiles characterizing the rings are calculated.

10.4.2. The Effect of Rigid External Currents

The effect of a set of rigid external current sources on MHD stability can be accounted for in a straightforward manner (Johnson *et al.*, 1969). The currents are assumed to arise from an instantaneously neutralized species of external current carriers. Thus, there is an external current $\mathbf{J}_e$ but no external charge density: $-en_e = 0$.

The relevant MHD equations are derived by first considering the two-fluid model given by (see Chapter 2)

$$\begin{aligned} m_i n \frac{d\mathbf{v}_i}{dt} &= en(\mathbf{E} + \mathbf{v}_i \times \mathbf{B}) - \nabla p_{ci} \\ 0 &= -en(\mathbf{E} + \mathbf{v}_e \times \mathbf{B}) - \nabla p_{ce} \end{aligned} \tag{10.53}$$

which apply to the core ions and core electrons respectively. Adding these equations yields

$$\rho \frac{d\mathbf{v}}{dt} = (\mathbf{J} - \mathbf{J}_e) \times \mathbf{B} - \nabla p_c \tag{10.54}$$

where $p_c = p_{ci} + p_{ce}$, $\mathbf{v} = \mathbf{v}_i$, and $\mathbf{J} = en(\mathbf{v}_i - \mathbf{v}_e) + \mathbf{J}_e$, and $\mathbf{J}_e$ is the external current. The electron equation can be written as

$$\mathbf{E} + \mathbf{v} \times \mathbf{B} = \frac{1}{en}[(\mathbf{J} - \mathbf{J}_e) \times \boldsymbol{B} - \nabla p_{ce}] \tag{10.55}$$

In the small gyroradius limit the right-hand side of Eq. (10.55) is negligible, implying that

$$\mathbf{E} + \mathbf{v} \times \mathbf{B} = 0 \tag{10.56}$$

Equations (10.54) and (10.56) are the MHD momentum equation and Ohm's law, including an internal current source $\mathbf{J}_e$.

The equilibrium is assumed to satisfy

$$\begin{aligned} (\mathbf{J} - \mathbf{J}_e) \times \boldsymbol{B} &= \nabla p_c \\ \nabla \times \mathbf{B} &= \mu_0 \mathbf{J} \\ \nabla \cdot \mathbf{B} &= 0 \end{aligned} \tag{10.57}$$

Following the formulation in Chapter 8, one can easily derive the normal mode equation for linear stability:

$$-\omega^2 \rho \boldsymbol{\xi} = \mathbf{F}(\boldsymbol{\xi}) \tag{10.58}$$

where

$$\begin{aligned} \mathbf{F}(\boldsymbol{\xi}) &= (\mathbf{J} - \mathbf{J}_e) \times \mathbf{B}_1 + \mathbf{J}_1 \times \mathbf{B} - \nabla p_{c1} \\ \mathbf{B}_1 &= \nabla \times (\boldsymbol{\xi}_\perp \times \mathbf{B}) \\ \mathbf{J}_1 &= \frac{1}{\mu_0} \nabla \times \nabla \times (\boldsymbol{\xi}_\perp \times \mathbf{B}) \\ p_{c1} &= -\boldsymbol{\xi}_\perp \cdot \nabla p_c - \gamma p_c \nabla \cdot \boldsymbol{\xi} \end{aligned} \tag{10.59}$$

The rigid ring assumption requires that $\mathbf{J}_e$ remain fixed at its equilibrium value during any MHD perturbation. It is for this reason that no $\mathbf{J}_{e1}$ terms appear in Eq. (10.59).

A simple calculation shows that the "standard" form of δW_F including external currents can be written as (Johnson *et al.*, 1969)

$$\delta W_F = \frac{1}{2} \int d\mathbf{r} [|\mathbf{Q}|^2/\mu_0 - \boldsymbol{\xi}_\perp^* \cdot (\mathbf{J} - \mathbf{J}_e) \times \mathbf{Q} + (\boldsymbol{\xi}_\perp \cdot \nabla p_c) \nabla \cdot \boldsymbol{\xi}_\perp^* + \gamma p_c |\nabla \cdot \boldsymbol{\xi}|^2] \tag{10.60}$$

As before, $\mathbf{Q} \equiv \mathbf{B}_1 = \nabla \times (\boldsymbol{\xi}_\perp \times \mathbf{B})$. Following the derivation in Section 8.8.3 one can obtain an "intuitive" form of δW_F which includes external currents. (See Problem 10.2) A short calculation gives

$$\delta W_F = \frac{1}{2} \int d\mathbf{r} \left[\frac{|\mathbf{Q}_\perp|^2}{\mu_0} + \frac{B^2}{\mu_0} |\nabla \cdot \boldsymbol{\xi}_\perp + 2\boldsymbol{\xi}_\perp \cdot \boldsymbol{\kappa}_T|^2 + \gamma p_c |\nabla \cdot \boldsymbol{\xi}|^2 - (J - J_e)_\| \boldsymbol{\xi}_\perp^* \times \mathbf{B} \cdot \mathbf{Q}_\perp - 2(\boldsymbol{\xi}_\perp \cdot \nabla p_c)(\boldsymbol{\xi}_\perp^* \cdot \boldsymbol{\kappa}_T) \right] \tag{10.61}$$

$$\boldsymbol{\kappa}_T = \boldsymbol{\kappa} + \frac{\mu_0}{2} \frac{\mathbf{b} \times \mathbf{J}_{e\perp}}{B}$$

The final step in this part of the analysis is the introduction of the eikonal representation for $\boldsymbol{\xi}_\perp$ corresponding to the ballooning mode analysis. This derivation closely parallels that given in Section 10.2. (See Problem 10.3.) The result is a formulation of δW_F that describes interchange and ballooning mode stability in the presence of fixed external current sources:

$$\delta W_2 = \frac{1}{2\mu_0}\int d\mathbf{r}\left[k_\perp^2\,|\mathbf{b}\cdot\nabla X|^2 - \frac{2\mu_0}{B^2}(\mathbf{b}\times\mathbf{k}_\perp\cdot\nabla p_c)(\mathbf{b}\times\mathbf{k}_\perp\cdot\boldsymbol{\kappa}_T)|X|^2\right] \tag{10.62}$$

A comparison of Eqs. (10.13) and (10.62) indicates that the external currents enter the ballooning mode analysis only as a modification to the curvature.

10.4.3. EBT Flux Coordinates

To proceed with the EBT stability analysis flux coordinates are introduced that reduce the ballooning mode δW_F to an explicitly one-dimensional form. The discussion below closely parallels that presented in Section 10.3 and in a sense is slightly redundant. However, the material is presented in a more practical form with additional details provided for those unfamiliar with flux coordinates.

For the present analysis it is convenient to introduce the flux function itself and arc length along the field as the flux coordinates. In the discussion that follows the coordinate transformation is defined and several important properties are derived. Also, a procedure is given for calculating the transformation in practical situations. Once established, the flux coordinates are introduced into the ballooning mode equation, giving rise to a one-dimensional variational formulation.

The standard procedures for calculating MHD equilibria lead to solutions for the flux function and magnetic fields in terms of the Eulerian coordinates (r, θ, z). For a straight EBT, θ is an ignorable coordinate and $B_z = B_z(r, z)$, $B_r = B_r(r, z)$ and $B_\theta = 0$. (The general analysis presented in Section 10.3 coincides with the case under consideration by setting the angle $\chi = \theta$.) The coordinates (r, z) are not very convenient for the ballooning mode analysis because the operator $\mathbf{B}\cdot\nabla = B_r(\partial/\partial r) + B_z(\partial/\partial z)$ appears to be two-dimensional. This difficulty is eliminated by introducing the flux ψ and the arc length l as new coordinates replacing r and z. If one wants to specify coordinates (r, z), this is equivalent to specifying the flux surface ψ and the arc length l along a given line lying on the surface. See Fig. 10.5. Observe that ψ is a radial-like coordinate while l is a longitudinal distance coordinate.

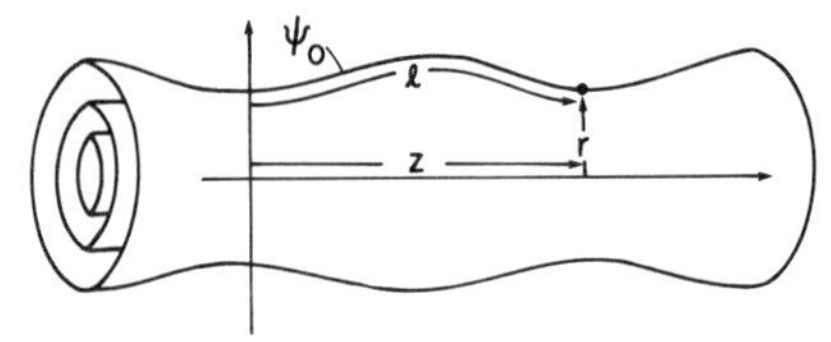

Figure 10.5. EBT geometry showing the flux coordinates ψ_0 and l and the usual cylindrical coordinates r and z.

The transformation of coordinates from (r, θ, z) to (ψ, θ, l) is defined by

$$\begin{aligned} r &= r(\psi, l) \\ z &= z(\psi, l) \\ \theta &= \theta \end{aligned} \tag{10.63}$$

The objective now is to determine the relations $r = r(\psi, l)$ and $z = z(\psi, l)$. The starting point assumes that the equilibrium fields and flux function are known in terms of Eulerian coordinates: $\psi(r, z)$, $B_r(r, z)$, $B_z(r, z)$ are specified. These quantities can be obtained numerically, or sometimes analytically, from the exact straight helical equilibrium equation (with helical wave number $l = 0$) given by Eq. (6.38). Alternatively, one can calculate B_r and B_z from the sideband expansion equation given by Eq. (7.41). Once the fields are known then ψ is obtained from the definition $\psi(r, z) = \int B_z(r', z)r'\, dr'$.

The assumption that $\psi(r, z)$ is known represents one of the desired transformation relations, although it is not in an easily usable form since a two-dimensional inversion is required. The relations $r(\psi, l)$, $z(\psi, l)$ are most easily obtained from the equations describing the field line trajectory

$$\frac{dr}{B_r} = \frac{dz}{B_z} = \frac{dl}{B} \tag{10.64}$$

where $dl = [(dr)^2 + (dz)^2]^{1/2}$ is the arc length along the field. These equations can be put in a form suitable for integration as follows:

$$\begin{aligned} \frac{dr}{dl} &= \frac{B_r(r, z)}{B(r, z)} \\ \frac{dz}{dl} &= \frac{B_z(r, z)}{B(r, z)} \end{aligned} \tag{10.65}$$

Without loss in generality the origins in z and l can be made to coincide: $z(l = 0) = 0$. The initial radial position of the field line is denoted by r_0: $r(l = 0) = r_0$. The quantity r_0 is an alternative flux surface label related directly to ψ. On a given flux surface ψ is by definition a constant so that $\psi = \psi(r_0, l = 0) = \text{const}$. This one-dimensional relation can be inverted to give $r_0 = r_0(\psi)$. If one now views r and z as functions of ψ and l then Eq. (10.65) can be written as

$$\begin{aligned} \frac{\partial r}{\partial l} &= \frac{B_r(r, z)}{B(r, z)}, \qquad r(\psi, 0) = r_0(\psi) \\ \frac{\partial z}{\partial l} &= \frac{B_z(r, z)}{B(r, z)}, \qquad z(\psi, 0) = 0 \end{aligned} \tag{10.66}$$

where the partial derivatives are at fixed ψ. Equation (10.66) describes a set of coupled nonlinear, one-dimensional equations for r and z as a function of l with ψ appearing as a parameter. These equations are in a convenient form for a numerical or, if one is fortunate, analytic solution. Hereafter, it is assumed that $r(\psi, l)$ and $z(\psi, l)$ are known functions.

Several important properties of the coordinates are now derived that are needed for the ballooning mode analysis. First, from the general definition of the flux function [Eq. (6.38)] it follows that

$$\mathbf{B} = \frac{1}{r}\nabla\psi \times \mathbf{e}_\theta \tag{10.67}$$

(For convenience a factor h has been absorbed into ψ.) This implies that

$$B = \frac{|\nabla\psi|}{r} \tag{10.68}$$

The second property concerns the operator $\mathbf{b} \cdot \nabla$, which can be expressed as

$$\mathbf{b} \cdot \nabla = \frac{1}{B}\left(B_r \frac{\partial}{\partial r} + B_z \frac{\partial}{\partial z}\right) \tag{10.69}$$

Along the field line coordinates given by Eq. (10.66) this reduces to

$$\mathbf{b} \cdot \nabla = \frac{\partial r}{\partial l}\frac{\partial}{\partial r} + \frac{\partial z}{\partial l}\frac{\partial}{\partial z} \tag{10.70}$$

or

$$\mathbf{b} \cdot \nabla = \frac{\partial}{\partial l} \tag{10.71}$$

In the flux coordinates, the operator $\mathbf{b} \cdot \nabla$ is explicitly one-dimensional. Third, note that in general l and ψ are not orthogonal coordinates: $\nabla l \cdot \nabla\psi \neq 0$. However, this adds no significant complications to the analysis. The last property of interest concerns the Jacobian of the transformation which is required in order to express the differential volume element in terms of the new coordinates. In Eulerian coordinates the volume element is given by $d\mathbf{r} = 2\pi r\, dr\, dz$ for systems with θ symmetry. In terms of the new coordinates it follows that

$$r\, dr\, dz = J\, d\psi\, dl \tag{10.72}$$

where

$$\frac{r}{J} = \begin{vmatrix} \partial\psi/\partial r & \partial\psi/\partial z \\ \partial l/\partial r & \partial l/\partial z \end{vmatrix} = \mathbf{e}_\theta \cdot \nabla l \times \nabla\psi \tag{10.73}$$

From Eq. (10.67) one finds

$$\mathbf{e}_\theta \cdot \nabla l \times \nabla\psi = r\mathbf{e}_\theta \cdot \nabla l \times (\mathbf{e}_\theta \times \mathbf{B}) = r\mathbf{B} \cdot \nabla l = rB \tag{10.74}$$

Thus $J = 1/B$ and

$$2\pi r\, dr\, dz = \frac{2\pi\, d\psi\, dl}{B} \tag{10.75}$$

10.4.4. The Ballooning Mode Equation

Using these relations it is now straightforward to simplify the ballooning mode equation for the axisymmetric EBT configuration. The first step is to calculate the eikonal function S which must satisfy $\mathbf{B} \cdot \nabla S = 0$. For shearless systems such as EBT, the difficulties in calculating S, alluded to in the general ballooning mode analysis, do not arise. The $\mathbf{B} \cdot \nabla S = 0$ condition is easily satisfied in flux coordinates by requiring $S(\psi, \theta, l) \rightarrow S(\psi, \theta)$. In addition, axisymmetry allows Fourier analysis with respect to θ. Hence, one can write

$$\begin{aligned} S &= m\theta + m\hat{S}(\psi) \\ \mathbf{k}_\perp &= \nabla S = m(k_\theta \mathbf{e}_\theta + k_n \mathbf{n}) \end{aligned} \tag{10.76}$$

where m is the azimuthal mode number. $\hat{S}(\psi)$ is an arbitrary function of ψ, $k_\theta = 1/r$, $k_n = rB\partial\hat{S}/\partial\psi$ and $\mathbf{n} = \nabla\psi/|\nabla\psi|$. The assumption $m \gg 1$ guarantees that S corresponds to rapid oscillations.

Since the θ dependence has been explicitly accounted for in S, the dependent variable X is a function only of ψ and l: $X = X(\psi, l)$. Equation (10.71) then implies that

$$\mathbf{b} \cdot \nabla X = \frac{\partial X}{\partial l} \tag{10.77}$$

Next, the vector $\mathbf{b} \times \mathbf{k}_\perp$ can be expressed as

$$\mathbf{b} \times \mathbf{k}_\perp = -\frac{1}{rB}\mathbf{k}_\perp \times (\nabla\psi \times \mathbf{e}_\theta) = -m(k_\theta \mathbf{n} - k_n \mathbf{e}_\theta) \qquad (10.78)$$

Using this relation it follows that

$$\begin{aligned} \mathbf{b} \times \mathbf{k}_\perp \cdot \nabla p_c &= -mk_\theta rB \frac{dp_c}{d\psi} \\ \mathbf{b} \times \mathbf{k}_\perp \cdot (2\boldsymbol{\kappa} + \mu_0 \mathbf{b} \times \mathbf{J}_{e\perp}/B) &= -mk_\theta \left(2\kappa_n - \frac{\mu_0 J_{\theta e}}{B}\right) \end{aligned} \qquad (10.79)$$

where $\kappa_n = \mathbf{n} \cdot \boldsymbol{\kappa}$.

These results are now substituted into the ballooning mode equation [Eq. (10.62)]. One finds

$$\begin{aligned} \delta W_2 = \frac{\pi m^2}{\mu_0} \int \frac{d\psi\, dl}{B} \Bigg[(k_\theta^2 + k_n^2) \left| \frac{\partial X}{\partial l} \right|^2 \\ - \frac{\mu_0 r k_\theta^2}{B} \frac{dp_c}{d\psi} \left(2\kappa_n - \frac{\mu_0 J_{\theta e}}{B}\right) |X|^2 \Bigg] \end{aligned} \qquad (10.80)$$

Equation (10.80) can be further simplified by noting the following points. The quantity k_n appears only in a positive term. Hence, δW_2 is minimized by choosing $k_n = 0$. Next, if one assumes that the hot electron pressure is isotropic, as in Section 7.4, then $J_{\theta e}$ can be expressed in terms of the ring pressure p_e: $J_{\theta e}\mathbf{e}_\theta \times \mathbf{B} = \nabla p_e$ or $J_{\theta e} = r\, dp_e/d\psi$. Finally, the only derivatives appearing on X are l derivatives. This implies that ψ appears only as a parameter and enables one to write δW_2 as follows (Nelson and Hedrick, 1979):

$$\delta W_2 = \frac{\pi m^2}{\mu_0} \int W(\psi)\, d\psi \qquad (10.81)$$

With the aid of the simplifications just discussed, $W(\psi)$ can be expressed as

$$W(\psi) = \oint dl \left[\frac{1}{r^2 B} \left(\frac{\partial X}{\partial l} \right)^2 - \frac{\mu_0 p_c'}{rB^2} \left(2\kappa_n - \frac{\mu_0 r p_e'}{B} \right) X^2 \right] \qquad (10.82)$$

where prime denotes differentiation with respect to ψ and without loss in generality X can be considered real, since all the coefficients are real.

Equation (10.82) is the desired form of the Energy Principle. The system is stable to localized internal pressure-driven modes if and only if $W(\psi) > 0$ on every flux surface. $W(\psi) < 0$ on any flux surface is a sufficient condition for instability. Observe that the effort required to test stability involves the independent evaluation of a one-dimensional Energy Integral on a series of separate flux surfaces. This represents an enormous savings over the fully coupled two-dimensional problem. This savings is a direct consequence of the localization assumption.

10.4.5. Interchange Stability

The stability of the EBT configuration, as described by Eq. (10.82), can be understood by separately investigating interchange and ballooning modes, with the aid of the expanded equilibrium solutions derived in Chapter 7. Consider first interchange stability. These perturbations "interchange" tubes of plasma and magnetic flux in such a way that the perturbed magnetic energy associated with line bending is zero. Recall that this is only possible in shearless systems. The interchange trial function assumes that X is constant along the field line under consideration, implying that $\mathbf{Q}_\perp = 0$. Upon setting $X(\psi, l) = X_0(\psi)$ in Eq. (10.82), one obtains a necessary condition for stability on each surface given by

$$-\mu_0 p_c' \oint \frac{dl}{rB^2}\left(2\kappa_n - \frac{\mu_0 r p_e'}{B}\right) > 0 \tag{10.83}$$

The quantity

$$U(\psi) = \oint \left(\frac{\kappa_n}{rB^2}\right) dl \tag{10.84}$$

represents the "average curvature" of the field line. In closed line systems without external current sources U must be positive on every field line to avoid interchange instabilities (assuming $p_c' < 0$, the normal situation).

The condition given by Eq. (10.83) can be evaluated analytically by making use of the expanded EBT equilibrium discussed in Section 7.4 plus two additional simplifications. First, the limit of a "straight EBT" corresponds to $r/R_0 \to 0$, implying that only Eq. (7.41) need be solved. Second, as is shown shortly, the ring β_e must be considered smaller, of order δ^2, in order to enter competitively with the curvature terms. Thus, in terms of the ordering parameter δ, defined by

$$\delta = \left.\frac{B_{\max} - B_{\min}}{B_{\max} + B_{\min}}\right|_{r=0} \tag{10.85}$$

the expanded equilibrium satisfies

$$\beta \approx \beta_c = \frac{2\mu_0 p_c}{B_0^2} \sim 1$$

$$\beta_e = \frac{2\mu_0 p_e}{B_0^2} \sim \delta^2 \qquad (10.86)$$

$$ha \sim 1$$

Straightforward application of the results in Section 7.4 shows that the equilibrium fields can be written as

$$\frac{B_{r1}}{B_z} = b_{r1} \sin hz$$

$$\frac{B_{z1}}{B_z} = b_{z1} \cos hz \qquad (10.87)$$

where $B_z(r) = B_0[1 - \beta_c(r)]^{1/2}$ is the zeroth-order toroidal field and

$$b_{r1}(r) = \frac{1}{1 - \beta_c} \frac{d\phi}{dr}$$

$$b_{z1}(r) = \frac{1}{1 - \beta_c} \left[h\phi - \frac{1}{2h(1 - \beta_c)} \frac{d\beta_c}{dr} \frac{d\phi}{dr} \right] \qquad (10.88)$$

Here, ϕ is a potential function related to V in Section 7.4 by $V = (i/2)\phi$. ϕ satisfies

$$\frac{1}{r} \frac{d}{dr} \left(\frac{r}{1 - \beta_c} \frac{d\phi}{dr} \right) - \frac{h^2 \phi}{1 - \beta_c} = 0$$

$$\phi(0) = \delta/h, \qquad \phi'(0) = 0 \qquad (10.89)$$

Similarly, the field line trajectory and flux function are given by

$$z = l + O(\delta^2)$$

$$r = r_0 - \left[\frac{b_{r1}(r_0)}{h} \right] \cos hl + O(\delta^2) \qquad (10.90)$$

$$\psi = \psi_0(r_0) + \left[\frac{r_0}{h(1 - \beta_c)^{1/2}} \frac{d\phi}{dr_0} \right] \cos hl + O(\delta^2)$$

where $\psi_0'(r_0) = r_0 B_z(r_0)$ and r_0 is used in place of ψ as a flux-surface label.

Finally, a short calculation making use of Eq. (7.30) shows that the normal curvature and the quantity rB^2 evaluated along the field line can be expressed as

$$\kappa_n|_l \approx \kappa_r|_l = \left\{hb_{r1}\cos hl + \frac{\partial}{\partial l}\left[\frac{B_{r2}}{B_z} - \frac{b_{r1}}{2}\left(b_{z1} - \frac{1}{h}\frac{db_{r1}}{dr}\right)\sin 2hl\right]\right\}_{r=r_0}$$

$$rB^2|_l = \left\{rB_z^2\left[1 + \left(2\frac{d\Lambda}{dr} + \frac{\Lambda}{r}\right)\cos hl\right]\right\}_{r=r_0} \tag{10.91}$$

with

$$\Lambda(r) = \frac{1}{(1-\beta_c)}\frac{d\phi}{dr} \tag{10.92}$$

These results are now substituted into Eq. (10.83). Another short calculation yields the following condition for interchange stability:

$$\frac{d\beta_c}{dr}\left[\frac{1}{2}\frac{d\beta_e}{dr} + \frac{(1-\beta_c)}{r}\frac{d}{dr}(r\Lambda^2)\right] > 0 \tag{10.93}$$

For simplicity, the subscript zero has been dropped from r.

From this condition emerges the following relatively simple picture of EBT stability. Assume $\beta_c' < 0$ for confinement. The term in the bracket of Eq. (10.93), which represents the negative of the combined average curvature of the plasma plus rings is positive (i.e., destabilizing) if $\beta_e' > 0$ since Λ is in general an increasing function of radius. The curvature contribution can be negative only if β_e' is sufficiently negative. Since the hot electron ring pressure is peaked off axis, the region from the axis to the maximum ring pressure is potentially unstable because of the average unfavorable curvature of the mirror field. To avoid such instabilities the EBT pressure profile must be flat (i.e., $\beta_c' = 0$) out to a radius somewhat past the peak of the ring pressure where the average curvature first becomes favorable. The entire pressure gradient of the plasma core must occur over the relatively narrow region of space where β_e' is sufficiently negative to create favorable curvature. Typical stable profiles are illustrated in Fig. 10.6.

A simple criterion that determines the minimum ring properties required to create a region of favorable curvature can be obtained as follows. First, assume that the hot electron ring profile is parabolic:

$$\beta_e = \beta_{e0}\left[1 - \left(\frac{r-a}{d}\right)^2\right] \tag{10.94}$$

where β_{e0} is the peak "beta," a is the radius where the ring is located and d is the half-width of the ring. Second, consider the "slow z" approxima-

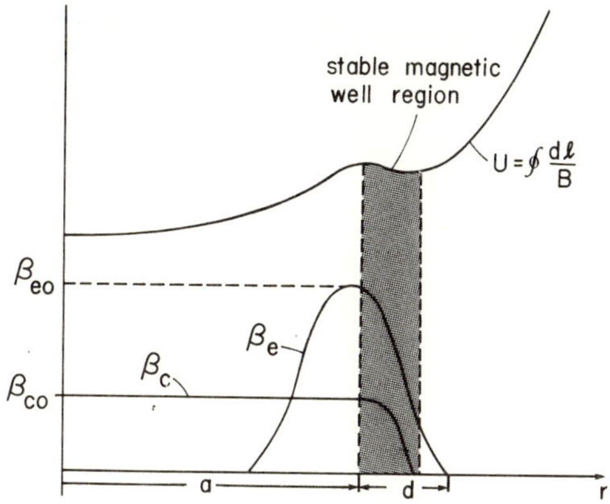

Figure 10.6. Stabilized EBT plasma pressure profile due to the magnetic well of the hot electron rings.

tion, $ha \ll 1$ and the limit of a thin ring, $d/a \ll 1$. Under these conditions, in the region of interest the functions ϕ and Λ are given by

$$\phi \approx \frac{\delta}{h}\left(1 + \frac{h^2r^2}{4}\right)$$
$$\Lambda \approx \frac{h\delta r}{2(1 - \beta_{c0})} \tag{10.95}$$

where β_{c0} is the nearly uniform peak "beta" of the plasma core. Substituting into Eq. (10.93) yields the following condition on the ring β for creating a region of average favorable curvature:

$$\beta_{e0} > \frac{3}{16}\frac{h^2(M-1)^2ad}{1-\beta_{c0}} \tag{10.96}$$

As in Section 7.4, $M = 1 + 2\delta$ is the vacuum mirror ratio and $(M - 1) \ll 1$. Observe that β_{e0} scales as δ^2 and that sufficiently large values of β_{e0} are required for stability.

A general comment concerning favorable and unfavorable curvature can be made by examining Eqs. (10.83) and (10.91). In configurations such as EBT the normal curvature κ_n, as given by Eq. (10.91), averages to zero along a field line. However, as seen in Eq. (10.83), the "average curvature" for stability is weighted by $1/rB^2$. This emphasizes the curvature in regions of low B. For EBT low B occurs in the bulge regions between the mirror coils where the curvature points inwards (i.e., is destabilizing). It is for this reason that an EBT without external current sources would be very unstable, even at zero β.

10.4.6. Ballooning Mode Stability

Once a region of average favorable curvature has been established, the next important question that arises is to determine the maximum value of β_{c0} that can be stably confined. The β_{c0} limit is a consequence of ballooning modes and is calculated as follows. Recall that the interchange trial function assumes $X(\psi, l) = X_0(\psi)$. In order to allow the perturbations to "balloon out" in the unfavorable curvature regions it is necessary to generalize the form of the trial function. For the expanded equilibrium as described by Eq. (10.86), (but not assuming the slow z or thin ring approximations) X is given by

$$X(\psi, l) = X_0(\psi)[1 + X_1(\psi) \cos hl] \tag{10.97}$$

where $|X_1|/|X_0| \sim \delta$. Upon substituting into Eq. (10.82) one obtains, after a straightforward calculation

$$\begin{aligned} \frac{W(\psi)}{2\pi R_0} = \frac{X_0^2}{2r^2 B} \bigg\{ & \frac{1}{(1-\beta_c)^2} \frac{d\beta_c}{dr} \left[\frac{1}{2} \frac{d\beta_e}{dr} + \frac{(1-\beta_c)}{r} \frac{d}{dr}(r\Lambda^2) \right] \\ & + h^2 X_1^2 - 2\left(\frac{\Lambda}{1-\beta_c} \frac{d\beta_c}{dr} \right) hX_1 \bigg\} \end{aligned} \tag{10.98}$$

Note that X_1 appears only algebraically. Thus, one can easily complete the square leading to the following expression for the minimizing X_1:

$$hX_1 = \frac{\Lambda}{1-\beta_c} \frac{d\beta_c}{dr} \tag{10.99}$$

Substituting back into Eq. (10.98) yields

$$\frac{d\beta_c}{dr} \left[\frac{1}{2} \frac{d\beta_e}{dr} + \frac{(1-\beta_c)}{r} \frac{d}{dr}(r\Lambda^2) - \Lambda^2 \frac{d\beta_c}{dr} \right] > 0 \tag{10.100}$$

Equation (10.100) represents the condition for stability against ballooning modes. A comparison of Eqs. (10.100) and (10.93) indicates the presence of an additional destabilizing term, proportional to $\Lambda^2 d\beta_c/dr$, for ballooning modes.

The β_{c0} limit can be evaluated analytically by using the same slow z and thin ring assumptions made in the previous section and ordering $\beta_{c0} \sim d/a \ll 1$. One obtains a local criterion for stability, valid in the region of average favorable curvature, given by

$$-\frac{d\beta_c}{dx} \leq \frac{3d}{a} \left(\frac{\beta_{e0}}{\beta_*} x - 1 \right), \qquad x_1 < x < 1 \tag{10.101}$$

where $r = a + dx$, $\beta_* = (3/16)h^2(M - 1)^2ad$, and $x_1 = \beta_*/\beta_{e0}$. Note that for a region of average favorable curvature $\beta_{e0} > \beta_*$ and $x_1 < x < 1$.

The maximum stable β_{c0} can be calculated by assuming equality in Eq. (10.101) and then integrating across the favorable curvature region. The result is

$$\beta_{c0} < \frac{3}{2}\left(\frac{d}{a}\right)\frac{(\beta_{e0} - \beta_*)^2}{\beta_{e0}\beta_*} \tag{10.102}$$

This result is in good agreement with the numerical studies carried out by Nelson and Hedrick (1979) and illustrated in Fig. 10.7. They found that for realistic configurations including finite mirror ratio, finite β, anisotropic hot electron rings and plasma compressibility, the threshold value of β_{e0} occurs between 0.1 and 0.15. Above this threshold the marginal value of β_{c0} increases rapidly and in the case of long electron rings a value of $\beta_{e0} = 0.2$ stabilizes the plasma core up to values of $\beta_{c0} \approx 0.3$. Also illustrated in Fig. 10.7 is the analytic result given by Eq. (10.102) for the parameters $\delta = 0.5$, $ha = 1.2$, $d/a = 0.26$, approximately corresponding to the case studied by Nelson and Hedrick. There is good qualitative and reasonable semiquantitative agreement in the results despite the many simplifications in the analytic theory.

One concludes that a combination of a sufficiently high ring β_e and a

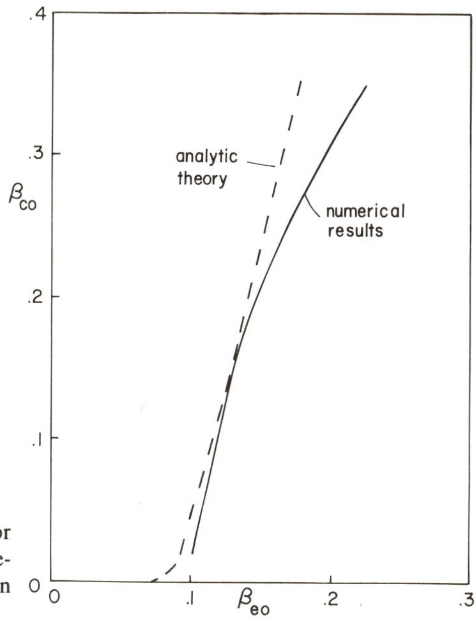

Figure 10.7. Critical plasma core β_{c0} for EBT stability. Shown are the analytic results and the numerical results from Nelson and Hedrick (1979).

flat plasma core pressure profile out to the favorable curvature region leads to a stable high-β_c EBT configuration. This is an optimistic result from the ideal MHD point of view.

10.4.7. Summary of EBT

In summary, high-β EBT profiles exist that are stable to all ideal MHD modes. The most dangerous perturbations correspond to interchange and ballooning modes. The stability is sensitively dependent upon the assumption that the hot electron rings are rigid with respect to MHD perturbations. Without this assumption, the system would be very unstable even at zero β. An interesting feature of the results is that finite values of β_{c0} can be stabilized by relatively low values of β_{e0} if the mirror ratio (i.e., β_*) is also small. (This optimistic result does not hold when the hot electron rings are treated with a self-consistent kinetic model. In this case one typically requires $\beta_{c0} \lesssim \beta_{e0}$). One characteristic feature of the stable profiles is that plasma pressure must be flat out to the region where the favorable curvature develops in order to prevent interchange instabilities. In a certain sense this requirement corresponds to "infinite" transport in the interior region since no gradients are allowed. Therefore, the overall transport is determined by the properties of the rings in the favorable curvature region and gives rise to an undesirable energy confinement scaling law $\tau_E \sim ad$ rather than $\tau_E \sim a^2$ as in the tokamak or RFP. This may indeed be the most important consequence of ideal MHD stability on the performance of EBT.

10.5. TOKAMAKS

10.5.1. Introduction

Toroidicity plays a very important role in determining the stability of tokamaks, particularly in the high $\beta \sim \epsilon$ regime. There are quantitative changes in the stability of current-driven modes as compared to the predictions of "straight tokamak" theory. More important, toroidicity leads to qualitative changes for pressure-driven modes; specifically, internal pressure-driven ballooning modes and external ballooning-kink modes set upper limits on the maximum value of β that can be stably confined.

To obtain a reasonably comprehensive picture of toroidal tokamak stability, one is required to investigate a number of different type modes. In order to accomplish this goal, the material has been organized as follows.

1. The beginning of Section 10.5 is concerned with interchange and ballooning modes in arbitrary tokamaks (i.e., no expansions in ϵ or β). As a preliminary, a short discussion is given of tokamak flux coordinates.

2. Once established, the flux coordinates are introduced into the general ballooning mode equation. The difficulties associated with the calculation of the eikonal function S in systems with shear are addressed and resolved. The result is an explicitly one-dimensional form of the ballooning mode potential energy.

3. The first application of the ballooning mode Energy Principle is to interchange perturbations. The analysis leads to the derivation of the Mercier criterion, a two-dimensional generalization of Suydam's criterion. This is an exact analytic criterion, necessary but not sufficient for stability, which makes important modifications to the one-dimensional criterion in the vicinity of the magnetic axis.

4. The second application of the ballooning mode Energy Principle is to ballooning modes themselves. By treating a simple model and making use of trial functions, it is possible to obtain an estimate of the β limit due to ballooning modes. Also, it is shown that under certain conditions there exists a "second region of stability" at high β, an enticing prospect for future tokamak operation.

5. Next, a short discussion is given of the stability properties of low-n internal kink modes. Although substantial analysis and numerical studies have been carried out, there is still uncertainty, as of this writing, in the overall picture.

6. The analysis continues with an investigation of external-kink modes. It is shown that toroidicity adds an important ballooning component to the external kink eigenfunction, ultimately leading to a β limit. This is often the most severe limit in the system. By making use of the simple surface current model it is also shown that q_* rather than q_a determines low-n external kink stability.

7. The results up to this point are largely analytical. As such it is prudent to compare these results with those obtained from extensive numerical studies. Furthermore, the numerical results provide a realistic treatment of noncircularity, finite aspect ratio, and general diffuse profiles. These studies have led to simple empirical scaling laws for the maximum stable β known as the Sykes limit and the Troyon limit.

8. Finally, a short discussion is given of the axisymmetric, $n = 0$ mode, the most unstable perturbation usually corresponding to a nearly rigid vertical shift. In the straight tokamak with circular cross section such modes (by symmetry) are neutrally stable. In a torus, however, if the vertical field is not properly designed, the plasma, particularly if it is elongated, has a tendency to drift upward. Stabilization of the $n = 0$ mode is discussed at the end of this section.

10.5.2. Tokamak Flux Coordinates

The first part of the tokamak stability analysis is concerned with interchange and ballooning modes. To carry out these investigations it is necessary to introduce a set of flux coordinates in order to reduce the ballooning mode δW_F into an explicitly one-dimensional energy integral.

As with EBT, the choice of flux coordinates is not unique and a number of different systems appear in the literature. Each of these systems uses the poloidal flux ψ as a "radial-like" coordinate. The "anglelike" coordinate, denoted here by χ, is where differences arise. Poloidal arc length, Hamada coordinates, and orthogonality are among the choices that have been made. For present purposes it is convenient to treat ψ and χ as the flux coordinates with χ an orthogonal anglelike variable. This leads to the simplest formulation of the ballooning mode δW_F.

The tokamak flux coordinates are defined as follows. Consider a general, axisymmetric torus in which no expansions are made in β or ϵ and which allows arbitrary shaped cross sections. Axisymmetry implies that the toroidal angle ϕ is an ignorable coordinate for the equilibrium. The flux coordinates represent a transformation of the poloidal variables (R, Z) or equivalently (r, θ). The full transformation is defined by

$$\begin{aligned} \psi &= \psi(R, Z) \\ \chi &= \chi(R, Z) \\ \phi &= \phi \end{aligned} \tag{10.103}$$

where ψ is the flux function satisfying the Grad–Shafranov equation and χ is an orthogonal poloidal anglelike variable. Once ψ is determined, either analytically or numerically, one can then evaluate the magnetic fields and curvature vector in the Eulerian (R, Z) system. These relations, derived in Section 6.2, are repeated here for convenience:

$$\begin{aligned} \mathbf{B} &= B_\phi \mathbf{e}_\phi + \mathbf{B}_p \\ B_\phi &= \frac{F(\psi)}{R} \\ \mathbf{B}_p &= \frac{1}{R}\nabla\psi \times \mathbf{e}_\phi \\ \boldsymbol{\kappa} &= \mathbf{b}\cdot\nabla\mathbf{b} \end{aligned} \tag{10.104}$$

Here, $\mathbf{b} = \mathbf{B}/B$.

The variable χ is determined from the orthogonality condition

$$\nabla\psi \cdot \nabla\chi = 0 \tag{10.105}$$

Also, since Eq (10.105) is homogeneous, one can without loss in generality normalize χ so that one complete poloidal circuit corresponds to a change in χ of 2π. For the analysis, what is actually required is the inverse transformation $R = R(\psi, \chi)$, $Z = Z(\psi, \chi)$. A practical method for calculating this transformation is discussed in Problem 10.5. Hereafter, it is assumed that the quantities given by Eq. (10.104) and the inverse transformation are known.

The next step is to derive the important properties of the transformation for the ballooning mode analysis. First, from Eq. (10.104) it follows that

$$B_p = \frac{|\nabla\psi|}{R} \tag{10.106}$$

Second, the transformation of the volume element assuming toroidal axisymmetry is given by

$$d\mathbf{r} = 2\pi R\, dR\, dZ = 2\pi J\, d\psi\, d\chi \tag{10.107}$$

where the Jacobian J satisfies

$$\frac{R}{J} = \begin{vmatrix} \partial\psi/\partial R & \partial\psi/\partial Z \\ \partial\chi/\partial R & \partial\chi/\partial Z \end{vmatrix} = \mathbf{e}_\phi \cdot \nabla\chi \times \nabla\psi \tag{10.108}$$

Using the definition of $\mathbf{B}_p$ in Eq. (10.104) one can write $\mathbf{e}_\phi \cdot \nabla\chi \times \nabla\psi = R\mathbf{B}_p \cdot \nabla\chi$. The Jacobian then has the form

$$JB_p = \frac{1}{\mathbf{b}_p \cdot \nabla\chi} = \mathbf{b}_p \cdot \frac{\partial\mathbf{r}}{\partial\chi} \tag{10.109}$$

where $\mathbf{b}_p = \mathbf{B}_p/B_p$ and $\mathbf{r}(\psi, \chi) = R(\psi, \chi)\mathbf{e}_R + Z(\psi, \chi)\mathbf{e}_Z$. The second form in Eq. (10.109) is the inverse representation of J and is easily computed as indicated in Problem 10.6 from $\mathbf{r} = \mathbf{r}(\psi, \chi)$.

Consider now the evaluation of $\mathbf{k}_\perp = \nabla S$ appearing in the general ballooning mode equation. Assume that a physically acceptable function $S = S(\psi, \phi, \chi)$ can be found satisfying $\mathbf{B} \cdot \nabla S = 0$. The difficulties of finding such a function are deferred until the next section. In terms of the flux coordinates, ∇S is given by

$$\nabla S = \frac{\partial S}{\partial\psi}\nabla\psi + \frac{\partial S}{\partial\chi}\nabla\chi + \frac{1}{R}\frac{\partial S}{\partial\phi}\mathbf{e}_\phi \tag{10.110}$$

It is convenient at this point to introduce a set of locally orthogonal unit vectors defined by $(\mathbf{n}, \mathbf{t}, \mathbf{b})$, where

$$\begin{aligned}\mathbf{n} &= \frac{\nabla\psi}{|\nabla\psi|}\\ \mathbf{t} &= \frac{B_\phi}{B}\mathbf{b}_p - \frac{B_p}{B}\mathbf{e}_\phi \\ \mathbf{b} &= \frac{B_p}{B}\mathbf{b}_p + \frac{B_\phi}{B}\mathbf{e}_\phi\end{aligned} \tag{10.111}$$

Note that $\mathbf{n}$ is a normal vector while $\mathbf{b}$ and $\mathbf{t}$ are tangential vectors lying in the flux surface parallel and perpendicular to the equilibrium magnetic field, respectively. These quantities can be derived from $\psi(R, Z)$ and hence are assumed known.

Using these definitions one can write $\mathbf{k}$ as follows:

$$\begin{aligned}\mathbf{k} &= k_n\mathbf{n} + k_t\mathbf{t}\\ k_n &= \mathbf{n}\cdot\nabla S = (\mathbf{n}\cdot\nabla\psi)\frac{\partial S}{\partial\psi}\\ k_t &= \mathbf{t}\cdot\nabla S = (\mathbf{t}\cdot\nabla\chi)\frac{\partial S}{\partial\chi} + (\mathbf{t}\cdot\mathbf{e}_\phi)\frac{1}{R}\frac{\partial S}{\partial\phi}\end{aligned} \tag{10.112}$$

Similarly, it can be shown that

$$\mathbf{b}\times\mathbf{k}_\perp = k_t\mathbf{n} - k_n\mathbf{t} \tag{10.113}$$

The last quantity to evaluate for the ballooning mode analysis is $\mathbf{b}\cdot\nabla X$. In general $X = X(\psi, \phi, \chi)$. However, for axisymmetric equilibria one can Fourier analyze with respect to ϕ: $\boldsymbol{\xi}(R, Z, \phi) = \boldsymbol{\xi}(R, Z)\exp(-in\phi)$ with n the toroidal mode number. The ϕ dependence is explicitly accounted for in the ballooning mode analysis by writing

$$S(\psi, \phi, \chi) = -n\phi + \tilde{S}(\psi, \chi) \tag{10.114}$$

No other ϕ dependence appears in the analysis. Hence $X(\psi, \phi, \chi) \to X(\psi, \chi)$. A simple calculation then shows that

$$\mathbf{b}\cdot\nabla X = \frac{1}{JB}\frac{\partial X}{\partial\chi} \tag{10.115}$$

These results can be combined and substituted into the general form of the ballooning mode energy integral given by Eq. (10.13). The result is

$$\delta W_2 = \frac{\pi}{\mu_0} \int d\psi \, W(\psi)$$

$$W(\psi) = \int_0^{2\pi} J \, d\chi \Big[(k_n^2 + k_t^2) \Big(\frac{1}{JB} \frac{\partial X}{\partial \chi} \Big)^2 - \frac{2\mu_0 R B_p}{B^2} \frac{dp}{d\psi} (k_t^2 \kappa_n - k_t k_n \kappa_t) X^2 \Big] \tag{10.116}$$

Here, the curvature has been decomposed as

$$\boldsymbol{\kappa} = \kappa_n \mathbf{n} + \kappa_t \mathbf{t} \tag{10.117}$$

where $\kappa_n = \mathbf{n} \cdot \boldsymbol{\kappa}$ is the normal curvature and $\kappa_t = \mathbf{t} \cdot \boldsymbol{\kappa}$ is the geodesic curvature. The χ integration corresponds to one complete poloidal circuit.

$W(\psi)$ as given by Eq. (10.116) is the one-dimensional form of the ballooning mode potential energy. A comparison of this relation with Eq. (10.80) for EBT shows a great similarity in form. The main differences are that there are no external current sources for the tokamak while the geodesic curvature is zero in the straight EBT.

10.5.3. The Tokamak Ballooning-Mode Equation: Shear versus Periodicity

The last step in the derivation of the general tokamak ballooning mode Energy Principle is the determination of the eikonal function $S(\psi, \phi, \chi)$. As is shown, important but subtle difficulties arise in the calculation of S. To clarify the issues the analysis is presented in the form of two parallel calculations. The first corresponds to the full toroidal problem carried out in flux coordinates (ψ, χ). The second corresponds to the infinite aspect ratio circular-cross-section limit (i.e., the straight cylinder) carried out in the familiar (r, θ) coordinates.

The analysis begins by recalling that S must have two properties: (1) it must produce very rapid oscillations perpendicular to the field and (2) it must be constant along the field, $\mathbf{B} \cdot \nabla S = 0$. For the general tokamak this last condition can be written as

$$\frac{B_\phi}{R} \frac{\partial S}{\partial \phi} + \frac{1}{J} \frac{\partial S}{\partial \chi} = 0 \tag{10.118}$$

In the cylindrical limit, $R \to R_0$, $\chi \to \theta$, $\psi(r, \theta) \to \psi(r)$ and the Jacobian reduces to $J = 1/(\mathbf{B}_p \cdot \nabla \chi) = r/B_\theta$. Thus, Eq. (10.118) simplifies to

$$\frac{B_\phi}{R_0}\frac{\partial S}{\partial \phi} + \frac{B_\theta}{r}\frac{\partial S}{\partial \theta} = 0 \tag{10.119}$$

As previously discussed, axisymmetry implies that Fourier analysis can be carried out with respect to ϕ: $S = -n\phi + \tilde{S}(\psi, \chi)$ or $S = -n\phi + \tilde{S}(r, \theta)$. For $n \gg 1$, S corresponds to rapid oscillating behavior. With this decomposition of S Eqs. (10.118) and (10.119) can be formally integrated as follows:

$$\begin{aligned} S &= n\left[-\phi + \int_{\chi_0}^{\chi} \frac{JB_\phi}{R}\, d\chi'\right] \\ S &= n[-\phi + q(r)(\theta - \theta_0)] \end{aligned} \tag{10.120}$$

where $q(r) = rB_\phi/R_0B_\theta$ is the safety factor and $\chi_0(\psi)$ and $\theta_0(r)$ are arbitrary integration functions.

The difficulties involved in finding a physically acceptable S are now apparent. The function S given by Eq. (10.120) guarantees that $\mathbf{B}\cdot\nabla S = 0$. However, to be physically acceptable, S must also correspond to periodic behavior in χ or θ. Focusing on the cylindrical case one notes that it is always possible to choose a resonant surface $r = r_0$ that is rational, thus satisfying the periodicity constraint: $nq(r_0) = m$. Even so, in systems with shear (i.e., $q' \neq 0$) the periodicity problem persists a short distance from the rational surface. Specifically, expanding S about $r = r_0$ one finds

$$S \approx n[-\phi + q(r_0)(\theta - \theta_0) + (r - r_0)q'(r_0)(\theta - \theta_0)] \tag{10.121}$$

Since $n \gg 1$ the term $n(r - r_0)q'(r_0)$ becomes finite even when $r - r_0$ is small.

A similar situation arises in the general case. If one expands about the rational surface $\psi = \psi_0$, then

$$S \approx n\left[-\phi + \int_{\chi_0}^{\chi}\left(\frac{JB_\phi}{R}\right)_{\psi_0} d\chi' + (\psi - \psi_0)\int_{\chi_0}^{\chi}\frac{\partial}{\partial\psi_0}\left(\frac{JB_\phi}{R}\right) d\chi'\right] \tag{10.122}$$

The condition that the surface be rational is given by

$$n\int_0^{2\pi}\left(\frac{JB_\phi}{R}\right)_{\psi_0} d\chi = 2\pi m \tag{10.123}$$

where the integration is around one poloidal circuit.

The basic incompatibility between shear and periodicity is fundamental and at first glance might appear to render the previously developed ballooning mode analysis irrelevant. However, an elegant resolution of these difficulties has been given by Connor, Hastie, and Taylor (1979). The end result is that from a practical point of view essentially all of the previous analysis remains formally intact although with important differences in the interpretation.

The basic idea is as follows. Consider the original plasma displacement appearing in the Energy Principle for axisymmetric tori: $\boldsymbol{\xi} = \exp(-in\phi)\boldsymbol{\xi}(\psi, \chi)$. Physically acceptable displacements must satisfy the poloidal periodicity constraint $\boldsymbol{\xi}(\psi, \chi) = \boldsymbol{\xi}(\psi, \chi + 2\pi)$. Connor *et al.* pointed out that a periodic $\boldsymbol{\xi}$ could be constructed from a series of nonperiodic functions, called quasimodes, defined over the infinite domain. In particular,

$$\boldsymbol{\xi}(\psi, \chi) = \sum_p \boldsymbol{\xi}_Q(\psi, \chi + 2\pi p) \tag{10.124}$$

where $\boldsymbol{\xi}_Q(\psi, \chi)$, $-\infty < \chi < \infty$ is the quasimode. The function $\boldsymbol{\xi}(\psi, \chi)$ is by definition periodic if $\boldsymbol{\xi}_Q$ satisfies appropriate convergence properties as $\chi \to \pm\infty$:

$$\boldsymbol{\xi}_Q(\psi, \chi \to \pm\infty) = 0 \tag{10.125}$$

An important property of the quasimode representation is that $\boldsymbol{\xi}_Q$ satisfies the same equation as $\boldsymbol{\xi}$; that is, the $\boldsymbol{\xi}$ appearing in δW satisfies

$$\mathbf{F}(\boldsymbol{\xi}) = 0 \tag{10.126}$$

where $\mathbf{F}$ is the MHD force operator. Since the equilibria of interest are periodic, this implies that $\mathbf{F}(\psi, \chi) = \mathbf{F}(\psi, \chi + 2\pi)$. As a consequence, application of the force operator $\mathbf{F}$ to Eq. (10.124) shows that

$$\begin{aligned} \mathbf{F}(\boldsymbol{\xi}) &= \sum_p \mathbf{F}(\psi, \chi)\boldsymbol{\xi}_Q(\psi, \chi + 2\pi p) \\ &= \sum_p \mathbf{F}(\psi, \chi + 2\pi p)\boldsymbol{\xi}_Q(\psi, \chi + 2\pi p) \end{aligned} \tag{10.127}$$

Thus, $\boldsymbol{\xi}_Q$ satisfies

$$\mathbf{F}(\boldsymbol{\xi}_Q) = 0 \tag{10.128}$$

If one takes Eq. (10.128) as the starting point, then by inspection all the previous analyses of the ballooning mode apply with two exceptions. First, the unknown X appearing in Eq. (10.116) is actually X_Q, the amplitude of the corresponding quasimode. Second, the domain of integration is no longer $0 < \chi < 2\pi$ but instead extends over the infinite domain

$$\int_0^{2\pi} J\, d\chi \to \int_{-\infty}^{\infty} J\, d\chi \tag{10.129}$$

The quasimode representation resolves the incompatibility between shear and periodicity. In this representation the eikonal function S corresponds to the quasimode ξ_Q and as such is no longer required to produce periodic behavior in χ. Consequently, the "nonperiodic" functions given by Eqs. (10.121) and (10.122) are valid, physically acceptable choices for S.

The ballooning mode analysis is now easily completed. For the cylindrical case, $\mathbf{k}_\perp = \nabla S$ is obtained from Eq. (10.121) and is given by

$$k_n = n[q'(r_0)(\theta - \theta_0)]$$
$$k_t = \left(\frac{nB}{R_0 B_\theta}\right)_{r_0} \tag{10.130}$$

where small terms of order $n(r - r_0)$ have been neglected in k_t. The corresponding value of $W(\psi)$, obtained from Eq. (10.116), can be written as

$$W(r_0) = \int_{-\infty}^{\infty} d\theta \left(\frac{B_\theta}{rB^2}\right)\left[(k_n^2 + k_t^2)\left(\frac{dX}{d\theta}\right)^2 + k_t^2\left(\frac{2\mu_0 r}{B^2}\frac{dp}{dr}\right)X^2\right] \tag{10.131}$$

Here, for simplicity the 0 subscript has been suppressed from r and use has been made of the relations $\kappa_n = -(B_\theta^2/rB^2)$, $\kappa_t = 0$ and $B_\theta = \psi'(r)/R_0$.

The analogous relations for the general case have the form

$$k_n = n\left[RB_p \int_{\chi_0}^{\chi} \frac{\partial}{\partial\psi}\left(\frac{JB_\phi}{R}\right) d\chi'\right]$$
$$k_t = \frac{nB}{RB_p} \tag{10.132}$$
$$W(\psi) = \int_{-\infty}^{\infty} J d\chi \left[(k_n^2 + k_t^2)\left(\frac{1}{JB}\frac{\partial X}{\partial \chi}\right)^2 - \frac{2\mu_0 RB_p}{B^2}\frac{dp}{d\psi}(k_t^2\kappa_n - k_t k_n \kappa_t)X^2\right]$$

All coefficients are evaluated on the resonant surface $\psi = \psi_0$, but for simplicity the zero subscript has been suppressed.

Equation (10.132) is the desired final form of the general tokamak ballooning mode Energy Principle. It is an explicitly one-dimensional formulation valid for arbitrary β and ϵ that can be solved separately on any given flux surface. In general, to evaluate W, one requires a solution to the Grad–Shafranov equation plus an evaluation of the inverse transformation. This defines all quantities except $\chi_0(\psi)$, which is a free constant associated with the "radial" wave number k_n. In principle, $W(\psi)$ on a given flux surface must be minimized over all χ_0. It often, but not always, turns out in practice that $\chi_0 = 0$ is the most unstable case. Finally, it is worth noting that Connor *et al.* have extended the ballooning mode analysis to include the $1/n$ corrections. They find that these corrections are stabilizing so that $n \to \infty$ results in the most severe β limits due to ballooning modes.

10.5.4. Interchange Stability: The Mercier Criterion

The first important application of the ballooning mode Energy Principle is to the problem of interchange stability. Recall that in a straight screw pinch with shear an analysis of interchange perturbations leads to Suydam's criterion, a necessary condition for stability. This condition has been generalized to axisymmetric toroidal geometry. The resulting criterion is known as Mercier's criterion (Mercier, 1960) and is valid for arbitrary β, ϵ, and cross section.

The derivation presented below follows the work of Connor *et al* (1979) and proceeds in two steps. First, the interchange condition is derived in the infinite aspect ratio cylindrical limit reproducing Suydam's criterion. The purpose of this calculation is to demonstrate how the ballooning mode formalism, in which the angle θ is the independent variable, is related to the standard Suydam analysis in which $x = r - r_0$ is the independent variable. This relationship is easiest to understand in the familiar (r, θ) cylindrical system.

In the second step, the calculation is repeated for the general axisymmetry torus. Conceptually the calculation is quite similar to the straight cylinder although the details are significantly more complicated because of the two-dimensionality of the equilibrium. The end result is the Mercier criterion. This criterion is applied to a simple tokamak configuration and the implications are discussed.

Cylindrical Limit: Suydam's Criterion. The ballooning mode Energy Principle corresponding to the cylindrical limit is given by Eq. (10.131). Upon substituting the values of k_n and k_t from Eq. (10.130) one obtains

$$\frac{W(r)}{W_0} = \int_{-\infty}^{\infty} d\theta \left[(\theta^2 + a_1)\left(\frac{dX}{d\theta}\right)^2 - D_s X^2 \right] \qquad (10.133)$$

where $W_0(r) = n^2 q'^2 B_\theta / rB^2$, $a_1(r) = B^2/R_0^2 B_\theta^2 q'^2$, and $D_s = -2\mu_0 p' q^2 / rB_z^2 q'^2$. Note that D_s is Suydam's parameter [Eq. (9.53)]. Also, because of cylindrical symmetry one can set $\theta_0 = 0$ without loss in generality.

Within the ballooning mode formalism the interchange stability condition is related to the behavior of X for large θ; specifically, if X oscillates as $\theta \to \infty$ the system is unstable. This can be seen as follows. The minimizing Euler–Lagrange equation determining X at large θ is given by

$$\frac{d}{d\theta}\left(\theta^2 \frac{dX}{d\theta}\right) + D_s X = 0 \tag{10.134}$$

The solution is $X = c_1\theta^{p_1} + c_2\theta^{p_2}$, where

$$p_{1,2} = -\tfrac{1}{2} \pm \tfrac{1}{2}(1 - 4D_s)^{1/2} \tag{10.135}$$

Oscillatory solutions occur for $D_s > 1/4$. One now constructs a trial function as shown in Fig. 10.8, where X satisfies Eq. (10.134) in regions I and III and $X = X_0 =$ const in region II. A simple calculation shows that

$$\frac{W}{W_0} = -D_s X_0^2(\theta_3 - \theta_2) \tag{10.136}$$

Since $D_s > 1/4$ for oscillatory behavior, $W < 0$, indicating instability. For $D_s < 1/4$ oscillatory solutions do not exist and the system is stable to interchange perturbations. The condition $D_s < 1/4$ is just Suydam's criterion.

It is interesting to observe that in the ballooning mode formalism one is concerned with the oscillatory behavior of X at large θ while in the standard Suydam analysis one is concerned with the oscillatory behavior of ξ at small $x = r - r_0$.

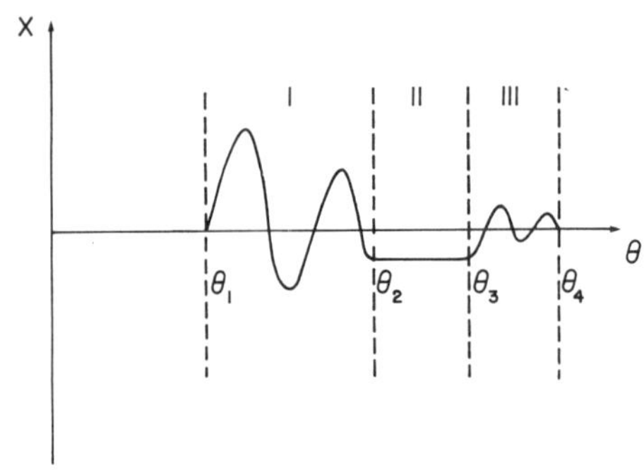

Figure 10.8. Trial function leading to violation of the Suydam criterion in the transform space.

The two analyses are actually exactly equivalent in that one is the Fourier transform of the other. For instance, in the standard Suydam's analysis, the Euler–Lagrange equation for the minimizing ξ is given by [Eq. (9.54)]

$$\frac{d}{dx}\left(x^2\frac{d\xi}{dx}\right) + D_s\xi = 0 \tag{10.137}$$

Upon introducing the Fourier transform

$$X(k) = \int_{-\infty}^{\infty} \xi(x)e^{ikx}\,dx \tag{10.138}$$

and assuming that ξ converges for large x, it can easily be shown that

$$\frac{d}{dk}\left(k^2\frac{dX}{dk}\right) + D_sX = 0 \tag{10.139}$$

One now identifies the eikonal function S with the total phase factor associated with $\xi: S = -n\phi + m\theta + kx$. Equation (10.121) then implies that $k = q'\theta$. Equation (10.139) reduces to

$$\frac{d}{d\theta}\left(\theta^2\frac{dX}{d\theta}\right) + D_sX = 0 \tag{10.140}$$

which is identical with Eq. (10.134).

Consequently, in the ballooning mode formalism the angle θ corresponds to the transform variable of the radius $x = r - r_0$ in real space. A standard result of Fourier transform theory is the equivalence of oscillatory behavior near $x \to 0$ in real space with oscillatory behavior near $\theta \to \infty$ in the transform space.

Toroidal Geometry: The Mercier Criterion. The derivation of the interchange stability criterion in toroidal geometry is conceptually quite similar to the cylindrical case. The behavior of the general ballooning mode equation is examined for large χ. Straightforward application of Newcomb's analysis (Section 9.4.4) indicates that the criterion for the onset of nonoscillatory behavior is a necessary condition for stability. Conversely, violation of this criterion is a sufficient condition for instability.

In effect the Mercier criterion can be viewed as a quantification of the convergence criterion for the validity of the ballooning mode formalism. If the Mercier criterion is violated the ballooning mode equation has improper convergence properties as $\chi \to \pm\infty$ and hence cannot be used. However, the system is already unstable to interchanges so one need proceed no further. If the Mercier criterion is satisfied the system is stable to interchange perturbations. In this case the convergence requirements on X as $\chi \to \pm\infty$ are satisfied and the ballooning mode equation can be used to test for ballooning stability.

The details of the derivation are more difficult than in the cylindrical case because of the complicated χ dependence of the coefficients. The critical feature of the analysis is the separation of the secular algebraic dependence of the coefficients from the periodic behavior. This allows a systematic asymptotic expansion for the solution as $\chi \to \infty$.

The derivation proceeds as follows. The minimizing Euler–Lagrange equation for the general ballooning mode Energy Principle [Eq. (10.132)] is given by

$$\begin{aligned} &\frac{\partial}{\partial \chi}\left(f\frac{\partial X}{\partial \chi}\right) - gX = 0 \\ &f = \frac{k_n^2 + k_t^2}{JB^2} \qquad\qquad (10.141) \\ &g = -\frac{2\mu_0 JRB_p}{B^2}\frac{dp}{d\psi}(k_t^2\kappa_n - k_t k_n \kappa_t) \end{aligned}$$

An examination of f and g indicates that secular algebraic terms (i.e., terms proportional to χ^ν as $\chi \to \infty$) appear only in the coefficient k_n; specifically, in the integral

$$\hat{\chi} \equiv \int_{\chi_0}^{\chi} \frac{\partial \hat{q}}{\partial \psi}\, d\chi' \qquad (10.142)$$

where $\hat{q}(\psi, \chi) = JB_\phi/R$. Note that from the definition of $q(\psi)$ [Eq. (6.25)] one finds that the average value of $\hat{q}$ over one period in χ satisfies

$$q(\psi) = \frac{1}{2\pi}\int_0^{2\pi} \hat{q}\, d\chi \qquad (10.143)$$

This implies that as $\chi \to \infty$, the function $\hat{\chi}(\chi, \psi) \to 2\pi(dq/d\psi)\chi$, explicitly exhibiting a linear secular dependence on χ.

The algebraic and periodic behavior of f and g are separated by rewriting Eq. (10.141) as follows:

$$\frac{\partial}{\partial \chi}\left[(a_0\hat{\chi}^2 + a_1)\frac{\partial X}{\partial \chi}\right] + \left(\frac{\partial c_0}{\partial \chi}\hat{\chi} + c_1\right)X = 0 \qquad (10.144)$$

Here, a_0, a_1, c_0, and c_1 are periodic functions of χ given by

$$\begin{aligned} a_0 &= \frac{R^2 B_p^2}{JB^2} \\ a_1 &= \frac{1}{JR^2 B_p^2} \\ c_0 &= \frac{\mu_0 p' F}{B^2} \\ c_1 &= \frac{2\mu_0 J p' \kappa_n}{RB_p} \end{aligned} \tag{10.145}$$

where $F(\psi) = RB_\phi$ and prime denotes $d/d\psi$. Also, in deriving the coefficient $\partial c_0/\partial\chi$ use has been made of the relation

$$\kappa_t = \left(\frac{1}{B}\frac{\partial B}{\partial \chi}\right)\mathbf{t}\cdot\nabla\chi = \frac{F}{JRB_pB^2}\left(\frac{\partial B}{\partial \chi}\right) \tag{10.146}$$

The goal now is to calculate the asymptotic solution for X as $\chi \to \infty$ and to determine the threshold for oscillatory behavior. The appropriate expansion for X can be written as

$$X = \hat{\chi}^\alpha\left[X_0 + \frac{X_1}{\hat{\chi}} + \frac{X_2}{\hat{\chi}^2} + \cdots\right] \tag{10.147}$$

where $X_j = X_j(\psi, \chi)$ are assumed to be periodic functions of χ with the same period as the equilibrium (2π). The parameter α is the indicial coefficient to be determined by solving the Euler–Lagrange equation. When α is imaginary or complex, X oscillates for large χ. If α is real no oscillatory solutions exist. Thus, the stability boundary defining the Mercier criterion corresponds to the transition value of α separating oscillatory and nonoscillatory behavior.

To determine α, Eq. (10.147) is substituted into Eq. (10.144). The resulting system of equations has the form of a polynomial series in $\hat{\chi}$. By equating the coefficients of descending powers of $\hat{\chi}$ to zero and making use of appropriate periodicity requirements one obtains a solution for α and the X_j. Specifically, to determine α the solution must be carried out to second order.

The leading-order equation, corresponding to the highest power of $\hat{\chi}$, $(\hat{\chi}^{\alpha+2})$, is given by

$$\frac{\partial}{\partial \chi}\left(a_0 \frac{\partial X_0}{\partial \chi}\right) = 0 \tag{10.148}$$

which has as its solution

$$X_0 = 1 \tag{10.149}$$

The first-order equation corresponding to $\hat{\chi}^{\alpha+1}$ can be written as

$$\frac{\partial}{\partial \chi}\left[a_0\left(\frac{\partial X_1}{\partial \chi} + \alpha \frac{\partial \hat{q}}{\partial \psi}\right)\right] + \frac{\partial c_0}{\partial \chi} = 0 \tag{10.150}$$

This equation can be integrated yielding

$$\frac{\partial X_1}{\partial \chi} + \alpha \frac{\partial \hat{q}}{\partial \psi} = \frac{K - c_0}{a_0} \tag{10.151}$$

where $K(\psi)$ is a free integration function. K is determined by requiring that X_1 be periodic. Therefore, integrating Eq. (10.151) over one period in χ gives

$$K = \langle c_0 \rangle + \alpha q' \langle a_0 \rangle \tag{10.152}$$

where

$$\langle Q \rangle \equiv \int_0^{2\pi} \frac{Q\, d\chi}{a_0} \Big/ \int_0^{2\pi} \frac{d\chi}{a_0} \tag{10.153}$$

Continuing to second order, one finds after a short calculation that the coefficient of $\hat{\chi}^{\alpha}$ can be expressed as

$$\frac{\partial}{\partial \chi}\left\{a_0\left[\frac{\partial X_2}{\partial \chi} + (\alpha - 1)\frac{\partial \hat{q}}{\partial \psi} X_1\right]\right\} + (\alpha + 1)a_0 \frac{\partial \hat{q}}{\partial \psi}\left(\frac{\partial X_1}{\partial \chi} + \alpha \frac{\partial \hat{q}}{\partial \psi}\right) + c_1 + \frac{\partial c_0}{\partial \chi} X_1 = 0 \tag{10.154}$$

The solution for X_2 is not explicitly required. The indicial coefficient α is determined as a periodicity constraint by integrating Eq. (10.154) over one period in χ, noting that the contribution due to the first term vanishes if X_2 is periodic. The result is

$$\int_0^{2\pi} d\chi\left[(\alpha + 1)a_0 \frac{\partial \hat{q}}{\partial \psi}\left(\frac{\partial X_1}{\partial \chi} + \alpha \frac{\partial \hat{q}}{\partial \psi}\right) + c_1 - c_0 \frac{\partial X_1}{\partial \chi}\right] = 0 \tag{10.155}$$

Substituting $\partial X_1/\partial \chi$ from Eq. (10.151) leads to the indicial equation

$$\alpha^2 + \alpha + D_M = 0 \tag{10.156}$$

where

$$D_M = \frac{1}{\langle a_0 \rangle^2 q'^2}\left[\langle a_0 c_1 \rangle + \left\langle c_0\left(c_0 - a_0 \frac{\partial \hat{q}}{\partial \psi}\right)\right\rangle - \langle c_0 \rangle\left\langle c_0 - a_0 \frac{\partial \hat{q}}{\partial \psi}\right\rangle\right] \tag{10.157}$$

The roots of the indicial equation are

$$\alpha = -\tfrac{1}{2} \pm \tfrac{1}{2}(1 - 4D_M)^{1/2} \tag{10.158}$$

The transition from oscillatory to nonoscillatory behavior occurs for $D_M = 1/4$. Consequently, the Mercier criterion for interchange stability is given by

$$D_M < 1/4 \tag{10.159}$$

The derivation is completed by rewriting D_M in terms of the physical variables as follows:

$$\begin{aligned} D_M &= \frac{\mu_0 p'}{q'^2}\left[2\left\langle\frac{RB_p\kappa_n}{B^2}\right\rangle + \left\langle\frac{\Lambda}{B^4}\right\rangle - \left\langle\frac{1}{B^2}\right\rangle\left\langle\frac{\Lambda}{B^2}\right\rangle\right]\Big/\left\langle\frac{R^2B_p^2}{JB^2}\right\rangle \\ \Lambda &= F\left(\mu_0 p'F - \frac{R^2B_p^2}{J}\frac{\partial \hat{q}}{\partial \psi}\right) \end{aligned} \tag{10.160}$$

and

$$\langle Q \rangle = \int_0^{2\pi}\left(\frac{QB^2}{R^2B_p^2}\right)J\,d\chi \Big/ \int_0^{2\pi}\left(\frac{B^2}{R^2B_p^2}\right)J\,d\chi \tag{10.161}$$

The Mercier criterion is a function only of the equilibrium quantities and can be tested separately, one flux surface at a time. It must be satisfied on each surface to guarantee interchange stability.

A comparison of Eqs. (10.160) and (10.83) indicates that toroidicity and shear add substantial complication to the interchange stability criterion. Even so, while neither necessary nor sufficient, it is still a desirable qualitative property that the quantity $\langle RB_p\kappa_n/B^2\rangle > 0$. Note that by transforming from orthogonal coordinates to arc length coordinates on any given surface (i.e., $J\,d\chi = dl_p/B_p = dl/B$) the following proportionality relations are obtained:

$$\left\langle\frac{RB_p\kappa_n}{B^2}\right\rangle \propto \int \frac{\kappa_n\,dl_p}{RB_p^2} \propto \int \frac{dl}{B^3}\frac{\partial}{\partial \psi}\left(p + \frac{B^2}{2\mu_0}\right) \tag{10.162}$$

The first relationship shows the similarity with EBT [Eq. (10.83)] while the second is closely related to the magnetic-well figure of merit introduced in Chapter 4 and further discussed in Section 10.3.

Application of Mercier's Criterion. Mercier's criterion is perhaps the simplest test of tokamak stability, although it still in general requires a numerical solution to the equilibrium Grad–Shafranov equation. There are, however, several simple analytic limits that shed some light on the requirements for Mercier stability. As might be expected, in the one-dimensional cylindrical limit the quantity D_M reduces to D_S, leading to Suydam's criterion

$$\left(\frac{rq'}{q}\right)^2 + \frac{8\mu_0 rp'}{B_\phi^2} > 0 \tag{10.163}$$

where prime denotes d/dr.

One of the better known toroidal limits (Ware and Haas, 1966; Shafranov and Yurchenko, 1968) corresponds to the case of a circular cross section, large aspect ratio tokamak with $\beta_p \sim 1$. The stability condition is given by

$$\left(\frac{rq'}{q}\right)^2 + 4r\beta'(1 - q^2) > 0 \tag{10.164}$$

where $\beta = 2\mu_0 p/B_0^2$. Note that this expression is very similar to Suydam's criterion for the straight tokamak except for the factor $(1 - q^2)$. This modification predicts that configurations with a negative pressure gradient, even near the axis, can be stabilized if $q > 1$. In contrast, the straight tokamak always predicts a small region of instability under similar circumstances.

The additional stability found in the torus is associated with favorable modifications to the average field line curvature; that is, on the inside of the torus ($\theta = \pi$) the toroidal field bends away from the plasma producing favorable curvature. The opposite is true on the outside ($\theta = 0$). To first order in ϵ the toroidal curvature averages to zero. However, there is a net favorable second-order contribution, which enters competitively with the poloidal field curvature. This can be seen qualitatively by noting that in the large aspect ratio limit

$$\boldsymbol{\kappa} \approx -\frac{B_\theta^2}{rB_0^2}\mathbf{e}_r - \frac{\mathbf{e}_R}{R} \tag{10.165}$$

The normal component of $\boldsymbol{\kappa}$ is approximately given by $\kappa_n \approx \mathbf{e}_r \cdot \boldsymbol{\kappa}$ and can be written as

$$\kappa_n \approx -\frac{B_\theta^2}{rB_0^2} - \frac{\cos\theta}{R_0}\left(1 - \frac{r}{R_0}\cos\theta + \cdots\right) \tag{10.166}$$

Observe that the magnitude of the toroidal curvature ($\sim 1/R_0$) is a factor $1/\epsilon$ larger than the poloidal curvature ($\sim B_\theta^2/rB_0^2$). However, after simple averaging over θ the contributions are competitive and are given by

$$\frac{1}{2\pi}\int_0^{2\pi} \kappa_n\, d\theta \approx \kappa_c(1 - q^2/2) \tag{10.167}$$

where $\kappa_c = -B_\theta^2/rB_0^2$ is the curvature in the straight cylinder.

The correct result, as given by Eq. (10.164), contains an extra factor of $q^2/2$ and requires a surprisingly lengthy calculation. All of the terms in the Mercier criterion must be maintained as well as the small toroidal shift of the flux surfaces. This more careful averaging with the correct weight factors indicates that there is a net favorable effect when $q > 1$.

For tokamaks, which are normally characterized by low β, the shear term in Mercier's criterion dominates over most of the profile. However, near the origin the shear is small, and it is here that the stability condition is most easily violated. With this insight, one concludes that the Mercier criterion essentially sets a limit on the minimum q_0.

Another important question concerning the Mercier criterion is the effect of noncircularity. This problem has been investigated by a number of authors, who examined the behavior of the Mercier criterion near the magnetic axis where the shear is small (Solov'ev *et al.*, 1969; Laval *et al.*, 1971; Lortz and Nuhrenberg, 1973; Mikhailovskii and Shafranov, 1974; and Mikhailovskii, 1974)]. Their results can be summarized as follows. For noncircular cross sections the on-axis stability criterion $1 < q_0^2$ is replaced by

$$1 < q_0^2\left\{1 - \frac{4}{1 + 3\kappa^2}\left[\frac{3}{4}\frac{\kappa^2 - 1}{\kappa^2 + 1}\left(\kappa^2 - \frac{2\delta}{\epsilon}\right) + \frac{(\kappa - 1)^2\beta_{p0}}{\kappa(\kappa + 1)}\right]\right\} \tag{10.168}$$

where κ is the elongation, β_{p0} is the poloidal β on axis, δ is the triangularity, and $\epsilon = r_0/R_0$ is the inverse aspect of the surface under consideration. The geometric interpretation of κ, δ, and ϵ is illustrated in Fig. 10.9. Observe the following points. For cross sections with no elongation ($\kappa = 1$), triangularity and finite β_{p0} have no effect on the

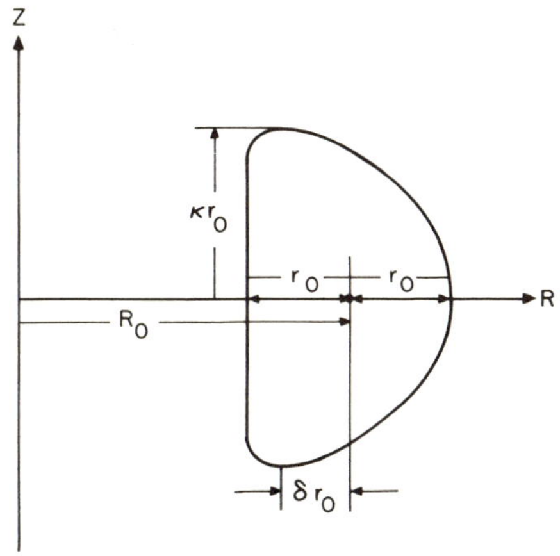

Figure 10.9. Geometry defining κ, δ, and r_0/R_0 in a noncircular tokamak. Note that $\delta > 0$ for an outward-pointing triangle.

stability boundary. Elongation by itself ($\kappa > 1$, $\delta = \beta_{p0} = 0$) is destabilizing. However, a combination of elongation and sufficient outward pointing triangularity ($\kappa > 1$, $\delta/\epsilon > \kappa^2/2$, $\beta_{p0} = 0$) is stabilizing; the allowable value of $1/q_0 \propto RJ_\phi/B_\phi$ on axis is increased. For any elongation the effect of β_{p0} is destabilizing. Also, as an academic point, Eq. (10.168) indicates that a flattened tokamak with inward pointing triangularity also improves stability.

The improved stability associated with elongated "D" shapes is related to the fact that for such cross sections a magnetic line has a relatively large fraction of its trajectory located in the favorable curvature region on the inside of the torus.

In summary, the Mercier criterion represents one limit on the maximum value of $J_\phi \sim 1/q$ on axis in a toroidal tokamak. For sufficiently high q_0 the average curvature of a field line becomes favorable and a negative pressure gradient on axis, which is always unstable in the straight tokamak, is stabilized. As with Suydam's criterion, a Mercier interchange perturbation by itself is unlikely to be important in an actual experiment. However, the presence of such instabilities implies that dangerous large-scale modes are also very likely to exist.

10.5.5. Ballooning Modes

An important application of the ballooning mode Energy Principle is to the problem of ballooning modes themselves. In general, this involves the solution of a second-order ordinary differential equation on each flux surface [Eq. (10.141)]. The coefficients depend upon the parameter χ_0 as well as the equilibrium, which is assumed known.

Stability can be ascertained in either of two ways. First, Newcomb's analysis can be applied. Thus, if on any flux surface a value of χ_0 is found such that the solution to Eq. (10.141) which is regular at $\chi \to -\infty$ has a zero crossing for $\chi < +\infty$, the system is unstable. A second approach is to introduce a convenient normalization and then compute the corresponding eigenvalue (as a function of χ_0). Stability is determined by the sign of the eigenvalue.

Analytic Model. Many numerical studies of ballooning mode stability have been carried out for realistic experimental situations. These results are summarized in Section 10.5.8. For present purposes, however, it is convenient to consider a very simple model which allows analytic insight into the behavior of ballooning modes. The model of interest is a large aspect ratio, circular cross section tokamak. The pressure profile has the

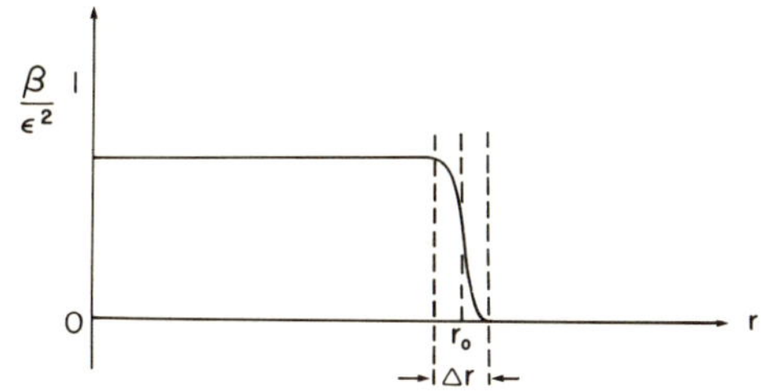

Figure 10.10. Simple $\beta(r)$ profile used in the analytic study of the ballooning mode.

special shape illustrated in Fig. 10.10. The crucial feature is that the average value of β is assumed small ($\beta \sim \epsilon^2$) although in a narrow layer $\Delta r \sim \epsilon$ about $r = r_0$ the gradient is larger by $1/\epsilon$: $r_0\beta' \sim \epsilon$.

This leads to a great simplification in the analysis. When $\beta \sim \epsilon^2$ the flux surface shifts are small and most quantities assume their cylindrical values. Even so, there remains a small region where $r_0\beta'$ is sufficiently large to drive ballooning instabilities. From a practical point of view this is a pessimistic profile since the overall average β is small while high β' ballooning modes can still be excited. Nevertheless, the model is quite reliable for developing qualitative insight because of its analytic simplicity.

The analysis proceeds with the reduction of the general Euler–Lagrange equation given by Eq. (10.141) or alternatively by Eq. (10.144). The coefficients a_0, a_1, c_0, c_1 are to be evaluated to leading order in the inverse aspect ratio expansion. With two exceptions, all quantities appearing in Eq. (10.144) can be replaced by their cylindrical values; the toroidal corrections are small by ϵ. Thus,

$$\begin{aligned}
\chi &\rightarrow \theta \\
R &\rightarrow R_0 \\
B_p &\rightarrow B_\theta(r) \\
J &\rightarrow r/B_\theta \\
B &\rightarrow B_0 \\
\frac{dp}{d\psi} &\rightarrow \frac{1}{rB_\theta}\frac{dp}{dr} \\
\kappa_n &\rightarrow -\frac{\cos\theta}{R_0}
\end{aligned} \tag{10.169}$$

Note that to leading order only the toroidal curvature enters κ_n. With these simplifications the coefficients a_0, a_1, and c_1 can be written as

$$
\begin{aligned}
a_0 &= \frac{R_0^2 B_\theta^3}{B_0^2 r} \\
a_1 &= \frac{1}{r B_\theta R_0^2} \\
c_1 &= -\left(\frac{2\mu_0 r p'}{R_0^2 B_\theta^3}\right) \cos\theta
\end{aligned}
\tag{10.170}
$$

The two exceptions mentioned above involve the evaluation of c_0 and $\hat{\chi}$. c_0 must be calculated to first order since the leading-order contribution is independent of θ and the quantity actually required is $\partial c_0/\partial\theta$. This evaluation is easily carried out by noting that, correct to first order, $B \approx B_\phi = F/R$. One finds

$$c_0 \approx \left(\frac{\mu_0}{F}\frac{dp}{d\psi}\right) R^2 \approx \left(\frac{\mu_0}{F}\frac{dp}{d\psi}\right)(R_0^2 + 2R_0 r \cos\theta) \tag{10.171}$$

Since $p(\psi)$ and $F(\psi)$ are constant on a flux surface, it follows that

$$\frac{\partial c_0}{\partial \chi} \approx -\left(\frac{2\mu_0 r p'}{R_0 B_0 B_\theta}\right) \sin\theta \tag{10.172}$$

where hereafter prime denotes (d/dr).

The evaluation of $\hat{\chi}$ is slightly subtle. Nominally, with the small shift assumption one might expect only the cylindrical contribution to appear. However, the integrand in $\hat{\chi}$ is proportional to the local shear $\partial\hat{q}/\partial\psi \propto \partial\hat{q}/\partial r$. The expansion of this quantity has the form $\partial\hat{q}/\partial r = q' + (K_1\psi_1 + K_2\psi_1' + K_3\psi_1'')\cos\theta$. The last three terms are the toroidal corrections expressed in terms of ψ_1, the first-order solution to the Grad–Shafranov equation. An examination of ψ_1 as given by Eq. (6.70) indicates that ψ_1 and ψ_1' involve only integrals of p'. Thus, these terms are of order ϵ and can be neglected. On the other hand, ψ_1'' has a contribution directly proportional to p' and hence is of order unity within the gradient layer. As a result $\hat{\chi}$ can be written as

$$\hat{\chi} \approx \frac{1}{R_0 B_\theta} \int_{\theta_0}^{\theta} d\hat{\theta} \left(q' - \frac{q\psi_1''}{R_0 B_\theta} \cos\hat{\theta}\right) \tag{10.173}$$

Within the layer ψ_1'' is easily evaluated from Eq. (6.70):

$$\psi_1'' \approx -\frac{2\mu_0 r p'}{B_\theta} \tag{10.174}$$

Thus, $\hat{\chi}$ is given by

$$\hat{\chi} = \frac{1}{R_0 B_\theta}\left[q'(\theta - \theta_0) + \frac{2\mu_0 rqp'}{R_0 B_\theta^2}(\sin\theta - \sin\theta_0)\right] \qquad (10.175)$$

These results are now substituted into Eq. (10.144) yielding

$$\frac{\partial}{\partial\theta}\left[(1 + \Lambda^2)\frac{\partial X}{\partial\theta}\right] + \alpha(\Lambda\sin\theta + \cos\theta)X = 0 \qquad (10.176)$$

where

$$\Lambda(\theta) = s(\theta - \theta_0) - \alpha(\sin\theta - \sin\theta_0) \qquad (10.177)$$

and s and α are two parameters given by

$$s = \frac{rq'}{q}$$

$$\alpha = -\frac{2\mu_0 r^2 p'}{R_0 B_\theta^2} = -q^2 R_0 \beta' \qquad (10.178)$$

Equation (10.176) is the desired form of the ballooning mode equation for the model equilibrium. It is considerably simpler than the starting equation and depends upon only two equilibrium quantities s and α. Note that s represents the average shear while α is a measure of the pressure gradient.

Numerical Solution. Before attempting to solve Eq. (10.176) analytically, it is helpful to present the results as obtained from a numerical solution. These are given in the form of an (s, α) diagram in Fig. 10.11. The two solid curves represent the numerically computed marginal stability boundaries. One branch separates the "first region of stability" from the unstable region and is consistent with the intuition developed for the behavior of ballooning modes. At a fixed value of shear the system is stable for small pressure gradients. As the pressure gradient (i.e., α) increases the perturbation develops a progressively larger ballooning component. Eventually, at sufficiently high α, the destabilizing contribution from the unfavorable curvature region overcomes the shear and the system becomes unstable. As expected, when the shear increases, the maximum allowable pressure gradient also increases.

Perhaps the most surprising feature of the (s, α) diagram is the second branch which separates the unstable region from the so-called "second region of stability." In this region sufficiently large values of the pressure gradient are stabilized. Interestingly, this region is most easily

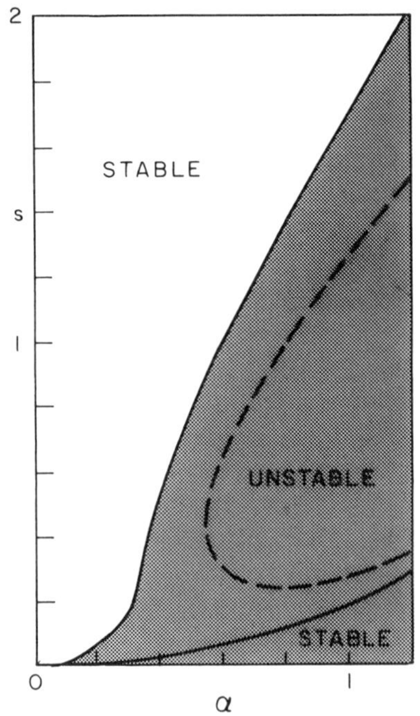

Figure 10.11. s vs. α marginal stability diagram for the analytic ballooning mode model. The region in the upper left corresponds to the first region of stability. The region in the lower right corresponds to the second region of stability. Courtesy J. Connor and M. Turner, Culham Laboratory, Great Britain.

accessible at low values of shear. The possibility of high-β operation in the second region of stability is an enticing prospect for the tokamak concept.

None of the insight thus far developed for ballooning modes has suggested the existence of such a region. To understand this behavior it is instructive to attempt an analytic solution to Eq. (10.176).

Analytic Solution. There is no simple analytic solution to Eq. (10.176). However, many of the main features can be ascertained by calculating the (s, α) diagram variationally. The Energy Principle corresponding to Eq. (10.176) is given by

$$W = \frac{1}{2\pi}\int_{-\infty}^{\infty} d\theta\left[(1 + \Lambda^2)\left(\frac{\partial X}{\partial \theta}\right)^2 - \alpha(\Lambda \sin\theta + \cos\theta)X^2\right] \quad (10.179)$$

Following the work of Coppi (1977) assume a simple trial function for X of the form

$$X = \begin{cases} 1 + \cos\theta, & -\pi < \theta < \pi \\ 0 & |\theta| > \pi \end{cases} \quad (10.180)$$

X is illustrated in Fig. 10.12. Observe that the trial function is largest

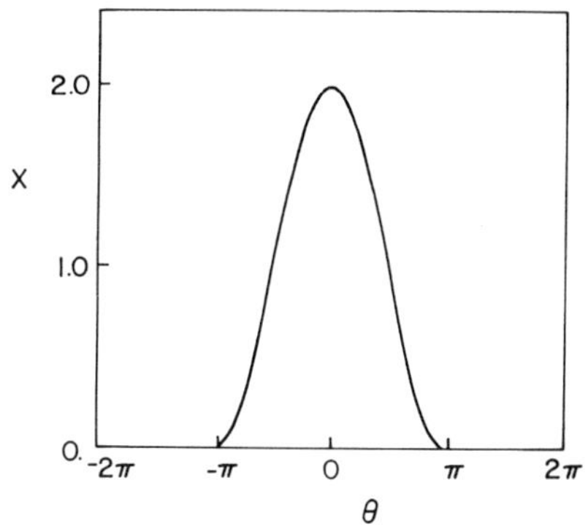

Figure 10.12. Simple trial function used in the analytic derivation of the ballooning mode criterion.

near $\theta = 0$ where the toroidal curvature is unfavorable. Conversely, $X = 0$ at $\theta = \pi$ where the curvature is favorable. Thus, the ballooning nature of the trial function is evident. Substituting X into Eq. (10.179) and assuming for simplicity that $\theta_0 = 0$ yields

$$W = 1.39s^2 - 2.17s\alpha + \alpha^2 - \alpha + 0.5 \qquad (10.181)$$

Setting $W = 0$ for marginal stability leads to a quadratic equation for s which has as its solution

$$s = 0.78\alpha \pm (0.72\alpha - 0.36 - 0.11\alpha^2)^{1/2} \qquad (10.182)$$

Equation (10.182) is illustrated as the dashed curve in Fig. 10.11. Except for small s and α the analytic solution is qualitatively similar to the numerical solution.

An examination of the individual terms contributing to Eq. (10.181) indicates that the second region of stability is closely related to a pressure-driven modulation of the local shear $\sim\partial\hat{q}/\partial\psi$. This can be seen by noting that the local shear is given by [see Eq. (10.173)]

$$\frac{r}{q}\frac{\partial\hat{q}}{\partial r} = s - \alpha\cos\theta \qquad (10.183)$$

where s represents the average shear and $\alpha\cos\theta$ represents the pressure-driven modulation. If the modulation is neglected, the local shear is equal to the average shear and $\Lambda \approx s(\theta - \theta_0)$. Upon repeating the variational analysis using this function for Λ and $\theta_0 = 0$ one obtains a simple relation for $\alpha = \alpha(s)$:

$$\alpha = \frac{0.5 + 1.39s^2}{1 + 0.83s} \qquad (10.184)$$

Equation (10.184) is illustrated in Fig. 10.13 along with the full

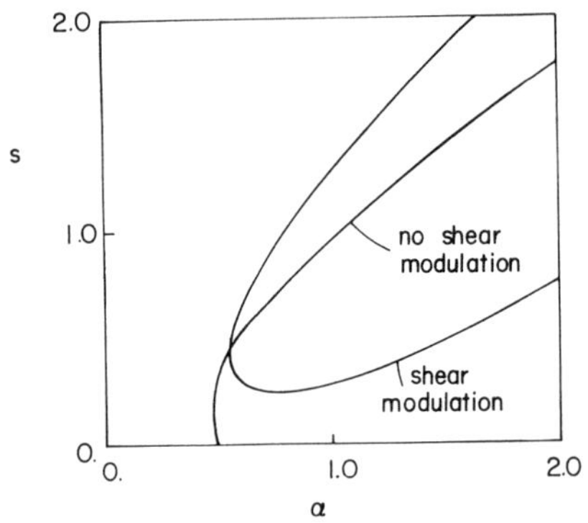

Figure 10.13. s vs. α diagram showing the importance of shear modulation on the second region of stability.

variational solution given by Eq. (10.182). Observe that without the pressure-driven modulation there is no second region of stability. For large pressure gradients relative to s, the local shear remains finite (i.e., does not cross zero) in the unfavorable curvature region and gives rise to a second region of stability at large α.

The results just presented are based on the trial function illustrated in Fig. 10.12. For comparison, an illustration is presented in Fig. 10.14 of a complete ballooning mode eigenfunction. Shown here are the quasi-mode $X_Q(\theta)$, $-\infty < \theta < \infty$, the corresponding eigenfunction $X(\theta)$, $-\pi \leq \theta \leq \pi$ as calculated from Eq. (10.124), and a sketch of a perturbed plasma surface exhibiting a ballooning instability.

Applications of the Ballooning Mode Stability Criterion. The ballooning mode stability criterion associated with the first region of stability sets a limit on the maximum allowable value of the pressure gradient, specifically on $|q^2R_0\beta'|$. For the simple model studied this limit only has an effect in the narrow gradient layer. For more general profiles, a qualitatively similar limit applies on each flux surface in the plasma. Consequently, if one integrates the critical pressure gradient over the entire profile there is a maximum allowable overall average β.

A simple estimate of the average β limit can be obtained by assuming that the ballooning stability criterion given by the (s, α) diagram applies over an entire diffuse pressure profile. Furthermore, it is assumed that the first region of stability boundary in the (s, α) diagram can be approximated by $\alpha = Ks$, where $K \approx 0.6$. This approximation is valid over most of the profile except near the origin. Thus, at every point

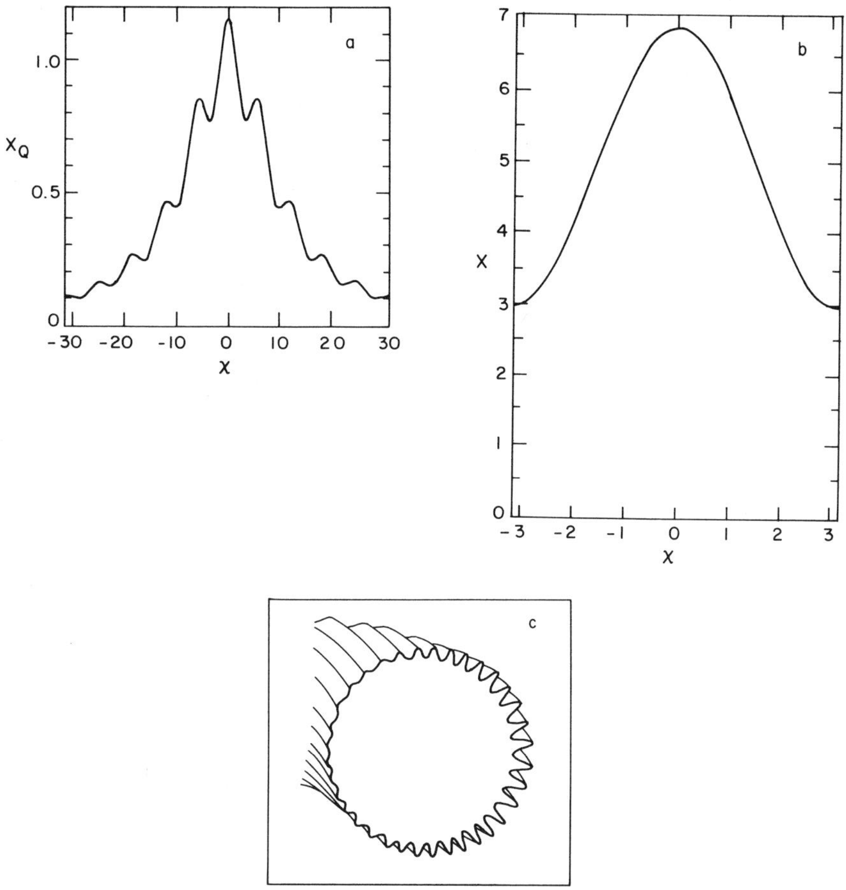

Figure 10.14. Ballooning mode perturbation (a) quasimode X_Q; (b) eigenfunction X; (c) ballooning mode displacement.

in the plasma the pressure gradient must satisfy

$$-R_0\beta' \leq K(rq'/q^3) \tag{10.185}$$

One now multiplies Eq. (10.185) by r^2 and integrates over the plasma volume. For a relatively flat q profile (i.e., broad current profile) given by

$$q = q_0(1 + r^3/r_0^3) \tag{10.186}$$

the right-hand term can be easily integrated. This leads to an average toroidal β limit of the form

$$\beta_t \leq \frac{K}{2}\frac{\epsilon}{q_0 q_a}\left(1 - \frac{q_0}{q_a}\right) \tag{10.187}$$

where $q_a = q_0(1 + a^3/r_0^3)$. Setting $K = 0.6$ and $q_0 = 1$ yields

$$\beta_t \leq 0.3\frac{\epsilon}{q_a}\left(1 - \frac{1}{q_a}\right) \tag{10.188}$$

For typical tokamak parameters ($\epsilon = 1/3$, $q_a = 3$) Eq. (10.188) predicts $\beta_t \leq 2\%$. This result is somewhat optimistic with respect to realistic diffuse profile numerical calculations. Still, it is a reasonable estimate considering the simplicity of the model.

As might be expected, noncircularity plays an important quantitative role in setting the value of the first region of stability β limit. Elongated, outward-pointing triangular cross sections, favorable for the Mercier criterion, are also favorable for raising the ballooning mode β limit, leading to values on the order of 10%. In addition, if the triangularity is further exaggerated so that the plasma assumes a "bean" shape, one finds from numerical calculations that there may be a path to the second region of stability that never crosses the unstable region. These issues are discussed in more detail in Section 10.5.8, which summarizes the numerical stability studies.

There is a final point worth noting with respect to optimizing the pressure and current profiles against first region of stability ballooning modes. Since shear has a strong stabilizing influence on these modes, one expects intuitively that the optimum pressure profile will have its largest gradients near the outer edge of the plasma where the shear is large. This insight implies broad pressure and current profiles, but must be viewed cautiously. As shown in the straight tokamak model, large current gradients near the plasma surface are likely to drive higher-m external kink modes unstable. Since there are opposing constraints for ballooning and kink modes, a completely optimized profile represents a compromise between these two effects.

In summary, ballooning modes associated with the first region of stability are pressure-driven instabilities localized in the region of unfavorable curvature on the outside of the torus. High shear is in general favorable for stability. These modes are important in that they ultimately set an upper limit on the maximum allowable average β which scales as $\beta_t \sim \epsilon/q_0 q_a$. The ballooning mode analysis also shows the existence of a second region of stability where sufficiently large pressure gradients are stabilized for low shear. This unanticipated result is a consequence of a pressure-driven modulation of the local shear and represents an enticing prospect for the tokamak concept.

10.5.6. Low-n Internal Modes

Although high-n ballooning modes lead to the lowest β_t limits for internal instabilities, it remains of interest to examine more general lower n number modes. The reason is that such modes set an upper limit on the toroidal current density on axis. Recall that in a straight tokamak $q_0 > 1$ is required for $n = 1$, $m = 1$ stability.

Since low n number modes have large-scale radial structure, the stability boundaries can often only be determined by numerical computation. However, an interesting analytic calculation has been carried out in the limit of a large aspect ratio, circular cross section tokamak by Bussac *et al.* (1975) which sheds considerable light on this problem. By applying the conventional tokamak expansion, $\beta_t \sim \epsilon^2$, $\beta_p \sim 1$, they derive an expression for δW as a function of n for the toroidal case. Since β is small their calculation can be viewed as a determination of the effects of toroidicity on the current density limit of low n number internal kinks.

After a rather involved calculation too complicated to reproduce here, Bussac *et al.* derive an expression for δW which can be written as

$$\frac{\delta W_F}{W_0} = \left(1 - \frac{1}{n^2}\right) W_c + \frac{1}{n^2} W_t \tag{10.189}$$

Here $W_0 = 2\pi^2 B_0^2 \xi_0^2 / \mu_0 R_0$ and

$$W_c = \int_0^{r_s} \left[r\beta' + \frac{n^2 r^2}{R_0^2} \left(1 - \frac{1}{nq}\right)\left(3 + \frac{1}{nq}\right)\right] r\, dr \tag{10.190}$$

is the contribution due to the straight tokamak using the standard $m = 1$ internal-mode eigenfunction [see Eq. (9.165) and Fig. 9.30] with r_s the singular surface defined by $nq(r_s) = 1$. The toroidal contribution W_t is in general quite complicated. However, in the limit of $|q_0 - 1| \ll 1$, $q_0 < 1$, and a parabolic current profile, a simple expression is obtained, given by

$$W_t \approx \frac{3n^2 r_s^4}{R_0^2}(1 - q_0)\left(\frac{13}{144} - \beta_p^2\right)$$

$$\beta_p = -\frac{R_0^2}{n^2 r_s^2} \int_0^{r_s} r^2 \beta'\, dr \tag{10.191}$$

Note that for $n \gg 1$ $\delta W_F / W_0 \approx W_c$ and stability is identical to that in the straight tokamak, requiring $nq_0 > 1$. For the most severe case, $n = 1$, the cylindrical contribution vanishes and $\delta W_F/W_0 = W_t$. Here, in contrast to the cylindrical result, the internal-kink mode is stable in the limit $\beta_p \to 0$ if $q_0 < 1$. However, in both cases increasing β_p is destabilizing and the toroidal calculation predicts instability for $\beta_p > \sqrt{13/12} \approx 0.3$.

Similar results have been found numerically (Sykes and Wesson, 1974; Kerner, 1976; Berger *et al.*, 1977). These calculations show that for small β_p there still exists a current-density limit but that it is associated with the $n = 2$, $m = 1$ mode; $q_0 > 0.5$. Consistent with the analytic calculation of Bussac *et al.*, the $n = 1$ mode is stabilized by toroidicity at low β_p. However, as β_p increases the $n = 1$, $m = 1$ mode is eventually excited. The numerical results show that this mode requires $q_0 > 1$ for stability.

The low β_p stabilization of the $n = 1$, $m = 1$ mode has been recently reexamined by Wesson (1985). It is shown that the stabilization is sensitively dependent upon the shape of the q profile near the axis. For parabolic profiles, as assumed by Bussac *et al.*, the mode is stabilized. For much flatter profiles, as observed in many tokamak experiments, the stabilization disappears and $q_0 > 1$ is required regardless of β_p.

Qualitatively this behavior can be understood in terms of the coupling of the straight tokamak modes. The $m = 2$, $n = 1$ mode is coupled to the $m = 1$, $n = 1$ mode when toroidicity is introduced. Since the $m = 2$, $n = 1$ mode is stable in the straight system there is a net stabilizing effect. When the q profile is parabolic, the shear is "high" producing a strong coupling. Hence, the mode is stabilized unless β_p is large enough. For very flat q profiles the shear is "low" and the coupling is weak. In this case instability persists even if $\beta_p = 0$.

Since the $n = 1$ internal kink depends crucially upon the existence of a $q = 1$ resonant surface, one does not expect noncircularity to significantly alter the $q_0 = 1$ stability condition for flat q profiles.

The $n = 1$ mode and resistive modifications play an important role in the operation of tokamaks. For realistic J and p_p profiles one cannot operate a tokamak with $q_0 < 1$ or else MHD instabilities would be excited. Interestingly, it is also quite difficult to operate with $q_0 > 1$ in most tokamaks even though such systems would be MHD stable. The reason is associated with the temperature dependence of the resistivity: $\eta \propto T_e^{-3/2}$. An initially peaked temperature profile is preferentially ohmically heated near the axis $r = 0$. Since $J \propto T^{3/2}$ in steady state, the current profile becomes increasingly peaked as the heating progresses. This peaking in J is equivalent to a decreasing value of $q_0 = 2B_0/\mu_0 r_0 J(0)$. Eventually q_0 decreases to unity and an internal $n = 1$, $m = 1$ mode is excited. The system then operates with a sequence of "sawtooth" relaxation oscillations about the value $q_0 = 1$. Consequently, from a practical operational point of view the low-n internal modes in a tokamak can be viewed as clamping the value of q on axis to $q_0 = 1$.

10.5.7. External Modes

As in the straight tokamak, external modes set severe limits on ideal MHD stability. Recall that in a straight tokamak stability is achieved by requiring $q_a \geq 1$ for the $m = 1$ mode and requiring a peaked current profile for the higher m modes. In a torus, there is a further limitation which sets a maximum value for β_t. This limit is usually more severe than that due to $n \to \infty$ internal ballooning modes.

In the regime of the ohmically heated tokamak $\beta \sim \epsilon^2$, ballooning effects are unimportant on external modes. Stability is identical to that in the straight tokamak. The new stability limit appears in the high $\beta \sim \epsilon$ regime and is associated with toroidal ballooning effects. In fact, for the most unstable modes the eigenfunction corresponds to a combination of ballooning and kinking. Thus, in contrast to $n \to \infty$ internal ballooning modes which are driven solely by the pressure gradient, external modes are driven by a combination of both the pressure gradient and the parallel current.

Another interesting feature of high-β external modes is the issue of q_a and q_* and their relationship to stability. Recall that in a low-β circular system $q_a = q_*$ and no distinction can be made. At high β where q_a and q_* no longer coincide it is shown that q_* is the critical parameter for stability.

Since external kinks are low-m-number modes with gross radial structure, most of the stability results thus far obtained have required numerical studies. These are summarized in Section 10.5.8. There is, however, one simple equilibrium model that gives surprisingly reliable information about the β limits. This model is known as the surface current model and its equilibrium and stability properties are discussed below (Freidberg and Haas, 1973).

Sharp Boundary Model. In the sharp boundary model, the shape of the plasma surface is prescribed and all currents are assumed to flow only on this surface. Within the plasma $\mathbf{J} = 0$ and $p = \text{const}$. Outside, the plasma is surrounded by a vacuum. The only requirements for equilibrium are $[\![\mathbf{n} \cdot \mathbf{B}]\!] = 0$ and $[\![p + B^2/2\mu_0]\!] = 0$ across the surface. Because of its simplicity the model can be solved exactly or in the context of either the ohmically heated or high-β expansion. For present purposes it is useful to focus attention on the high-β expansion. In this case it can be shown that for a circular cross section the poloidal field just outside the plasma is given by (see Problem 10.8)

$$b_\theta(\theta) \equiv \frac{B_\theta(a, \theta)}{\epsilon B_0} = \frac{1}{q_*}\left(\frac{\pi}{2E}\right)[1 - k^2 \sin^2(\theta/2)]^{1/2} \qquad (10.192)$$

where $q_* = 2\pi^2 B_0/\mu_0 R_0 I_0$, $E(k)$ is the complete elliptic integral of the second kind and $k \sim 1$ is a parameter related to $\beta_t \equiv 2\mu_0 p/B_0^2$ as follows:

$$\frac{\beta_t q_*^2}{\epsilon} = \left(\frac{\pi k}{4E}\right)^2 \qquad (10.193)$$

As β_t varies, the shape of the cross section is held fixed. If the current I_0 (i.e., q_*) is also held fixed there is an equilibrium limit corresponding to $k \to 1$ given by

$$\begin{aligned} \frac{\beta_t}{\epsilon} &\leq \frac{\pi^2}{16 q_*^2} \\ \epsilon\beta_p &\leq \frac{\pi^2}{16} \end{aligned} \qquad (10.194)$$

In analogy with the uniform current equilibrium (see Section 6.7) a crude approximation to flux conservation can be made by requiring q_a to be held fixed rather than q_* as β_t is varied. In this case there is no equilibrium limit and as $\beta_t \to \infty$

$$\begin{aligned} \frac{1}{q_*} &\to \frac{4}{\pi}\left(\frac{\beta_t}{\epsilon}\right)^{1/2} \\ \epsilon\beta_p &\to \frac{\pi^2}{16} \end{aligned} \qquad (10.195)$$

Thus, despite its simplicity, the surface current equilibrium has many features in common with more complicated diffuse high-β equilibria.

The stability of the system is determined by an analysis of the Extended Energy Principle $\delta W = \delta W_F + \delta W_S + \delta W_V$ (see Section 8.8.2). An advantage of the surface current model is that after Fourier analysis with respect to ϕ, the stability problem is reduced to a minimization with respect to a *single, one-dimensional, scalar* quantity $\xi(\theta)$. The procedure can be carried out for arbitrary shape cross sections and represents a substantial saving over the general diffuse case where the unknown perturbations are two-dimensional functions of r and θ.

Some insight into the nature of the high-β instabilities can be ascertained by examining δW_S the only source of instability in the surface current model. For the circular-cross-section case a simple calculation shows that

$$\frac{\delta W_S}{W_0} = -\frac{1}{2\pi}\int_0^{2\pi} d\theta\, |\xi|^2[b_\theta^2 + (\beta_t/\epsilon)\cos\theta] \tag{10.196}$$

where $W_0 = 2\pi^2\epsilon^2 B_0^2 R_0/\mu_0$. Note that there are two contributions to drive instabilities. The first, proportional to B_θ^2, represents the usual destabilizing effect associated with the parallel currents (i.e., the kink term). The second contribution represents the effect of the toroidal field curvature (i.e., the pressure-driven term). The $\cos\theta$ dependence reflects the fact that the toroidal curvature is unfavorable on the outside of the torus while favorable on the inside. Indeed, it is just this effect which causes the eigenfunction to balloon. Observe also that for the low-β ordering, $\beta \sim \epsilon^2$, the ballooning contribution is negligible.

The stability analysis is carried out by Fourier analyzing the normal component of plasma displacement, $\xi(\theta) = \exp(-in\phi)\sum_m' \xi_m \exp(im\theta)$, and then minimizing over the coefficients ξ_m. Both the plasma and vacuum contributions can be converted to surface integrals. These are then expressed in terms of the ξ_m through the solution to Laplace's equation for the perturbed scalar magnetic potentials and application of the boundary conditions given by Eq. (8.86).

After a lengthy calculation an expression for δW is obtained which can be written as follows (Freidberg and Haas, 1973):

$$\frac{\delta W}{n^2 W_0} = \boldsymbol{\xi}\cdot\mathsf{W}\cdot\boldsymbol{\xi} \tag{10.197}$$

where $\boldsymbol{\xi}$ is a vector whose components consist of the amplitudes ξ_m and W is a matrix which has the form

$$\mathsf{W} = \mathsf{W}^{(1)} + \frac{\mathsf{W}^{(2)}}{nq_*} + \frac{\mathsf{W}^{(3)}}{n^2q_*^2} \tag{10.198}$$

The matrices $\mathsf{W}^{(1)}$, $\mathsf{W}^{(2)}$, and $\mathsf{W}^{(3)}$ are functions only of the parameter k and are given by

$$\begin{aligned} W_{mp}^{(1)} &= \frac{2\delta_{m-p}}{|m|} \\ W_{mp}^{(2)} &= -\frac{\pi}{2E}\left(\frac{m}{|m|} + \frac{p}{|p|}\right)G_{mp} \\ W_{mp}^{(3)} &= \frac{\pi^2}{4E^2}\left[-\left(1-\frac{k^2}{2}\right)\delta_{m-p} \right. \\ &\quad \left. - \frac{3k^2}{8}(\delta_{m-p-1} + \delta_{p-m-1}) + \sum_l{}' |l|\, G_{lm}G_{lp}\right] \end{aligned} \tag{10.199}$$

Here G_{lm} is defined as

$$G_{lm} = \frac{1}{\pi}\int_0^{\pi} d\theta[1 - k^2\sin^2(\theta/2)]^{1/2}\cos(l - m)\theta \qquad (10.200)$$

Results. There is an important general property (valid for diffuse profiles as well) that can be inferred from W despite its complexity. This property is a consequence of the fact that the matrix elements are functions only of two parameters nq_* and k. Equivalently, from Eq. (10.193) the two parameters can be considered as nq_* and $\beta_t/\epsilon n^2$. A diffuse profile formulation has the same property although additional shape factors are also present. Assume now that a point of marginal stability has been found corresponding to a certain value of nq_* and $\beta_t/\epsilon n^2$. Clearly then, $n = 1$ is the most unstable external mode. For the marginal state $n = 1$ leads to the largest value of q_* and the smallest value of β_t/ϵ.

The scaling of W given by Eq. (10.198) also applies to ballooning modes except that in this case $\mathsf{W}^{(1)} = \mathsf{W}^{(2)} = 0$. The only parameters that appear are $\beta_t q_*^2/\epsilon$ plus shape factors. Recall from the ballooning mode analysis that α scales as $\beta q^2/\epsilon$ and s is independent of the magnitude of I_0 or β; that is s is a shape factor.

The conclusion is that ballooning instabilities in themselves do not in general set limits on β_t or I_0, but only on the ratio $\beta_t q_*^2/\epsilon$ (plus shape factors). In contrast, external kinks set individual limits on both β_t/ϵ and q_*.

The marginal stability boundary for the surface current model is determined by setting $n = 1$ and varying β_t/ϵ until the lowest eigenvalue of W crosses zero. This is a simple numerical procedure. The stability curve for the circular cross section high-β tokamak is illustrated in Fig. 10.15. Plotted here is the critical β_t/ϵ versus $q_* \sim 1/I_0$. Also illustrated is the equilibrium limit $\beta_t q_a^2/\epsilon = \pi^2/16$ corresponding to $q_a = \infty$. Observe that for low $q_* < 1$ the plasma is unstable for any β_t. When $q_* > 1$ the plasma becomes stable and the critical β_t increases from zero. At $q_* \approx 1.7$ the stability curve intersects the equilibrium curve predicting a maximum critical $\beta_t/\epsilon \approx 0.21$. At higher values of q_* the plasma is always stable, but these lower currents support increasingly smaller β_t's in toroidal equilibrium.

An interesting point to note is that the system is unstable along the equilibrium boundary for $q_* < 1.7$; that is, even though $q_a = \infty$, the kink mode is unstable. One thus concludes that q_* rather than q_a is the critical parameter describing the stability of external modes.

The structure of a marginally stable eigenfunction is illustrated in

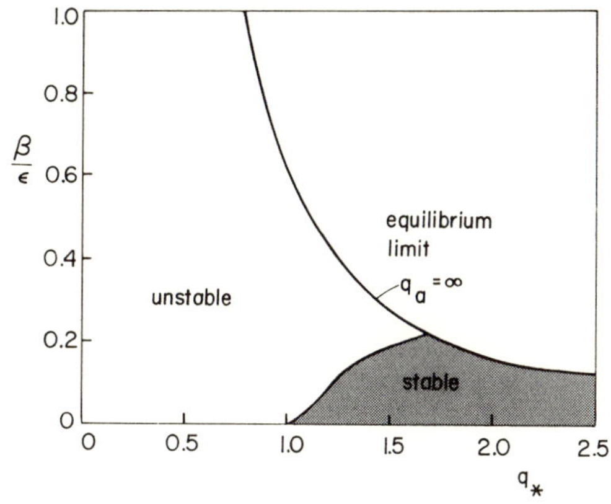

Figure 10.15. Stability boundary for external modes in a high-β tokamak described by the surface current model [from D'Ippolito *et al.* (1978)].

Fig. 10.16. Shown here is the unperturbed plasma surface (the dashed curve) and the perturbed surface (the solid curve) at four different ϕ locations. The ballooning-kink nature of the eigenfunction is apparent.

The β_t limit predicted by the surface current model is optimistic by a factor of about 3 compared to realistic diffuse profile numerical calculations. As might be expected noncircularity and profile shaping make important quantitative differences in the β_t limit. Elongated D-shaped cross sections are favorable for stability while large current gradients near

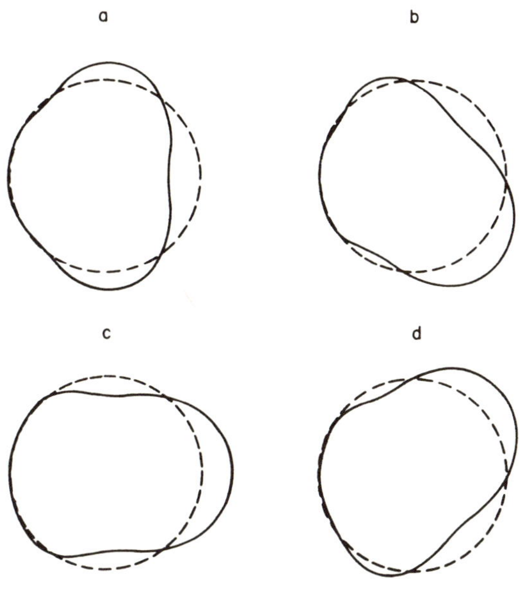

Figure 10.16. Marginally stable eigenfunction for the sharp boundary model at four different ϕ values. The major axis is to the left. The dashed and solid curves represent the unperturbed and perturbed surfaces respectively. [From Freidberg and Grossmann, (1975).]

the plasma edge are unfavorable. The numerical results are summarized in the next section. As a guideline one should note that it is relatively difficult to design a tokamak that is stable to external kinks at $\beta_t = 5\%$ without a conducting wall.

In summary, the surface current model demonstrates that external ballooning-kink modes in a tokamak require both a current limit $q_* > 1$ and a pressure limit $\beta_t/\epsilon < 0.21$ for stability. Numerical studies indicate that from a theoretical point of view these modes appear to be the most dangerous ideal MHD instabilities. However, their behavior is strongly dependent on external conditions so that the presence of a metal limiter or quasiconducting vacuum chamber near the plasma could have a substantial effect on the actual critical β_t achievable in any experiment. These problems require further investigation.

10.5.8. Numerical Results: The Sykes Limit, the Troyon Limit

Over the past decade rather sophisticated numerical codes have been developed to study the problem of MHD equilibrium and stability in tokamaks. These codes contain an equilibrium Grad–Shafranov solver valid for arbitrary β, ϵ, and cross section. They can treat either fixed or free boundary configurations. General profiles are allowed and usually $p(\psi)$, and either $F(\psi)$, $q(\psi)$, or $\langle J_\phi(\psi)\rangle$, is specified. Once an equilibrium is established the following stability tests can be made: (1) Mercier stability, (2) high-n ballooning modes, (3) low-n internal modes, and (4) external ballooning-kink modes. Such studies, while time consuming, provide a complete and detailed picture of tokamak stability. They are extremely helpful in the design of new experiments and in the interpretation and analysis of existing experimental data. These studies also play a major role in the determination of optimized configurations and are ultimately able to make quantitative predictions for the maximum β_t and I_0 that can be stably maintained in MHD equilibrium.

In recent years many numerical studies have been merged, providing a relatively simple picture of tokamak stability. This section summarizes these results, which have been separated into three main areas: ballooning mode studies, full stability studies, and second region of stability studies.

Ballooning Mode Studies. The first major topic of interest concerns first region of stability ballooning mode studies. A number of independent results were unified and expanded in the optimization studies of Sykes *et al.* (1983). In this work an attempt is made to calculate the tokamak configuration with the maximum stable β_t against ballooning modes. The q profile always has $q_0 = 1.05$ to ensure stability against low-n internal kinks. Different q_* profiles are examined by varying $I_0 \propto 1/q_*$. For a given I_0 the $q(\psi)$ profile is held fixed as the pressure is iteratively maximized. For a given shaped cross section the optimized state is obtained when the pressure profile is essentially marginally stable to ballooning modes on every surface and satisfies the Mercier criterion on axis. This optimization procedure is then carried out for several different shaped cross sections.

The end result is remarkably simple in form. The optimized β_t for all cases studied is accurately represented by the Sykes limit

$$\beta_t = 0.044\left(\frac{I_0}{aB_0}\right) \tag{10.201}$$

where the units are $I(MA)$, $a(m)$, $B_0(T)$. An alternate form of Eq. (10.201) can be written as

$$\beta_t = 0.22\left(\frac{\epsilon\kappa}{q_*}\right) \tag{10.202}$$

Here $\epsilon = a/R_0$, $q_* = 2B_0A/\mu_0R_0I_0$, A is the cross-sectional area, and $\kappa = A/\pi a^2$ is the elongation.

Observe the following points. Equation (10.201) implies that high β_t's are achieved by operating at high I_0/B_0. Clearly though, from a physical point of view one would not expect this relation to hold as $I_0 \to \infty$, $B_0 \to 0$. In practice two effects limit the maximum ratio of I_0/B_0. First, as I_0/B_0 increases the flux surfaces near the axis become highly elongated causing violation of the Mercier criterion. As discussed previously, triangularity helps alleviate this problem, although not indefinitely. Second, for large I_0/B_0 external kinks are excited, although this possibility is not included in the ballooning mode analysis. Recent studies by Yamazaki *et al.* (1985) show that for small to moderate triangularities within a "pure" ballooning mode optimization the curve of β_t versus I_0/aB_0 peaks and then decays for large I_0/aB_0, but usually at higher values than one would anticipate for the onset of external kinks. Even so, the linear relation between β_t and I_0 is valid over quite a wide practical parameter range, although one must ensure a sufficiently triangular shape so that β_t is not degraded by Mercier instabilities.

Another point worth noting is that the form of β_t given by Eq. (10.202) indicates that the optimum β_t scales as $1/q_*$. From the discussion accompanying Eq. (10.198), a scaling $\beta_t \sim 1/q_*^2$ would be expected. In fact there is an implied quadratic scaling in the Sykes limit, but this has been masked because q_0 has been held fixed at a value near unity; that is, a comparison of Eq. (10.202) with Eq. (10.187) suggests that the Sykes limit may actually scale as $\beta_t \sim \epsilon\kappa/q_0 q_*$.

Finally, Eq. (10.202) shows that β_t increases linearly with κ at fixed q_*. The critical question is how low q_* can become while still maintaining Mercier stability (and external kink stability in more general analysis). The results of Sykes *et al.* demonstrate that with moderate triangularity ($\delta = 0.24$) the scaling persists down to values of $q_* = 1.4$ for $\kappa = 2$ with no apparent saturation in sight. This is just an alternate restatement of the problem of how large a value of I_0/B_0 is allowed. As an example, if one considers a tokamak with an elongation $\kappa = 2$, aspect ratio $\epsilon = 1/3$, operating at $q_* = 1.5$ then the critical $\beta_t = 10\%$. If experimentally achievable, this is an acceptably high value of β_t to be of interest to fusion power reactors.

In summary, the optimized β_t against first region of stability ballooning modes is given by Eq. (10.202). The absolute maximum value of β_t depends upon how high κ and how low q_* can be made. One limit is due to Mercier stability, and strong triangularity is required to delay the onset of these modes. The other limit is due to external kinks, although this is not included in the ballooning mode studies.

Full Stability Studies. A rather complete set of stability studies has been carried out by Troyon *et al.* (1984). These studies have a number of similarities with the ballooning mode calculations of Sykes *et al.* (1983). In particular, an optimized β_t is obtained by varying the p and q profiles as well as the shape of the cross section. Also, the optimized profiles are stable against ballooning modes and the Mercier criterion. The important additional feature, one which requires substantial numerical effort, is the inclusion of low-n kink modes. Consequently, the final optimized configurations obtained by Troyon *et al.* are stable against the Mercier criterion, ballooning modes, the $n = 1$ internal kink, and low-n external ballooning-kink modes, corresponding to an essentially complete stability analysis.

The numerical studies indicate that elongated, outward pointing triangular cross sections are favorable for stability. When the results are examined Troyon *et al.* also obtain a remarkably simple relation for the optimized β_t limit given by

$$\beta_t = 0.028\left(\frac{I_0}{aB_0}\right) \tag{10.203}$$

or equivalently

$$\beta_t = 0.14\left(\frac{\epsilon\kappa}{q_*}\right) \tag{10.204}$$

This limit, known as the Troyon limit, is identical in form to the ballooning criterion given by Eq. (10.201) except that the numerical coefficient is reduced from 0.044 to 0.028.

Over most of the regimes of interest the $n = 1$ internal kink is the most unstable mode setting a slightly stricter limit on β_t than that due to ballooning modes. The optimized profiles have q_0 slightly greater than unity to satisfy the Mercier criterion. This automatically guarantees stability for the $n = 1$ internal kink. The pressure profile is broad and the q profile is flat over most of the cross section; q rises rapidly near the plasma surface. The current density is zero at the plasma edge to suppress higher n external kinks.

The one crucial difference between the ballooning mode and full stability studies is the lowest allowable value of q_*. Recall that in the Sykes limit the scaling $\beta_t \sim 1/q_*$ shows no indication of a minimum q_* over the range of parameters investigated. In the Troyon limit, however, the scaling $\beta_t \sim 1/q_*$ breaks down once $q_a \gtrsim 2$; that is, when a $q = 2$ resonant surface appears in the vacuum, the system becomes unstable to an external kink mode. Consequently, the lowest allowable value of q_* is that value which corresponds to $q_a \approx 2$. In general the relation between q_a and q_* is a complicated function of elongation, triangularity, aspect ratio, β_t, and current and pressure profiles. For a large aspect ratio, low β_t circular tokamak, $q_a = q_*$. For the optimized profiles found by Troyon *et al.* $q_* = 1.5$ when $\kappa \approx 1.6$. Using this value, one finds for the maximum β_t

$$\beta_t|_{\kappa=1.6} = 0.15\epsilon \tag{10.205}$$

There are two points requiring caution in the use of Eq. (10.205). First, consider the effects of elongation. For fixed profiles $q_a \propto q_*\kappa$ at large κ. Thus, if q_a is held fixed to avoid a $q_a = 2$ resonant surface in the vacuum, then $q_* \propto 1/\kappa$ and, from Eq. (10.204), $\beta_t \propto \kappa^2$. One infers that substantial gains in β_t would follow from elongation. However, keep in mind that the $q_a \approx 2$ requirement has been obtained for moderate elongations: $1 < \kappa < 1.8$. At large elongations the critical q_a for the onset of kink modes may increase with κ. For instance, above a certain elongation a $q_a = 3$ surface could trigger instabilities implying that $q_a = q_a(\kappa)$. Since there is a one-to-one equivalence between q_a and q_*, there is in general a minimum $q_* = q_*(\kappa)$. There is some indication that $q_* \propto \kappa$ for large elongations implying a saturation of β_t. This important question is still under investigation.

The second point of caution concerns the criterion $q_a = 2$ for the onset of kink instabilities for moderate elongations. This condition appears incompatible with the surface current results which predict kink instabilities for $q_a \to \infty$ near the equilibrium β_t limit. The resolution of this point is that the "high β" corresponding to the Troyon limit is still sufficiently low that the system is far from its equilibrium β_t limit. Thus, for practical purposes the β_t dependence between q_a and q_* can be neglected and the one-to-one relation between these two quantities is a purely geometric one depending upon the shape of the cross section and the current profile. However, to allow for the possibility of a separatrix such as in a diverted plasma, one should treat q_* rather than q_a as the more basic parameter describing kink stability.

In summary the $n = 1$ external ballooning-kink mode sets the most severe β_t limit in a tokamak as given by Eq. (10.203). The value of q_0 must be slightly greater than unity to satisfy the Mercier criterion and the $n = 1$ internal kink condition. The maximum value of β_t occurs for the lowest allowable value of q_* which is, in general, a function of κ as set by external kinks.

Second Region of Stability Studies. The last topic of interest concerns the second region of stability. Recall that the analytic studies of ballooning modes presented in Section 10.5.5 lead to an (s, α) diagram for marginal stability (Fig. 10.11). The diagram consists of two branches, one of which predicts stability for sufficiently large values of the pressure gradient (i.e., α). This is the second region of stability. The goal of the numerical studies carried out by Grimm *et al.* (1984) is the discovery of configurations that lie entirely within this region and are accessible in a practical experiment.

The work of Grimm *et al.*, as well as others, demonstrates that tokamaks with a "bean-shaped" cross section meet these goals. A typical bean-shaped tokamak is illustrated in Fig. 10.17. The crucial feature is the indentation of the cross section on the inside of the torus. A sequence of such configurations has been tested for MHD stability. Several classes of modes are considered, although most of the study is focused on ballooning instabilities.

One interesting set of numerical calculations considers profiles with $q_0 = 1.03$, $q_a = 4.2$, and $\kappa = 1.4$. The marginal stability diagrams obtained by Grimm *et al.* are illustrated in Fig. 10.18. Plotted here are curves of β_t versus indentation i for various aspect ratios, where $i \equiv d/2a$ (see Fig. 10.17). The value of β_t corresponds to a $p(\psi)$ profile in which the most unstable flux surface is marginally stable to the ballooning mode stability criterion.

Observe the following points. For a fixed aspect ratio there exists a first region of stability at low values of β_t for any indentation. Above a critical indentation $i = i_c$, there are no ballooning instabilities for any values of β_t considered. For smaller indentations $i < i_c$ a second region of

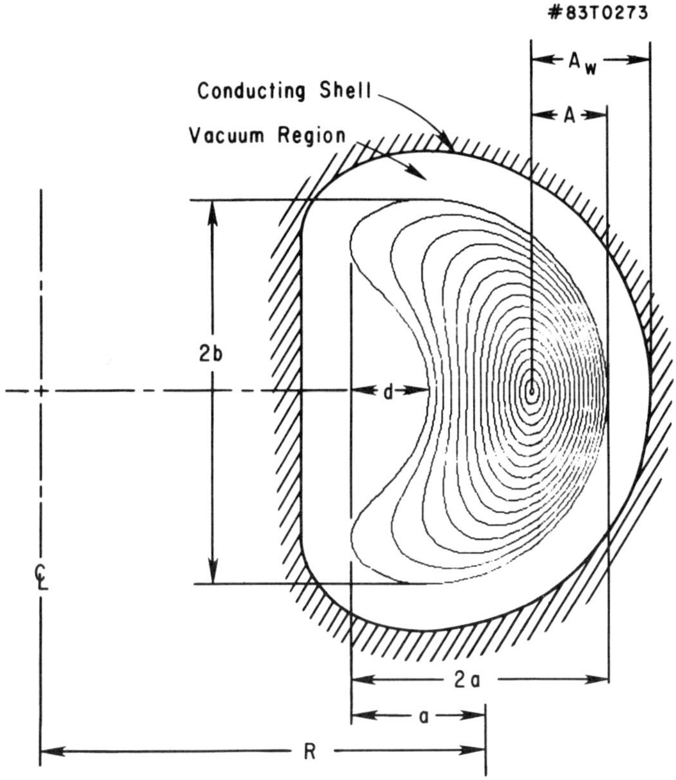

Figure 10.17. Numerically computed equilibrium demonstrating a "bean-shaped" cross section [from Grimm *et al.* (1984)].

stability is found at sufficiently large β_t. As a general trend, access to the second region is easiest at large aspect ratios. Specifically, while the boundary for the first region of stability is a weak function of ϵ, the minimum β_t for the second region is a strong increasing function of ϵ. The studies of Grimm *et al.* also show that relatively high values of $q_a \gtrsim 4$ are desirable for access to the second region. At low q_a, the stability boundary does not turn back at high β_t and the second region of stability vanishes. However, the β_t limit for the first stability region increases.

Using these results one can visualize a practical experimental situation which might gain access to the second region of stability. Consider a bean-shaped experiment as shown in Fig. 10.17. The bean shape is created by adding a "pusher coil" in the region of indentation. As the current in the pusher coil increases, the indentation increases. Assume now that the plasma is initially in a low-β_t ohmic state without any indentation; that is, a "D" shape. The current in the pusher coil is

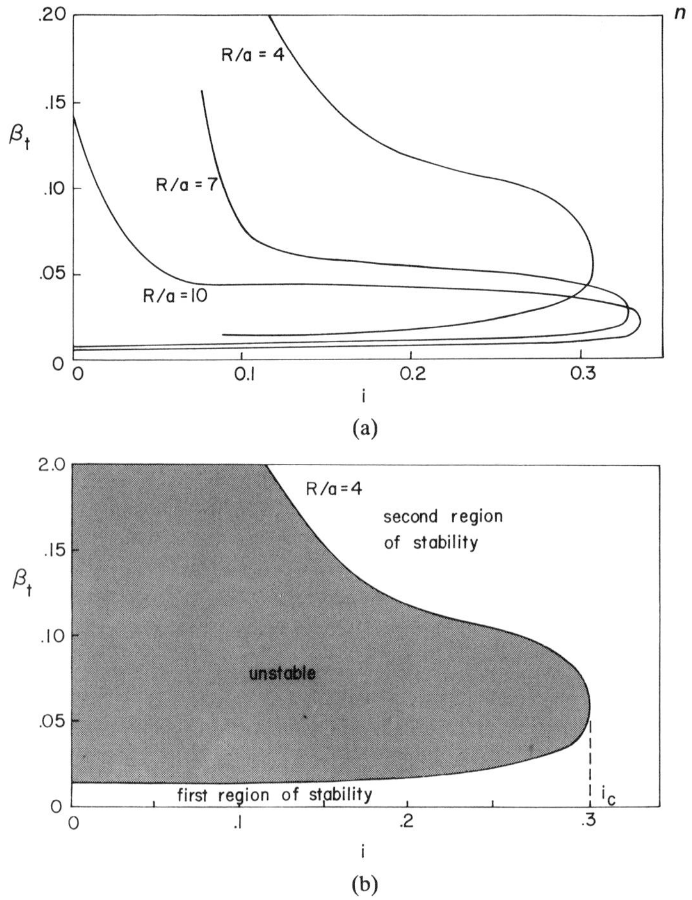

Figure 10.18. Stability diagrams of β_t vs. indentation i in bean-shaped tokamaks (a) as a function of aspect ratio and (b) explicitly showing the second region of stability [from Grimm *et al.* (1984)].

increased so that the configuration evolves along the path illustrated in Fig. 10.19. Once the critical indentation is exceeded, auxiliary power is applied raising the plasma β_t. At sufficiently high β_t the current in the pusher coil is decreased and the plasma enters the second region of stability. This procedure will be tested experimentally on the Princeton Bean Experiment (PBX). It is important to be able to reduce the pusher current once high β_t is achieved since it is quite difficult technologically to achieve and maintain an indentation on the order of $i \sim 0.3$. Further studies involving profile and cross-sectional optimization are underway to

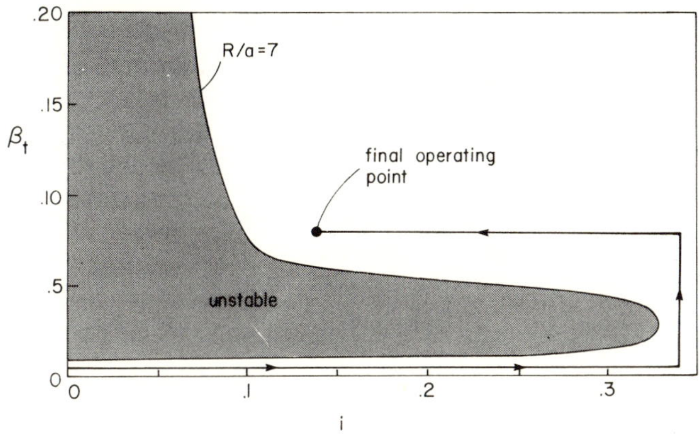

Figure 10.19. Possible experimental path to gain access to the second region of stability [from Grimm *et al.* (1984)].

reduce the critical indentation for entry into the second region of stability.

In summary it may be possible to achieve high-β_t operation in a tokamak by means of the second region of stability. Such operation requires a bean-shaped cross section, large aspect ratio, $q_0 \approx 1$, and $q_a \approx 4$.

10.5.9. $n = 0$ Axisymmetric Modes

The final class of toroidal instability to be considered is the axisymmetric mode. This perturbation corresponds to an $n = 0$ external mode which at marginal stability can be viewed equivalently as a neighboring equilibrium state. The most unstable perturbation usually corresponds to a nearly rigid vertical shift of the plasma (i.e., there is a large $m = 1$ component to the eigenfunction). It is important to stabilize these modes since they represent a macroscopic motion of the plasma towards the wall of the discharge chamber.

The stability of axisymmetric modes is directly coupled to toroidicity and noncircularity. This can be seen by examining δW for the $m = 1$, $n = 0$ external mode in a straight circular tokamak [see Eq. (9.170)]. As might be expected because of symmetry, one finds $\delta W = 0$, indicating neutral stability. Toroidicity and noncircularity in general require relatively complicated two-dimensional calculations. There is, however, one simple model that can be analyzed, illustrating the basic nature of the axisymmetric instability. This model, first investigated by Osovets (1959) and later by Yoshikawa (1964) and Mukhovatov and Shafranov (1971), is discussed below.

The calculation treats the plasma as a thin ($a/R_0 \ll 1$) current-carrying loop of wire embedded in an externally applied vertical field. For simplicity the effects of plasma pressure and the internal magnetic flux are neglected. The object of the calculation is to determine the appropriate constraints on the shape of the vertical field to provide stability against rigid vertical and horizontal displacements. Clearly, a pure uniform vertical field $\mathbf{B}_v = B_v \mathbf{e}_z$, $B_v =$ const, would not be adequate, since by symmetry the system would be only neutrally stable to vertical displacements.

The calculation proceeds by introducing a potential $\phi(R, Z)$, such that the equilibrium forces acting on the plasma are given by $\mathbf{F}(R, Z) = -\nabla\phi$. Equilibrium occurs at the point R_0, Z_0 where $F_R(R_0, Z_0) = F_Z(R_0, Z_0) = 0$. The condition that the plasma then be stable to a rigid shift in either the vertical or horizontal direction is that $\partial F_Z(R_0, Z_0)/\partial Z_0 < 0$ and $\partial F(R_0, Z_0)/\partial R_0 < 0$ respectively; that is, stability occurs when the restoring force is in the opposite direction of the displacement. Note that because the plasma is a perfect conductor the calculation must be carried out under the constraint that the poloidal flux contained within the current loop be conserved under either plasma displacement.

For the simple model under consideration, the appropriate equilibrium potential and the poloidal flux contained within the current loop

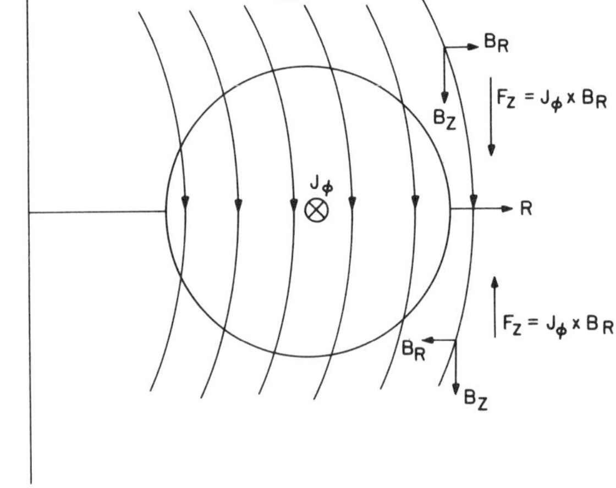

Figure 10.20. Geometry for calculating $n = 0$, axisymmetric mode stability.

are given by (see Fig. 10.20)

$$\phi(R,Z) = \tfrac{1}{2}LI^2$$

$$\psi_p(R, Z) = LI - 2\pi \int_0^R B_Z(R', Z)R'\, dR' \qquad (10.206)$$

Here, $L(R) = \mu_0 R[\ln(8R/a) - 2]$ is the external inductance associated with the toroidal current I and B_Z and B_R are the components of the externally applied vertical field.

The equilibrium forces, calculated from ϕ can be written as

$$F_Z = -LI\frac{\partial I}{\partial Z}$$

$$F_R = -LI\frac{\partial I}{\partial R} - \frac{I^2}{2}\frac{dL}{dR} \qquad (10.207)$$

From ψ_p = const it then follows that the equilibrium relation, $\mathbf{F}(R_0, Z_0) = 0$ reduces to

$$B_R(R_0, Z_0) = 0 \qquad (10.208)$$

$$B_Z(R_0, Z_0) = \frac{I}{4\pi R_0}\frac{dL}{dR_0} \qquad (10.209)$$

Note that Eq. (10.209) is identical to those terms in the Shafranov shift associated with the external poloidal flux [see Eq. (6.91)].

Consider now stability against a rigid vertical shift. Using the fact that $\nabla \times \mathbf{B} = 0$ for the externally applied vertical field one can easily calculate $\partial F_Z(R_0, Z_0)/\partial Z_0$. One finds

$$\frac{\partial F_Z}{\partial Z_0} = -\frac{I^2}{2R_0}\frac{dL}{dR_0}n \qquad (10.210)$$

where

$$n(R_0, Z_0) = -\left(\frac{R}{B_z}\frac{\partial B_z}{\partial R}\right)_{R_0, Z_0} \qquad (10.211)$$

is known as the field index. The condition for vertical stability is thus given by

$$n > 0 \qquad (10.212)$$

This condition can easily be understood from Fig. 10.20. If the directions of the vertical field and toroidal current are as shown then (1) the vertical field produces an inward force for toroidal equilibrium and (2) a small upward shift of the plasma gives rise to a downward $\mathbf{J} \times \mathbf{B}$

force, of magnitude $2\pi RIB_R$, which is in the direction to restore equilibrium. Thus, the curvature of vertical field shown in Fig. 10.20 is the appropriate one for stability, and corresponds to the condition $n > 0$.

Consider now horizontal stability which is only slightly more complicated. In this case, a straightforward calculation yields the following expression for $\partial F_R(R_0, Z_0)/\partial R_0$:

$$\frac{\partial F_R}{\partial R_0} = \frac{I^2}{2R_0}\frac{dL}{dR_0}\left[n - 1 - \frac{1}{2}\frac{d\ln L}{d\ln R_0} + \frac{1}{2}\frac{d\ln(dL/dR_0)}{d\ln R_0}\right] \quad (10.213)$$

In the simple limit $\ln(8R_0/a) \gg 1$ stability is achieved when

$$n < \tfrac{3}{2} \quad (10.214)$$

This situation can be understood as follows. Assume the plasma is given a small outward shift. At constant flux, the outward hoop force per unit length on the plasma decreases because of the smaller current at the larger major radius. However, because of its larger radius of curvature, the vertical field decreases with increasing R. Hence, the toroidal restoring force also decreases at the larger major radius. The stability condition given by Eq. (10.214) expresses the requirement that the curvature of the vertical field be sufficiently weak so that the toroidal restoring force decays at a slower rate than that of the hoop force, thus providing a net inward stabilizing force.

The results of this simple model are in good agreement with more exact numerical calculations (Wesson and Sykes, 1974; Johnson *et al.*, 1977). In general, for circular-cross-section plasmas axisymmetric stability is relatively easy to achieve, requiring only a modest shaping of the vertical fields. The situation is more serious for elongated tokamaks. In such configurations there is a strong tendency for the plasma to be unstable to vertical shifts as the elongation increases. A number of calculations which derive a more accurate form of the vertical stability condition and relate the elongation b/a to n indicate that only rather small elongations, $b/a \lesssim 1.25$, can be stable to vertical shifts when no conducting wall is present (Laval *et al.*, 1974; Laval and Pellat, 1973; Rebhan, 1975; Haas, 1975). Also, with no wall the critical elongation does not depend very strongly on β_p or the current profile. If a conducting wall is allowed even a moderate distance away, $a_{\text{wall}}/a_{\text{plasma}} \sim 1.5$–$2.0$ there is a substantial improvement in vertical stability. However, the necessity of such a wall in a reactor is a disadvantage. Even more important, such a wall would certainly be resistive on the reactor time scale, implying the requirement of a feedback system for vertical positioning. This would be in addition to the horizontal feedback system required for centering the plasma column along the major radius.

It is currently thought that through a combination of careful design and modest active or passive feedback, elongations on the order of $\kappa = 1.6$–2.0 should be readily attainable in actual experiments.

In summary, $n = 0$ axisymmetric modes can lead to potentially serious instabilities in a tokamak. For circular cross sections a moderate shaping of the vertical field should provide stability. For noncircular tokamaks, vertical instabilities produce important limitations on the maximum achievable elongations. Even moderate elongations require a conducting wall or a feedback system for vertical stability. In practice values of $\kappa \sim 1.6$ should be possible.

10.5.10. Overview of the Tokamak

By combining the equilibrium and stability results one obtains the following overview of the tokamak concept with respect to ideal MHD behavior.

The tokamak is a relatively low-β axisymmetric toroidal configuration. It has a large component of toroidal magnetic field which is required for stability. Radial pressure balance is provided by the poloidal field (i.e., as in a Z pinch) in the low-β ohmically heated regime characterized by $\beta_t \sim \epsilon^2$, $\beta_p \sim 1$, $q \sim 1$. In contrast, in the high-β regime, $\beta_t \sim \epsilon$, $\beta_p \sim 1/\epsilon$, $q \sim 1$, radial pressure balance is provided by the toroidal field (i.e., as in a θ pinch). For both regimes toroidal force balance is achieved by the interaction of the toroidal current with an externally applied vertical field. Since the highest values of β and toroidal current scale with ϵ for a fixed toroidal field, high-performance tokamaks require tight aspect ratio, somewhat of a technological disadvantage from the reactor viewpoint.

There are two classes of ideal MHD instabilities which define important boundaries for tokamak operation. These involve a current limit which is found in both straight and toroidal systems and a β limit which is a consequence of toroidicity.

The most important current limitations result from the $n = 1$ internal kink, the $n \to \infty$ Mercier condition, the $m = 1$, $n = 1$ Kruskal–Shafranov external kink, and $m \geq 2$ low-β external kinks. The $n = 1$, $m = 1$ internal kink requires $q_0 > 1$ for stability. This mode, and its nonlinear evolution (including resistivity), is believed to be responsible for the sawtooth oscillations observed in tokamak experiments. The Mercier criterion for interchange stability is most difficult to satisfy near the magnetic axis where the shear is small. To prevent such instabilities the profiles must satisfy $q_0 \gtrsim 1$. The exact threshold depends upon the cross section. Elongated, outward pointing triangular shapes are favorable for stability. The most well-known current limit arises from the $m = 1$, $n = 1$ external kink. This leads to the Kruskal–Shafranov condition $q_a > 1$. Stability against higher m external kinks requires a current profile which decays to zero sufficiently rapidly near the edge of the plasma. Typically, this implies $q_a/q_0 \gtrsim 2\text{–}3$.

In practice when plasma resistivity is included the high-m external modes generalize to include $m \gtrsim 2$ internal resistive tearing modes. These instabilities can cause major disruptions and terminate the discharge if the current becomes too large. The requirements $q_0 \gtrsim 1$ for ideal internal modes and $q_a/q_0 \gtrsim 2\text{–}3$ for tearing and external kink modes imply that $q_a \sim 2\text{–}3$ for most operation. Hence, in general tokamaks do not reach the $q_a = 1$ Kruskal–Shafranov limit.

The overall situation with regard to MHD current limitations is that most tokamaks operate in a regime that is stable, or at least marginally stable to ideal modes, but with some finite level of resistive MHD turbulence. Although there are regions of parameter space where even resistive modes can be stabilized, these are not very attractive because they correspond to low values of the current limit; that is, the benefits of higher current outweigh the enhanced transport due to weak resistive modes at the optimum operating conditions.

Turning to β limits, the most important instabilities are internal $n \to \infty$ ballooning modes and low-n, high-β external ballooning-kink modes. The $n \to \infty$ ballooning modes correspond to perturbations that are enhanced in the regions of unfavorable curvature. This instability sets a limit on the maximum allowable pressure gradient on any given flux surface (the first region of stability). When integrated over the entire profile, the ballooning mode criterion leads to a maximum volume averaged toroidal β of the form $\beta_t \lesssim K\epsilon\kappa/q_0q_*$. The ballooning mode itself does not actually yield a direct absolute limit on β_t since the mode does not set a minimum value for q_*. As with the Mercier criterion an outward-pointing triangular cross section is favorable for stability. The ballooning mode stability analysis also indicates the possibility of achieving high β in a tokamak by means of operating in the second region of stability. Such operation will likely require a "bean-shaped" cross section. The low-n, high-β external ballooning-kink modes usually set the most severe limits on β_t in a tokamak. The stability criterion is of the form $\beta_t \lesssim K\epsilon\kappa/q_0q_*$, which is similar to that deduced from $n \to \infty$ internal ballooning modes. However, the constant K is somewhat smaller for the external modes. Also there is a minimum value of q_* corresponding to the onset of the external kink. Several tokamaks with large auxiliary power supplies have reached the external ballooning-kink β

limit. There is usually a degradation of confinement time as β increases, and, as of this writing (1986), it has not been possible to exceed that MHD β_t limit: that is, the maximum β_t appears to represent a real experimental limitation on tokamak operation.

These stability results serve as valuable guidelines for tokamak operation and are summarized as follows:

Current Limitations

$n = 1$ internal kink	$q_0 > 1$
Mercier criterion	$q_0 \gtrsim 1$
$n = 1$, $m = 1$ Kruskal–Shafranov limit	$q_a > 1$
$m \geq 2$ low-β external kinks	$q_a/q_0 \gtrsim 2$–3

β Limitations

$n \to \infty$ ballooning modes	$\beta_t < 0.22\epsilon\kappa/q_0 q_*$
low-n external ballooning-kink modes	$\beta_t < 0.14\epsilon\kappa/q_0 q_*$
	$q_a > 2$

There is one further, indirect constraint on tokamak operation resulting from MHD requirements that should be noted. The stability requirement $q \gtrsim 1$ combined with technological constraints limiting the maximum B_ϕ set an upper limit on the toroidal current. Theoretical calculations show that this current can joule heat a plasma to several keV. However, it is quite difficult, perhaps impossible to reach the ignition temperature. As a result, the standard operation of present experiments and future tokamak reactors will likely require some form of auxiliary heating such as neutral beams or rf power.

In summary, ideal MHD equilibrium and stability requirements have given rise to important constraints on the operation of a tokamak. Nevertheless, the available stable operating space is sufficiently broad and flexible that tokamaks are currently the most successful experimental configuration and the most promising concept on the path to a fusion reactor.

10.6. STELLARATORS

10.6.1. Introduction

The investigation of the stability properties of stellarators is one of the most challenging problems in MHD theory. The challenge arises because of the wide variety of stellarator configurations, each a fully three-dimensional system.

The "simplest" configuration is the "standard" stellarator, which is characterized by a single helicity field with $l \geq 2$. Such configurations have a circular magnetic axis: $R = R_a$, $Z = 0$. More sophisticated systems have a helical magnetic axis: $R = R_a + \Delta_h \cos(\theta - N\phi)$, $Z = \Delta_h \sin(\theta - N\phi)$. These stellarators have a large $l = 1$ component in the helical field system. The early figure-eight stellarator and recent heliac are configurations of this type. Reactor extrapolations indicate that helical axis stellarators, as they are currently envisaged, will lead to large systems, thus placing them in the role of a backup concept. Perhaps the most advanced configuration is the Drakon, a system consisting of two long parallel cylindrical sections connected together at each end by "half a stellarator." Each end section, called a CREL, is specially designed so that zero local parallel currents flow in the straight sections. Parallel currents would quickly lead to loss of single-particle confinement. At present, no Drakon experiments have yet been built.

In this section the stability analysis is focused on the standard stellarator. A general discussion is presented of the various possible MHD instabilities and which in particular are expected to be the most dangerous. One important conclusion is that it is necessary to have a magnetic well in order to achieve stability at high values of β.

With this insight, a discussion is presented of the different methods proposed to create a magnetic well. Included are the effects of a vertical field, plasma β, helical sideband fields and toroidicity. It is shown that it is indeed possible to create a magnetic well capable of stabilizing high values of $\beta \sim 0.1$.

The final topic concerns the application of MHD theory to the design of an actual experiment. Both the equilibrium and stability requirements are considered simultaneously. A series of extensive and sophisticated three-dimensional MHD numerical calculations provides the basis for the design of the ATF torsatron at the Oak Ridge National Laboratory. These studies represent an excellent example of the interaction of MHD theory with practical experimental design (ATF Proposal, 1983; Carreras *et al.*, 1983).

10.6.2. General Description of Stellarator Instabilities

In general an MHD equilibrium is subject to current-driven and pressure-driven instabilities whose mode structure corresponds to either internal or external perturbations. To understand stellarator stability it is useful to review briefly tokamak stability as a point of reference. In a tokamak the most dangerous current-driven modes all involve low-n long-wavelength perturbations. These modes set limits on the current density on axis (the $m = 1$ internal kink), the current profile (the $m \geq 2$ external kinks), and the total plasma current (the $m = 1$ external kink). The most dangerous modes driven by the pressure gradient are $n \to \infty$ interchanges and ballooning modes with the latter setting important limits on β_t. The external ballooning-kink modes also set limits on β_t. However, the mode structure is predominantly kink-like in that only low-n values are typically involved.

Consider now the stellarator. One major reason for interest in this configuration, as well as other three-dimensional systems, is the potential for steady state operation made possible by the absence of any requirements that a net ohmic current flow in the plasma. Consequently, in a current-free stellarator one expects that low-n current-driven modes should play a minor role in comparison to the tokamak. This is indeed the case.

The most dangerous MHD instabilities in a stellarator are driven by the pressure gradient. In analogy with the tokamak the severest limits (theoretically) arise from $n \to \infty$ interchanges and ballooning modes. There is, however, an interesting contrast. In a tokamak, ballooning modes set more stringent stability limits than interchanges; the opposite is true for most stellarator configurations thus far investigated. One can see this intuitively by assuming that the analytic ballooning mode theory for the tokamak (presented in Section 10.5.5) is qualitatively similar for the stellarator. For the analytic model the local shear is given by [Eq. (10.183)]

$$\frac{r}{q}\frac{\partial \hat{q}}{\partial r} = s - \alpha \cos \theta \tag{10.215}$$

where $s = rq'(r)/q(r)$ is the average shear and $\alpha = -q^2 R_0 d\beta(r)/dr$ is a measure of the pressure gradient. Note that s is positive in a tokamak. As the pressure gradient α increases from zero the magnitude of the local shear first decreases in the region of unfavorable toroidal curvature, $\theta = 0$. This behavior is partially responsible for the first region of instability. At sufficiently large $\alpha \gg s$ the magnitude of the local shear again increases with α in the region of unfavorable curvature, leading to the second region of stability.

In a stellarator, s is negative since ι is in general an increasing function of radius. Thus, as α increases the local shear monotonically increases in the region of unfavorable toroidal curvature $\theta = 0$. The local shear is small at $\theta = \pi$, but here the toroidal curvature is favorable. The situation is clearly more desirable in a stellarator than in a tokamak; the local shear has the maximum increase in the region where it is most needed, the region of unfavorable toroidal curvature. It is suggestive to reexamine the $s - \alpha$ diagram for the tokamak illustrated in Fig. 10.11. To the extent that a similar diagram applies to the stellarator, one sees that a stellarator operates in the region $\alpha > 0$, $s < 0$, entirely in the "second region of stability." In fact, for a well-designed stellarator there is no "first region of instability" and such configurations are stable to ballooning modes up until the equilibrium β_t limit.

This insight implies that the most dangerous modes in a stellarator are interchanges, which essentially measure the average curvature. Even without a magnetic well one might think that there is sufficient shear to provide stability against interchanges. A simple calculation presented in the next section shows that this is only possible at very low $\beta_t \lesssim 0.01$. To satisfy the interchange condition at larger $\beta_t \sim 0.05$–0.1 requires a configuration with a favorable magnetic well. Methods of creating such a well are also discussed in the next section.

In summary the most dangerous instabilities in the standard stellarator are $n \to \infty$ interchanges. These modes can be stabilized at large $\beta_t \sim 0.1$ by means of a magnetic well. Such a system is capable of stable operation up to the equilibrium β_t limit.

10.6.3. Creating a Magnetic Well in a Stellarator

In this section various methods of creating a vacuum magnetic well in a stellarator are analyzed. The discussion in based on the figure of merit introduced in Chapter 4 and repeated here for convenience:

$$W = \frac{2V}{\langle B^2 \rangle} \frac{d}{dV} \frac{\langle B^2 \rangle}{2} \tag{10.216}$$

where $\langle Q \rangle$ is defined as

$$\langle Q \rangle = \lim_{L \to \infty} \left[\int_0^L \frac{Q\,dl}{B} \bigg/ \int_0^L \frac{dl}{B} \right] \tag{10.217}$$

Recall that while W is a useful figure of merit, it is neither necessary nor sufficient for stability. A more complete stability analysis (see, for example, Shafranov, 1983) indicates that additional terms of comparable magnitude appear in the stability critierion. These calculations are somewhat involved and are beyond the scope of the textbook. For simplicity, the analysis presented is restricted to the evalution of W. Even so, the results obtained are qualitatively correct when compared to the more exact theory.

The discussion proceeds as follows. The analysis is focused on the standard $l = 2$ stellarator, the configuration that has received the most attention experimentally. The first system studied is the large aspect ratio, pure single helicity, $l = 2$ stellarator. In this limit toroidicity has no effect on the magnetic well. The corresponding value of $W < 0$ indicates a hill as opposed to a well. A simple estimate is presented that demonstrates that shear can only stabilize low values of β_t in the presence of a magnetic hill. Based on this result, a number of additional fields are added to the basic $l = 2$ configuration (e.g., a vertical field, helical sideband fields). W is recalculated showing that under a variety of conditions it is possible to create a magnetic well.

Pure $l = 2$ Stellarator. Within the context of the stellarator expansion, the magnetic well given by Eq. (10.216) reduces to (see Problem 7.6)

$$W(r) = -r\frac{d}{dr}\langle (r_0/R_0)\cos\theta_0 + |\nabla_\perp V|^2 + h^2|V|^2\rangle \tag{10.218}$$

where

$$\langle Q\rangle = \int_0^{2\pi}\frac{Qr_0\,d\theta_0}{\partial\psi/\partial r_0}\bigg/\int_0^{2\pi}\frac{r_0\,d\theta_0}{\partial\psi/\partial r_0} \tag{10.219}$$

and r is the flux surface label: $r_0 = r_0(\theta_0, r)$. The flux function ψ satisfies Eq. (7.88) while V is the potential function describing the vacuum helical field

$$V = \sum_l V_l I_l(hr)e^{il\theta} \tag{10.220}$$

Here the V_l are a measure of the helical field amplitudes.

For a pure $l = 2$ system in the limit of large aspect ratio, the flux surfaces are concentric circles; $r_0(\theta_0, r) = r = \text{const}$. The rotational transform and magnetic well are given by [see Eq. (7.100) and Problem 7.6]

$$\begin{aligned}\iota &= 2lh^3R_0V_l^2\left[\frac{1}{u}\left(\frac{I_lI_l'}{u}\right)\right]' \\ W &= -h^2V_l^2u(I_l'^2 + I_l^2/u^2 + I_l^2)'\end{aligned} \tag{10.221}$$

where $u = hr$, $I_l = I_l(u)$, and $h = N/R_0$. Note that $W < 0$, indicating a magnetic hill.

One can now ask whether high shear is capable of stabilizing interchanges in the presence of a magnetic hill. In the special limit under consideration the Princeton group (Sinclair *et al.*, 1965) has calculated a simple form for the interchange stability criterion. It has the appearance of a generalization to Suydam's criterion and is given by

$$\left(\frac{r}{\hat{\iota}}\frac{d\hat{\iota}}{dr}\right)^2 - 4r\frac{d\beta}{dr}W_n > 0 \tag{10.222}$$

Here $W_n = (R_0/r\hat{\iota})^2 W$ is a normalized form of the magnetic well. The critical β_t that can be stabilized by shear is estimated by using the small argument expansion for the Bessel functions:

$$\begin{aligned}\hat{\iota} &\approx \frac{h^3 R_0 V_2^2}{4}\left(1 + \frac{u^2}{2}\right)\\ s &\approx \frac{u^2 + (7/24)u^4}{1 + u^2/2}\\ W &\approx -\frac{h^2 V_2^2}{4}\left(u^2 + \frac{3}{4}u^4\right)\end{aligned} \tag{10.223}$$

Equation (10.222) is now assumed to be marginally satisfied. This yields an expression for β' given by

$$\frac{d\beta}{dr} = -\frac{h^6 V_2^2}{16}\left[\frac{1 + (7/12)h^2r^2}{1 + (3/4)h^2r^2}\right]r^3 \approx \frac{h^6 V_2^2 r^3}{16} \tag{10.224}$$

Multiplying by r^2 and integrating over the plasma volume, one obtains the following value of β_t:

$$\beta_t = \frac{h^6 a^4 V_2^2}{96} = \frac{\hat{\iota}_0}{3}\left(\frac{a}{R_0}\right)\left(\frac{\hat{\iota}_a - \hat{\iota}_0}{2\hat{\iota}_0}\right)^{3/2} \tag{10.225}$$

where $\hat{\iota}_0 = h^3 R_0 V_2^2/4$ and $\hat{\iota}_a = \hat{\iota}_0(1 + h^2a^2/2)$. For a high shear system with $\hat{\iota}_0 = 0.5$, $\hat{\iota}_a = 1$, and $R_0/a = 8$, then $\beta_t = 0.0074$.

A high shear, pure $l = 2$ stellarator is capable of stable operation only at very low values of β_t in the absence of a magnetic well.

Creating a Magnetic Well. To improve the stability of the standard $l = 2$ stellarator it is necessary to create a magnetic well. The existence of a well is strongly coupled to the shift of the flux surfaces. Assume that small additional fields are superimposed on the basic $l = 2$ system. The specific nature of these fields will be discussed shortly. Assume that the additional fields cause a shift in flux surfaces described by

$$r_0(\theta_0, r) = r + \Delta(r)\cos\theta_0 \tag{10.226}$$

As $\beta \to 0$ the dominant change in the magnetic well caused by the shift appears in the $(r_0/R_0)\cos\theta_0$ toroidicity term [Eq. (10.218)]. After expanding $\psi(r_0, \theta_0) \approx \psi_0(r) + \psi_0'(r)\cos\theta_0 + \psi_1(r)\cos\theta_0$ and using the definition $\Delta \equiv -\psi_1/\psi_0'$, one obtains a modified form of the magnetic well given by

$$W(r) \approx -\frac{h^2V_2^2}{4}\left(h^2r^2 + \frac{3}{4}h^4r^4\right) - \frac{1}{2R_0}\left[\frac{(r^2\Delta)'}{r}\right]' \tag{10.227}$$

The first term is the previously calculated pure $l = 2$ contribution while the second term is the correction due to the flux surface shift.

If Δ is a decreasing function of radius then W develops a stabilizing well. This corresponds to the situation where the inner flux surfaces are shifted outward along the major radius with respect to the outer surfaces.

The shift Δ consists of several contributions: $\Delta = \Delta_v + \Delta_\beta + \Delta_{l+1} + \Delta_{l-1}$. Δ_v is the contribution due to an externally applied vertical field. Δ_β is the shift due to the small, but nonzero plasma pressure. $\Delta_{l\pm1}$ are the contributions due to externally applied helical sideband fields. Δ_v and Δ_p have already been calculated in Chapter 7. $\Delta_{l\pm1}$ are easily evaluated by assuming the applied vacuum helical field is of the form

$$V = V_2I_2(u)e^{2i\theta} + V_3I_3(u)e^{3i\theta} + V_1I_1(u)e^{i\theta} \tag{10.228}$$

where $V_3/V_2 \sim V_1/V_2 \ll 1$.

The flux function ψ which determines Δ is expanded as $\psi(r, \theta) = \psi_0(r) + \psi_1(r)\cos\theta$ where

$$\psi_0'(r) = \frac{r}{R_0}\,\hat{\iota}(r) \tag{10.229}$$

with $\hat{\iota}$ the vacuum $l = 2$ transform [Eq. (10.221)] and ψ_1 satisfying the first-order contribution to Eq. (7.88). The solution for $\psi_1(r)$ is given by

$$\begin{aligned}\psi_1 = {} & r\frac{B_v}{B_0} + r\int_r^\infty \frac{dx}{x^3}\int_0^x \frac{y^2\beta'(y)\,dy}{\hat{\iota}(y)} \\ & + 2hV_1V_2\left(\frac{I_1I_2' + 2I_1'I_2}{u}\right) + 2hV_3V_2\left(\frac{3I_3I_2' + 2I_3'I_2}{u}\right)\end{aligned} \tag{10.230}$$

Combining these results leads to the following expression for $\Delta = -\psi_1/\psi_0'$:

$$\begin{aligned}\Delta_v &= -\frac{R_0B_v}{\hat{\iota}B_0} \\ \Delta_\beta &= -\frac{R_0}{\hat{\iota}}\int_r^\infty \frac{dx}{x^3}\int_0^x \frac{y^2\beta'(y)\,dy}{\hat{\iota}(y)} \\ \Delta_{l+1} &\approx -\left(\frac{hV_3}{4V_2}\right)r^2 \\ \Delta_{l-1} &\approx -\frac{2V_1}{hV_2}\left(1 - \frac{h^2r^2}{8}\right)\end{aligned} \tag{10.231}$$

Δ_3 and Δ_1 have been calculated using the small ha expansion.

Having evaluated Δ observe the following points. An externally applied vertical field of proper magnitude and sign can produce a magnetic well, but only if the shear is nonzero; that is, if ι = const (i.e., zero shear) then Δ_v = const and as Eq. (10.227) indicates, a constant shift has no effect on W. If $\beta' < 0$ then Δ_β is a decreasing function of radius corresponding to a stabilizing contribution to $W(r)$. Similarly, with a proper choice of sign in either helical sideband field, a magnetic well can be created.

In practical situations the vertical field has the most flexibility. Most stellarators have a set of external vertical field coils whose current can be adjusted to produce a magnetic well. Helical sideband fields can be installed initially by winding the coils in a nonsinusoidal manner. However, once installed they cannot be varied. Separate sideband coils require additional space and cost and for those reasons are not practical. The plasma contribution is always stabilizing but is not effective at low values of β_t. At higher β_t the plasma makes an important stabilizing contribution to $W(r)$.

An important point to observe concerns the radial dependence of the magnetic well. While the methods just described can produce a stabilizing shift, the original destabilizing helical contribution is a rapidly increasing function of r, particularly in high shear systems. In practical systems it is often only possible to create a well over the inner portion of the profile. At larger r, the well vanishes and one must rely solely on shear for stability.

Once a strong magnetic well is established the system is stable against interchange perturbations. If properly designed the configuration will also be stable against ballooning and kink modes as previously discussed. Such a configuration should allow stable operation up to the equilibrium limit. The most practical method for creating a well in an $l = 2$ system is by a combination of high shear and an external vertical field.

10.6.4. Design of a Stellarator Experiment

The magnetic well provides useful insight into the stability properties of a stellarator, but is not accurate or detailed enough to be used in the design of an actual experiment. Although sophisticated numerical tools have been developed in recent years to test MHD equilibrium and stability in three-dimensional systems, they are quite time consuming to run. Thus, for a stellarator configuration potentially characterized by the dimensionless quantities l, N, a/R_0, β_t, V_l/a, V_{l+1}/a, V_{l-1}/a, B_v/B_0, the parameter space is so large that it is not practical to obtain an optimized design solely by parameter scans.

In this section it is shown how the simple analytic theory and the sophisticated numerical tools can be combined to yield an optimized design for a stellarator experiment. The particular example considered is the ATF torsatron at Oak Ridge National Laboratory (ATF Proposal, 1983; Carreras *et al.*, 1983).

General Considerations. The primary objective of the ATF experiment is the design and operation of a stellarator with the maximum β_t capability. Based on the previous discussion, good stability requires a configuration with moderate to high shear and a strong magnetic well. The shear is also important in that it raises the equilibrium β_t limit and creates relatively robust flux surfaces not susceptible to large island formation.

The magnetic configuration must also provide good single-particle confinement. This in general also requires robust flux surfaces and moderate to large rotational transform. The combined requirements of shear and transform indicate that an $l = 2$ system with many helical periods is a good choice. Furthermore, substantial helical sideband fields are not desirable because they introduce symmetry-breaking components into the configuration. This can lead to a deterioration of single-particle confinement and stochastic field line behavior.

The ATF plasma must have a reasonably large minor radius to provide an adequate energy confinement time due to transport. The basic scale of the experiment is set by economic considerations and leads to $R_0 \approx 2$ m. One goal of the design is to achieve a configuration with $a \gtrsim 0.2$ m when $R_0 \approx 2$ m.

The experiment should not require too tight an aspect ratio. While tight aspect ratio has favorable physics properties (e.g., large a at fixed R_0, strong MHD stabilizing effect due to toroidicity), it is difficult to achieve technologically. The coil aspect ratio must satisfy $R/a_c \gtrsim 3$–4 where a_c is the coil radius.

The device should be flexible in order to vary the basic properties of the magnetic geometry, such as shear and magnetic well, over a wide range to test the MHD β_t limits. This is accomplished by means of a vertical field circuit capable of being programmed during the discharge evolution.

Lastly, the experiment must be relatively easy to construct, be able to withstand the mechanical forces produced by moderate magnetic fields, and provide good diagnostic access. These criteria lead to the choice of a torsatron configuration. Its advantages outweigh the loss in flexibility of varying the ratio of helical to toroidal field characteristic of all torsatrons.

In summary, general considerations suggest an ATF design corresponding to a moderate shear $l = 2$ torsatron with major radius $R_0 \approx 2$ m. The system should include a flexible vertical field circuit but no substantial helical sideband fields. There are two important quantities as yet undetermined: the coil aspect ratio a_c/R_0 and the number of helical periods N. Recall that once a_c/R_0 and N are specified, this leads to a unique rotational transform profile in a torsatron. In practice it is convenient to treat Na_c/R_0 and a_c/R_0 as the undetermined parameters. The first quantity, Na_c/R_0, is chosen to optimize the equilibrium vacuum flux surfaces. The second quantity, a_c/R_0, is then chosen to optimize the MHD stability limits. This two-parameter optimization is well within the capabilities of the numerical codes. A summary of the results is given below.

Flux Surface Optimization. The first step in the optimization procedure is the discovery of configurations with robust vacuum flux surfaces and moderate values of shear. The numerical studies indicate that the most sensitive parameter affecting the vacuum surfaces is the pitch of the helical coils, defined as $P = Na_c/lR_0$.

A sequence of calculations has been carried for various values of P at fixed coil aspect ratio. One such sequence is illustrated in Fig. 10.21 for fixed $R_0/a_c = 4.4$. The first curve corresponds to $P = 1.14$, $N = 10$. Observe that this configuration has high rotational transform but low shear. Also, the last well-defined flux surface has a small minor radius of approximately $a \approx 0.1$ m.

At the largest value of the pitch $P = 1.71$, $N = 15$, corresponding to the third curve, there is a good shear and a large plasma radius $a \approx 0.3$ m. However, over most of the region where the plasma is located the transform is quite small: $\iota \sim 0.1$–0.2.

The intermediate case $P = 1.37$, $N = 12$ appears to be the best compromise between plasma size, shear, and transform. On this basis a value of $P = 1.4$ is chosen as the optimal value. In the discussion of the stability analysis that follows, all numerical scans are carried out assuming a fixed value $P = 1.4$. For this value the ratio of plasma radius to coil radius is found numerically as $a/a_c \approx 0.6$.

Equilibrium and Stability Optimization. The remaining parameter to be determined is the aspect ratio, or equivalently at fixed pitch, the number of helical periods N. The optimum value of N is a balance between two competing effects. First, equilibrium β_t limits are highest in large aspect ratio systems (more closely resembling a straight cylinder). At fixed pitch $P = Na_c/lR_0 = \text{const}$. This implies large N. Second,

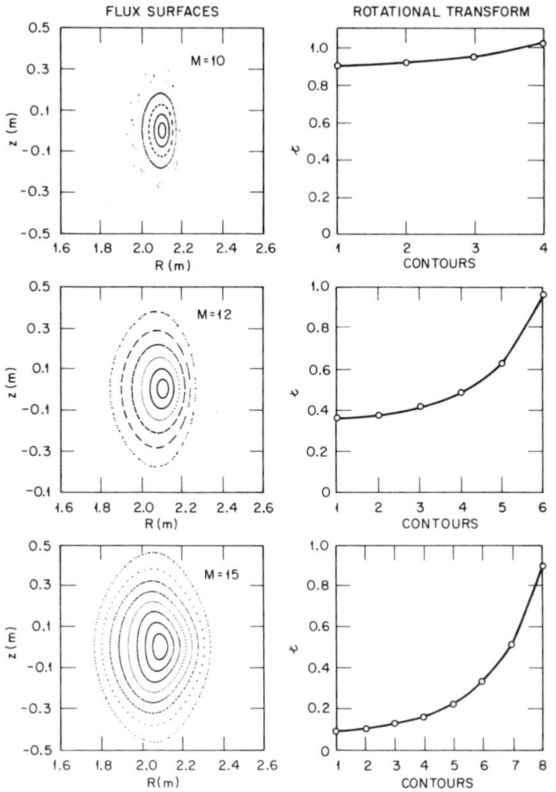

Figure 10.21. Vacuum magnetic surfaces for torsatrons of different pitch with fixed coil aspect ratio $R_0/a_c = 4.4$ [from Carreras *et al.* (1983)].

stability β_t limits are highest in tight aspect ratio systems where the stabilizing influence of toroidicity is most pronounced. This implies that low N is desirable. The optimal N balances these two effects, each of which is discussed below.

The equilibrium β_t limit in a stellarator has been derived in Chapter 7 [Eq. (7.122)] and is given by

$$\beta_t < \iota_a^2(a/R_0) \tag{10.232}$$

In a large aspect ratio system the rotational transform is equal to the rotational transform per helical period multiplied by the number of helical periods. For a torsatron, the transform per period is a function only of the pitch P. Hence $\iota_a = Nf(P)$. This yields the following scaling for the equilibrium β_t limit as a function of N at fixed P:

$$\beta_t < [lPf^2(a/a_c)]N \propto N \tag{10.233}$$

The plasma to coil ratio a/a_c is also a function only of P and for ATF $a/a_c \approx 0.6$. As stated the equilibrium β_t increases with N.

The stability analysis of the ATF configuration is focused on the internal $n = 1$ mode. While this mode is theoretically not as dangerous as the $n \to \infty$ interchanges, it is the most dangerous mode observed experimentally. The goal is to discover configurations that have robust $n = 1$ stability properties, not depending sensitively on large, perhaps fragile, flux surface shifts.

A series of $n = 1$ stability studies has been performed using the stellarator expansion model for different values of N. The pitch is held constant and the vertical field is adjusted so that the shift of the vacuum magnetic surfaces is zero. For large aspect ratio one would expect such a system to possess an unstable magnetic hill generated by the $l = 2$ helical field. This is indeed the case.

As the aspect ratio decreases, toroidicity becomes more important. At finite a/R_0 toroidicity creates a vacuum magnetic well over the inner portion of the profile even in the absence of a vertical field induced flux surface shift. Furthermore, at higher values of β_t the plasma itself produces an important stabilizing contribution to the magnetic well by means of the β_t-induced shift.

The numerical studies suggest a simple guideline to provide good $n = 1$ stability. Two requirements must be satisfied: (1) the configuration must contain a magnetic well over a portion of the profile and (2) the $n = 1$ resonant surfaces must lie within the magnetic well region. These requirements place strong constraints on the allowable values of N.

The situation is illustrated in Fig. 10.22. The curve $\hat{\iota} = \hat{\iota}_a$ represents the transform on the surface as a function of N for fixed pitch $P = 1.4$. The large aspect ratio analytic scaling $\hat{\iota}_a \propto N$ is in good agreement. The

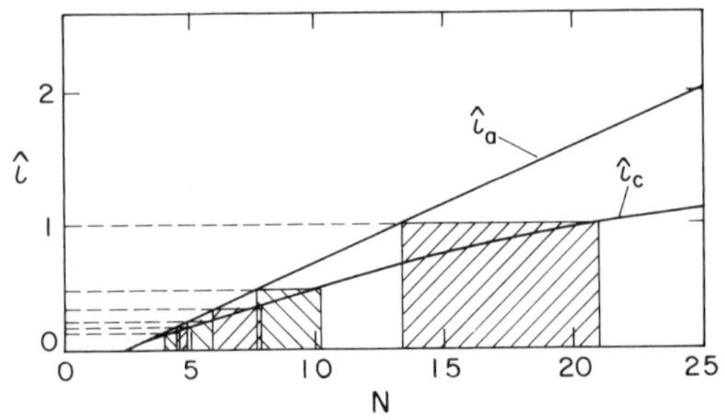

Figure 10.22. Diagram showing those values of N for which the $n = 1$ resonant surfaces lie in the region of a stabilizing magnetic well. The unshaded region is stable. [From Carreras *et al.* (1983).]

curve $\iota = \iota_c$ corresponds to that value of radius r_c where the magnetic well (including β_t effects) switches from favorable to unfavorable; $W > 0$ for $0 < r < r_c$, $W < 0$ for $r_c < r < a$ and $\iota_c = \iota(r_c)$. This curve is computed numerically as a function of N. The dashed lines in Fig. 10.22 represent the $n = 1$ resonant surfaces $\iota = n/m = 1, 1/2, 1/3, \ldots$. The shaded areas correspond to configurations where an $n = 1$ resonant surface lies in a region of unfavorable magnetic well, $W < 0$. Observe that there is a desirable gap of stability between $N = 10 - 13$.

The detailed numerical investigations of MHD equilibrium and $n = 1$ stability are combined and summarized in Fig. 10.23. Shown here are curves of peak β_0 vs. N corresponding to the equilibrium and stability boundaries. The equilibrium curve represents the highest numerically converged value of β_0. Presumably the "true equilibrium" limit where a separatrix moves onto the plasma surface is only slightly higher. The $n = 1$ stability curve represents the numerically determined boundary based on the stellarator expansion model. Observe that the highest value of β_0 achievable is about $\beta_0 = 0.18$ and is equivalent to $\beta_t \approx 0.07$–0.1 depending on the pressure profile. This occurs for $N = 12$. Based on these results, the ATF torsatron has been designed with $N = 12$.

Further stability studies have been carried out examining the $n \to \infty$ interchange mode and low-n-number external modes. With a proper choice of magnetic well created by the vertical field for small and intermediate radii and moderate shear at large radii these modes are also fully stabilized in the ATF configuration.

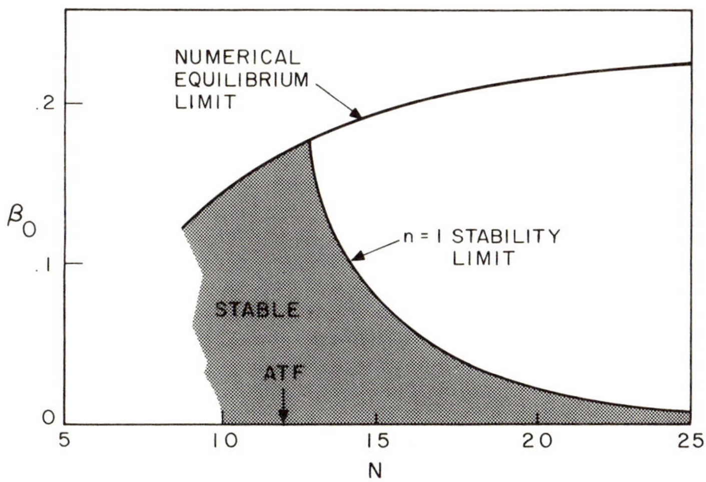

Figure 10.23. Combined equilibrium and $n = 1$ stability limits as functions of N for the ATF studies (from the ATF Proposal, Oak Ridge National Laboratory).

The studies of Carreras *et al.* demonstrate how MHD theory can be applied in a practical manner to the design of stellarator experiments. The final ATF design obtained by MHD optimization is one that exhibits no ideal MHD instabilities and is capable of operating up to the equilibrium β_t limit. For the optimized parameters this limit is quite significant, $\beta_t \sim 0.07$–0.10 and represents a very favorable result for the stellarator concept.

10.7. SUMMARY

In Chapter 10 the Energy Principle has been applied to the problem of MHD stability in multidimensional geometries. A summary of these results is given below.

General Ballooning Mode Equation. One important difference between cylindrical and toroidal configurations is the following. In cylindrical systems symmetry implies that the curvature is unidirectional and only a function of r. In toroidal systems there are in general alternating regions of favorable and unfavorable curvature. The $1/R$ dependence of the toroidal field produces favorable curvature on the inside of torus $\theta = \pi$ and unfavorable curvature on the outside $\theta = 0$.

These alternating regions allow the possibility of ballooning modes, perturbations which attempt to minimize δW by concentrating in the unfavorable region while not causing significant line bending. The most dangerous modes correspond to $n \to \infty$ and are highly localized about a flux surface. By exploiting the high-n limit and the mode localization one can reduce the multidimensional Energy Principle to a one-dimensional differential equation along the field line on each flux surface. This greatly reduces the effort required to test stability.

Relation between Magnetic Well and Average Curvature. An analysis of the general ballooning mode Energy Principle provides justification for the use of the magnetic well W as a figure of merit. It is shown that the justification arises from an examination of interchange perturbations in closed-line systems. In the limit $\beta \to 0$ the interchange condition reduces to a test of the average curvature along a field line. The magnetic well is shown to be identical to the average curvature in this limit. A positive well is a necessary condition for stability in low-β closed-line systems.

Finite-β modifications complicate the criterion. A simple correction is added to the vacuum well which cancels the plasma diamagnetic effect.

The resulting criterion is neither necessary nor sufficient for stability but serves as a useful figure of merit. The magnetic well is also used in toroidal systems with shear. Here too it serves as a figure of merit and is neither necessary nor sufficient for stability even in the limit $\beta \to 0$.

EBT Stability. The EBT configuration is the first multidimensional configuration tested for stability. Since toroidicity is required primarily to close the field lines in equilibrium, but does not play a major role in stability, the analysis is carried out for a straight two-dimensional model

In EBT $J_{\parallel} = 0$ implying that kink modes are not important. The most dangerous instabilities are $n \to \infty$ interchanges and ballooning modes. A crucial feature of EBT is the requirement that the hot electron rings be considered rigid during any MHD motions. This is equivalent to assuming the rings behave as fixed external sources of current.

The analysis shows that the hot electrons provide a magnetic well over a narrow region in r centered about the ring: $a < r < a + d$. For $r < a$ the bumpy field has average unfavorable curvature and produces a magnetic hill. Stable EBT profiles are constructed by requiring the plasma core pressure $p_c(r)$ to be flat out to $r \approx a$ where the magnetic well first develops. Since $p_c'(r) = 0$ for $0 < r < a$, interchanges and ballooning modes are not excited in this region. The entire pressure gradient occurs in the narrow well region. The analysis shows that a sufficiently high value of ring β_c is required to provide a magnetic well capable of stabilizing significant values of the core β_c. In theory quite high values of $\beta_c \sim 0.2$–0.3 are stabilized by this procedure. One difficulty with EBT is that all the gradients and hence transport occurs in the narrow region corresponding to the well. This leads to an unfavorable scaling of transport for large devices; $\tau \sim ad/D^2$ rather than $\tau \sim a^2/D$.

Tokamak Stability. Toroidicity makes important modifications to the stability predictions of the straight tokamak. There are quantitative changes in kink stability requirements and qualitative changes due to pressure-driven modes.

In a low-β straight tokamak only current-driven modes are important. Of the internal modes $n = 1$, $m = 1$ is the most dangerous instability and is associated with sawtooth oscillations. The stability condition is $q_0 > 1$. In a torus the situation is more complicated and depends sensitively on the shape of the q profile near the axis. For a parabolic q profile the $n = 1$ mode is stable down to $q_0 = 1/2$ as $\beta_p \to 0$. Only when $\beta_p \gtrsim 0.3$ does stability again require $q_0 > 1$. On the other hand, for a very flat q profile the stability boundary is $q_0 > 1$ even as $\beta_p \to 0$. Low-β external kink stability is similar in straight and toroidal tokamaks. A value $q_a > 1$ is required to avoid the $n = 1$, $m = 1$ Kruskal–Shafranov mode and $q_a/q_0 \gtrsim 2.5$–3 to avoid high m-number external kinks.

For straight tokamaks pressure-driven modes do not lead to any limits on β_t. In contrast, for toroidal tokamak interchanges, ballooning modes and ballooning-kink modes play an important role in stability. The toroidal interchange condition is given by the Mercier criterion, a two-dimensional generalization of Suydam's criterion. The $n \to \infty$ interchanges are most important near the magnetic axis where the shear is small. A circular-cross-section torus develops a magnetic well when $q_0 > 1$. With such a well the plasma is stable to localized interchanges over the entire profile. A combination of elongation and outward triangularity improves interchange stability and allows operation at values of $q_0 < 1$.

The $n \to \infty$ ballooning modes are more important than interchanges in a tokamak and set important limits on the maximum allowable pressure gradient, and hence β_t. High shear and elongated outward-pointing triangular cross sections are favorable for raising the β_t limit associated with the first region of stability. The numerical studies of Sykes *et al.* show that for optimized profiles the β_t limit due to ballooning modes is given by $\beta_t = 0.22\epsilon\kappa/q_*$. For realistic experimental cases $\beta_t \sim 0.1$.

An interesting discovery from the ballooning mode analysis is the existence of a second region of stability for large pressure gradients and low shear (equivalent to $\beta_t \gtrsim 0.15$ at large aspect ratio). The stability is a consequence of a high value of local shear caused by strong β modifications to the flux surfaces. Plasmas with bean-shaped cross sections appear to have the widest access to the second region of stability and represent an interesting prospect for high-β operation in a tokamak reactor.

External kink modes are strongly affected by plasma pressure in a toroidal tokamak. As β_t increases, the unstable perturbations develop a strong ballooning component acquiring the appearance of a ballooning-kink mode. These instabilities usually set the most severe β_t limits on tokamak operation. A combination of elongation and outward triangularity improves stability. The studies of Troyan *et al.* indicate that the maximum β_t limit in a profile optimized tokamak, including ballooning and kink modes is given by $\beta_t = 0.14\epsilon\kappa/q_*$. Here, in contrast to ballooning modes alone, q_* must exceed $q_{*\min}$, where $q_{*\min}$ corresponds to the kink limit for low β_t moderately elongated cross sections. For practical cases ballooning-kink modes limit $\beta_t \sim 0.06$ unless a conducting wall or other stabilizing influence is introduced.

In summary, while a number of unstable modes are possible in a tokamak, regions of stability can be found by a combination of cross-sectional shaping, tight aspect ratio, wall stabilization, sufficiently low current, and sufficiently low β_t. The maximum stable values of β_t lie in the range $\beta_t \sim 0.05$–0.1, which is perhaps slightly low but still well within the regime of reactor interest.

Stellarator Stability. The stellarator is the most complicated of the configurations investigated. It is a truly three-dimensional system with a wide variety of options. The analytic discussion is focused on providing qualitative insight into the stability of an $l = 2$ stellarator.

In current-free stellarators kink modes do not play a major role and stability is dominated by pressure-driven instabilities. An interesting contrast is that ballooning modes are unimportant in stellarators compared to tokamaks. This is a consequence of the fact that $s < 0$ in a stellarator. Consequently as β_t increases the local shear increases in regions of local unfavorable curvature and decreases in regions of favorable curvature, a desirable feature for stability. The opposite is true in a tokamak. Consequently interchanges are the most unstable perturbations in a stellarator.

Interchanges are stabilized by a combination of shear and magnetic well. Shear alone is capable of stabilizing only low values of $\beta_t \lesssim 0.01$, indicating the necessity for a magnetic well. The toroidal shift of the flux surfaces plays a crucial role in developing a magnetic well. Helical sideband fields, plasma pressure, and a vertical field in a system with a shear are each capable of transforming the basic $l = 2$ helical magnetic hill into a magnetic well. At low β, a vertical field in a system with shear is the most practical method. At high β_t, the pressure-induced shift makes an important stabilizing contribution.

The analytic insight has been combined with a series of sophisticated numerical calculations by Carreras *et al.* leading to the design of the ATF experiment. The ATF configuration is a torsatron optimized with respect to the quality of the vacuum surfaces, equilibrium β_t limits, and MHD stability limits. The optimized configuration should be capable of stable operation up to the equilibrium limit, which is approximately $\beta_t \sim 0.07$–0.1. This is indeed a favorable result.

The studies show how MHD considerations play a very important role in the conceptual design of fusion experiments. They also show how a combination of analytic insight and sophisticated numerical studies can be combined in a practical manner to provide detailed design optimization.

REFERENCES

ATF Proposal, J. F. Lyon *et al.* (1983). Oak Ridge National Laboratory Report No. ORNL/TM-8496.

Berger, D., R. Gruber, and F. Troyon (1977). In *Controlled Fusion and Plasma Physics,* Proceedings of the Eighth European Conference, Prague, Vol. I, p. 51.

Bussac, M. N., R. Pellat, D. Edery, and J. L. Soule (1975). *Phys. Rev. Lett.* **35,** 1638.

Carreras, B. A., H. R. Hicks, J. A. Holmes, V. E. Lynch, L. Garcia, J. H. Harris, T. C. Hender and B. F. Masden (1983). *Phys. Fluids* **26,** 3569.

Connor, J. W., R. J. Hastie, and J. B. Taylor (1979). *Proc. R. Soc. London, Ser. A.* **365,** 1.

Coppi, B. (1977). *Phys. Rev. Lett.* **39,** 939.

D'Ippolito, D. A., J. P. Freidberg, J. P. Goedbloed, and J. Rem (1978). *Phys. Fluids* **21,** 1600.

Freidberg, J. P., and F. A. Haas (1973). *Phys. Fluids* **16,** 1909.

Freidberg, J. P., and W. Grossmann (1975). *Phys. Fluids* **18,** 1494.

Grimm, R. C., M. S. Chance, A. M. M. Todd, J. Manickam, M. Okabayashi, W. M. Tang, R. L. Dewar, H. Fishman, S. L. Mendesohn, and D. A. Monticello (1984). "MHD Stability Properties of Bean-Shaped Tokamaks", Princeton Plasma Physics Laboratory Report 2090.

Haas, F. A. (1975). *Nucl. Fusion* **15,** 407.

Johnson, J. L., M. S. Chance, J. M. Greene, R. C. Grimm *et al.* (1977). In *Plasma Physics and Controlled Nuclear Fusion Research 1976,* IAEA, Vienna, Vol. II, 395.

Johnson, J. L., R. M. Kulsrud, and K. E. Weimer (1969). *Plasma Phys.* **11,** 463.

Kerner, W. (1976). Princeton University, PPL Report MATT-1229.

Laval, G., H. Luc, E. K. Maschke, C. Mercier, and R. Pellat (1971). In *Plasma Physics and Controlled Nuclear Fusion Research 1970,* IAEA, Vienna, Vol. II, p. 507.

Laval, G., and R. Pellat (1973). In *Controlled Fusion and Plasma Physics,* Proceedings of the Sixth European Conference, Moscow, Vol. II, p. 640.

Laval, G., R. Pellat, and J. L. Soule (1974). *Phys. Fluids* **17,** 835.

Lortz, D., and J. Nuhrenberg (1973). *Nucl. Fusion* **13,** 821.

Mercier, C. (1960). *Nucl. Fusion* **1,** 47.

Mikhailovskii, A. B. (1974). *Nucl. Fusion* **14,** 483.

Mikhailovskii, A. B., and V. D. Shafranov (1974). *Sov. Phys.—JETP* **39,** 88.

Mukhovatov, V. S., and V. D. Shafranov (1971). *Nucl. Fusion* **11,** 605.

Nelson, D. B., and C. L. Hedrick (1979). *Nucl. Fusion* **19,** 283.

Osovets, S. M. (1959). In *Plasma Physics and the Problems of Controlled Thermonuclear Reactions,* Pergamon, Oxford, Vol. 2, p. 322.

Rebhan, E. (1975). *Nucl. Fusion* **15,** 277.

Shafranov, V. D. (1966). In *Review of Plasma Physics,* edited by M. A. Leontovich, Consultants Bureau, New York, Vol. III.

Shafranov, C. D., and E. I. Yurchenko (1968). *Sov. Phys.—JETP* **26,** 682.

Shafranov, V. D. (1983). *Phys. Fluids* **26,** 357.

Sinclair, R. M., S. Yoshikawa, W. L. Harries, K. M. Young, K. E. Weimer, and J. L. Johnson (1965). *Phys. Fluids* **8,** 118.

Solov'ev, L. S., V. D. Shafranov, and E. I. Yurchenko (1969). In *Plasma Physics and Controlled Nuclear Fusion Research 1968,* IAEA, Vienna, Vol. I, p. 173.

Sykes, A., and J. A. Wesson (1974). *Nucl. Fusion* **14,** 645.

Sykes, A., M. F. Turner, and S. Patel (1983). In *Controlled Fusion and Plasma Physics,* Eleventh European Conference, Aachen, W. Germany, p. 363.

Troyon, F., R. Gruber, H. Saurenmann, S. Semenzato, and S. Succi (1984). *Plasma Phys.* **26,** 209.

Ware, A. A., and F. A. Haas (1966). *Phys. Fluids* **9,** 956.
Wesson, J. A. (1985). In the 12th European Conference on Controlled Fusion and Plasma Physics, Budapest, Hungary.
Wesson, J. A., and A. Sykes (1974). In *Plasma Physics and Controlled Nuclear Fusion Research 1974*, IAEA, Vienna, Vol. I, p. 449.
Yamazaki, K., T. Amano, H. Naitou, Y. Hamada, and M. Azumi (1985). *Nucl. Fusion* **25,** 1543.
Yoshikawa, S. (1964). *Phys. Fluids* **7,** 278.

PROBLEMS

10.1. This problem demonstrates that any divergence-free vector field can be written in the form $\mathbf{B} = \nabla\alpha \times \nabla\beta$. Consider two functions $\hat{\alpha}(\mathbf{r})$, $\hat{\beta}(\mathbf{r})$ such that the magnetic lines lie in the contours $\hat{\alpha} = \text{const}$ and $\hat{\beta} = \text{const}$. The intersection of two contours $\hat{\alpha} = \alpha_0$ and $\hat{\beta} = \beta_0$ defines a given magnetic line.

a. Show that $\nabla\hat{\alpha} \times \nabla\hat{\beta}$ points in the direction of the magnetic field at every point in space.

b. Assuming (a) to be true, one can write the magnetic field as $\mathbf{B} = A(\mathbf{r})\nabla\hat{\alpha} \times \nabla\hat{\beta}$. Here, $A(\mathbf{r})$ is a scalar function adjusted appropriately so that $\mathbf{B}$ has the correct magnitude. Show that the condition $\nabla \cdot \mathbf{B} = 0$ implies that $A(\mathbf{r}) = A(\hat{\alpha}, \hat{\beta})$.

c. Define new coordinates α and β as follows:

$$\alpha = \hat{\alpha}$$

$$\beta = \int_0^{\hat{\beta}} A(\hat{\alpha}, \beta')\, d\beta'$$

Show that

$$\mathbf{B} = \nabla\alpha \times \nabla\beta$$

10.2. Consider an ideal MHD plasma in the presence of a fixed external current density source $\mathbf{J}_e$. The linearized stability problem for this system is given by Eqs. (10.57) and (10.58). Following the derivation in Section 8.8.3, show that δW_F can be written in the following intuitive form:

$$\delta W_F = \frac{1}{2}\int d\mathbf{r}\left[\frac{|\mathbf{Q}_\perp|^2}{\mu_0} + \frac{B^2}{\mu_0}|\nabla \cdot \boldsymbol{\xi}_\perp + 2\boldsymbol{\xi}_\perp \cdot \boldsymbol{\kappa}_c|^2 + \gamma p_c\, |\nabla \cdot \boldsymbol{\xi}|^2 \right.$$
$$\left. - (\mathbf{b} \cdot \mathbf{J}_c)\boldsymbol{\xi}_\perp^* \times \mathbf{b} \cdot \mathbf{Q}_\perp - 2(\boldsymbol{\xi}_\perp \cdot \nabla p_c)(\boldsymbol{\xi}_\perp^* \cdot \boldsymbol{\kappa}_c)\right]$$

where p_c is the plasma pressure, $\mathbf{J}_c = \mathbf{J} - \mathbf{J}_e$ is the plasma current, and

$$\boldsymbol{\kappa}_c = \boldsymbol{\kappa} + \frac{\mu_0}{2}\left(\frac{\mathbf{b} \times \mathbf{J}_{e\perp}}{B}\right)$$

10.3. Consider the form of δW_F given in Problem 10.2. Introduce an eikonal representation for $\boldsymbol{\xi}_\perp$ and show that in the limit $\mathbf{k}_\perp \to \infty$, δW_F can be written as

$$\delta W_F = \frac{1}{2\mu_0}\int d\mathbf{r}\left[k_\perp^2\,|\mathbf{b}\cdot\nabla X|^2 - \frac{2\mu_0}{B^2}(\mathbf{b}\times\mathbf{k}_\perp\cdot\nabla p_c)(\mathbf{b}\times\mathbf{k}_\perp\cdot\boldsymbol{\kappa}_c)\,|X|^2\right]$$

The derivation closely parallels that given in Section 10.2.

10.4. In this problem the effects of compressibility on the ballooning mode stability of closed-line systems is investigated. For simplicity the analysis is carried out for a straight EBT system using the flux coordinates introduced in Section 10.4.3. However, the calculation can easily be generalized to arbitrary three-dimensional closed-line systems.

a. Review the discussion in Section 8.9. Show that the terms giving rise to a zeroth-order contribution in δW proportional to $\mathbf{k}_\perp$ in the eikonal representation are given by

$$\delta W_0 = \frac{1}{2}\int \frac{d\theta\,d\psi\,dl}{B}\left[\frac{B^2}{\mu_0}|\nabla\cdot\boldsymbol{\xi}_\perp + 2\boldsymbol{\xi}_\perp\cdot\boldsymbol{\kappa}_c|^2 + \gamma p_c\,|\langle\nabla\cdot\boldsymbol{\xi}_\perp\rangle|^2\right]$$

where p_c is the plasma core pressure, $\boldsymbol{\kappa}_c = \boldsymbol{\kappa} + \mu_0\mathbf{b}\times\mathbf{J}_e/2B$ and

$$\langle F\rangle = \oint\frac{F\,dl}{B}\Big/\oint\frac{dl}{B}$$

b. Following the discussion in Section 10.2 write $\boldsymbol{\xi}_\perp = \boldsymbol{\eta}_\perp\exp(iS)$, where $\mathbf{B}\cdot\nabla S = 0$ and $\mathbf{k}_\perp = \nabla S$. Expand $\boldsymbol{\eta}_\perp = \boldsymbol{\eta}_{\perp 0} + \boldsymbol{\eta}_{\perp 1} + \cdots$. Minimize δW_0 and show that the first nonvanishing contribution is given by

$$\delta W_0 = \frac{1}{2}\int d\theta\,d\psi\,W_0$$

where

$$W_0 = \oint\frac{dl}{B}\left[\frac{B^2}{\mu_0}(U + 2\boldsymbol{\eta}_{\perp 0}\cdot\boldsymbol{\kappa}_c)^2 + \gamma p_c\langle U\rangle^2\right]$$

Here $U = i\mathbf{k}_\perp\cdot\boldsymbol{\eta}_{\perp 1} + \nabla\cdot\boldsymbol{\eta}_{\perp 0}$ and $\boldsymbol{\eta}_{\perp 0} = (X/B)(\mathbf{b}\times\mathbf{k}_\perp)$.

c. Note the $\boldsymbol{\eta}_{\perp 1}$ appears only in W_0 and not in the remainder of δW. Minimize W_0 with respect to U. A convenient way to carry out the minimization is to introduce the variable

$$V(l) = \int_0^l \frac{U\,dl}{B}$$

It immediately follows that

$$U = B\frac{\partial V}{\partial l}$$

and

$$\langle U\rangle = V(L)/K_1$$

where

$$K_n = \oint \frac{dl}{B^n}$$

Show that the minimized W_0 can be written as

$$W_0 = \frac{4\gamma p_c K_1^2 \langle \boldsymbol{\eta}_{\perp 0} \cdot \boldsymbol{\kappa}_c \rangle^2}{\mu_0 \gamma p_c K_3 + K_1}$$

d. Show that the full contribution to W (for ballooning modes) is given by

$$W = \frac{4\gamma p_c K_1^2}{\mu_0 \gamma p_c K_3 + K_1} \left\langle \frac{\kappa_{cn} X}{rB} \right\rangle^2 + \oint dl \left[\frac{1}{\mu_0 r^2 B} \left(\frac{\partial X}{\partial l} \right)^2 - 2 \frac{dp_c}{d\psi} \frac{\kappa_{cn} X^2}{rB^2} \right]$$

where $\kappa_{cn} = \mathbf{n} \cdot \boldsymbol{\kappa}_c$ is the normal curvature.

e. Consider the interchange perturbation $X = 1$. Show that the stability condition can be expressed

$$-2 \frac{dp_c}{d\psi} \left\langle \frac{\kappa_{cn}}{rB} \right\rangle + \left[\frac{4\gamma p_c}{1 + \mu_0 \gamma p_c \langle 1/B^2 \rangle} \right] \left\langle \frac{\kappa_{cn}}{rB} \right\rangle^2 > 0$$

Compare this result with Eq. (10.83).

10.5. This problem demonstrates a practical procedure for calculating the orthogonal flux coordinates used in the tokamak ballooning mode analysis. Assume that the flux function $\psi(R, Z)$ and its R and Z derivatives have been found either numerically or analytically by solving the Grad–Shafranov equation. The goal is to determine the coordinates $R = R(\psi, \chi)$, $Z = Z(\psi, \chi)$, where χ is an anglelike variable with period 2π, orthogonal to ψ: $\nabla\psi \cdot \nabla\chi = 0$

a. Write the total differentials for $d\psi$ and $d\chi$. Show that for orthogonal coordinates, R and Z satisfy the equations

$$\frac{\partial R}{\partial \psi} = \frac{1}{(\nabla\psi)^2} \frac{\partial \psi}{\partial R}$$

$$\frac{\partial Z}{\partial \psi} = \frac{1}{(\nabla\psi)^2} \frac{\partial \psi}{\partial Z}$$

b. The equations in part (a) represent a set of coupled, nonlinear ordinary differential equations for R and Z with χ appearing as a parameter. R and Z can be obtained by a simple numerical integration once appropriate starting conditions in ψ and a value of χ are specified. The starting conditions are slightly subtle, but can be obtained as follows. Expand all quantities about the magnetic axis: $R = R_a + x$, $Z = y$, $\psi = (\psi_{RR} x^2 + \psi_{ZZ} y^2)/2$. Here, ψ_{RR} and ψ_{ZZ} are constants, equal to the derivatives evaluated on the magnetic axis and the arbitrary constant in ψ has been chosen so that $\psi(R_a, 0) = 0$. Show that the equations in the vicinity of the axis can be written as

$$\frac{\partial X}{\partial \psi} = 2\psi_{RR}\frac{X}{X+Y}$$

$$\frac{\partial Y}{\partial \psi} = 2\psi_{ZZ}\frac{Y}{X+Y}$$

where $X = \psi_{RR}^2 x^2$, $Y = \psi_{ZZ}^2 y^2$.

c. Solve the equations in part (b) for X and Y. Show that

$$X = F(\chi)Y^{\alpha}$$

$$X + \alpha Y = 2\psi_{RR}\psi$$

where $F(\chi)$ is an arbitrary function and $\alpha = \psi_{RR}/\psi_{ZZ}$. For elongated flux surfaces $\alpha > 1$.

d. The function $F(\chi)$ is not unique; that is, there are an infinite number of ways to label the orthogonal angle coordinate χ. A convenient choice, satisfying the 2π periodicity constraint and possessing the conventional properties that $\chi = 0$ when $y = 0$ and $\chi = \pi/2$ when $x = 0$ is given by

$$F(\chi) = K\cot^2\chi$$

The constant K is arbitrary and its value is of no great significance. Different K values correspond to different labels of the constant χ surfaces as χ varies from 0 to 2π. However, the shape of the constant χ surfaces remains unchanged. Choose $K = \psi_{RR}^2/\psi_{ZZ}^{2\alpha}$. Show that the proper starting conditions for small ψ can be written as a set of coupled transcendental algebraic equations given by

$$\frac{y^2}{\alpha^{1/2}} + y^{2\alpha}\alpha^{1/2}\cot^2\chi = \frac{2\psi}{(\psi_{RR}\psi_{ZZ})^{1/2}}$$

$$x^2 = y^{2\alpha}\cot^2\chi$$

e. Sketch the $\psi = \text{const}$ and $\chi = \text{const}$ contours in the vicinity of the origin. To calculate the orthogonal coordinates $R(\psi, \chi)$, $Z(\psi, \chi)$ one specifies a value of χ and a small value of ψ (near the magnetic axis). The equations in part (d) are then solved simultaneously for y and x. These values are used as starting conditions for the basic equations in part (a). This process is then repeated for a sequence of χ values.

10.6. Consider the transformation of coordinates $(R, Z) \to (\psi, \chi)$ used in the discussion of axisymmetric tori in Section 10.5.2. First carry out the transformation in the forward direction

$$R\,dR\,dZ = J\,d\psi\,d\chi$$

Next carry out the transformation in the reverse direction

$$d\psi\,d\chi = J'R\,dR\,dZ$$

By definition $J = 1/J'$. Show that [Eq. (10.109)]

$$JB_p = \frac{1}{\mathbf{b}_p \cdot \nabla\chi} = \mathbf{b}_p \cdot \frac{\partial \mathbf{r}}{\partial \chi}$$

where $\mathbf{r} = R(\psi, \chi)\mathbf{e}_R + Z(\psi, \chi)\mathbf{e}_Z$ and $\mathbf{b}_p = \mathbf{B}_p/B_p$.

10.7. In Section 10.5.5 the maximum value of β_t allowed by ballooning instabilities was investigated using a simple analytic model. This problem reexamines the analysis using a slightly more realistic model. As in Section 10.5.5 assume the entire profile is determined by operation on the marginal stability boundary of the ballooning mode. For simplicity assume the boundary is given by $\alpha = Ks$ where $K \approx 0.6$.

a. Show that the critical β_t can be written as

$$\beta_t = \frac{\epsilon K}{2}\left[-\frac{1}{q_a^2} + 3\int_0^1 \frac{x^2}{q^2}dx\right]$$

where $x = r/a$.

b. Assume the q profile has the form

$$q(x) = q_0(1 + \lambda x^3)^\nu$$

Find the relationship between λ and ν such that $J(a) = 0$.

c. Calculate $\beta_t = (\epsilon K/2)F(q_a, q_0)$. Set $q_0 = 1$ and plot $F(q_a)$ versus $1/q_a$. Compare the dependence of F on $1/q_a$ with the predictions of Sykes and Troyan: $F \propto 1/q_a$. Note that there is quite good scaling agreement over the experimentally interesting range $2 < q_a < 6$.

10.8. a. Write down the general pressure-balance jump condition for the sharp boundary surface current model of an axisymmetric tokamak. Derive an analytic expression for $B_p = B_p(\theta)$ corresponding to an arbitrary plasma cross section $r = r_p(\theta)$. Do not make a large aspect ratio or low-β expansion.

b. Using the high-β tokamak ordering for a circular cross section plasma show that $B_p(\theta)$ reduces to

$$\frac{B_p(\theta)}{\epsilon B_0} = \frac{1}{q_*}\left(\frac{\pi}{2E}\right)[1 - k^2 \sin^2(\theta/2)]^{1/2}$$

where $q_* = 2\pi a^2 B_0/\mu_0 R_0 I_0$, $E(k)$ is the complete elliptic function of the second kind and k is a parameter related to $\beta_t = 2\mu_0 p/B_0^2$ as follows: $\beta_t q_*^2/\epsilon = (\pi k/4E)^2$.

c. Make a sketch of β_t/ϵ versus q_* corresponding to the equilibrium limit. Calculate q_a. On the β_t/ϵ versus q_* diagram sketch the curve corresponding to $q_a = \text{const} = 2$. This curve represents, in an approximate sense, the trajectory of a flux conserving tokamak evolution.

10.9. Show that the surface term in δW for an arbitrary three-dimensional sharp boundary equilibrium can be written as

$$\delta W_S = \frac{1}{2}\int dS\, |\mathbf{n}\cdot\boldsymbol{\xi}|^2[2p\langle\kappa_n\rangle + \langle B^2/\mu_0\rangle(\hat{\kappa}_n - \kappa_n)]$$

where $\langle\kappa_n\rangle = (\hat{\kappa}_n + \kappa_n)/2$ is the average curvature across the surface and $\langle B^2\rangle = (\hat{B}^2 + B^2)/2$ is the average magnetic energy. Note that the first term is analogous to the pressure-driven contribution in δW proportional to the product of the pressure gradient and the curvature. The second term is analogous to the current-driven contribution in δW; the quantity $\hat{\kappa}_n - \kappa_n$ is nonzero only when there is a parallel current.

10.10. In this problem the high-β tokamak expansion is used to reduce δW_F to a two-dimensional form involving only a single scalar unknown. Furthermore, the final form of δW_F is valid for arbitrary m and n modes; no restriction to short wavelengths is required as in the ballooning mode analysis.

a. Using the high-β tokamak expansion show that to leading order, δW_F is minimized (to zero) by choosing $\nabla\cdot\boldsymbol{\xi}_\perp = 0$. From this result it follows that $\boldsymbol{\xi}_\perp$ can be written as

$$\boldsymbol{\xi}_\perp = \mathbf{e}_\phi \times \nabla[X(r, \theta)\exp(-in\phi)]$$

Here, $X(r, \theta)$ is a two-dimensional scalar function representing the plasma displacement.

b. With this representation for $\boldsymbol{\xi}_\perp$ show that the first nonzero contribution to δW_F is of order ϵ^2 and is given by

$$\delta W_F \approx \pi R_0 \int r\,dr\,d\theta\bigg[\frac{1}{\mu_0}|\nabla_\perp(\mathbf{B}\cdot\nabla X)|^2 + 2\frac{dp}{d\psi}(\mathbf{B}_p\cdot\nabla_\perp X)(\mathbf{e}_Z\cdot\nabla_\perp X^*)$$
$$+ J_\parallel \mathbf{e}_\phi\cdot(\nabla_\perp X^*)\times\nabla_\perp(\mathbf{B}\cdot\nabla X)\bigg]$$

where $\mathbf{e}_Z = \mathbf{e}_r\sin\theta + \mathbf{e}_\theta\cos\theta$, $\nabla_\perp = \mathbf{e}_r\partial/\partial r + (\mathbf{e}_\theta/r)\partial/\partial\theta$, and $\mathbf{B}\cdot\nabla X = (-inB_0/R_0 + \mathbf{B}_p\cdot\nabla_\perp)X$.

c. Now consider ballooning modes in the context of the high-β tokamak form of δW_F. Introduce an eikonal representation for X: $X = \hat{X}(r, \theta)\exp[iS_\perp(r, \theta)]$ where $\mathbf{B}_p\cdot\nabla_\perp S_\perp - nB_0/R_0 = 0$ and $n \to \infty$. Derive the ballooning mode equation following the procedure given in Section 10.2. Show that the resulting equation agrees with Eq. (10.13) in the limit of the high-β expansion. Thus, for this system it makes no difference whether one expands in ϵ and then takes the ballooning limit or vice versa; the result is the same.

10.11. Consider a large aspect ratio torsatron in which the magnetic fields are created by a surface current located at $r = a_c$ given by

$$\mathbf{K} = K_0[1 + \cos(l\theta - N\phi)](\cos\alpha\,\mathbf{e}_\theta + \sin\alpha\,\mathbf{e}_\phi)$$

where $\tan\alpha = lR_0/Na_c$. Derive a formula for the rotational transform at $r = a$.

Appendices

APPENDIX A. SELF-ADJOINTNESS OF THE FORCE OPERATOR F

The goal of Appendix A is to show that the force operator $\mathbf{F}$ is self-adjoint; that is

$$\int \boldsymbol{\eta} \cdot \mathbf{F}(\boldsymbol{\xi})\, d\mathbf{r} = \int \boldsymbol{\xi} \cdot \mathbf{F}(\boldsymbol{\eta})\, d\mathbf{r} \tag{A.1}$$

where $\boldsymbol{\xi}$ and $\boldsymbol{\eta}$ are two arbitrary vectors satisfying the boundary condition $\mathbf{n} \cdot \boldsymbol{\xi} = \mathbf{n} \cdot \boldsymbol{\eta} = 0$ on the surface. This corresponds to the perfectly conducting wall boundary condition.

The integrand can be written as

$$\boldsymbol{\eta} \cdot \mathbf{F}(\boldsymbol{\xi}) = \boldsymbol{\eta} \cdot \left[\frac{1}{\mu_0}(\nabla \times \mathbf{B}) \times \mathbf{Q} + \frac{1}{\mu_0}(\nabla \times \mathbf{Q}) \times \mathbf{B} + \nabla(\boldsymbol{\xi} \cdot \nabla p + \gamma p \nabla \cdot \boldsymbol{\xi})\right] \tag{A.2}$$

with $\mathbf{Q} = \nabla \times (\boldsymbol{\xi} \times \mathbf{B})$. The last term is integrated by parts yielding

$$\boldsymbol{\eta} \cdot \mathbf{F}(\boldsymbol{\xi}) = \boldsymbol{\eta} \cdot \left[\frac{1}{\mu_0}(\nabla \times \mathbf{B}) \times \mathbf{Q} + \frac{1}{\mu_0}(\nabla \times \mathbf{Q}) \times \mathbf{B} + \nabla(\boldsymbol{\xi} \cdot \nabla p)\right] - \gamma p(\nabla \cdot \boldsymbol{\xi})(\nabla \cdot \boldsymbol{\eta}) \tag{A.3}$$

One now writes $\boldsymbol{\xi} = \boldsymbol{\xi}_\perp + \xi_\parallel \mathbf{b}$, $\boldsymbol{\eta} = \boldsymbol{\eta}_\perp + \eta_\parallel \mathbf{b}$. The term in the square bracket in Eq. (A.3) has no parallel component; specifically,

a. $$\frac{1}{\mu_0}\mathbf{B} \cdot (\nabla \times \mathbf{B}) \times \mathbf{Q} = -\mathbf{Q} \cdot \mathbf{J} \times \mathbf{B} = -\mathbf{Q} \cdot \nabla p = \nabla \cdot [\nabla p \times (\boldsymbol{\xi} \times \mathbf{B})] = -\nabla \cdot [(\boldsymbol{\xi} \cdot \nabla p)\mathbf{B}] \tag{A.4}$$

b. $$\mathbf{B} \cdot \nabla(\boldsymbol{\xi} \cdot \nabla p) = \nabla \cdot [(\boldsymbol{\xi} \cdot \nabla p)\mathbf{B}]$$

Clearly, the parallel component cancels. Consequently one finds

$$\boldsymbol{\eta} \cdot \mathbf{F}(\boldsymbol{\xi}) = -\gamma p(\nabla \cdot \boldsymbol{\xi})(\nabla \cdot \boldsymbol{\eta}) + I$$

where I is a function only of the perpendicular components of $\boldsymbol{\xi}$ and $\boldsymbol{\eta}$:

$$I(\boldsymbol{\xi}_\perp, \boldsymbol{\eta}_\perp) = \boldsymbol{\eta}_\perp \cdot \left[\frac{1}{\mu_0}(\nabla \times \mathbf{B}) \times \mathbf{Q} + \frac{1}{\mu_0}(\nabla \times \mathbf{Q}) \times \mathbf{B} + \nabla(\boldsymbol{\xi}_\perp \cdot \nabla p)\right] \tag{A.5}$$

The last term is now integrated by parts and the first two terms rewritten using standard vector identities:

$$I = \frac{1}{\mu_0}\boldsymbol{\eta}_\perp \cdot [\mathbf{Q}\cdot\nabla\mathbf{B} + \mathbf{B}\cdot\nabla\mathbf{Q} - \nabla(\mathbf{B}\cdot\mathbf{Q})] - (\boldsymbol{\xi}_\perp\cdot\nabla p)\nabla\cdot\boldsymbol{\eta}_\perp \quad \text{(A.6)}$$

The three terms inside the square brackets in Eq. (A.6) are expanded as follows:

$$\begin{aligned}
\boldsymbol{\eta}_\perp\cdot(\mathbf{Q}\cdot\nabla\mathbf{B}) &= \boldsymbol{\eta}_\perp\cdot[(\mathbf{B}\cdot\nabla\boldsymbol{\xi}_\perp)\cdot\nabla\mathbf{B} - (\boldsymbol{\xi}_\perp\cdot\nabla\mathbf{B})\cdot\nabla\mathbf{B}] - B^2(\boldsymbol{\eta}_\perp\cdot\boldsymbol{\kappa})\nabla\cdot\boldsymbol{\xi}_\perp \\
\boldsymbol{\eta}_\perp\cdot(\mathbf{B}\cdot\nabla\mathbf{Q}) &= \mathbf{B}\cdot\nabla(\boldsymbol{\eta}_\perp\cdot\mathbf{Q}) - \mathbf{Q}\cdot(\mathbf{B}\cdot\nabla\boldsymbol{\eta}_\perp) \\
&= \nabla\cdot[(\boldsymbol{\eta}_\perp\cdot\mathbf{Q})\mathbf{B}] - (\mathbf{B}\cdot\nabla\boldsymbol{\xi}_\perp - \boldsymbol{\xi}_\perp\cdot\nabla\mathbf{B} \\
&\quad - \mathbf{B}\nabla\cdot\boldsymbol{\xi}_\perp)\cdot(\mathbf{B}\cdot\nabla\boldsymbol{\eta}_\perp) \\
&= -(\mathbf{B}\cdot\nabla\boldsymbol{\xi}_\perp)\cdot(\mathbf{B}\cdot\nabla\boldsymbol{\eta}_\perp) + (\boldsymbol{\xi}_\perp\cdot\nabla\mathbf{B})\cdot(\mathbf{B}\cdot\nabla\boldsymbol{\eta}_\perp) \\
&\quad - B^2(\boldsymbol{\eta}_\perp\cdot\boldsymbol{\kappa})\nabla\cdot\boldsymbol{\xi}_\perp \\
-\boldsymbol{\eta}_\perp\cdot\nabla(\mathbf{B}\cdot\mathbf{Q}) &= -\nabla\cdot[(\mathbf{B}\cdot\mathbf{Q})\boldsymbol{\eta}_\perp] + (\mathbf{B}\cdot\mathbf{Q})\nabla\cdot\boldsymbol{\eta}_\perp \\
&= -B^2(\nabla\cdot\boldsymbol{\xi}_\perp)(\nabla\cdot\boldsymbol{\eta}_\perp) - [\boldsymbol{\xi}_\perp\cdot\nabla B^2/2 + B^2(\boldsymbol{\xi}_\perp\cdot\boldsymbol{\kappa})]\nabla\cdot\boldsymbol{\eta}_\perp
\end{aligned} \quad \text{(A.7)}$$

In the second and third terms, full divergence contributions have been dropped since they integrate to zero. Combining terms one finds

$$\begin{aligned}
\boldsymbol{\eta}\cdot\mathbf{F}(\boldsymbol{\xi}) = &-\frac{B^2}{\mu_0}(\nabla\cdot\boldsymbol{\xi}_\perp)(\nabla\cdot\boldsymbol{\eta}_\perp) - \frac{1}{\mu_0}(\mathbf{B}\cdot\nabla\boldsymbol{\xi}_\perp)\cdot(\mathbf{B}\cdot\nabla\boldsymbol{\eta}_\perp) \\
&- \gamma p(\nabla\cdot\boldsymbol{\xi})(\nabla\cdot\boldsymbol{\eta}) \\
&- \left[\boldsymbol{\xi}_\perp\cdot\nabla\left(p + \frac{B^2}{2\mu_0}\right) + \frac{B^2}{\mu_0}\boldsymbol{\xi}_\perp\cdot\boldsymbol{\kappa}\right]\nabla\cdot\boldsymbol{\eta}_\perp \\
&- 2\frac{B^2}{\mu_0}(\boldsymbol{\eta}_\perp\cdot\boldsymbol{\kappa})\nabla\cdot\boldsymbol{\xi}_\perp \\
&+ R
\end{aligned} \quad \text{(A.8)}$$

where

$$\mu_0 R = \boldsymbol{\eta}_\perp\cdot[(\mathbf{B}\cdot\nabla\boldsymbol{\xi}_\perp)\cdot\nabla\mathbf{B} - (\boldsymbol{\xi}_\perp\cdot\nabla\mathbf{B})\cdot\nabla\mathbf{B}] + (\boldsymbol{\xi}_\perp\cdot\nabla\mathbf{B})\cdot(\mathbf{B}\cdot\nabla\boldsymbol{\eta}_\perp) \quad \text{(A.9)}$$

The middle line is simplified by noting that $\boldsymbol{\xi}_\perp\cdot\nabla(p + B^2/2\mu_0) = (B^2/\mu_0)(\boldsymbol{\xi}_\perp\cdot\boldsymbol{\kappa})$. The quantity R can be rewritten by using the two identities

$$\begin{aligned}
\nabla\cdot\{[\boldsymbol{\eta}_\perp\cdot(\boldsymbol{\xi}_\perp\cdot\nabla\mathbf{B})]\mathbf{B}\} &= (\mathbf{B}\cdot\nabla\boldsymbol{\eta}_\perp)\cdot(\boldsymbol{\xi}_\perp\cdot\nabla\mathbf{B}) \\
&\quad + \boldsymbol{\eta}_\perp\cdot(\mathbf{B}\cdot\nabla\boldsymbol{\xi}_\perp)\cdot\nabla\mathbf{B} + \boldsymbol{\eta}_\perp\cdot(\mathbf{B}\boldsymbol{\xi}:\nabla\nabla)\mathbf{B} \\
\boldsymbol{\eta}_\perp\cdot(\boldsymbol{\xi}_\perp\cdot\nabla)(\mathbf{B}\cdot\nabla\mathbf{B}) &= \boldsymbol{\eta}_\perp\cdot(\boldsymbol{\xi}_\perp\cdot\nabla\mathbf{B})\cdot\nabla\mathbf{B} + \boldsymbol{\eta}_\perp\cdot(\mathbf{B}\boldsymbol{\xi}:\nabla\nabla\mathbf{B})
\end{aligned} \quad \text{(A.10)}$$

To within a divergence term which integrates to zero, R is given by

$$R = -\frac{1}{\mu_0}\boldsymbol{\eta}_\perp \cdot (\boldsymbol{\xi}_\perp \cdot \nabla)(\mathbf{B} \cdot \nabla \mathbf{B}) = -(\boldsymbol{\eta}_\perp \boldsymbol{\xi}_\perp : \nabla\nabla)\left(p + \frac{B^2}{2\mu_0}\right) \quad \text{(A.11)}$$

The final result is

$$\begin{aligned}\int \boldsymbol{\eta} \cdot \mathbf{F}(\boldsymbol{\xi})\, d\mathbf{r} = -\int d\mathbf{r}\bigg[&\frac{1}{\mu_0}(\mathbf{B} \cdot \nabla \boldsymbol{\xi}_\perp) \cdot (\mathbf{B} \cdot \nabla \boldsymbol{\eta}_\perp) + \gamma p(\nabla \cdot \boldsymbol{\xi})(\nabla \cdot \boldsymbol{\eta}) \\ &+ \frac{B^2}{\mu_0}(\nabla \cdot \boldsymbol{\xi}_\perp + 2\boldsymbol{\xi}_\perp \cdot \boldsymbol{\kappa})(\nabla \cdot \boldsymbol{\eta}_\perp + 2\boldsymbol{\eta}_\perp \cdot \boldsymbol{\kappa}) \qquad \text{(A.12)} \\ &- \frac{4B^2}{\mu_0}(\boldsymbol{\xi}_\perp \cdot \boldsymbol{\kappa})(\boldsymbol{\eta}_\perp \cdot \boldsymbol{\kappa}) + (\boldsymbol{\eta}_\perp \boldsymbol{\xi}_\perp : \nabla\nabla)\left(p + \frac{B^2}{2\mu_0}\right)\bigg]\end{aligned}$$

which is clearly a self-adjoint form by inspection.

APPENDIX B. EVALUATION OF δW_F FOR THE GENERAL SCREW PINCH

The starting point for the calculation is the intuitive form of δW_F with the plasma compressibility term set to zero [see Eq. (9.33)]:

$$\delta W_F = \frac{1}{2}\int d\mathbf{r}\left[\frac{|\mathbf{Q}_\perp|^2}{\mu_0} + \frac{\mathbf{B}^2}{\mu_0}|\nabla\cdot\boldsymbol{\xi}_\perp + 2\boldsymbol{\xi}_\perp\cdot\boldsymbol{\kappa}|^2 - 2(\boldsymbol{\xi}_\perp\cdot\nabla p)(\boldsymbol{\kappa}\cdot\boldsymbol{\xi}_\perp^*) - J_\parallel(\boldsymbol{\xi}_\perp^*\times\mathbf{b})\cdot\mathbf{Q}_\perp\right] \tag{B.1}$$

The plasma displacement is written as

$$\boldsymbol{\xi}_\perp = \xi\mathbf{e}_r + \boldsymbol{\eta}\mathbf{e}_\eta \tag{B.2}$$

with

$$\begin{aligned}\boldsymbol{\eta} &= (\xi_\theta B_z - \xi_z B_\theta)/B\\ \mathbf{e}_\eta &= (B_z\mathbf{e}_\theta - B_\theta\mathbf{e}_z)/B\end{aligned} \tag{B.3}$$

The individual terms required for the integrand can be easily evaluated

$$\begin{aligned}\mathbf{Q}_\perp &= \nabla\times(\boldsymbol{\xi}_\perp\times\mathbf{B})_\perp\\ &= iF\xi\mathbf{e}_r + \left\{iF\boldsymbol{\eta} + \xi\left[\frac{B_z'B_\theta}{B} - \frac{rB_z}{B}\left(\frac{B_\theta}{r}\right)'\right]\right\}\mathbf{e}_\eta\\ \boldsymbol{\kappa} &= \mathbf{b}\cdot\nabla\mathbf{b} = -\frac{B_\theta^2}{rB^2}\mathbf{e}_r\\ \nabla\cdot\boldsymbol{\xi}_\perp + 2\boldsymbol{\xi}_\perp\cdot\boldsymbol{\kappa} &= \frac{(r\xi)'}{r} - 2\frac{B_\theta^2}{rB^2}\xi + i\frac{G}{B}\eta\\ \mu_0 J_\parallel &= \mu_0\mathbf{J}\cdot\mathbf{b} = \frac{B_z}{rB}(rB_\theta)' - \frac{B_\theta B_z'}{B}\end{aligned} \tag{B.4}$$

where $F = kB_z + mB_\theta/r$ and $G = mB_z/r - kB_\theta$. These expressions are substituted into Eq. (B.1). One obtains

$$\begin{aligned}\frac{\delta W_F}{2\pi R_0} &= \frac{\pi}{\mu_0}\int_0^a W(r)r\,dr\\ W(r) &= F^2|\xi|^2 + \left|iF\eta + \xi\left[\frac{B_z'B_\theta}{B} - \frac{rB_z}{B}\left(\frac{B_\theta}{r}\right)'\right]\right|^2\\ &\quad + B^2\left|\frac{(r\xi)'}{r} - \frac{2B_\theta^2}{rB^2}\xi + i\frac{G}{B}\eta\right|^2 + \frac{2\mu_0 p' B_\theta^2}{rB^2}|\xi|^2\\ &\quad - \mu_0 J_\parallel\left\{iF(\xi\eta^* - \xi^*\eta) - |\xi|^2\left[\frac{B_z'B_\theta}{B} - \frac{rB_z}{B}\left(\frac{B_\theta}{r}\right)'\right]\right\}\end{aligned} \tag{B.5}$$

Although this expression is quite complicated, observe that η appears only algebraically. After a short calculation, the terms explicitly containing η can be grouped together as follows:

$$\begin{aligned} W_\eta &= k_0^2 B^2 |\eta|^2 + 2\frac{ikBB_\theta}{r}(\eta\xi^* - \eta^*\xi) + \frac{iGB}{r}[\eta(r\xi^*)' - \eta^*(r\xi)'] \\ &= \left| ik_0 B\eta + \frac{2kB_\theta}{rk_0}\xi + \frac{G}{rk_0}(r\xi)' \right|^2 - \left| \frac{2kB_\theta}{rk_0}\xi + \frac{G}{rk_0}(r\xi)' \right|^2 \end{aligned} \tag{B.6}$$

where $k_0^2 = k^2 + m^2/r^2$. Since η appears only in a positive term, δW_F is minimized by choosing

$$\eta = \frac{i}{rk_0^2 B}[G(r\xi)' + 2kB_\theta\xi] \tag{B.7}$$

With this choice for η, $W(r)$ can be written as

$$W(r) = A_1\xi'^2 + 2A_2\xi\xi' + A_3\xi^2 \tag{B.8}$$

Note that without loss in generality, ξ can be considered as a real quantity since A_1, A_2, and A_3 are real coefficients. A short calculation shows that A_1, A_2, and A_3 are given by

$$\begin{aligned} A_1 &= \frac{F^2}{k_0^2} \\ A_2 &= \frac{1}{rk_0^2}\left(k^2B_z^2 - \frac{m^2B_\theta^2}{r^2}\right) \\ A_3 &= F^2 + \frac{2\mu_0 p' B_\theta^2}{rB^2} + \frac{B^2}{r^2}\left(1 - 2\frac{B_\theta^2}{B^2}\right)^2 - \frac{1}{r^2k_0^2}(G + 2kB_\theta)^2 \\ &\quad + \frac{2B_\theta B_z}{rB}\left[\frac{B_\theta B_z'}{B} - \frac{rB_z}{B}\left(\frac{B_\theta}{r}\right)'\right] \end{aligned} \tag{B.9}$$

The middle term in Eq. (B.8) can be integrated by parts. This leads to the following form for δW_F:

$$\frac{\delta W_F}{2\pi^2 R_0/\mu_0} = \int_0^a (f\xi'^2 + g\xi^2)\,dr + \left[\frac{k^2r^2B_z^2 - m^2B_\theta^2}{k_0^2 r^2}\right]_a \xi^2(a) \tag{B.10}$$

Here,

$$\begin{aligned} f &= rA_1 = \frac{rF^2}{k_0^2} \\ g &= rA_3 - (rA_2)' \end{aligned} \tag{B.11}$$

After a straightforward, but tedious calculation, g can be written as

$$g = \frac{2k^2}{k_0^2}(\mu_0 p)' + \left(\frac{k_0^2 r^2 - 1}{k_0^2 r^2}\right) rF^2 + \frac{2k^2}{rk_0^4}\left(kB_z - \frac{mB_\theta}{r}\right)F \quad \text{(B.12)}$$

This completes the derivation of δW_F.

APPENDIX C. FULL EIGENMODE EQUATION FOR THE GENERAL SCREW PINCH

The eigenmode equation for the general screw pinch is obtained from the reduction of

$$\begin{aligned} -\omega^2\rho\boldsymbol{\xi} &= \mathbf{J}_1 \times \mathbf{B} + \mathbf{J} \times \mathbf{B}_1 - \nabla p_1 \\ &= \frac{1}{\mu_0}(\mathbf{B}\cdot\nabla\mathbf{B}_1 + \mathbf{B}_1\cdot\nabla\mathbf{B}) - \nabla\left(p_1 + \frac{\mathbf{B}_1\cdot\mathbf{B}}{\mu_0}\right) \end{aligned} \tag{C.1}$$

where

$$\mathbf{B}_1 = \nabla \times (\boldsymbol{\xi}_\perp \times \mathbf{B}) \tag{C.2}$$

$$p_1 = -\boldsymbol{\xi}\cdot\nabla p - \gamma p \nabla\cdot\boldsymbol{\xi} \tag{C.3}$$

For the cylindrically symmetric system it is convenient to decompose the vector components as follows:

$$\begin{aligned} \boldsymbol{\xi} &= \xi\mathbf{e}_r + \eta\mathbf{e}_\eta + \xi_\|\mathbf{b} \\ \mathbf{B}_1 &= \tilde{B}_r\mathbf{e}_r + \tilde{B}_\eta\mathbf{e}_\eta + \tilde{B}_\|\mathbf{b} \\ \mathbf{b} &= (B_\theta\mathbf{e}_\theta + B_z\mathbf{e}_z)/B \\ \mathbf{e}_\eta &= (B_z\mathbf{e}_\theta - B_\theta\mathbf{e}_z)/B \end{aligned} \tag{C.4}$$

All perturbed quantities are assumed to vary as $Q_1 = Q_1(r)\exp[i(m\theta + kz)]$.

Expansion of Eq. (C.2) yields expressions for $\tilde{B}_r$, $\tilde{B}_\eta$, $\tilde{B}_\|$ in terms of ξ, η, $\xi_\|$:

$$\begin{aligned} \tilde{B}_r &= iF\xi \\ \tilde{B}_\eta &= iF\eta + \xi\left[\frac{B_\theta B_z'}{B} - \frac{rB_z}{B}\left(\frac{B_\theta}{r}\right)'\right] \\ \tilde{B}_\| &= -iG\eta - \frac{B}{r}(r\xi)' + \frac{\xi}{B}\left(\mu_0 p' + \frac{2B_\theta^2}{r}\right) \end{aligned} \tag{C.5}$$

where $F = kB_z + mB_\theta/r$ and $G = mB_z/r - kB_\theta$. Similarly, expansion of Eq. (C.3) leads to

$$p_1 = -p'\xi - \gamma p\left[\frac{(r\xi)'}{r} + i\frac{G}{B}\eta + i\frac{F}{B}\xi_\|\right] \tag{C.6}$$

Equations (C.5) and (C.6) are substituted into the momentum equation [Eq. (C1)]. Here, it is also convenient to separate the equation into $(\mathbf{e}_r, \mathbf{e}_\eta, \mathbf{b})$ vector components. If one forms the dot product of

Eq.(C.1) with **b**, then a short calculation gives

$$\left(\frac{\gamma p}{B^2}F^2 - \omega^2\rho\right)\xi_\parallel + \left(\frac{\gamma p F G}{B^2}\right)\eta = \left(\frac{i\gamma p F}{rB}\right)(r\xi)' \tag{C.7}$$

Similarly, upon forming the dot product of Eq. (C.1) with $\mathbf{e}_\eta$, one finds after another short calculation

$$\left(\frac{\gamma p F G}{B^2}\right)\xi_\parallel + \left(\frac{k_0^2 B^2}{\mu_0} + \frac{\gamma p}{B^2}G^2 - \omega^2\rho\right)\eta$$
$$= \left[\frac{iG(\gamma p + B^2/\mu_0)}{rB}\right](r\xi)' + \left(\frac{2ikBB_\theta}{\mu_0 r}\right)\xi \tag{C.8}$$

where $k_0^2 = k^2 + m^2/r^2$.

Equations (C.7) and (C.8) should be viewed as two simultaneous equations for $\xi_\parallel$ and η. Solving these equations yields

$$\xi_\parallel = -\left(\frac{i\gamma p F}{r\rho BD}\right)\left[(\omega^2 - \omega_a^2)(r\xi)' + \left(\frac{2kGB_\theta}{\mu_0\rho}\right)\xi\right] \tag{C.9}$$

$$\eta = -\left(\frac{i}{r\rho BD}\right)\left[G\left(\gamma p + \frac{B^2}{\mu_0}\right)(\omega^2 - \omega_h^2)(r\xi)' + \frac{2kB^2B_\theta}{\mu_0}(\omega^2 - \omega_g^2)\xi\right] \tag{C.10}$$

where

$$\begin{aligned}
\omega_a^2 &= \frac{F^2}{\mu_0\rho}\\
\omega_h^2 &= \left(\frac{V_s^2}{V_s^2 + V_a^2}\right)\omega_a^2\\
\omega_g^2 &= \left(\frac{V_s^2}{V_a^2}\right)\omega_a^2\\
D &= (\omega^2 - \omega_f^2)(\omega^2 - \omega_s^2)\\
\omega_{f,s}^2 &= \tfrac{1}{2}[k_0^2(V_s^2 + V_a^2)][1 \pm (1 - \alpha^2)^{1/2}]\\
\alpha^2 &= \frac{4V_s^2\omega_a^2}{k_0^2(V_s^2 + V_a^2)^2}, \qquad 0 < \alpha^2 < 1
\end{aligned} \tag{C.11}$$

and

$$\begin{aligned}
V_s^2 &= \gamma p/\rho\\
V_a^2 &= B^2/\mu_0\rho
\end{aligned} \tag{C.12}$$

The last step in the calculation is to substitute into the radial component of the momentum equation, given by

$$\rho(\omega^2 - \omega_a^2)\xi = \frac{2B_\theta}{\mu_0 r B}(B_z\tilde{B}_\eta + B_\theta \tilde{B}_\|) + \frac{d}{dr}\left(p_1 + \frac{B\tilde{B}_\|}{\mu_0}\right) \tag{C.13}$$

After a straightforward and slightly tedious calculation, the separate terms contributing to Eq. (C.13) can be written as

$$\begin{aligned} p_1 + \frac{B\tilde{B}_\|}{\mu_0} = &-\left[\frac{\rho(V_s^2 + V_a^2)}{r}\right]\left[\frac{(\omega^2 - \omega_a^2)(\omega^2 - \omega_h^2)}{(\omega^2 - \omega_f^2)(\omega^2 - \omega_s^2)}\right](r\xi)' \\ &+ 2\left[\frac{B_\theta^2}{\mu_0 r^2} - \left(\frac{kGB_\theta}{\mu_0 r^2}\right)\frac{(V_a^2 + V_s^2)(\omega^2 - \omega_h^2)}{(\omega^2 - \omega_f^2)(\omega^2 - \omega_s^2)}\right](r\xi) \end{aligned} \tag{C.14}$$

$$\begin{aligned} \frac{2B_\theta}{\mu_0 r B}(B_z\tilde{B}_\eta + B_\theta\tilde{B}_\|) = &-2\left[\frac{B_\theta^2}{\mu_0 r^2} - \left(\frac{kGB_\theta}{\mu_0 r^2}\right)\frac{(V_a^2 + V_s^2)(\omega^2 - \omega_h^2)}{(\omega^2 - \omega_f^2)(\omega^2 - \omega_s^2)}\right](r\xi)' \\ &+ \left[\left(\frac{4k^2B_\theta^2V_a^2}{\mu_0 r^3}\right)\frac{(\omega^2 - \omega_g^2)}{(\omega^2 - \omega_f^2)(\omega^2 - \omega_s^2)} \right. \\ &\left. - \left(\frac{B_\theta^2}{r^2}\right)'\right](r\xi) \end{aligned} \tag{C.15}$$

Substituting into Eq. (C.13) leads to the desired form of the normal-mode eigenvalue equation

$$\frac{d}{dr}\left[A\frac{d}{dr}(r\xi)\right] - C(r\xi) = 0 \tag{C.16}$$

where

$$A = \left[\frac{\rho(V_s^2 + V_a^2)}{r}\right]\frac{(\omega^2 - \omega_a^2)(\omega^2 - \omega_h^2)}{(\omega^2 - \omega_f^2)(\omega^2 - \omega_s^2)} \tag{C.17}$$

$$\begin{aligned} C = &-\frac{\rho}{r}(\omega^2 - \omega_a^2) + \left(\frac{4k^2V_a^2B_\theta^2}{\mu_0 r^3}\right)\frac{(\omega^2 - \omega_g^2)}{(\omega^2 - \omega_f^2)(\omega^2 - \omega_s^2)} \\ &+ \frac{d}{dr}\left[\frac{B_\theta^2}{\mu_0 r^2} - \left(\frac{2kB_\theta G}{\mu_0 r^2}\right)\frac{(V_a^2 + V_s^2)(\omega^2 - \omega_h^2)}{(\omega^2 - \omega_f^2)(\omega^2 - \omega_s^2)}\right] \end{aligned} \tag{C.18}$$

APPENDIX D. TAYLOR'S MINIMUM ENERGY STATES

D.1. Ideal MHD Constraint

Consider the minimization of the energy

$$W = \frac{1}{2\mu_0}\int B^2\, d\mathbf{r} \tag{D.1}$$

subject to the constraints $\nabla \cdot \mathbf{B} = 0$ and $\mathbf{E} + \mathbf{v} \times \mathbf{B} = 0$. Perfect conductivity requires any magnetic field variation to be coupled to the motion of the plasma by the relation

$$\frac{\partial \mathbf{B}}{\partial t} = \nabla \times (\mathbf{v} \times \mathbf{B}) \tag{D.2}$$

As shown in Chapter 3, Eq. (D.2) corresponds to the constraint that the magnetic field lines are "frozen into the plasma." When varying $\mathbf{B}$ in the energy minimization, only those states are allowed that preserve the initial topology of the system; no coalescing or tearing of field lines can occur.

The energy is minimized as follows. As the magnetic field is varied by an amount $\delta\mathbf{B}$, this is coupled to the plasma motion by the relation

$$\delta\mathbf{B} = \nabla \times (\boldsymbol{\xi} \times \mathbf{B}) \tag{D.3}$$

where $\delta\mathbf{v} \equiv \partial\boldsymbol{\xi}/\partial t$ and it is assumed that the background state is static. Furthermore, Eq. (D.3) guarantees that $\nabla \cdot \delta\mathbf{B} = 0$ automatically for all allowable plasma displacements $\boldsymbol{\xi}$. Upon minimizing W with respect to $\boldsymbol{\xi}$ one finds

$$(\delta \mathrm{W})_{\xi} = \frac{1}{\mu_0}\int \mathbf{B} \cdot \nabla \times (\boldsymbol{\xi} \times \mathbf{B})\, d\mathbf{r} \tag{D.4}$$

Integrating by parts yields

$$(\delta W)_{\xi} = -\frac{1}{\mu_0}\int \boldsymbol{\xi} \cdot (\mathbf{B} \times \nabla \times \mathbf{B})\, d\mathbf{r} \tag{D.5}$$

Thus, the energy is minimized if $\mathbf{B} \times \nabla \times \mathbf{B} = 0$ or

$$\begin{aligned} \nabla \times \mathbf{B} &= \lambda(\psi)\mathbf{B} \\ \mathbf{B} \cdot \nabla\lambda &= 0 \end{aligned} \tag{D.6}$$

The second relation is a consequence of $\nabla \cdot \mathbf{B} = 0$. When perfect conductivity is introduced as a constraint, the energy is minimized by a force-free field. Note that one does not in general have the option to set $\lambda(\psi) = 0$. The quantity $\lambda(\psi)$ is determined by the initial topology of the system and is preserved during the evolution by the perfect conductivity assumption.

An important result in the calculation of minimum energy states concerns the relationship between perfect conductivity and the plasma helicity defined by

$$K(\psi, t) = \int_{\psi} \mathbf{A} \cdot \mathbf{B}\, d\mathbf{r} \tag{D.7}$$

where $\mathbf{A}$ is the vector potential and the integration is over the volume enclosed by any arbitrary flux surface ψ. Taylor has shown that $K(\psi, t)$ is a constant of motion on every flux surface in an ideal MHD plasma. Furthermore, the constancy of K is equivalent to the perfect conductivity constraint on the allowable class of $\delta\mathbf{B}$ variations.

That K is a constant of the motion follows from the easily derivable relation (see Problem 9.8).

$$\frac{dK}{dt} = -2\int_{\psi} \mathbf{E} \cdot \mathbf{B}\, d\mathbf{r} + \int_{S} \mathbf{n} \cdot \mathbf{A} \times (\mathbf{E} + \mathbf{v} \times \mathbf{B})\, dS \tag{D.8}$$

For ideal MHD, $\mathbf{E} + \mathbf{v} \times \mathbf{B} = 0$ and $dK/dt = 0$ on every flux surface. The equivalence of $K = \text{const}$ and the perfect conductivity constraint on $\delta\mathbf{B}$ follows from minimizing

$$W = \frac{1}{2\mu_0}\left[\int d\mathbf{r}(B^2 - \lambda\mathbf{A} \cdot \mathbf{B}) + \int d\psi\left(\lambda \frac{dK}{d\psi}\right)\right] \tag{D.9}$$

where the first integration is over the entire volume, and the second integration covers all flux surfaces within the system. The constant helicity constraint has been introduced by the technique of Lagrange multipliers. Note that the multiplier $\lambda = \lambda(\psi)$ since K is a constant on every flux surface. If W is varied with respect to λ in the transformed coordinate system $(r, \theta, z) \to (\psi, \theta, z)$ then one obtains

$$(\delta W)_\lambda = \int d\psi\, \delta\lambda\left(\frac{dK}{d\psi} - \int J\, d\theta\, dz\, \mathbf{A} \cdot \mathbf{B}\right) \tag{D.10}$$

where $J = r/(\partial\psi/\partial r)$ is the Jacobian of the transformation. Setting $(\delta W)_\lambda = 0$ for arbitrary $\delta\lambda(\psi)$ yields the original constraint

$$K(\psi) = \int_0^{\psi} d\psi' \int (\mathbf{A} \cdot \mathbf{B}) J\, d\theta\, dz = \int_{\psi} \mathbf{A} \cdot \mathbf{B}\, d\mathbf{r} \tag{D.11}$$

If one now writes $\mathbf{B} = \nabla \times \mathbf{A}$, $\delta\mathbf{B} = \nabla \times \delta\mathbf{A}$, then the constraint

$\nabla \cdot \mathbf{B} = \nabla \cdot \delta\mathbf{B} = 0$ is automatically satisfied. The variation of W with respect to $\mathbf{A}$ is given by

$$(\delta W)_{\mathbf{A}} = \frac{1}{2\mu_0} \int_{\psi} d\mathbf{r}[2\mathbf{B} \cdot \nabla \times \delta\mathbf{A} - \lambda(\mathbf{B} \cdot \delta\mathbf{A} + \mathbf{A} \cdot \nabla \times \delta\mathbf{A})] \quad \text{(D.12)}$$

Integrating by parts yields

$$(\delta W)_{\mathbf{A}} = \frac{1}{\mu_0} \int_{\psi} d\mathbf{r}\delta\mathbf{A} \cdot [\nabla \times \mathbf{B} - \lambda\mathbf{B}] \quad \text{(D.13)}$$

Setting $(\delta W)_{\mathbf{A}} = 0$ for arbitrary $\delta\mathbf{A}$ gives $\nabla \times \mathbf{B} = \lambda\mathbf{B}$, which is identical to Eq. (D.6). Hence, holding K constant on every flux surface is equivalent to the perfect conductivity constraint.

D.2. Taylor's Helicity Constraint

Taylor conjectures that the ideal MHD constraint is unnecessarily restrictive in describing the evolution of RFPs. Instead, in a realistic system with dissipation only the helicity of the entire system K_a is conserved. The quantity K_a is given by

$$K_a = \int \mathbf{A} \cdot \mathbf{B}\, d\mathbf{r} \quad \text{(D.14)}$$

where the integration is over the entire plasma volume. The minimum energy state is obtained by varying

$$W = \frac{1}{2\mu_0}\left[\int (B^2 - \mu\mathbf{A} \cdot \mathbf{B})\, d\mathbf{r} + \mu K_a\right] \quad \text{(D.15)}$$

where the multiplier μ = const, since there is only one global constraint. Variation of W with respect to μ gives

$$(\delta W)_{\mu} = \delta\mu\left(K_a - \int \mathbf{A} \cdot \mathbf{B}\, d\mathbf{r}\right) \quad \text{(D.16)}$$

which leads to the original constraint [Eq. (D.14)]. Variation of W with respect to $\mathbf{A}$ yields

$$(\delta W)_{\mathbf{A}} = \frac{1}{2\mu_0} \int [2\mathbf{B} \cdot \nabla \times \delta\mathbf{A} - \mu(\delta\mathbf{A} \cdot \mathbf{B} + \mathbf{A} \cdot \nabla \times \delta\mathbf{A})]\, d\mathbf{r} \quad \text{(D.17)}$$

Upon integrating by parts one finds

$$(\delta W)_{\mathbf{A}} = \frac{1}{\mu_0} \int \delta\mathbf{A} \cdot [\nabla \times \mathbf{B} - \mu\mathbf{B}]\, d\mathbf{r} \quad \text{(D.18)}$$

Setting $(\delta W)_{\mathbf{A}} = 0$ for arbitrary $\delta\mathbf{A}$ leads to the equation describing the Taylor minimum energy state:

$$\nabla \times \mathbf{B} = \mu \mathbf{B} \tag{D.19}$$

D.3. Solutions for the Minimum Energy States

The solutions for the minimum energy states are obtained by solving Eq. (D.19). This can be conveniently carried out for the straight cylinder by operating on Eq. (D.19) with $\mathbf{e}_z \cdot \nabla \times$ and using the relation $\nabla \cdot \mathbf{B} = 0$. The result is a single second-order equation for B_z given by

$$\nabla^2 B_z + \mu^2 B_z = 0 \tag{D.20}$$

The general solution is given by

$$B_z = \sum_{m,k} a_{m,k} J_m(\alpha r) \cos(m\theta + kz) \tag{D.21}$$

where $\alpha = (\mu^2 - k^2)^{1/2}$. B_r and B_θ are found from the r and θ components of $\nabla \times \mathbf{B} = \mu\mathbf{B}$. If one writes

$$\begin{aligned} B_r &= \sum_{m,k} G_{m,k}(r) \sin(m\theta + kz) \\ B_\theta &= \sum_{m,k} H_{m,k}(r) \cos(m\theta + kz) \end{aligned} \tag{D.22}$$

then a simple calculation shows that $G_{m,k}$ and $H_{m,k}$ satisfy

$$\begin{aligned} kG_{m,k} - \mu H_{m,k} &= a_m \alpha J'_m(\alpha r) \\ \mu G_{m,k} - kH_{m,k} &= -\frac{m a_m}{r} J_m(\alpha r) \end{aligned} \tag{D.23}$$

Here a prime denotes differentiation with respect to the argument. Solving for $G_{m,k}$ and $H_{m,k}$, one finds

$$\begin{aligned} G_{m,k} &= -\frac{a_m}{\alpha^2}\left[\frac{m\mu}{r} J_m + k\alpha J'_m\right] \\ H_{m,k} &= -\frac{a_m}{\alpha^2}\left[\mu\alpha J'_m + \frac{mk}{r} J_m\right] \end{aligned} \tag{D.24}$$

Equations (D.21), (D.22), and (D.24) represent the general solutions for the cylindrical minimum energy states.

D.4. Evaluation of W(K)

The evaluation of $W(K_a)$ is obtained as follows. For any minimum energy state, substitution of Eq. (9.152) into the definition of K_a yields

$$K_a = \frac{1}{\mu}\int \mathbf{B}\cdot[\mathbf{B} - B_0 J_0(\mu a)\mathbf{e}_z]\, d\mathbf{r} \tag{D.25}$$

Upon integration one obtains

$$\hat{K} = \frac{\hat{W}}{x} - \frac{J_0(x)}{J_1(x)} \tag{D.26}$$

where $x = \mu a$, $\hat{W} = 2a^2\mu_0 W/R_0\psi^2$ and $\hat{K} = aK_a/R_0\psi^2$.

Equation (D.26) constitutes one relationship for $\hat{W}$ and $\hat{K}$. For the $m = 0$ cylindrical solution, the second relation follows from the direct evaluation of W:

$$W = \frac{2\pi^2 R_0}{\mu_0}\int B_0^2[J_0^2(\mu r) + J_1^2(\mu r)]r\, dr \tag{D.27}$$

or

$$\hat{W} = \frac{x^2}{J_1^2}\left(J_0^2 + J_1^2 - \frac{J_0 J_1}{x}\right) \tag{D.28}$$

Substituting $\hat{W}$ into Eq. (D.26) gives

$$\hat{K} = \frac{x}{J_1^2}\left(J_0^2 + J_1^2 - \frac{2J_0 J_1}{x}\right) \tag{D.29}$$

Equations (D.28) and (D.29) are two parametric equations in x specifying the relation $\hat{W} = \hat{W}(\hat{K})$ for the $m = 0$ solution.

For the $m = 0,1$ solution the second relation between $\hat{W}$ and $\hat{K}$ follows from the requirement that $B_r(a, \theta, z) = 0$ at the conducting wall. This condition is easily obtained from Eq. (9.150) and can be written as

$$\hat{k} y J_1'(y) + x J_1(y) = 0 \tag{D.30}$$

where $\hat{k} = ka$ and $y = (x^2 - \hat{k}^2)^{1/2}$. Directly substituting into the expression for W for the $m = 0,1$ solution yields information about the helical amplitude c, but does not provide an independent relationship between $\hat{W}$ and $\hat{K}$. Consequently, $\hat{W}(\hat{K})$ is determined by Eqs. (D.26) and (D.30) for the $m = 0,1$ state.

Index